Lecture Notes in Computer Science 10988

Commenced Publication in 1973
Founding and Former Series Editors:
Gerhard Goos, Juris Hartmanis, and Jan van Leeuwen

More information about this series at http://www.springer.com/series/7409

Yi Cai · Yoshiharu Ishikawa
Jianliang Xu (Eds.)

Web and Big Data

Second International Joint Conference, APWeb-WAIM 2018
Macau, China, July 23–25, 2018
Proceedings, Part II

 Springer

Editors
Yi Cai
South China University of Technology
Guangzhou
China

Jianliang Xu
Hong Kong Baptist University
Kowloon Tong, Hong Kong
China

Yoshiharu Ishikawa
Nagoya University
Nagoya
Japan

ISSN 0302-9743   ISSN 1611-3349   (electronic)
Lecture Notes in Computer Science
ISBN 978-3-319-96892-6   ISBN 978-3-319-96893-3   (eBook)
https://doi.org/10.1007/978-3-319-96893-3

Library of Congress Control Number: 2018948814

LNCS Sublibrary: SL3 – Information Systems and Applications, incl. Internet/Web, and HCI

This Springer imprint is published by the registered company Springer Nature Switzerland AG
The registered company address is: Gewerbestrasse 11, 6330 Cham, Switzerland

Preface

This volume (LNCS 10987) and its companion volume (LNCS 10988) contain the proceedings of the second Asia-Pacific Web (APWeb) and Web-Age Information Management (WAIM) Joint Conference on Web and Big Data, called APWeb-WAIM. This joint conference aims to attract participants from different scientific communities as well as from industry, and not merely from the Asia Pacific region, but also from other continents. The objective is to enable the sharing and exchange of ideas, experiences, and results in the areas of World Wide Web and big data, thus covering Web technologies, database systems, information management, software engineering, and big data. The second APWeb-WAIM conference was held in Macau during July 23–25, 2018. As an Asia-Pacific flagship conference focusing on research, development, and applications in relation to Web information management, APWeb-WAIM builds on the successes of APWeb and WAIM: APWeb was previously held in Beijing (1998), Hong Kong (1999), Xi'an (2000), Changsha (2001), Xi'an (2003), Hangzhou (2004), Shanghai (2005), Harbin (2006), Huangshan (2007), Shenyang (2008), Suzhou (2009), Busan (2010), Beijing (2011), Kunming (2012), Sydney (2013), Changsha (2014), Guangzhou (2015), and Suzhou (2016); and WAIM was held in Shanghai (2000), Xi'an (2001), Beijing (2002), Chengdu (2003), Dalian (2004), Hangzhou (2005), Hong Kong (2006), Huangshan (2007), Zhangjiajie (2008), Suzhou (2009), Jiuzhaigou (2010), Wuhan (2011), Harbin (2012), Beidaihe (2013), Macau (2014), Qingdao (2015), and Nanchang (2016). The first joint APWeb-WAIM conference was held in Bejing (2017). With the fast development of Web-related technologies, we expect that APWeb-WAIM will become an increasingly popular forum that brings together outstanding researchers and developers in the field of the Web and big data from around the world. The high-quality program documented in these proceedings would not have been possible without the authors who chose APWeb-WAIM for disseminating their findings. Out of 168 submissions, the conference accepted 39 regular (23.21%), 31 short research papers, and six demonstrations. The contributed papers address a wide range of topics, such as text analysis, graph data processing, social networks, recommender systems, information retrieval, data streams, knowledge graph, data mining and application, query processing, machine learning, database and Web applications, big data, and blockchain. The technical program also included keynotes by Prof. Xuemin Lin (The University of New South Wales, Australia), Prof. Lei Chen (The Hong Kong University of Science and Technology, Hong Kong, SAR China), and Prof. Ninghui Li (Purdue University, USA) as well as industrial invited talks by Dr. Zhao Cao (Huawei Blockchain) and Jun Yan (YiDu Cloud). We are grateful to these distinguished scientists for their invaluable contributions to the conference program. As a joint conference, teamwork was particularly important for the success of APWeb-WAIM. We are deeply thankful to the Program Committee members and the external reviewers for lending their time and expertise to the conference. Special thanks go to the local Organizing Committee led by Prof. Zhiguo Gong.

Thanks also go to the workshop co-chairs (Leong Hou U and Haoran Xie), demo co-chairs (Zhixu Li, Zhifeng Bao, and Lisi Chen), industry co-chair (Wenyin Liu), tutorial co-chair (Jian Yang), panel chair (Kamal Karlapalem), local arrangements chair (Derek Fai Wong), and publicity co-chairs (An Liu, Feifei Li, Wen-Chih Peng, and Ladjel Bellatreche). Their efforts were essential to the success of the conference. Last but not least, we wish to express our gratitude to the treasurer (Andrew Shibo Jiang), the Webmaster (William Sio) for all the hard work, and to our sponsors who generously supported the smooth running of the conference. We hope you enjoy the exciting program of APWeb-WAIM 2018 as documented in these proceedings.

June 2018   Yi Cai
                                                                  Jianliang Xu
                                                            Yoshiharu Ishikawa

Organization

Organizing Committee

Honorary Chair

Lionel Ni   University of Macau, SAR China

General Co-chairs

Zhiguo Gong	University of Macau, SAR China
Qing Li	City University of Hong Kong, SAR China
Kam-fai Wong	Chinese University of Hong Kong, SAR China

Program Co-chairs

Yi Cai	South China University of Technology, China
Yoshiharu Ishikawa	Nagoya University, Japan
Jianliang Xu	Hong Kong Baptist University, SAR China

Workshop Chairs

Leong Hou U	University of Macau, SAR China
Haoran Xie	Education University of Hong Kong, SAR China

Demo Co-chairs

Zhixu Li	Soochow University, China
Zhifeng Bao	RMIT, Australia
Lisi Chen	Wollongong University, Australia

Tutorial Chair

Jian Yang   Macquarie University, Australia

Industry Chair

Wenyin Liu   Guangdong University of Technology, China

Panel Chair

Kamal Karlapalem   IIIT, Hyderabad, India

Publicity Co-chairs

An Liu	Soochow University, China
Feifei Li	University of Utah, USA

Wen-Chih Peng   National Taiwan University, China
Ladjel Bellatreche   ISAE-ENSMA, Poitiers, France

Treasurers

Leong Hou U   University of Macau, SAR China
Andrew Shibo Jiang   Macau Convention and Exhibition Association,
                               SAR China

Local Arrangements Chair

Derek Fai Wong   University of Macau, SAR China

Webmaster

William Sio   University of Macau, SAR China

Senior Program Committee

Bin Cui   Peking University, China
Byron Choi   Hong Kong Baptist University, SAR China
Christian Jensen   Aalborg University, Denmark
Demetrios   University of Cyprus, Cyprus
   Zeinalipour-Yazti
Feifei Li   University of Utah, USA
Guoliang Li   Tsinghua University, China
K. Selçuk Candan   Arizona State University, USA
Kyuseok Shim   Seoul National University, South Korea
Makoto Onizuka   Osaka University, Japan
Reynold Cheng   The University of Hong Kong, SAR China
Toshiyuki Amagasa   University of Tsukuba, Japan
Walid Aref   Purdue University, USA
Wang-Chien Lee   Pennsylvania State University, USA
Wen-Chih Peng   National Chiao Tung University, Taiwan
Wook-Shin Han Pohang   University of Science and Technology, South Korea
Xiaokui Xiao   National University of Singapore, Singapore
Ying Zhang   University of Technology Sydney, Australia

Program Committee

Alex Thomo   University of Victoria, Canada
An Liu   Soochow University, China
Baoning Niu   Taiyuan University of Technology, China
Bin Yang   Aalborg University, Denmark
Bo Tang   Southern University of Science and Technology, China
Zouhaier Brahmia   University of Sfax, Tunisia
Carson Leung   University of Manitoba, Canada
Cheng Long   Queen's University Belfast, UK

Chih-Chien Hung	Tamkang University, China
Chih-Hua Tai	National Taipei University, China
Cuiping Li	Renmin University of China, China
Daniele Riboni	University of Cagliari, Italy
Defu Lian	Big Data Research Center, University of Electronic Science and Technology of China, China
Dejing Dou	University of Oregon, USA
Dimitris Sacharidis	Technische Universität Wien, Austria
Ganzhao Yuan	Sun Yat-sen University, China
Giovanna Guerrini	Università di Genova, Italy
Guanfeng Liu	The University of Queensland, Australia
Guoqiong Liao	Jiangxi University of Finance and Economics, China
Guanling Lee	National Dong Hwa University, China
Haibo Hu	Hong Kong Polytechnic University, SAR China
Hailong Sun	Beihang University, China
Han Su	University of Southern California, USA
Haoran Xie	The Education University of Hong Kong, SAR China
Hiroaki Ohshima	University of Hyogo, Japan
Hong Chen	Renmin University of China, China
Hongyan Liu	Tsinghua University, China
Hongzhi Wang	Harbin Institute of Technology, China
Hongzhi Yin	The University of Queensland, Australia
Hua Wang	Victoria University, Australia
Ilaria Bartolini	University of Bologna, Italy
James Cheng	Chinese University of Hong Kong, SAR China
Jeffrey Xu Yu	Chinese University of Hong Kong, SAR China
Jiajun Liu	Renmin University of China, China
Jialong Han	Nanyang Technological University, Singapore
Jianbin Huang	Xidian University, China
Jian Yin	Sun Yat-sen University, China
Jiannan Wang	Simon Fraser University, Canada
Jianting Zhang	City College of New York, USA
Jianxin Li	Beihang University, China
Jianzhong Qi	University of Melbourne, Australia
Jinchuan Chen	Renmin University of China, China
Ju Fan	Renmin University of China, China
Jun Gao	Peking University, China
Junhu Wang	Griffith University, Australia
Kai Zeng	Microsoft, USA
Kai Zheng	University of Electronic Science and Technology of China, China
Karine Zeitouni	Université de Versailles Saint-Quentin, France
Lei Zou	Peking University, China
Leong Hou U	University of Macau, SAR China
Liang Hong	Wuhan University, China
Lianghuai Yang	Zhejiang University of Technology, China

Lisi Chen   Wollongong University, Australia
Lu Chen   Aalborg University, Denmark
Maria Damiani   University of Milan, Italy
Markus Endres   University of Augsburg, Germany
Mihai Lupu   Vienna University of Technology, Austria
Mirco Nanni   ISTI-CNR Pisa, Italy
Mizuho Iwaihara   Waseda University, Japan
Peiquan Jin   University of Science and Technology of China, China
Peng Wang   Fudan University, China
Qin Lu   University of Technology Sydney, Australia
Ralf Hartmut Güting   Fernuniversität in Hagen, Germany
Raymond Chi-Wing Wong   Hong Kong University of Science and Technology,
                            SAR China
Ronghua Li   Shenzhen University, China
Rui Zhang   University of Melbourne, Australia
Sanghyun Park   Yonsei University, South Korea
Sanjay Madria   Missouri University of Science and Technology, USA
Shaoxu Song   Tsinghua University, China
Shengli Wu   Jiangsu University, China
Shimin Chen   Chinese Academy of Sciences, China
Shuai Ma   Beihang University, China
Shuo Shang   King Abdullah University of Science and Technology,
                            Saudi Arabia
Takahiro Hara   Osaka University, Japan
Tieyun Qian   Wuhan University, China
Tingjian Ge   University of Massachusetts, Lowell, USA
Tom Z. J. Fu   Advanced Digital Sciences Center, Singapore
Tru Cao   Ho Chi Minh City University of Technology, Vietnam
Vincent Oria   New Jersey Institute of Technology, USA
Wee Ng   Institute for Infocomm Research, Singapore
Wei Wang   University of New South wales, Australia
Weining Qian   East China Normal University, China
Weiwei Sun   Fudan University, China
Wen Zhang   Wuhan University, China
Wolf-Tilo Balke   Technische Universität Braunschweig, Germany
Wookey Lee   Inha University, South Korea
Xiang Zhao   National University of Defence Technology, China
Xiang Lian   Kent State University, USA
Xiangliang Zhang   King Abdullah University of Science and Technology,
                            Saudi Arabia
Xiangmin Zhou   RMIT University, Australia
Xiaochun Yang   Northeast University, China
Xiaofeng He   East China Normal University, China
Xiaohui (Daniel) Tao   The University of Southern Queensland, Australia
Xiaoyong Du   Renmin University of China, China
Xike Xie   University of Science and Technology of China, China

Xin Cao	The University of New South Wales, Australia
Xin Huang	Hong Kong Baptist University, SAR China
Xin Wang	Tianjin University, China
Xingquan Zhu	Florida Atlantic University, USA
Xuan Zhou	Renmin University of China, China
Yafei Li	Zhengzhou University, China
Yanghua Xiao	Fudan University, China
Yanghui Rao	Sun Yat-sen University, China
Yang-Sae Moon	Kangwon National University, South Korea
Yaokai Feng	Kyushu University, Japan
Yi Cai	South China University of Technology, China
Yijie Wang	National University of Defense Technology, China
Yingxia Shao	Peking University, China
Yongxin Tong	Beihang University, China
Yu Gu	Northeastern University, China
Yuan Fang	Institute for Infocomm Research, Singapore
Yunjun Gao	Zhejiang University, China
Zakaria Maamar	Zayed University, United Arab of Emirates
Zhaonian Zou	Harbin Institute of Technology, China
Zhiwei Zhang	Hong Kong Baptist University, SAR China

Keynotes

Graph Processing: Applications, Challenges, and Advances

Xuemin Lin

School of Computer Science and Engineering,
University of New South Wales, Sydney
lxue@cse.unsw.edu.au

Abstract. Graph data are key parts of Big Data and widely used for modelling complex structured data with a broad spectrum of applications. Over the last decade, tremendous research efforts have been devoted to many fundamental problems in managing and analyzing graph data. In this talk, I will cover various applications, challenges, and recent advances. We will also look to the future of the area.

Differential Privacy in the Local Setting

Ninghui Li

Department of Computer Sciences, Purdue University
ninghui@cs.purdue.edu

Abstract. Differential privacy has been increasingly accepted as the de facto standard for data privacy in the research community. Recently, techniques for satisfying differential privacy (DP) in the local setting, which we call LDP, have been deployed. Such techniques enable the gathering of statistics while preserving privacy of every user, without relying on trust in a single data curator. Companies such as Google, Apple, and Microsoft have deployed techniques for collecting user data while satisfying LDP. In this talk, we will discuss the state of the art of LDP. We survey recent developments for LDP, and discuss protocols for estimating frequencies of different values under LDP, and for computing marginal when each user has multiple attributes. Finally, we discuss limitations and open problems of LDP.

Big Data, AI, and HI, What is the Next?

Lei Chen

Department of Computer Science and Engineering, Hong Kong University
of Science and Technology
leichen@cse.ust.hk

Abstract. Recently, AI has become quite popular and attractive, not only to the academia but also to the industry. The successful stories of AI on Alpha-go and Texas hold 'em games raise significant public interests on AI. Meanwhile, human intelligence is turning out to be more sophisticated, and Big Data technology is everywhere to improve our life quality. The question we all want to ask is "what is the next?". In this talk, I will discuss about DHA, a new computing paradigm, which combines big Data, Human intelligence, and AI. First I will briefly explain the motivation of DHA. Then I will present some challenges and possible solutions to build this new paradigm.

Contents – Part II

Data Mining and Application

Contents – Part I

Demo Papers

Database and Web Applications

Fuzzy Searching Encryption with Complex Wild-Cards Queries on Encrypted Database

He Chen[1], Xiuxia Tian[2(✉)], and Cheqing Jin[1]

[1] School of Data Science and Engineering, East China Normal University,
Shanghai, China
watch_ch@163.com, cqjin@dase.ecnu.edu.cn
[2] College of Computer Science and Technology,
Shanghai University of Electric Power, Shanghai, China
xxtian@fudan.edu.cn

Abstract. Achieving fuzzy searching encryption (FSE) can greatly enrich the basic function over cipher-texts, especially on encrypted database (like CryptDB). However, most proposed schemes base on centralized inverted indexes which cannot handle complicated queries with wild-cards. In this paper, we present a well-designed FSE schema through Locality-Sensitive-Hashing and Bloom-Filter algorithms to generate two types of auxiliary columns respectively. Furthermore, an adaptive rewriting method is described to satisfy queries with wild-cards, such as percent and underscore. Besides, security enhanced improvements are provided to avoid extra messages leakage. The extensive experiments show effectiveness and feasibility of our work.

Keywords: Fuzzy searching encryption · Wild-cards searching
CryptDB

1 Introduction

Cloud database is a prevalent paradigm for data outsourcing. In consideration of data security and commercial privacy, both individuals and enterprises prefer outsourcing them in encrypted form. CryptDB [21] is a typical outsourced encrypted database (OEDB) which supports executing SQL statements on cipher-texts. Its transparency essentially relies on the design of splitting attributions and rewriting queries on proxy middle-ware. Under this proxy-based encrypted framework, several auxiliary columns are extended with different encryptions and query semantics are preserved through modifying or appending SQL statements.

Supported by the National Key Research and Development Program of China (No. 2016YFB1000905), NSFC (Nos. 61772327, 61532021, U1501252, U1401256 and 61402180), Project of Shanghai Science and Technology Committee Grant (No. 15110500700).

To enrich basic functions on cipher-texts, searchable symmetric encryption (SSE) is proposed for keyword searching with encrypted inverted indexes [4,13,22,24], and then dynamic SSE (DSSE) achieves alterations on various centralized indexes to enhance applicability [2,11,12,14]. Besides, the studies about exact searching with boolean expressions are extended in this field to increase accuracy [3,10]. Furthermore, the researches of similar searching among documents or words are widely discussed through introducing locality sensitive hashing algorithms [1,7–9,15,18,19,23,25–27]. However, these proposed schemes are not applicable to OEDB scenario because of the centralized index design and cannot handle complex fuzzy searching with wild-cards.

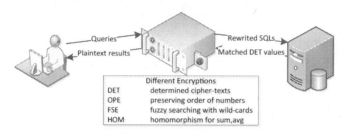

Fig. 1. The client-proxy-database framework synthesizes various encryptions together, such as the determined encryption (DET) preserves symmetric character for en/decryption, the order-preserving encryption (OPE) persists order among numeric values, the fuzzy searching encryption (FSE) handles queries on text, and the homomorphic encryption (HOM) achieves aggregation computing.

Therefore, it is meaningful and necessary to achieve fuzzy searching encryption over outsourced encrypted database. As shown in Fig. 1, the specific framework accomplishes transparency and homomorphism by rewriting SQL statements on auxiliary columns. In this paper, we focus on resolving the functionality of 'like' queries with wild-cards ('%' and '_'). Our contributions are summarized as follows:

- We propose a fuzzy searching encryption with complex wild-cards queries on encrypted database which extends extra functionality for the client-proxy-database framework like CryptDB.
- We present an adaptive rewriting method to handle different query cases on two types of auxiliary columns. The formal column works for similar searching by locality sensitive hashing and the latter multiple columns work for maximum substring matching by designed bloom-filter vectors.
- We evaluate the efficiency, correctness rate and space overhead by adjusting the parameters in auxiliary columns. Besides, security enhanced improvements are provided to avoid extra messages leakage. The extensive experiments also indicate the effectiveness and feasibility of our work.

The rest of paper is organized as follows. Section 2 discusses the related work and Sect. 3 introduces some basic concepts and definitions. Section 4 describes our schema including initialization of auxiliary columns, adaptive rewriting queries and security enhanced improvements. Section 5 presents the experiments and a brief conclusion is given in Sect. 6.

2   Related Work

In recent years, many proposed schemes have been attempting to achieve fuzzy searching encryption with helps of similarity [1,8,9,15,23,25–27]. The most of them introduce locality sensitive hashing (LSH) to map similar items together and bloom-filter to change the method of measuring. Wang et al.'s work [23] was one of the first works to present fuzzy searching. They encode every words in each file into same large bloom-filter space as a vector and evaluate similarity of target queries by computing the inner product for top-k results among vectors. Kuzu et al.'s work [15] generates similar feature vectors by embedding keyword strings into the Euclidean space which approximately preserves the relative edit distance. Fu et al.'s work [8] proposes an efficient multi-keyword fuzzy ranked search schema which is suitable for common spelling mistakes. It benefits from counting uni-gram among keywords and transvection sorting to obtain ranked candidates. Wang et al.'s work [26] generates a high-dimensional feature vector by LSH to support large-scale similarity search over encrypted feature-rich multimedia data. It stores encrypted inverted file identifier vectors as indexes while mapping similar objects into same or neighbor keyword-buckets by LSH based on Euclidean distance. In contrast to sparse vectors from bi-gram mapping, their work eliminates the sparsity and promotes the correctness as well. However, there are many problems in existing schemes including the insufficient metric conversion, the coarse-grained similarity comparison, the extreme dependency of assistant programs and the neglect about wild-card queries.

Meanwhile, the proposal of CryptDB [21] has attracted world-wide attention because they provide a practical way to combine various attribution-preserving encryptions over encrypted database. Then many analogous researches [5,16,17, 20] study its security definitions, feasible frameworks, extensible functions and optimizations. Chen et al. [5] consider these encrypted database as a client-proxy-database framework and presents symmetric column for en/decryption and auxiliary columns for supporting executions. This framework helps execute SQL statements directly over cipher-texts through appending auxiliary columns with different encryptions. It also benefits from the transparency of en/decryption processes and combines various functional encryptions together. Therefore, it is meaningful to achieve efficient fuzzy searching with complex wild-cards queries on proxy-based encrypted database.

3   Preliminaries

3.1   Basic Concepts

A. N-gram. In the fields of computational linguistics and probability, the n-gram method is proposed for measurement by generating a contiguous sequence of items from given strings. Essentially, it converts texts to fragments sets for vectorization while preserving some connotative connections. As shown in Table 1, various n-gram methods are utilized to preserve different implicit inner relation from origin strings.

Table 1. Various n-gram forms in our scheme

N-gram methods	Value	Description
String	secure	The original keyword
Counting uni-gram [8]	s1, e1, c1, u1, r1, e2	Preserve repetitions
Bi-gram	#s, se, ec, cu, ur, re, e#	Preserve adjacent letters
Tri-gram	sec, ecu, cur, ure	Preserve triple adjacent letters
Prefix and suffix	@s, e@	Beginning and ending of sentence

In general, bi-gram is the most common converting method which maintains the connotative information between adjacent letters. However, each change of single letter will double influence bi-gram results and cause reduction of matching probability. The counting uni-gram preserves repetitions and benefits on letter-confused comparison cases, such as misspelling of a letter, missing or adding a letter and reversing the order of two letters. However, it reduces the degree of constraint along with increasing false positives. The tri-gram is a more strict method which only suits the specific scene like existing judgment. The prefix and suffix preserve the beginning and ending of data to meet edge-searching.

B. Bloom-Filter. The Bloom-filter is a compact structure reflecting whether specific elements exist in prepared union. In our schema, we introduce this algorithm to judge existence about maximized substring fragments and represent the sparse vector through decimal numbers in separated columns. Given words fragments set $S = \{e_1, \ldots, e_{\#e}\}$, a bloom-filter maps each element e_i into a same l-bit sparse array by k independent hash functions. Positive answer is provided only if all bits of matched positions are true.

C. Locality Sensitive Hashing. The locality sensitive hashing (LSH) algorithm helps reduce the dimension of high-dimensional data. In our schema, we introduce this algorithm to map similar items together with high probability. Besides, the specific manifestation of the algorithm is different under different measurement standards. However, there is no available method for levenshtein distance among text. So that a common practice is converting texts to fragment sets with n-gram methods.

Definition 1 (Locality sensitive hashing). *Given a distance metric function D, a hash function family $\mathcal{H} = \{h_i : \{0,1\}^d \rightarrow \{0,1\}^t | i = 1, \ldots, M\}$ is (r_1, r_2, p_1, p_2)-sensitive if for any $s, t \in \{0,1\}^d$ and any $h \in \mathcal{H}$ satisfies:*
if $D(s,t) \leq r_1$ then $Pr[h_i(p) = h_i(q)] \geq p_1$;
if $D(s,t) \geq r_2$ then $Pr[h_i(p) = h_i(q)] \leq p_2$.

For nearest neighbor searching, $p_1 > p_2$ and $r_1 < r_2$ is needed. Practically, feasible permutations are generated through surjective hashing functions with our security parameter λ. And the minhash algorithm helps map fragment sets of every separated words which achieves similar searching.

3.2   Functional Model

Let $D = (d_1, \ldots, d_{\#D})$ be sensitive row data (each line contains some words respectively, as $d_i = \bigcup_{j=1}^{|d_i|} w_j^i$) and $C = \{c_{det}, c_{lsh}, c_{bf}\}$ be the corresponding cipher-texts. Two types of indexing methods are enforced: the first one achieves similar searching among words through dimension reductions with locality sensitive hashing (let m be the dimension of LSH, n be the tolerance and $\mathcal{L}$ represents its conversion); the last one achieves maximum substring matching through bit operation with bloom-filter (let l be the length of vector space, k be the amount of hashing functions and $\mathcal{B}$ represents its conversion). We consider LSH tokens set $T_i = \mathcal{L}_m^n(\bigcup_{j=1}^{|d_i|} \mathcal{G}_{ss}(w_j^i))$ be the elementary ciphers for c_{lsh}, and BF vector $V_i = \mathcal{B}_l^k(\bigcup_{j=1}^{|d_i|} \mathcal{G}_{msm}(w_j^i))$ be the ciphers of whole continuous sequence for c_{bf}. Besides, $\mathcal{G}$ represents n-gram methods for similar searching or maximum substring matching.

Definition 2 (Fuzzy searching encryption). *A proxy-based encrypted database implements fully fuzzy searching with rewriting SQL statements through the following polynomial-time algorithms:*

$(K_{det}, \mathcal{L}_m^n, \mathcal{B}_l^k) \leftarrow$ KeyGen(λ, m, n, l, k): Given security parameter λ, dimension m of LSH and tolerance n, vector length l of BF and hash amount k, it outputs a primary key K_{det} for determining encryption, $\mathcal{L}_m^n$ for LSH, $\mathcal{B}_l^k$ for BF. The security parameter λ helps initialize the hash functions and randomization processes.

$(c_{det}, T_i, V_i) \leftarrow$ Index($d_i, \mathcal{L}_m^n, \mathcal{B}_l^k$): Given the LSH function $\mathcal{L}_m^n$ and the BF function $\mathcal{B}_l^k$, the plain-text d_i is encrypted to determined cipher-texts c_{det}, ciphers T_i for similar searching and ciphers V_i for maximum substring matching respectively.

$(c_{det}||T_i||V_i) \leftarrow$ Trapdoor(*expression*): Given the query expression analyzed from 'like' clause, the adaptive rewriting method help generate representing elements out of different considerations with wild-cards condition. The determined cipher-texts would return in next step over encrypted database and K_{det} helps decryption.

As shown in definition of fuzzy searching encryption, we mainly emphasize transformation processes like building, indexing and executing. There exist other functional methods such as updating, deleting to achieve dynamically of our schema. It is applicable for outsourced encrypted database through rewriting SQL statements including 'create', 'insert', 'select' and so on.

3.3   Security Notions

Our security definition follows the widely-accepted security frameworks in this field [6, 12, 15, 22]. It is summarized in fuzzy query over encrypted database that the overall security relies on the cryptographic assurance of indexes and trapdoors. In our schema, we store extra functional ciphers as indexes and rewrite queries as trapdoors. The security guarantee means there is no additional information leaked other than the functional results of fuzzy query.

4 Proposed Fuzzy Searching Encryption

4.1 Two Types of Functional Auxiliary Columns

The multiple-attributions-splitting design in cloud database synthesizes various encryptions to preserve query semantics. As shown in Table 2, two types of auxiliary columns (c-LSH and c-BF) are appended on cloud database along with a symmetrical determined column (DET).

Table 2. Storage pattern of multiple functional columns in database

c_{det}	c_{lsh}(m = 4, wid = 2)	$c_{bf}(1)$	...	$c_{bf}(\lceil\frac{l}{32}\rceil)$
0x1234 ("I love apple")	19030024, 01000409, 00020412	1077036627	...	1957741388
0x3456 ("lave banana")	01000409, 00020303	1079642851	...	625017556
0x5678 ("I love coconut")	19030024, 01000409, 06000700	1626500087	...	1687169793

This schema aims at handling queries with wild-cards on cipher-texts. So that several appended columns could store different functional ciphers with various encryptions, such as determination (DET) of data for equality, locality sensitive hashing (LSH) of words fragments for similar searching, bloom-filter (BF) among lines for maximum substring matching.

A. c-LSH. The c-LSH column, which stores the locality sensitive hashing values of each sentence, represents a message digest after dimensionality reduction. It

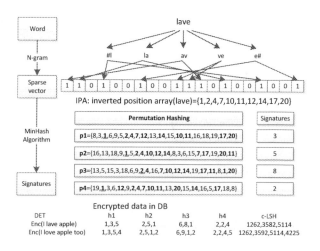

Fig. 2. A sample with bi-gram method (counting uni-gram as well) to show transforming process: (1) split sentences in line to multiple words; (2) transport a word to fragments with n-gram and build inverted position array; (3) execute dimension reduction with LSH and get m features; (4) link features to a token for each word; (5) combine tokens in line with comma.

helps map similar items together with probability which equals to the jaccard distance between their inverted position arrays (IPA for short).

During transforming process, n-gram methods are utilized (such as bi-gram and counting uni-gram) for dividing texts into fragments and finally to sparse vectors (IPA for short). As shown in Fig. 2, the transforming process maps every rows to separate signature collections by steps. This process changes measurement from levenshtein distance on texts to jaccard similarity on IPAs. So that the particular minhash algorithm could reduce the dimensions of numeric features for each subject (words). Finally, each word is converted to a linked sequence as a token and the c-LSH stores tokens set with comma to represent data of whole line.

B. c-BF. The multiple auxiliary c-BF columns, which represents macroscopic bloom-filter spaces for each row, are implemented on several 'bigint' (32-bit) columns. The database will return the DET ciphers where all c-BF columns cover the target sequences through native bit arithmetic operation '&'. Briefly, these columns are proposed for maximum substring matching which is a supplement to the c-LSH column above.

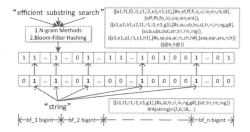

Fig. 3. The maximum substring matching over c-BF vectors which are stored in multiple 'bigint' auxiliary columns separately. After mapping fragments from whole sentence to vectors, queries execute with bit matching.

During mapping process, we respectively generate vectors for each row through bloom-filter hashing with following n-gram methods: bi-gram, tri-gram, prefix and suffix. These auxiliary columns are designed for substring matching so that the implicit information need be maximally persisted from origin strings. Through matching fragments between target bit vector and stored separated 'bigint' numbers, we could obtain all matched rows as shown in Fig. 3.

To meet application scenarios of inextensible cloud database, we accomplish operations completely through rewriting SQL statements by native bit arithmetic operation over multiple auxiliary columns, such as *select m_det from t where m_bf0&1=1 and m_bf1&3=3*. We experiment the connection between length of the sparse vector and correct rate of maximum substring matching in Sect. 5.

4.2 Adaptive Rewriting Method over Queries with Wild-Cards

In SQL, wild-card characters are used in 'like' expression: the percent sign '%' matches zero or more characters and the underscore '_' matches a single character. Usually the former symbol is a coarse-grained comparable delimiter and the latter could be tolerated by locality sensitive hashing in slightly different cases. So we construct an adaptive rewriting method over queries with wild-cards as shown in Fig. 4.

We consider three basic cases according to the number of percent signs to meet indivisible string fragments. Furthermore in every basic case, we also divide three sub-cases according to the number of underscore to benefit from different auxiliary columns. Besides, each query text is considered as whole word and substring while experiment exhibits the optimal selection.

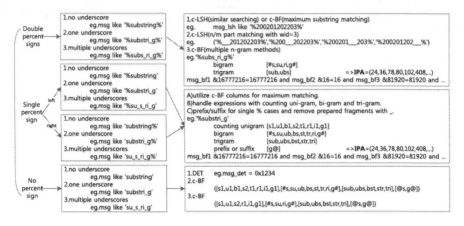

Fig. 4. Adaptive rewriting method over queries with wild-cards. We consider percent sign as a coarse-grained separator and few underscore could be tolerated according to similarity.

Firstly, the double percent signs case means that user attempts finding rows which contains the given string. Because the LSH function could tolerate small differences naturally, the sub-case with no underscore could accomplish similar searching among whole words. We achieve the one underscore sub-case with part matching method. This clever trick helps adjust fineness of similar searching as shown in Fig. 5. The multiple underscores sub-case is achieved by maximum substring matching on c-BF columns with bloom-filter.

Secondly, the single percent sign case need to consider prefix and suffix. The occurs of this type of queries reflect more detailed information and we match them all as substrings with maximum degree of constraint through various N-gram forms on c-BF column. Meanwhile, the prefix and suffix help preserve beginning and ending information of whole sentences in row. During splitting process, every fragments with underscore would be abandoned and the rest part would be mapped to the sparse bloom filter space which represented by IPA.

Thirdly, in the last no percent sign case, the user might already obtain most of target information and attempt to match specific patterns with underscore. Besides no underscore sub-case could be treated as determining equality operation, the maximum substring matching on c-BF column could meet the rest sub-cases' requirements.

Additionally, the tolerance parameter n is proposed as a flexible handler under the dimension m of locality sensitive hashing auxiliary column. Briefly, every features of word are set as fixed-length numbers which is filled by zero in basic scheme. As a linked string with all m features, the token could be converted to different variants where some feature parts replacing with underscores. We joint every possible cases together for database searching with keyword 'or' through a called bubble function as shown in Fig. 5.

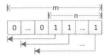

Fig. 5. The part matching method represents the adjustable fineness in c-LSH column with the tolerance parameter n and the LSH dimension m. For instance, let $m = 4, n = 3$ and the target feature set be $\{1, 2, 3, 4\}$, therefore the candidate set is $\{\_234, 1\_34, 12\_4, 123\_\}$ by this method.

The adaptive rewriting method helps generate trapdoor queries to meet the wild-cards fuzzy searching encryption in database through similar searching on c-LSH column and maximum substring matching on c-BF columns.

4.3   LSH-Based Security Improvements

The security of our schema relies on three parts. The symmetric cryptography algorithm guarantees the security on determining column and the divided bloom-filter vectors are presented by unidentifiable hashing ciphers. However, the content in c-LSH column might leak some extra information such as sizes and sequences of plain-texts. We present three improvements to enhance security and an integrated algorithm as followed.

A. Linking Features Without Padding
In basic scheme, we pad each feature with zero by the upper limit *wid* which benefits selecting process. To enhance security, we cancel the zero padding before linking features to a token. Meanwhile, the part matching method is also changed to an analogous bi-gram form. For instance, a secure enhanced part matching method is *select m_det where m_lsh like '%ab%' or m_lsh like '%bc%'* where *a,b,c* are multiple features of a word. We discuss the validity with experiments.

B. Modifying Sequences of Tokens
Each line of c-LSH auxiliary column stores a tokens set for whole sentence. Therefore, the sequences of tokens might exhibit the relevancy among words

to malicious attacker. To overcome this leakage, we modify the sequences randomly by hashing permutations. Additionally, we implement the permutation function with $\mathcal{P} : y = a * x + b \mod c$ where a is relatively-prime to c. This improvement protects the relation between invisible words and specific tokens. Since the matching only demands on existing rather than order, so this sequence modification helps for security protection.

C. Appending Confusing Tokens

The tokens sets in row leak the size of words. Appending tokens is a practical way for security, but what kind of token content should be added is the target of our discussion. The first way is appending repeated tokens from itself. It is simple and effective, but it only improves limited security. The second way is appending a little random tokens. Because of sparsity and randomization, few random tokens might not change the matching results. The third way is appending tokens combined from separated features among this tokens set. This way also influences the matching precision and increases proportion of false positive. Actually, these ways help greatly enhance security despite of disturbances.

D. Integrated Security Enhanced Algorithm

We present an integrated algorithm for security enhancement which combines all above implementations. As shown in Algorithm 1, this algorithm transforms the tokens set in each row to an security enhanced one. It helps prevent information leakage from c-LSH column.

Algorithm 1. Security Enhanced Improvements

Input: $token_m[\#word_i]$ which represents the m-dimensional tokens set of line i, wid be width of feature with zero padding, $amount$ be the lower bound for appending tokens

Output: an optimal security enhanced set $e\_token[amount]$

1 Let t represent token and each t can be split into m features by wid;
2 Generate a permutation function with $\mathcal{F} : y = a * x + b \mod c$ where $c = amount$ and $(a, c) = 1$;
3 Let $c = 0$ be the count for permutation;
4 **foreach** t in $token_m[\#word_i]$ **do**
5   Generate a temporary string et;
6   **for** $int\ j=0;\ j<m;\ j++$ **do**
7   Remove the zero prefix of $t.substr(j * wid, (j + 1) * wid)$;
8   Link it to et;
9     $e\_token[\mathcal{F}(c + +)]=et$;
10 **while** $c < amount$ **do**
11   Generate a temporary string et; **for** $int\ k=0;\ k<m;\ k++$ **do**
12   Get a feature $token_m[random()].substr(k * wid, (k + 1) * wid)$;
13   Remove the zero prefix and link it to et;
14     $e\_token[\mathcal{F}(c + +)]=et$;
15 return $e\_token[amount]$;

---

5   Performance Evaluation

In this section, we evaluate the performance of our work. Firstly, we discuss the effect of different n-gram methods about matching accuracy in c-LSH column. Secondly, we discuss the effect of bloom-filter length on collision degree and space usage of maximum substring matching. Thirdly, we discuss performance of adaptive rewriting method. Finally, we compare execution efficiency and space occupancy among efficient proposed schemes. The proposed scheme is implemented in Core i5-4460 3.20 GHz PC with 16 GB memory, and the used datasets include 2000 TOEFL words, the leaked user data of CSDN and the reuters news.

Manifestations of Different N-gram Methods on c-LSH Column. Utilizing bi-gram and counting uni-gram, we achieve similar searching on c-LSH column by introducing minhash algorithm based on the jaccard distance of fragments set. Intuitively, every change of character would greatly influence the corresponding fragments union over bi-gram method. So we introduce the counting uni-gram method to balance this excessiveness relativity. In this experiment, we evaluate the performance of these two n-gram methods and the combined one respectively.

The dataset we used is a 2000 TOEFL words set and we construct three variants of them to reveal the efficiency about LSH-based similar searching under different N-gram methods. The ways getting variants include appending a letter in the middle or in one side for every words, such as 'word' into 'words','wosrd','sword'. We calculate the average matched rows to reflect the searching results.

As shown in Fig. 6, we choose $m = 4, 6, 8$ to reveal matched numbers through part matching method with n. And the accuracy rate has a big promotion when n is larger than half of m. Besides, the combined method performs well when $m \geq 6$. It is reflected about the variation trend of accuracy that the amount of false positive reduces while the correct items remain unchanged.

The Bloom-Filter Length on c-BF Columns. In second experiment, we valuate collision accuracy and space occupancy under impacts of bloom-filter length and hashing function amount on c-BF columns when executing maximum substring matching. In detail, we attempt to find out an appropriate setting about the number of hashing function and the vector length of our bloom-filter structures.

The dataset we used is a leaked accounts set about CSDN, one of the most famous technical forum websites in China, and contains user name, password and e-mail. To guarantee effectiveness and avoid collisions, we change the vector length and keep the sparsity in several degrees such as half, quarter, one-sixth and one-eighth. Meanwhile different amount of hashing functions in bloom-filter influence accuracy and collision.

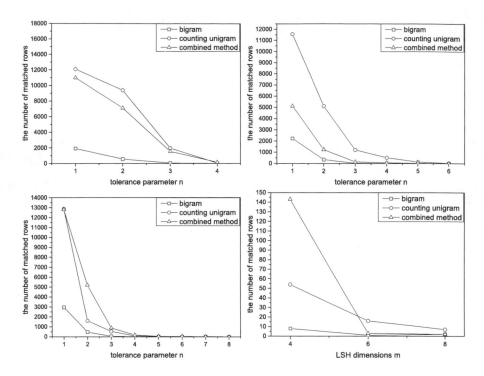

Fig. 6. The matching size under the tolerance parameter n and the LSH dimension m over variants of TOEFL words set. The first three graphs show the performance of different n-gram methods about part-matching respectively. The last graph shows the LSH dimension only complete-matching when $m = n$.

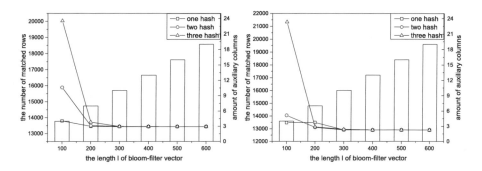

Fig. 7. The experiments show performance of maximum substring searching under different bloom-filter length l and different hashing amount k. We utilize fifty thousand rows of leaked CSDN account data and set several degrees of sparsity about bloom-filter vector while each row contains 50 characters. The left graph shows matching sizes of substring '163.com' on $\lceil \frac{l}{32} \rceil$ auxiliary columns when we build indexes under different length of bloom-filter. And the right graph represents 'qq.com'.

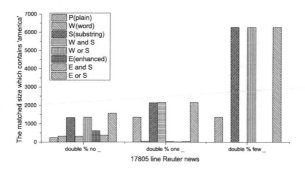

Fig. 8. The experiment shows the performance of adaptive rewriting method under different combinations, and reveals the most qualified modes for each fuzzy searching cases. Some expressions are used, such as '%america%', '%am_rica%', '%am_ri_a%'.

Because the bloom-filter length l corresponds to the amount of c-BF columns, this experiment discuss relations between matching accuracy and space occupancy under different amount of bloom-filter hashing functions. As shown in Fig. 7, the amount of matching size drops rapidly in the first place and then gets stable when sparsity is close to one-sixth.

The Performance of Adaptive Rewriting Method. This experiment aims at verifying effectiveness of the adaptive rewriting method. After auxiliary columns storing values as indexes, the 'like' clauses with wild-cards are analyzed by an adaptive rewriting method and rewritten to trapdoors. In this experiment, we consider the content of expression as a word or substring for comparison, and execute different types of queries with basic and security enhanced schemes respectively.

The dataset we used is Reuters-21578 news of 1987 [28]. In this experiment, we mainly discuss the double '%' cases because the other single '%' and no '%' cases carry out analogous steps. The only difference is that these cases additionally consider the prefix and suffix.

As shown in Fig. 8, we compare the matched size under different combinations. We also execute the origin SQL statements on extra stored plain-text column for contrast. It helps find the best combination modes under various wild-cards cases. We accomplish this experiment with the sparsity of c-BF columns being one-sixth and the dimension of c-LSH column being six. The graph shows that 'W and S' is fit for double '%' no '_' and double '%' one '_' cases while 'S' is fit for double '%' few '_' case. Besides, we discuss the performance of LSH-based security enhanced method and the graph confirms its feasibility.

Performance Comparison Among Proposed Schemes. In this section, we compare the efficiency of proposed schemes about inserting and selecting data. In general, the inserting process involves generating indexing values in auxiliary columns, and the selecting process involves decrypting determined cipher-texts.

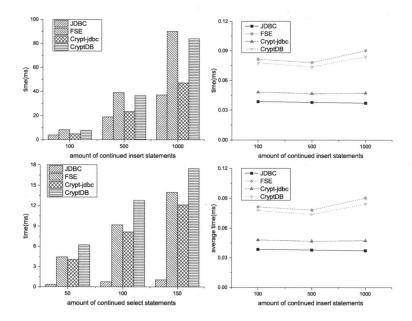

Fig. 9. This experiment show the execution efficiency among proposed schemes.

As shown in Fig. 9, our schema verifies this point and performs well comparing to normal JDBC, Crypt-jdbc and CryptDB.

6   Conclusion

This paper investigates the problem about fuzzy searching encryption with complex wild-cards queries on proxy-based encrypted database, then gives a practical schema with two types of auxiliary columns and rewriting SQL statements. Besides, security enhanced implementations and extensive experiments show the effectiveness. In future, the serialization and compression of functional ciphertexts would be studied to reduce space overhead.

References

1. Boldyreva, A., Chenette, N.: Efficient fuzzy search on encrypted data. In: Cid, C., Rechberger, C. (eds.) FSE 2014. LNCS, vol. 8540, pp. 613–633. Springer, Heidelberg (2015). https://doi.org/10.1007/978-3-662-46706-0_31
2. Cash, D., Jaeger, J., Jarecki, S., Jutla, C., Krawczyk, H., Rou, M.C., Steiner, M.: Dynamic searchable encryption in very-large databases: data structures and implementation. In: Network and Distributed System Security Symposium (2014)
3. Cash, D., Jarecki, S., Jutla, C., Krawczyk, H., Roşu, M.-C., Steiner, M.: Highly-scalable searchable symmetric encryption with support for boolean queries. In: Canetti, R., Garay, J.A. (eds.) CRYPTO 2013. LNCS, vol. 8042, pp. 353–373. Springer, Heidelberg (2013). https://doi.org/10.1007/978-3-642-40041-4_20

4. Chase, M., Kamara, S.: Structured encryption and controlled disclosure. In: Abe, M. (ed.) ASIACRYPT 2010. LNCS, vol. 6477, pp. 577–594. Springer, Heidelberg (2010). https://doi.org/10.1007/978-3-642-17373-8_33

5. Chen, H., Tian, X., Yuan, P., Jin, C.: Crypt-JDBC model: optimization of onion encryption algorithm. J. Front. Comput. Sci. Technol. **11**(8), 1246–1257 (2017)

6. Curtmola, R., Garay, J., Kamara, S., Ostrovsky, R.: Searchable symmetric encryption: improved definitions and efficient constructions. In: ACM Conference on Computer and Communications Security, pp. 79–88 (2006)

7. Fan, K., Yin, J., Wang, J., Li, H., Yang, Y.: Multi-keyword fuzzy and sortable ciphertext retrieval scheme for big data. In: 2017 IEEE Global Communications Conference, GLOBECOM 2017, pp. 1–6. IEEE (2017)

8. Fu, Z., Wu, X., Guan, C., Sun, X., Ren, K.: Toward efficient multi-keyword fuzzy search over encrypted outsourced data with accuracy improvement. IEEE Trans. Inf. Forensics Secur. **11**(12), 2706–2716 (2017)

9. Hahn, F., Kerschbaum, F.: Searchable encryption with secure and efficient updates. In: ACM SIGSAC Conference on Computer and Communications Security, pp. 310–320 (2014)

10. Jho, N.S., Chang, K.Y., Hong, D., Seo, C.: Symmetric searchable encryption with efficient range query using multi-layered linked chains. J. Supercomput. **72**(11), 1–14 (2016)

11. Kamara, S., Papamanthou, C.: Parallel and dynamic searchable symmetric encryption. In: Sadeghi, A.-R. (ed.) FC 2013. LNCS, vol. 7859, pp. 258–274. Springer, Heidelberg (2013). https://doi.org/10.1007/978-3-642-39884-1_22

12. Kamara, S., Papamanthou, C., Roeder, T.: Dynamic searchable symmetric encryption. In: ACM Conference on Computer and Communications Security, pp. 965–976 (2012)

13. Kurosawa, K., Ohtaki, Y.: UC-secure searchable symmetric encryption. In: Keromytis, A.D. (ed.) FC 2012. LNCS, vol. 7397, pp. 285–298. Springer, Heidelberg (2012). https://doi.org/10.1007/978-3-642-32946-3_21

14. Kurosawa, K., Sasaki, K., Ohta, K., Yoneyama, K.: UC-secure dynamic searchable symmetric encryption scheme. In: Ogawa, K., Yoshioka, K. (eds.) IWSEC 2016. LNCS, vol. 9836, pp. 73–90. Springer, Cham (2016). https://doi.org/10.1007/978-3-319-44524-3_5

15. Kuzu, M., Islam, M.S., Kantarcioglu, M.: Efficient similarity search over encrypted data. In: IEEE International Conference on Data Engineering, pp. 1156–1167 (2012)

16. Lesani, M.: MrCrypt: static analysis for secure cloud computations. ACM SIGPLAN Not. **48**(10), 271–286 (2013)

17. Li, J., Liu, Z., Chen, X., Xhafa, F., Tan, X., Wong, D.S.: L-EncDB: a lightweight framework for privacy-preserving data queries in cloud computing. Knowl.-Based Syst. **79**, 18–26 (2015)

18. Liu, Z., Li, J., Li, J., Jia, C., Yang, J., Yuan, K.: SQL-based fuzzy query mechanism over encrypted database. Int. J. Data Wareh. Min. (IJDWM) **10**(4), 71–87 (2014)

19. Liu, Z., Ma, H., Li, J., Jia, C., Li, J., Yuan, K.: Secure storage and fuzzy query over encrypted databases. In: Lopez, J., Huang, X., Sandhu, R. (eds.) NSS 2013. LNCS, vol. 7873, pp. 439–450. Springer, Heidelberg (2013). https://doi.org/10.1007/978-3-642-38631-2_32

20. Popa, R.A., Li, F.H., Zeldovich, N.: An ideal-security protocol for order-preserving encoding. In: IEEE Symposium on Security and Privacy, pp. 463–477 (2013)

21. Popa, R.A., Redfield, C.M.S., Zeldovich, N., Balakrishnan, H.: CryptDB: protecting confidentiality with encrypted query processing. In: ACM Symposium on Operating Systems Principles, SOSP 2011, Cascais, Portugal, October, pp. 85–100 (2011)
22. Song, D.X., Wagner, D., Perrig, A.: Practical techniques for searches on encrypted data. In: IEEE Symposium on Security and Privacy, p. 44 (2000)
23. Wang, B., Yu, S., Lou, W., Hou, Y.T.: Privacy-preserving multi-keyword fuzzy search over encrypted data in the cloud. In: 2014 Proceedings of IEEE INFOCOM, pp. 2112–2120 (2014)
24. Wang, C., Cao, N., Li, J., Ren, K.: Secure ranked keyword search over encrypted cloud data. In: IEEE International Conference on Distributed Computing Systems, pp. 253–262 (2010)
25. Wang, J., Ma, H., Tang, Q., Li, J., Zhu, H., Ma, S., Chen, X.: Efficient verifiable fuzzy keyword search over encrypted data in cloud computing. Comput. Sci. Inf. Syst. $10(2)$, 667–684 (2013)
26. Wang, Q., He, M., Du, M., Chow, S.S.M., Lai, R.W.F., Zou, Q.: Searchable encryption over feature-rich data. IEEE Trans. Dependable Secur. Comput. $PP(99)$, 1 (2016)
27. Wei, X., Zhang, H.: Verifiable multi-keyword fuzzy search over encrypted data in the cloud. In: International Conference on Advanced Materials and Information Technology Processing (2016)
28. Wiki: Reuters. http://www.research.att.com/~lewis

Towards Privacy-Preserving Travel-Time-First Task Assignment in Spatial Crowdsourcing

Jian Li[1], An Liu[1(✉)], Weiqi Wang[1], Zhixu Li[1], Guanfeng Liu[1], Lei Zhao[1], and Kai Zheng[2]

[1] School of Computer Science and Technology, Soochow University, Suzhou, China
anliu@suda.edu.cn
[2] University of Electronic Science and Technology of China, Chengdu, China

Abstract. With the ubiquity of mobile devices and wireless networks, spatial crowdsourcing (SC) has gained considerable popularity and importance as a new tool of problem-solving. It enables complex tasks at specific locations to be performed by a crowd of nearby workers. In this paper, we study the privacy-preserving travel-time-first task assignment problem where tasks are assigned to workers who can arrive at the required locations first and no private information are revealed to unauthorized parties. Compared with existing work on privacy-preserving task assignment, this problem is novel as tasks are allocated according to travel time rather than travel distance. Moreover, it is challenging as secure computation of travel time requires secure division which is still an open problem nowadays. Observing that current solutions for secure division do not scale well, we propose an efficient algorithm to securely calculate the least common multiple (LCM) of every workers speed, based on which expensive division operation on ciphertexts can be avoided. We formally prove that our protocol is secure against semi-honest adversaries. Through extensive experiments over real datasets, we demonstrate the efficiency and effectiveness of our proposed protocol.

Keywords: Spatial crowdsourcing · Privacy-preserving
Task assignment

1 Introduction

Thanks to the ubiquitous wireless networks and powerful mobile devices, spatial crowdsourcing has gained considerable popularity and importance as a new tool of problem-solving. It can be applied to simple tasks such as photo-taking where people act as sensors, or to complex tasks such as handyman service where people work as intelligent processing units. As an emerging crowdsourcing mode, spatial crowdsourcing differs from other crowdsourcing modes in that people in spatial crowdsourcing, also known as workers, must physically move to certain places to perform those spatial tasks. Recently years have witnessed an upsurge of interest

© Springer International Publishing AG, part of Springer Nature 2018
Y. Cai et al. (Eds.): APWeb-WAIM 2018, LNCS 10988, pp. 19–34, 2018.
https://doi.org/10.1007/978-3-319-96893-3_2

in spatial crowdsourcing applications in daily life, ranging from local search-and-discovery (e.g., Foursquare) to home repair and refresh (e.g., TaskRabbit).

A typical workflow of spatial crowdsourcing consists of four steps: task/worker registration, task assignment, answer aggregation, and quality control [1]. Among them, task assignment focuses on allocating a set of tasks to a set of workers according to a set of constraints such as location, time, and budget. Typically finding an optimal assignment subject to multiple constraints is NP-hard, which calls for efficient yet effective algorithms. Based on specific optimization goals, a variety of approaches have been proposed, for example, to maximize the total number of completed tasks [2], to maximize the number of tasks performed by a single worker [3] and to maximize the reliability-and-diversity score of assignments [4].

The problem of task assignment becomes even tougher when privacy issues are taken into account. It is not hard to see that the data used for decision making in task assignment is usually private and thus need to be kept secret due to the lack of trust among workers, task requesters, and the spatial crowdsourcing server. To achieve privacy, these private data should be protected by for example encryption using mature cryptographical algorithms or perturbation using emerging privacy-preserving techniques. However, the noise introduced by these mechanisms will decrease significantly the utility of the data and sometimes even will make the data useless. It is therefore more challenging to deal with task assignment with the extra privacy constraint.

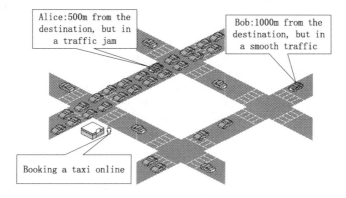

Fig. 1. Spatial crowdsourcing where travel time is more important than travel distance

The above hard problem has been studied by several work recently [5–8]. A common strategy of task assignment adopted by these work is travel-distance-first, that is, a task will be assigned to the worker who has the shortest travel distance to its location. This strategy is simple but sometimes is unreasonable in practice as it is common for some workers to move faster than others. Consider a simple example where a user wants to request a car through a spatial crowdsourcing platform (e.g., Uber). As shown in Fig. 1, two workers (i.e., drivers) named Alice and Bob are available when the task is issued, and their distances to the user are 500 m and 1,000 m, respectively. Using the aforementioned strategy, the spatial crowdsourcing server will assign this task to Alice as she is nearer to the user. As shown in the figure, however, Alice is in a traffic jam. On the other hand, Bob is in a smooth traffic and he can arrive at the users location before Alice. This simple example motivates us to consider, travel-time-first, a more effective strategy when allocating tasks to workers in practice.

In this paper, we propose a privacy-preserving task assignment protocol for spatial crowdsourcing platforms taking travel-time-first strategy, that is, workers who can first arrive at the location of a given task have priority over others. While it is more effective than travel-distance-first in practice, travel-time-first makes privacy-preserving task assignment more challenging due to the required division operation involved in the computation of travel time. For every user, his/her location and speed are both private and should be protected. How to perform division efficiently and accurately on encrypted or perturbed data is still an open problem. In [9], the authors propose a protocol for secure division based on ElGamal cryptosystem. However, this protocol does not scale well and cannot be applied to large spatial crowdsourcing system for the key length should be set large enough to avoid computation overflow and this will introduce prohibitive computation cost. To overcome this weakness, we transform the secure division problem into a secure least common multiple (LCM) problem. We propose an efficient way to calculate the LCM securely. Through extensive experiments, we demonstrate the feasibility and efficiency of our solution.

The remainder of this paper is organized as follows: Sect. 2 discusses related work. Problem definition and background knowledge are presented in Sect. 3. Section 4 introduces our approach in details. Section 5 analyzes the security and complexity of our approach theoretically. Section 6 evaluates our approach on real datasets. Section 7 concludes the paper.

2   Related Work

To be consistent with our contributions, we only review the works that are relevant to task assignment and privacy-preserving. Kazemi and Shahabi [2] propose several solutions to maximize the overall number of assigned tasks under the constraints of workers. Similarly, The assignment protocol proposed by [10] is to assign the time-constrained and multi-skill-required spatial tasks with dynamically moving workers. In [11], Zheng et al. take workers' rejection into consideration and try to maximize workers' acceptance in order to improve the system throughput. Tong et al. [12] devise efficient algorithms with provable competitive radio with online dynamic scenarios. And in [13], Tong et al. propose an online task assignment framework based on offline guidance to maximize the task allocation while maintaining the efficient task assignment. In [14], Gao et al. design a two-level-based framework to recommend suitable teams to accomplish a task. However, these works are all based on a pre-condition that workers do not refuse to disclose their private information to the SC platform that is hard to achieve in reality. Our work focuses on privacy-preserving during an execution of task assignment.

In recent years, the public concern over privacy has stimulated lots of research efforts in privacy-preserving. A location based query solution is proposed by Paulet et al. [15] that employs two protocols that enables a user to privately determine and acquire location data. In [16], Liu et al. propose an efficient approach to protecting mutual privacy in location-based queries by performing two rounds of oblivious transfer (OT) extension on two small key sets. A solution built on the Paillier public-key cryptosystem is presented by Yi et al. [17] for mutual privacy-preserving kNN query with fixed k and is extended in [18] where k is dynamic. Unfortunately, these solutions where workers location are private data of the SC platform are not suitable for our framework for workers location should be known to the SC platform in a secret way. Also, in [19], Sun et al. focus on the privacy-preserving task assignment in SC by presenting an approach where location privacy of workers can be protected in a k-annoymity manner. In [5], To et al. propose a framework for protecting location privacy of workers participating in SC tasks without protecting task location. Liu et al. [20] propose an efficient solution to securely compute the similarity between two encrypted trajectories without revealing nothing about the trajectories. However, their protocols also cannot be applied to our framework for they have too heavy computation cost to solve large task assignment problems.

3   Problem and Preliminary

In this section, we first present some definitions used in our work and then briefly introduce some cryptosystems based on which our protocol is built.

3.1   Problem Definitions

Definition 1 *(Spatial Task). A spatial task, denoted as T, is a task to be performed l_T.*

Definition 2 *(Workers). Let $W = \{w_1, \cdots, w_n\}$ be a set of n workers. Each worker w has an ID id_w, a location l_w, a constant speed s_w, and an acceptance rate AR_w which is the probability that he/she accepts a task assigned to him/her.*

As mentioned in the introduction, we mainly consider travel-time-first, a new task assignment strategy in privacy-preserving spatial crowdsourcing. Ideally, we only need to find a worker $w \in W$ who can first arrive at l_T and then assign T to w. This works if the worker is certain to accept the assigned task, but sometimes it is not. Therefore we consider a more general case where every worker w has an acceptance ratio denoted as AR_w for an assignment, and we need to ensure the probability that a task T is accepted by at least one worker is larger than a given threshold α_T. In this case, we need to find a set of workers $U \subset W$ rather than a single worker. It is easy to see that the probability that T is accepted by at least one worker in U is $\alpha_U = 1 - \prod_{w \in U}(1 - AR_w)$. Hence the travel-time-first task assignment problem can be formalized as follows:

Definition 3 *(Travel-time-first Task Assignment Problem). Given a set of workers W, a task T and its acceptance threshold α_T, the travel-time-first task assignment assigns task T to a set of workers $U \subset W$ such that:*

$$\frac{d\left(l_i, l_T\right)}{s_l} \leq \frac{d\left(l_j, l_T\right)}{s_j} \quad and \quad \alpha_T \leq \alpha_U \tag{1}$$

for $\forall i \in U$ and $\forall j \in W \setminus U$.

Privacy-preserving means all the private data should be hidden from unauthorized parties in the procedure of task assignment. To accurately define the ability of unauthorized parties, we adopt a typical adversary model, i.e., the semi-honest model [21]. Specifically, all parties in this model are assumed to be semi-honest, that is, they follow a given protocol exactly as specified, but may try to learn as much as possible about other parties private input from what they see during the protocols execution. This can be formally defined by the real-ideal paradigm as follows: for all adversaries, there exists a probabilistic polynomial-time simulator, so that the view of the adversary in the real world and the view of the simulator in the ideal world are computationally indistinguishable. Specifically, the security of a protocol Π is defined as follows:

Definition 4. *Let $p_i(1 \leq i \leq n)$ be n parties involved in a protocol Π. For $p_i(1 \leq i \leq n)$, its view, private input and extra knowledge it can infer during an execution of P_i are defined as V_i, X_i and K_i respectively. A protocol Pi has a strong privacy guarantee, that is, p_i cannot learn any knowledge except the final output of p_i, if these exists a probabilistic polynomial-time simulator P_i such that:*

$$P_i(X_i, \Pi(X_1, \cdots, X_n), K_i)_{X_1, \cdots, X_n} \equiv V_i(X_1 \cdots, X_n)_{X_1 \cdots, X_n} \tag{2}$$

and $K_i = \emptyset$, where $\equiv$ means computational indistinguishability. However, this strong guarantee cannot be achieved sometimes for $K_i \neq \emptyset$. If $K_i \neq \emptyset$, Π is said to be privacy-preserving with K_i disclosure against p_i in the sense that it reveals no more knowledge than K_i and the final output to p_i.

Now we are ready to define the problem of privacy-preserving travel-time-first task assignment as follows:

Definition 5 *(Privacy-preserving Travel-time-first Task Assignment Problem). Given a set of workers W, a task T and its acceptance threshold α_T, the travel-time-first task assignment assigns task T to a set of workers $U \subset W$ such that Eqs. (1) and (2) hold.*

3.2   Cryptosystems

The privacy-preserving property of our protocol is built on several well-known cryptosystems: PRG [22], Paillier [23] and ElGamal [24]. The details of PRG, Paillier and ElGamal can be found in the given references and all of them are proved to be secure. Here we only emphasize some important properties of these cryptosystems.

PRG can be implemented by using a one-way hash function denoted as G_k. For Paillier, its encryption and decryption are denoted as E_p and D_p, respectively. For ElGamal, its encryption and decryption are denoted as E_e and D_e, respectively. The important properties of Paillier and ElGamal are listed as follows:

Homomorphic Properties of Paillier: Given m_1 and m_2 are two messages, we have:

$$E_p(m_1)E_p(m_2) = E_p(m_1 + m_2). \tag{3}$$

$$E_p(m)^k = E_p(km). \tag{4}$$

Commutative-Like Property of ElGamal: Given a message m, we have:

$$E_e^{h_a}(E_e^{h_b}(m)) = E_e^{h_b}(E_e^{h_a}(m)). \tag{5}$$

4   Proposed Privacy-Preserving Framework

In this section, we will introduce our privacy-preserving framework in details and explain how to get LCM in a safe and secret way by AP encryption strategy.

4.1   Framework Overview

As Fig. 2 shows, our proposed framework consists of six stages, namely Initialization, Distance, LCM, Time, Comparison and Verification respectively. Different colors mean different stages.

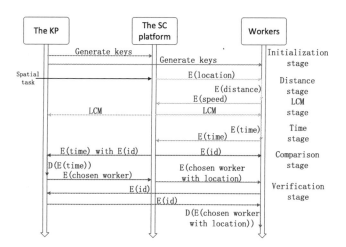

Fig. 2. Framework overview

4.2   Detailed Framework

Next, every stage in details is described in this subsection.

Initialization Stage. Firstly, the KP generates a pair of keys for Paillier. Then the KP keeps the encryption key public and the decryption key private respectively because the computations of the SC platform and workers are based on encrypted data while the KP has to decrypt data to find the chosen worker. Besides, the KP generates a cyclic group G for ElGamal based on which the KP and all workers generate their own pair of keys and keep them secret.

Distance Stage. Given a spatial task T, the SC platform encrypts task location $l_T(x_T, y_T)$ with the encryption key of Paillier by calculating $E_p(x_T^2 + y_T^2)$, $E_p(x_T)$ and $E_p(y_T)$. Then these three ciphertexts are sent to all workers. Without holding the decryption key of Paillier, every worker w_i can calculate the encrypted square of the distance based on the Euclidean distance and the homomorphic properties shown in Eqs. 3 and 4 as follows:

$$E_p(d^2(l_T, l_i)) = E_p(x_T^2 + y_T^2)E_p(x_T)^{-2x_i}E_p(y_T)^{-2y_i}E_p(x_i^2 + y_i^2) \qquad (6)$$

It should be noted that it also works when every worker encrypts location and the SC platform calculates $E_p(d^2(l_T, l_i))$. However, it will cost much more computing resources for every worker can calculate in parallel. That is to say, our proposed method is good for reducing the computation cost of the SC platform.

LCM Stage. At first, we explain why we need to get the LCM of all worker's speed. As defined in Definition 3, our framework prefers the worker who has the shortest travel time. To this end, we have to face division operation on ciphertexts which is still an open problem nowadays during the computation of travel time. Though we cannot solve the problem of division operation, a transformation can be employed to avoid the division operation based on the following lemma:

Lemma 1. *Let $W = \{w_1, \cdots, w_n\}$ be a set of n workers, $D = \{d_1, \cdots, d_n\}$ be the distance between task location and the worker w_i, S_{lcm} be the LCM of every worker's speed s_i and $s_i' = S_{lcm}/s_i$ where $1 \le i \le n$. So for any two different workers $w_i, w_j \in W$, if $d_i s_i' < d_j s_j'$ holds then we must infer $d_i/s_i < d_j/s_j$.*

Proof. $d_i s_i' < d_j s_j' \Longleftrightarrow d_i s_i'/S_{lcm} < d_j s_j'/S_{lcm} \Longleftrightarrow d_i/s_i < d_j/s_j$.

Deforming the formula of travel time can help us avoid the division operation over ciphertexts, which is the reason why we need to get the LCM. Note that the product of all speeds is not suitable here for it may cause the overflow of the multiplication of all speeds [9]. The process of calculating the LCM by AP encryption strategy in a safe and secret way will be introduced in the next subsection. In the end, the SC platform will inform the KP and all workers of the LCM.

Time Stage. Upon receiving the LCM S_{lcm}, every worker w_i can calculate an equivalent encrypted travel time t_i' to replace real encrypted travel time t_i based on the Lemma 1 where $t'_i = d(l_i, l_T)s'_i$ and $t_i = d(l_i, l_T)/s_i$. For no worker holds the decryption key of Paillier, homomorphic properties of Paillier are used again as follows:

$$E_p(t'^2_i) = E_p((d(l_t, l_i)s'_i)^2) = E_p((d(l_t, l_i)S_{lcm}/s_i)^2) = E_p(d^2(l_T, l_i))^{(S_{lcm}/s_i)^2} \tag{7}$$

where $E_p\left(d^2(l_t, l_i)\right)$ is calculated by Eq. 6 and s_i is the speed of worker w_i. Then the worker sends the encrypted equivalent travel time with his own ID to the SC platform for comparison in the form of $(i, E_p(t'^2_i))$.

Comparison Stage. When receiving the list of $(i, E_p(t'^2_i))$, the SC platform adopts a PRG G_k to encrypt the ID of workers as $(G_k(i), E_p(t'^2_{G_k(i)}))$ for the protection of workers especially the chosen worker. Then the SC platform sends the list $(G_k(i), E_p(t'^2_{G_k(i)}))$ to the KP and sends every $G_k(i)$ to the corresponding worker w_i. With the decryption key of Paillier, the KP can decrypt $E_p(t'^2_{G_k(i)})$ to obtain the $t'^2_{G_k(i)}$ and the real travel time $t_{G_k(i)}$ can be computed by $\sqrt{\frac{tt'^2_{G_k(i)}}{S^2_{lcm}}}$ where S_{lcm} is achieved in the LCM. And then the KP can easily find the chosen worker who has the shortest $t_{G_k(i)}$. Then, the ID of the chosen worker $G_k(i^*)$ is encrypted by ElGamal, whose output is $E^{KP}_e(G_k(i^*))$. At last, the KP sends $E^{KP}_e(G_k(i^*))$ to the SC platform. This encrypting operation is essential because the SC platform can infer that who is the chosen worker from $G_k(i^*)$. However, when AR is not always 100%, we will return a set of chosen workers instead of a chosen worker.

Verification Stage. To ensure only the chosen worker can learn the true task location, the SC platform hides the true task location by encrypting $E^{KP}_e(G_k(i^*))$ and l_T as follows:

$$E(l_T) = h\left(E^{KP}_e(G_k(i^*))\right) \oplus l_T \tag{8}$$

where function h is a length-match hash function which is used shorten a long bit-string and it is proved to be semantically secure. We perform exclusive-OR on the l_T and the output of function h because an important property of exclusive-OR is $a \oplus b \oplus a = b$. Based on this property, only the chosen worker w^*_i can infer the true task location by $l_s = E(l_s) \oplus h(E^{KP}_e(G_k(i^*)))$. The detailed procedure is as follows:

With their own ElGamal, every worker encrypts their own encrypted ID $G_k(i)$ received in the comparison stage as $E^{w_i}_e(G_k(i))$ and sends it to the KP. For all ElGamals are based on the same cyclic group G, commutative-like encryption can be implemented by $E^{KP}_e(E^{w_i}_e(G_k(i))) = E^{w_i}_e(E^{KP}_e(G_k(i)))$ with the same random number for the consistence of E^{KP}_e and the result is sent back to workers. Every worker w_i can decrypt it by the decryption key of his own ElGamal and get $E^{KP}_e(G_k(i))$. It is obvious that only the chosen worker can infer $E^{KP}_e(G_k(i^*))$ and thus infer the true task location.

Algorithm 1. Calculating LCM

Input: the maximal speed S_{max}, the speed s_i of every worker $w_i (1 \leq i \leq n)$
Output: the LCM of all speeds S_{lcm}

1: The SC platform and all workers perform the same exclusion algorithm on S_{max} to get a same list L of 2-tuples $<p, c_p>$ where p is a prime meeting $p \leq S_{max}$ and c_p is the maximal times of p meeting $p^{c_p} \leq S_{max}$.
2: Every worker w_i computes his own factorization F_i of s_i by Pollard's rho algorithm.
3: AP performs $\sum_{p \in P} p*(c_p+1)$ key generations and assigns these secrets respectively
4: **for** each prime p in L **do**
5:     **for** number $k(0 \leq k \leq c_p)$ **do**
6:         Every worker w_i generates his own flag data $f[k]$, encrypts it by
7:         the assigned AP secrets and sends it to the SC platfrom.
8: $S_{lcm} = 1$
9: **for** each prime p in L **do**
10:     **for** number $k(c_p \geq k \geq 0)$ **do**
11:         The SC platform decrypts the sum of all $f[k]$, denoted as H.
12:         **if** $H > 0$ **then**
13:             $S_{lcm} = S_{lcm} * p^H$
14:             break
15: **return** S_{lcm}

4.3   Calculating LCM

To compute the LCM securely, we adopt an aggregation protocol denoted as AP [25] which can calculate the sum of multiple messages in a privacy-preserving manner. It works as follows:

Key Generation: Let S be a set of nc random numbers where n is the number of workers and c is a random number. Then, divide S into n random disjoint subsets S_i with c numbers and define $M = 2^{\lceil \log_2 n\Delta \rceil}$ where Δ is maximum value of workers's data. At last, send k_i to w_i and the sum k_0 to the SC platform where $k_i = (\sum_{s' \in S_i} s') \mod M$ and $k_0 = (\sum_{s' \in S} s') \mod M$.

Encryption E_a: For each worker w_i, he encrypt data m_i by computing:

$$c_i = (k_i + m_i) \mod M \tag{9}$$

Encryption D_a: The SC platform can decrypt the sum by computing:

$$S(\sum_{i=1}^{n} m_i) = (\sum_{i=1}^{n} c_i - k_0) \mod M \tag{10}$$

Based on a credible assumption that the maximal worker's speed is limited and known to all, we explain the Algorithm 1 as follow: In line 1 and 2, exclusion algorithm is performed to get the list L of 2-tuples $<p, c_p>$ whose complexity is $O(n \log(\log n))$. For example, our maximal speed is 10. Then 3 is one prime where $3 < 10$, and its maximal times is 2 for $3^2 \leq 10$. So the tuple $<3, 2>$ will be inserted into the list. Besides, every worker calculates the factorization F_i of his own speed s_i by Pollard's rho algorithm whose complexity is $O(n^{\frac{1}{4}})$. For example, the factorization F of a worker($s_i = 6$) is $F = 2 * 3$ for $6 = 2 * 3$. Based on the list L, the AP generates $\sum_{p \in L} p * (c_p + 1)$ different keys for same key may disclose workers' speed in line 3. In line 4 to 7, each worker w_i generates his flag data $f[k](k \in [0, c_p])$ as follows:

$$f[k] = \begin{cases} 1, & AT[p] = k \\ 0, & otherwise \end{cases} \tag{11}$$

where $AT[p]$ is the appearance times of p in the corresponding F_i. Then, encrypts and sends flag data. In the above examples, when $p = 3$, this worker ($s_i = 6$) generates these flag data $f[0] = 0, f[1] = 1, f[2] = 0$. In line 9 to 14, the LCM is computed by $S_{lcm} = \prod_{p \in L} p^H$. For example, the factorization of another worker($s_i = 9$) is $3 * 3$. If $p = 3$, this worker generates flag data $f[0] = 0, f[1] = 0, f[2] = 1$. So the maximal times of 3 is 2 for the decrypted sum of $f[2]$ meets the condition in line 12. Meanwhile, the maximal times of $2, 5, 7$ are $1, 0, 0$ respectively. So $S_{lcm} = 2^1 * 3^2 * 5^0 * 7^0 = 18$ will be returned.

5   Security and Complexity Analysis

Denoting the LCM stage as $E_a(s_i)$ and $D_a(S_{lcm})$, we will prove the security and complexity of our framework next.

5.1   Security Analysis

Theorem 1. *Our framework is allowed to be privacy-preserving with $K_0 = S_{lcm}, K_{-1} = \left\{ S_{lcm}, t_{G_k(i)} \right\}$ and $K_i = S_{lcm}(1 \leq i \leq n)$ extra knowledge.*

Proof. We firstly consider the SC platform w_0 with $K_0 = S_{lcm}$. Then the view is $V_0 = \left\{ E_e^{KP}(G_k(i^*)), S_{lcm}, E_a(s_j), E_p(t_j'^2) \right\} (1 \leq j \leq n)$. There is a probabilistic polynomial-time simulator P_0 that generates $V_0' = \left\{ E_e^{KP}(x_1), S_{lcm}, E_a(y_i), E_p(z_i) \right\}$ where x_1 is random number from a cyclic group G, $y_i(1 \leq i \leq n)$ are random numbers distributed in $\mathbb{Z}$ and $z_i(1 \leq i \leq n)$ are random numbers uniformly distributed in $\mathbb{Z}_N$. As Paillier, ElGamal and AP are all secure, it is clear that $V_0 \equiv V_0'$.

Next we analyze every worker w_i with $K_i = S_{lcm}$. There is a probabilistic polynomial-time simulator P_i to simulate worker w_i's view. However, There are two kinds of workers to be analyzed. The difference between them is that only the chosen worker can infer the chosen ID is his ID. For the chosen worker w_i^*, his view is $V_{i^*} = \left\{ G_k(i), i*, S_{lcm}, E_p(x_T^2 + y_T^2), E_p(x_T), E_p(y_T) \right\}$. So simulator P_{i^*} generates $V_{i^*}' = \left\{ g, i*, S_{lcm}, E_p(x_1), E_p(x_2), E_p(x_3) \right\}$ where $x_i(i = 1, 2, 3)$ are random numbers uniformly distributed in $\mathbb{Z}_N$ and g is a random element uniformly distributed over $\{0, 1\}^\lambda$. For others, the view for them is $V_i = \{G_k(i), E_{re}(E_e^{w_i}(G_k(i^*))), S_{lcm}, E_p(x^2 + y^2), E_p(x), E_p(y)\}$ and simulator P_i generates $V_i' = \{g, E_{re}(E_e^{w_i}(y)), S_{lcm}, E_p(x_1), E_p(x_2), E_p(x_3)\}$ where x_i and g are the same as V_{i*}' and y is a random number from G. Based on the semantic security of Paillier, ElGamal and PRG, we can easily verify that $V_i \equiv V_i'(1 \le i < n)$.

Finally, we analyze the KP w_{-1} with $K_{-1} = \{S_{lcm}, t_{G_k(i)}\}$ $(1 \le i \le n)$. The view of the KP is $V_{-1} = \{S_{lcm}, t_{G_k(i)}, E_e^{w_i}(G_k(i))\}$ $(1 \le i \le n)$. There is a probabilistic polynomial-time simulator P_{-1} that generates $V'_{-1} = \{S_{lcm}, t_{x_i}, E_e^{w_i}(x_i)\}$ where $x_i(1 \le i \le n)$ are random numbers uniformly distributed in G. Due to the semantic security of ElGamal, $V_{-1} \equiv V'_{-1}$ is clearly true.

Based on the above proofs, our framework is secure with K disclosure where K has neglected effects on individual privacy.

5.2 Complexity Analysis

In our framework, every worker computes and communicates in parallel. To this end, we only need to consider one user. Ignoring some cheap operations, the computation and communication cost are summarized in Table 1 where $L_i(i = p, e)$ is the key size of encryption strategy, e is modular exponentiation and $+, -$ means sending and receiving. Note that ElGamal encryption and communicative-like encryption is two and three times longer than L_e. Due to the size of ciphertext by Paillier and ElGamal are larger than plaintext and the ciphertext by AP, we exclude the latter two from communication cost. In the situation when the AR is not always 100%, the KP needs $|W^*|E_e$ instead of $1E_e$ in computation cost and the communication cost changes from $|2L_e|$ to $2|W^*|L_e$ during the comparison stage.

Table 1. Computation and communication cost

	Computation cost			Communication cost		
	The SC platform	The KP	Workers	The SC platform	The KP	Workers
Distance	$3E_p$	0	$1E_p + 2e$	$+3L_p$	0	$-3L_p$
LCM	D_a	0	E_a	0	0	0
Time	0	0	$3e$	$-L_p$	0	$+L_p$
Comparison	$nPRG$	$nD_p + 1E_c$	0	$+nL_p - 2L_c$	$-nL_p + 2L_c$	0
Verification	0	nE_c	$E_r + D_c$	0	$-2nL_c + 3nL_e$	$+2L_e - 3L_e$

6 Experiment Study

In the first subsection, we introduce our experiment settings and evaluation criteria. Then we show and analyze the experiment results in the second subsection.

6.1   Experiment Settings

We conduct our experiments on an area in Pennsylvania of Gowalla dataset with latitude from 39.804250 to 41.787732 and longitude from -80.418515 to -75.189944 with 3036 workers.

Three criteria are introduced to evaluate our proposed framework, namely computing time, travel distance, and worker number respectively. For computing time, we compare our framework with Liu et al.'s framework [9] for all of them are based on the public-key cryptosystems. In these two frameworks, it is meaningless to take the computing time of the SC platform and the KP into consideration because we pay more attention on the workers computing time in the task assignment and these two parties are the same in these two frameworks. For travel distance and worker number, we compare our framework with To et al.'s framework [7] for Liu et al.'s framework has the same values as ours in travel distance and worker number. Tables 2 and 3 summarize the parameters in these two comparisons.

Table 2. Computing time

Parameters	Default	Range	Description
W	200	100, 200, 300, 400, 500	The number of workers
S_{max}	10	5, 10, 15, 20, 15	The maximal speed

Table 3. Travel distance and worker number

Parameters	Default	Range	Description
AR_{max}	0.6	0.2, 0.4, 0.6, 0.8, 1.0	The maximal AR
α	0.9	0.8, 0.85, 0.9, 0.95, 0.99	The expected rate of a task
ϵ	0.6	0.2, 0.4, 0.6, 0.8, 1.0	The privacy budget of To et al.'s framework

6.2   Performance Analysis

Computing Time. In the computing time comparison, two key sizes (1024 and 2048) of Paillier and ElGamal are considered in our framework and Liu et al.'s framework.

Firstly, we study the effect of S_{max}. As described in Fig. 3, no matter what key size is adopted, our framework has much shorter average computing time than Liu et al.'s framework which means tasks can be assigned more quickly and thus improve the service quality of all platforms. Also, there is a fault of Liu et al.'s framework where S_{max} is 10 when key size is 1024 because when S_{max} is larger than 10, theirs framework based on the product of all speeds will face the overflow of product. Meanwhile, our framework can support these calculations

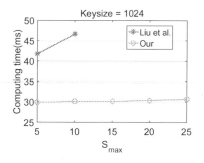
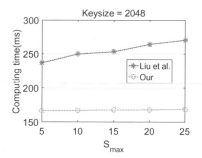

Fig. 3. Effect of S_{max}

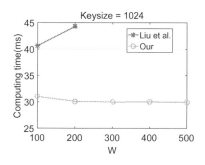
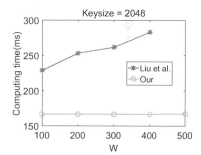

Fig. 4. Effect of W

for our framework is based on the LCM of all speeds. Note that there is still a fault where S_{max} is 100 in Liu et al.'s framework when key size is 2048 which is not shown in the Fig. 3. That is to say, the most important meaning for our framework is to break through the speed limitation of Liu et al.'s framework. Moreover, within our expectations, the computing time of Liu et al.'s framework increases as S_{max} grows while ours is a constant for the same reason as before.

Next, the effect of W is evaluated. Similar performance trend can be observed in Fig. 4 where the larger W is, the computing time grows. In addition, there are two obvious faults in Fig. 4 where W are 200 and 400 when key sizes are 1024 and 2048 respectively for the same reason as first part. Also, our framework has much shorter computing time than Liu et al.'s framework. Based on the LCM, our framework can be applied to more workers and a bigger speed.

Travel Time and Worker Number. In the travel time and worker number comparison, two functions are used to change the AR of every worker (Linear and Zipf). As To et al.'s framework does not consider the speed of workers, we set the speed of all workers is 1.

Firstly, we investigate the effect of AR_{max}. As depicted in Fig. 5, our framework has much shorter travel distance and smaller number of notified workers than To et al.'s framework because theirs is to choose some grid cells which

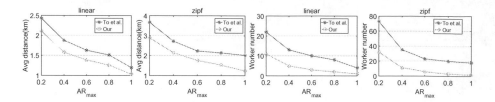

Fig. 5. Effect of AR_{max}

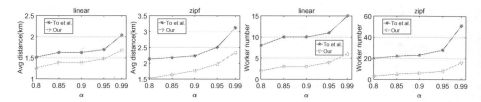

Fig. 6. Effect of α

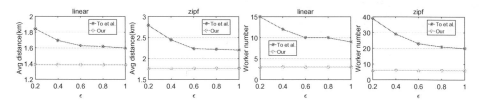

Fig. 7. Effect of ϵ

contains a number of workers. Some of them may be far away from task location. Yet, our framework is to visit the worker sorted by travel distance. In addition, the travel distance and worker number of our framework decrease when AR_{max} increases for a larger AR_{max} means workers are more willing to achieve this task.

Secondly, we study the effect of α. Figure 6 shows that our framework is much better than To et al.'s framework for the same reason as before. Also the travel distance and worker number of our framework grow with α increases for a larger α means a task has a higher expected rate to be accepted and thus more workers are required to accomplish the task.

At last, we assess the effect of ϵ. The higher ϵ is, the weaker privacy guarantee To et al.'s framework has. As expected, the change of ϵ only affects To et al.'s framework for ours is stable which is shown in Fig. 7. Also, with ϵ increases, the travel distance and worker number of their framework decreases by sacrificing of privacy. But ours still works better than theirs even in weakest privacy guarantee.

7   Conclusion

In this paper, we have identified a new task assignment strategy, travel-time-first, when allocating workers to tasks in spatial crowdsourcing. We have presented an efficient privacy-preserving task assignment protocol for this new strategy. The proposed protocol scales well because the expensive secure division operation is replaced by the secure least common multiple (LCM) computation, for which we have designed an efficient algorithm based on data aggregation. We have theoretically proved that our approach is secure against semi-honest adversaries. We have conducted extensive experiments on real-world datasets. Experimental results have shown that our protocol is efficient and effective.

Acknowledgement. Research reported in this publication was partially supported Natural Science Foundation of China (Grant Nos. 61572336, 61632016, 61572335).

References

1. Chen, L., Shahabi, C.: Spatial crowdsourcing: challenges and opportunities. IEEE Data Eng. Bull. **39**(4), 14–25 (2016)
2. Kazemi, L., Shahabi, C.: GeoCrowd: enabling query answering with spatial crowdsourcing. In: SIGSPATIAL, pp. 189–198 (2012)
3. Deng, D., Shahabi, C., Demiryurek, U.: Maximizing the number of worker's self-selected tasks in spatial crowdsourcing. In: SIGSPATIAL, pp. 324–333 (2013)
4. Cheng, P., Lian, X., Chen, Z., Fu, R., Chen, L., Han, J., Zhao, J.: Reliable diversity-based spatial crowdsourcing by moving workers. PVLDB **8**(10), 1022–1033 (2015)
5. To, H., Ghinita, G., Fan, L., Shahabi, C.: Differentially private location protection for worker datasets in spatial crowdsourcing. TMC **16**(4), 934–949 (2017)
6. Liu, B., Chen, L., Zhu, X., Zhang, Y., Zhang, C., Qiu, W.: Protecting location privacy in spatial crowdsourcing using encrypted data. In: EDBT, pp. 478–481 (2017)
7. To, H., Ghinita, G., Shahabi, C.: A framework for protecting worker location privacy in spatial crowdsourcing. PVLDB **7**(10), 919–930 (2014)
8. Liu, A., Li, Z., Liu, G., Zheng, K., Zhang, M., Li, Q., Zhang, X.: Privacy-preserving task assignment in spatial crowdsourcing. J. Comput. Sci. Technol. **32**(5), 905–918 (2017)
9. Liu, A., Wang, W., Shang, S., Li, Q., Zhang, X.: Efficient task assignment in spatial crowdsourcing with worker and task privacy protection. GeoInformatica **22**(2), 335–362 (2018)
10. Cheng, P., Lian, X., Chen, L., Han, J., Zhao, J.: Task assignment on multi-skill oriented spatial crowdsourcing. TKDE **28**(8), 2201–2215 (2016)
11. Zheng, L., Chen, L.: Maximizing acceptance in rejection-aware spatial crowdsourcing. TKDE **29**(9), 1943–1956 (2017)
12. Tong, Y., She, J., Ding, B., Wang, L., Chen, L.: Online mobile micro-task allocation in spatial crowdsourcing. In: ICDE, pp. 49–60 (2016)
13. Tong, Y., Wang, L., Zhou, Z., Ding, B., Chen, L., Ye, J., Xu, K.: Flexible online task assignment in real-time spatial data. PVLDB **10**(11), 1334–1345 (2017)
14. Gao, D., Tong, Y., She, J., Song, T., Chen, L., Xu, K.: Top-k team recommendation and its variants in spatial crowdsourcing. Data Sci. Eng. **2**(2), 136–150 (2017)
15. Paulet, R., Kaosar, M.G., Yi, X., Bertino, E.: Privacy-preserving and content-protecting location based queries. TKDE **26**(5), 1200–1210 (2014)
16. Liu, S., et al.: Efficient query processing with mutual privacy protection for location-based services. In: Navathe, S.B., Wu, W., Shekhar, S., Du, X., Wang, X.S., Xiong, H. (eds.) DASFAA 2016. LNCS, vol. 9643, pp. 299–313. Springer, Cham (2016). https://doi.org/10.1007/978-3-319-32049-6_19

17. Yi, X., Paulet, R., Bertino, E., Varadharajan, V.: Practical k nearest neighbor queries with location privacy. In: ICDE, pp. 640–651 (2014)
18. Yi, X., Paulet, R., Bertino, E., Varadharajan, V.: Practical approximate k nearest neighbor queries with location and query privacy. TKDE **28**(6), 1546–1559 (2016)
19. Sun, Y., Liu, A., Li, Z., Liu, G., Zhao, L., Zheng, K.: Anonymity-based privacy-preserving task assignment in spatial crowdsourcing. In: Bouguettaya, A., et al. (eds.) WISE 2017. LNCS, vol. 10570, pp. 263–277. Springer, Cham (2017). https://doi.org/10.1007/978-3-319-68786-5_21
20. Liu, A., Zheng, K., Li, L., Liu, G., Zhao, L., Zhou, X.: Efficient secure similarity computation on encrypted trajectory data. In: ICDE, pp. 66–77 (2015)
21. Goldreich, O.: The Foundations of Cryptography - Volume 2, Basic Applications. Cambridge University Press, Cambridge (2004)
22. Reddaway, S.: Pseudo-random number generators. US, pp. 57–67 (1974)
23. Paillier, P.: Public-key cryptosystems based on composite degree residuosity classes. In: Stern, J. (ed.) EUROCRYPT 1999. LNCS, vol. 1592, pp. 223–238. Springer, Heidelberg (1999). https://doi.org/10.1007/3-540-48910-X_16
24. ElGamal, T.: A public key cryptosystem and a signature scheme based on discrete logarithms. IEEE Trans. Inf. Theory **31**(4), 469–472 (1985)
25. Li, Q., Cao, G., La Porta, T.F.: Efficient and privacy-aware data aggregation in mobile sensing. TDSC **11**(2), 115–129 (2014)

Plover: Parallel In-Memory Database Logging on Scalable Storage Devices

Huan Zhou[1], Jinwei Guo[1], Ouya Pei[2], Weining Qian[1(⊠)], Xuan Zhou[1], and Aoying Zhou[1]

[1] School of Data Science and Engineering, East China Normal University, Shanghai 200062, China
{zhouhuan,guojinwei}@stu.ecnu.edu.cn,
{wnqian,xzhou,ayzhou}@dase.ecnu.edu.cn
[2] Northwestern Polytechnical University, Xian 710072, China
oypei@mail.nwpu.edu.cn

Abstract. Despite the prevalence of multi-core processors and large main memories, most in-memory databases still universally adopt a centralized ARIES-logging with a single I/O channel, which can be a serious bottleneck. In this paper, we propose a parallel logging mechanism, named `Plover` for in-memory databases, which utilizes the partial order property of transactions' dependencies and allows for concurrent logging in scalable storage devices. To further alleviate the performance overheads caused by log partitioning, we present a workload-aware log partitioning scheme to minimize the number of cross-partition transactions, while maintaining load balance. As such, `Plover` can scale well with the increasing number of storage devices and extensive experiments show that `Plover` with workload-aware partitioning can achieve 2× speedup over a centralized logging scheme and more than 42% over `Plover` with random partitioning.

Keywords: In-memory database · Parallel logging · Scalability

1 Introduction

The advent of multi-core processors makes low-speed disk a major performance bottleneck. Owing to the increasing size of main memory, many databases can host the entire data set in main memory to reduce disk I/Os. Unfortunately, to ensure the durability of transactions, in-memory systems have to flush logs to permanent storage regularly. Using a single disk as the permanent storage is not performant, due to its limited I/O bandwidth. Meanwhile, these systems still rely heavily on a centralized ARIES-style [1] logging mechanism to guarantee the global order of log entries. Since the total order property of logging implies the dependencies among transactions, databases can be reconstructed correctly in accordance of the order of log entries after failure recovery. However, contentions for the centralized log buffer and limited synchronous I/Os still exist, which may become a major overhead as system load increases.

© Springer International Publishing AG, part of Springer Nature 2018
Y. Cai et al. (Eds.): APWeb-WAIM 2018, LNCS 10988, pp. 35–43, 2018.
https://doi.org/10.1007/978-3-319-96893-3_3

In this paper, we propose a parallel logging mechanism for the in-memory database called `Plover`, which utilizes partial order of transactions' dependencies. The key idea is to employ distributed logging instead of centralized logging to mitigate the contention on the centralized data structure, and to use scalable storage devices to increase the I/O bandwidth. Implementing such a distributed logging is not trivial, due to two main challenges: (1) how to preserve the temporal order among log entries; (2) how to distribute the log entries across the storage devices. To address the first challenge, we use a global sequence number to identify the partial order of log entries, and a persistent group commit method to ensure all log entries of a transaction are persistent before committing. To simplify the implementation and accelerate the recovery process, we adopt tuple-level distributed logging, which partitions log entries by tuples. However, this leads to the second challenge: cross-partition transactions and workload skew, which may significantly deteriorate the performance. To resolve the potential defects, we propose a workload-aware log partitioning scheme, which applies a graph partitioning algorithm to find workload balanced partitions, while minimizing the number of distributed transactions. Finally, we demonstrate that `Plover` can achieve linear scalability with an increasing number of storage devices. In TATP and TPC-C, `Plover` with workload-aware log partitioning outperformed centralized logging by a factor of 2× and `Plover` using random partitioning by a factor of 1.42× on two storage devices.

2   Background and Related Work

Centralized Logging. To recover data from failures, a database system needs to leverage logging mechanism to guarantee atomicity and durability for transactions. For a in-memory database, the ARIES logging ensures that all REDO log entries are organized in a global order and a transaction can be committed only if all of its log entries have been persisted. The log sequence number (LSN)—which is unique and monotonically increasing— can be used to guarantee the global order of log entries. More specifically, the procedure of logging is described as follows:

(1) Log entry insertion. Before copying the log entry to the centralized log buffer, the transaction must acquire an LSN and claim the buffer space it will eventually fill with the intended log entry by a lock or a mutex. The lock or mutex will be released once the transaction finishes copying the log entry.

(2) Log entry persistence. The logging subsystem appends the log entries cached in log buffer to the log file in a single storage device. This can ensure that the entries are consecutive in the log file.

(3) Transaction committing. The transaction can commit safely after the log entries whose LSNs are less than or equal to those of its own log entries are persisted in the storage device.

However, with the CPU cores increases in a single machine, centralized logging is becoming a main bottleneck, especially in main-memory database systems, where logging is the only source of synchronous I/Os. Traditional centralized logging faces the following challenges: (1) upper limit of generating LSNs; (2) log buffer contention; and (3) limited synchronous I/Os.

Related Work. To improve the scalability of centralized logging, there have been active researches on above bottlenecks to develop new logging protocols. To alleviate the contention of allocation of LSNs, Kim et al. [3] presented a latch-free approach and Jung et al. [10] designed a concurrent data structure to ensure the global order of log entries; to improve the performance of log insertion, Johnson et al. [2] proposed a scalable logging with decoupling log inserts method so that log entries of different transactions can be copied into the log buffer in parallel; to eliminate the cost of synchronous log writes, most databases provided asynchronous commit strategy but at expense of durability. And there have been active researches to develop new logging protocols [7,9] based on the arrival of non-volatile memory (NVM) technology; to eliminate the limited I/O bandwidth of single storage device, Zheng et al. [6] implemented a transaction-level distributed logging mechanism with multiple storage devices and Wang et al. [8] proposed a universal distributed logging mechanism on multiple NVMs.

To the best of our knowledge, there are not works that can address all the issues we proposed. Therefore, we design a novel parallel logging mechanism, which utilizes partial order property of transactions' dependencies and adopts multiple log buffers and storage devices.

3   Parallel Logging

Overview. Plover aims at providing excellent performance and scalability for transaction logging, by leveraging distributed logging and multiple permanent storage devices. In our approach, the distributed logging is partitioned under tuple level, each log partition is processed by a dedicated logger thread and all of the log partitions can be accessed by all the worker threads. As modifications from a transaction may be written into many log partitions, there are two main challenges: (1) how to identify transaction dependencies for log entries over multiple log partitions; (2) how to protect committed work for a transaction. To tackle the two challenges, we prefer to employ a *global sequence number* (GSN), and propose a variant of group commit method, *persistent group commit*. The GSN provides a partial order based on logical clock [4] and guarantees the transaction dependencies among log entries over multiple log buffers. And a transaction can not safely commit until all of its log entries, along with all the log entries that logically precede them, have become persistent. Therefore, the persistent group commit starts a daemon thread to periodically monitor the submission of all logger threads and ensures that transactions can correctly commit.

Normal Processing. Next, we detailedly describe the logging processing of transactions (t × 6, t × 7, t × 8) in Plover with two log buffers (partitions) *partition A, B*, as illustrated in Fig. 1.

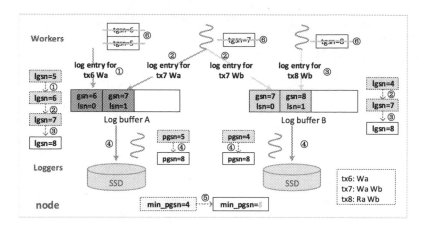

Fig. 1. Overview of parallel logging for main-memory database.

(1) **Log entry generation.** When a transaction is ready to commit, a worker thread generates a corresponding number of log entries based on the data partitions modified by the transaction. For the running transactions, $t \times 6$ and $t \times 8$ severally have a log entry, but $t \times 7$ produces two log entries.

(2) **Log entry insertion.** Before writing the generated log entries into matched log partitions, the worker thread needs to assign a GSN for all the log entries. The value of GSN is also maintained in each transaction (t_GSN) and each log partition (l_GSN). Computing a GSN should get the l_GSNs of corresponding partitions and set the value as $max(\text{l\_GSN}_i) + 1$, where i is the serial number of corresponding partitions. For tx6 which updates tuple a, it only acquires the l_GSN of partition A ($\text{l\_GSN}_a = 5$) and assigns its GSN as $\text{l\_GSN}_a + 1 = 6$, as step ①. For $t \times 7$ which modifies tuple a and b, it must get the l_GSN of partition A and B and computes its GSN as $max(\text{l\_GSN}_a = 6, \text{l\_GSN}_b = 4) + 1 = 7$, as step ②. To guarantee the true-dependency (RAW) and anti-dependency (WAR) among transactions, we also consider the case that read and write operation of a transaction across over multiple partitions. For $t \times 8$, although it only modifies tuple b, it also needs to acquire $\text{l\_GSN}_a = 7$, $\text{l\_GSN}_b = 7$ and sets its GSN as 8, as step ③. In addition to the GSN, each log entry also stores a LSN, which is used to indicate the space of an individual log buffer. Moreover, to further improve performance, we release the buffer latch once a transaction have obtained the GSN so that many worker threads can copy log entries in parallel.

(3) **Log entry persistence.** When many log entries are accumulated in log buffers, each logger thread triggers group commit to force them into disk within a single I/O, and then updates its thread-local variable (pgsn) as the GSN of the last log entry that have been persistent, as step ④. Subsequently, the persistent group commit daemon examines the pgsn of all logger threads and computes the smallest pgsn as min_pgsn, as step ⑤. The min_pgsn represents the upper bound of persistent log entries and transactions whose t_GSN $\leq$ min_pgsn can

be allowed to commit, as step ⑥. If a logger thread takes too long to update its `pgsn` (perhaps because of the corresponding partition accessed by a long read-only transaction), the persistent group commit daemon updates the logger thread's `pgsn` as the maximum value among all the `pgsn` of logger threads.

4 Recovery

Checkpoint. To accelerate data recovery from a failure, the in-memory database mandates a periodic checkpoint of its state during normal processing. In our `Plover`, the checkpoint is also partitioned according to tuples. Each checkpoint partition relates to a log partition and is processed by a dedicated checkpointer thread. When launching a new checkpoint, a checkpoint manager records the current `min_pgsn` as `c_GSN` which indicates the timestamp for a consistent snapshot, and then starts up n checkpointer threads, where n is the number of storage devices. Each checkpointer thread stores the consistent snapshot into m checkpoint files and reports to the checkpoint manager. At last, the manager writes the `c_GSN` and checkpoint metadata into a special file.

Failure Recovery. `Plover` masks outages by loading the most recent checkpoints (checkpoints recovery) and then repaying the log entries in log files (log recovery). In checkpoints recovery phase, a recovery manager thread acquires the newest metadata and `c_GSN`, where `c_GSN` denotes the starting point for log recovery, and then initiates m ∗ n threads to recovery all the checkpoint files in parallel. In log recovery phase, all the recovery threads are used to replay the log entries whose GSNs are larger than `c_GSN` and less than `r_GSN`. The `r_GSN` is the latest `min_pgsn` at the database crash, which written into a storage device by the persistent group commit daemon during transaction processing.

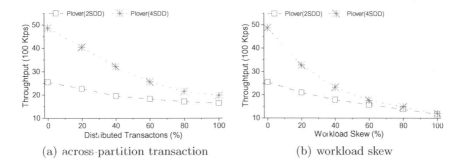

(a) across-partition transaction   (b) workload skew

Fig. 2. Impact of distributed transactions and workload skew on throughput.

5 Workload-Aware Log Partitioning

Performance Issues. Recall that the normal processing of our parallel logging, we find that the execution of a transaction is closely related to log partitioning

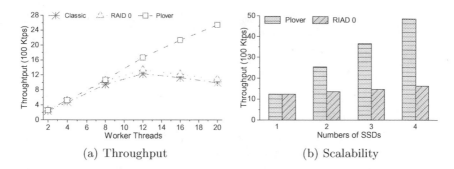

Fig. 3. Performance of parallel logging when running the microbenchmark.

and application workloads. Therefore, there are two subtle performance pitfalls: distributed transactions and workload skew. For the distributed transactions, as their GSN generation involves multiple log buffers, it increases computing overhead and reduces parallelism for logging processing. For the workload skew, it causes a log partition to suffer significant contention and excessive I/O overhead. As shown in Fig. 2, we explore the impact of distributed transaction and workload skew on throughput. We perform Plover with 2 and 4 log partitions (referred as 2SDD and 4SDD) respectively in microbenchmark and the experimental setup is shown in Sect. 6.

Partitioning Design. To solve the problems mentioned above, we implement a workload-aware log partitioning in our distributed logging. Firstly, we model the workload as a graph, $G = (V, E)$, where each vertex $v \in V$ represents a tuple, and the edge $e_{ij} \in E$ between v_i and v_j represents the connected tuples accessed by a same transaction. Each edge is associated with an edge weight w_e which accounts for the frequency of the transactions. After establishing the graph, we use a *k-way balanced min-cut partitioning* [5] to split the graph into k non-overlapping partitions such that the number of distributed transactions is minimized, while keeping the partitions within a constant factor perfectly balanced. To achieve the workload evenly across partitions, we consolidate the tuple size and access frequencies as a *factor* and assign the factor to each vertex.

6   Evaluation

Experimental Setup. All of our experiments are run on a single machine with two Intel Xeon E5-2630 (a total of 20 physical cores). The machine is equipped with 268GB DRAM and 4 pieces of SATA SSDs. We implemented a transactional logging prototype Plover in Java and each thread combines a database worker thread with a workload generator in our implementation. We compare the performance of our parallel logging equipped with multiple SSDs (referred to as *plover*) with two approaches: centralized logging with a single SSD (*classic*) and centralized logging equipped with RAID 0 (*raid0*). And

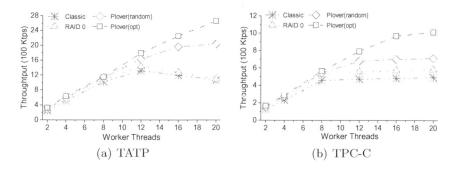

Fig. 4. Performance of parallel logging with workload-aware log partitioning.

Table 1. Recovery performance.

Variant	Checkpoint recovery time (seconds)	Log recovery time (seconds)	Total time (seconds)
Classic	67.7	163.5	231.2
RAID0	37.9	87.9	125.8
Plover	34.8	82.6	117.4

then we conduct experiment on the logging with the proposed workload-aware partitioning (*plover(opt)*) and with random partitioning (*plover(random)*). We run a microbenchmark which models a single write transaction with a 100 bytes log entry, TATP (Insert Call Forwarding) and TPC-C (New Order) on all system variants. For each benchmark and variant, each point reported in all graphs is the average throughput of three consecutive 120 s runs.

Effectiveness of Parallel Logging. We first compare the throughput and scalability with `Plover`, `Classic` and `RAID0` in microbenchmark. Figure 3 illustrates the experimental results.

Throughput. In this experiment, `Plover` and `RAID0` are equipped with two SSDs. In Fig. 3(a), as we increase the number of worker threads, the throughput of both `Classic` and `RAID0` rises steadily at first, but dramatically decreases when the number is larger than 12. However, `Plover` achieves linear scalability up to 20 threads. Owing to two logging simultaneously, `Plover` avoids the intensive contention of centralized logging and improves near 2× better performance in terms of peak throughput than `Classic` and `RAID0`.

Scalability. As shown in Fig. 3(b), `Plover` scales effectively as we increase the number of SSD drivers. The performance of `Plover` is proportional to the number of SSDs, but for `RAID0`, the non-linear speed-up is due to contention on the centralized log buffer.

Overall Performance. Next, we evaluate the performance of our parallel logging with the workload-aware log partitioning scheme in diverse workloads.

For TATP, both random partitioning and our approach can perfectly avert distributed transactions. But the random partitioning may suffers workload-skew. Hence, as shown in Fig. 4(a), `Plover` with the workload-ware partitioning `plover(opt)` has the best performance as increasing the number of worker threads, which improves 2× better peak throughput than `Classic` and `RAIDO`, and increases performance more than 30 % compared with `plover(random)`.

For TPC-C, "New Order" produces a variable-sized log entry, about from 800 byte to 2250 byte. The larger size per log entry makes the peak throughput of `Classic` and `RAIDO` quickly become saturated as growing the number of worker threads, as shown in Fig. 4(b). And the throughput of `plover(random)` does not further increase when the number of worker threads is larger than 12. That is because there are distributed transactions and workload skew in the random log partitioning. But our proposed scheme, `plover(opt)` achieves the best performance, which improves the peak throughput by factor of 2× over `Classic` and `RAIDO`, and more than 42% over `plover(random)`.

Recovery. To investigate the effectiveness of our logging for recovery, we use the microbenchmark without distributed transactions and workload skew. When the system fails, we acquire 28 GB checkpoints and 54 GB log files. In this experiment, `Plover` and `RAIDO` are equipped with two SSDs. As shown in Table 1, `Classic` has the largest total recovery time. This is because all of the checkpoint files and log files are stored in a single storage device and the limited I/O bandwidth seriously reduces its recovery performance. Owing to the parallel load, `RAIDO` and `Plover` can respectively improve the recovery time by a factor of 1.83× and 1.97× speedup over `Classic`.

7   Conclusion

In this paper, we introduce a parallel logging in the main memory database named `Plover`, which replaces the centralized log buffer with multiple tuple-level distributed log buffers and allows log entries to be simultaneously forced into multiple storage devices. Our distributed logging relies on a logical global sequence number to identify the uniqueness of log entries and a persistent group commit method to ensure a transaction can be safely committed. We also analyze the impacts of distributed transactions and workload skew on performance and present a workload-aware log partitioning scheme based on a graph-partitioning algorithm to produce high-quality partitions. Our experimental evaluations demonstrate that `Plover` can provide linear scalability with the growing number of storage devices and the increasing number of worker threads. Due to the parallel design, our approach significantly alleviates the contention of centralized logging and the limitation of single I/O bandwidth.

Acknowledgement. This work is partially supported by National High-tech R&D Program (863 Program) under grant number 2015AA015307, National Key Research & Development Program of China (No. 2018YFB1003400), National Science Foundation of China under grant numbers 61702189, 61432006 and 61672232.

References

1. Mohan, C., Haderle, D., Lindsay, B., Pirahesh, H., Schwarz, P.: ARIES: a transaction recovery method supporting fine-granularity locking and partial rollbacks using write-ahead logging. TODS **17**(1), 94–162 (1992)
2. Johnson, R., Pandis, I., Stoica, R., Manos, A.: Aether: a scalable approach to logging. VLDB **3**(1–2), 681–692 (2010)
3. Kim, K., Wang, T.Z., Johnson, R., et al.: Ermia: fast memory-optimized database system for heterogeneous workloads. In: SIGMOD, pp. 1675–1687 (2016)
4. Lamport, L.: Time, clocks, and the ordering of events in a distributed system. Commun. ACM **21**(7), 558–565 (1978)
5. Andreev, K., Racke, H.: Balanced graph partitioning. Theor. Comput. Syst. **39**(6), 929–939 (2006)
6. Zheng, W.T., Tu, S.: Fast databases with fast durability and recovery through multicore parallelism. In: OSDI, pp. 465–477 (2014)
7. Huang, J., Schwan, K., Qureshi, M.K.: NVRAM-aware logging in transaction systems. PVLDB **8**(4), 389–400 (2014)
8. Wang, T.Z., Johnson, R., Stoica, R., Manos, A.: Scalable logging through emerging non-volatile memory. PVLDB **7**(10), 865–876 (2014)
9. Arulraj, J., Perron, M., Pavlo, A.: Write-behind logging. PVLDB **10**(4), 337–348 (2016)
10. Jung, H., Han, H., Kang, S.: Scalable database logging for multicores. PVLDB **11**(2), 135–148 (2017)

Inferring Regular Expressions
with Interleaving from XML Data

Xiaolan Zhang[1,2], Yeting Li[1,2], Fei Tian[3], Fanlin Cui[1,2], Chunmei Dong[1,2],
and Haiming Chen[1(✉)]

[1] State Key Laboratory of Computer Science, Institute of Software,
Chinese Academy of Sciences, Beijing 100190, China
{zhangxl,liyt,cuifl,dongcm,chm}@ios.ac.cn
[2] University of Chinese Academy of Sciences, Beijing, China
[3] University of Science and Technology of China, Hefei, China
tf4811@mail.ustc.edu.cn

Abstract. Document Type Definition (DTD) and XML Schema Definition (XSD) are two popular schema languages for XML. However, many XML documents in practice are not accompanied by a schema, or by a valid schema. Therefore, it is essential to devise efficient algorithms for schema learning. Schema learning can be reduced to the inference of restricted regular expressions. In this paper, we first propose a new subclass of restricted regular expressions called *Various CHAin Regular Expression with Interleaving (VCHARE)*. Then based on single occurrence automaton (SOA) and maximum independent set (MIS), we introduce an inference algorithm *GenVCHARE*. The algorithm has been proved to infer a descriptive generalized *VCHARE* from a set of given sample. Finally, we conduct a series of experiments based on our data set crawled from the Web. The experimental results show that *VCHARE* can cover more content models than other existing subclasses of regular expressions. And, based on the data sets of *DBLP*, regular expressions inferred by *GenVCHARE* are more accurate and concise compared with other existing methods.

1 Introduction

Document Type Definition (DTD) and XML Schema Definition (XSD) are two popular schema languages for XML recommended by World Wide Web Consortium (W3C) [31]. The presence of a schema has numerous advantages such as data processing, automatic data integration, static analysis of transformations and so on [2, 11, 20, 22–24, 28]. Besides, the existence of schemas is necessary when integrating (meta) data through schema matching [30] and in the area of generic model management [3, 26]. However, many XML documents are not accompanied by a (or valid) schema in practice. A survey [19] shows that XML

H. Chen—Work supported by the National Natural Science Foundation of China under Grant No. 61472405.

documents on the Web which have schema definitions only account for 24.8% in 2013, of which the proportion of valid schemas is only about 8.9%. Therefore, it is essential to devise algorithms for schema inference. And schema inference can be reduced to learning restricted regular expressions from a set of given sample [6,8,16].

Gold [18] proposed a classical language learning model (*learning in the limit or explanatory learning*) and pointed out that the class of regular expressions cannot be learnable from positive examples only. Furthermore, Bex et al. proved in [4] that even the class of deterministic regular expressions is too rich to be learnable from positive data. Consequently, researchers have turned to study the restricted subclasses of regular expressions [27].

The popular existing subclasses of regular expressions used in XML such as SORE [6], CHARE (Simplified CHARE) [6], eSimplified CHARE [12], Simple regular expression (CHARE) [5], eCHARE [25] were renamed as in the brackets and analyzed together in [21]. These subclasses are all based on standard regular expressions. In data-centric applications using XML, there may be no order constraint among siblings [1]. However, the relative order within siblings may be still important. In [9], Ciucanu and Staworko proposed two schema formalisms for unordered XML: *disjunctive multiplicity expressions (DME)* and *disjunction-free multiplicity expressions (ME)* where the relative order among siblings was ignored. These two formalisms do not support the concatenation within siblings. For example, $E_1 = (a|b)^+\&c$ is a DME and $E_2 = a\&b^*\&c^?$ is an ME. But $E_3 = (a^+b^?)\&c^*$ does not satisfy both two formalisms. Peng and Chen in [29] also focused on the unordered relation among siblings and proposed *SIRE*. *SIRE* supports the concatenation operation within siblings. Therefore E_3 is a *SIRE*. However, *SIRE* does not support union operation. In [17], Ghelli et al. proposed a restricted subclass defined by grammar $T ::= \varepsilon | a^{[m,n]} | T+T | T \cdot T | T\&T$ where $m \in N \setminus \{0\}$ and $n \in N \setminus \{0\} \cup \{*\}$. For this subclass, counters (repetition operation) can only occur as a constraint for terminal symbols of strings in $L(T)$. For example, $E_4 = a^?(b|c|d)^*$ is not allowed.

In this paper, we focus on learning a restricted deterministic regular expression considering interleaving from a set of given positive examples. We propose a new subclass named as *Various CHAin Regular Expression with Interleaving (VCHARE)*. *VCHARE* supports union, concatenation and interleaving operators together. For example, $E_5 = a^*\&b^+\&c^?$ and $E_6 = (a|b^?)(c^*d^?|e^*)^+$ are both *VCHAREs*.

As for learning algorithms for XML data, Bex et al. [6,7] proposed two inference algorithms *RWR* and *CRX* for SOREs and its Simplified CHAREs, respectively. Freydenberger and Kötzing [13] proposed another two inference algorithms *Soa2Chare* and *Soa2Sore* based on *Single Occurrence Automaton* (SOA) for Simplified CHAREs and SOREs, respectively. These two algorithms can infer descriptive generalized regular expressions (explained below) while *RWR* and *CRX* can not. Ciucanu and Staworko introduced an algorithms for *DME* based on *max clique* [9]. Peng and Chen [29] proposed an approximation algorithm and heuristic solution to infer a descriptive generalized *SIRE*.

The concept of *descriptive generalization* [14], is different from Gold-style language learning. Gold-style learners are required to infer an exact description for the target language in a class. But descriptive generalization views the hypothesis space and the space of target language as distinct. Here is a formal explanation. For a class $\mathcal{D}$ of language representation mechanisms (e.g., a class of automata, regular expressions, or grammars), a representation $\alpha \in \mathcal{D}$ is called $\mathcal{D}$-descriptive for a set of given sample S if the language of α is an inclusion-minimal generalization of S. It means that there is no $\beta \in \mathcal{D}$ such that $S \subseteq \mathcal{L}(\beta) \subset \mathcal{L}(\alpha)$.

In present paper, the inference algorithm (*GenVCHARE*) is also based on the concept of descriptive generalization which aims to infer descriptive generalized *VCHAREs* for a set of given sample S. The main idea of *GenVCHARE* is based on SOA and *Maximum Independent Set* (MIS). We first construct an SOA for S. Then replace each non-trival strongly connected component (NTSCC) by the return value of *RepairRE()* as one new node. Next, assign each node a level number. Finally, all nodes of each level will be converted to one or more chain factors.

The main contributions of this paper are listed as follows.

- We propose a subclass of restricted regular expressions named as *Various CHAin Regular Expression with Interleaving (VCHARE)*.
- We design an inference algorithm *GenVCHARE* to infer descriptive generalized *VCHAREs*.
- We analyze the coverage proportion of *VCHARE* compared with other subclasses based on the real-world data set. Based on the data sets (*DBLP*), we compare the inferred results with other inferrence methods. The experimental results shows that regular expressions inferred by *GenVCHARE* are more accurate.

This paper is organized as follows. In Sect. 2 introduces some basic definitions. Section 3 is the inference algorithm *GenVCHARE*. Section 4 gives the experiments. Conclusions are drawn in Sect. 5.

2   Preliminaries

Definition 1. *Regular Expression with Interleaving.* *Let Σ be a finite alphabet. Σ^* is the set of all strings over Σ. A regular expression with interleaving over Σ is inductively defined as follows: ε or $a \in \Sigma$ is a regular expression where $a \in \Sigma$. For any regular expressions E_1 and E_2, the disjunction $E_1|E_2$, the concatenation $E_1 \cdot E_2$, the interleaving $E_1 \& E_2$, or the Kleene-Star E_1^* is also a regular expression. The language generated by E is defined as follows: $L(\emptyset) = \emptyset$; $L(\varepsilon) = \{\varepsilon\}$; $L(a) = \{a\}$; $L(E_1^*) = L(E)^*$; $L(E_1 E_2) = L(E_1)L(E_2)$; $L(E_1|E_2) = L(E_1) \cup L(E_2)$; $L(E_1 \& E_2) = L(E_1 E_2) \cup L(E_2 E_1)$. $E^?$ and E^+ are used as abbreviations of $E + \epsilon$ and EE^*, respectively.*

In the specification of XSD, the interleaving operator is used in the form of $a_1^{c_1} \& a_2^{c_2} \& \cdots \& a_n^{c_n}$ where $a_i \in \Sigma$ and $c_i \in \{1, ?, +, *\}$. For $a, b \in \Sigma$, $x, y \in \Sigma^*$, we have $a \& \epsilon = \epsilon \& a = a$ and $ax \& by = a(x \& by) \cup b(ax \& y)$.

Let S be the set of given sample. $POR(S)$ is the set of all partial order relations of each string in S. Using $POR(S)$, we can compute the Constraint Set (CS) and Non-Constraint Set (NCS) for S by the following formula.

1. $CS(S) = \{< a_i, a_j > \mid < a_i, a_j > \in POR(S)$, and $< a_j, a_i > \in POR(S)\}$;
2. $NCS(S) = \{< a_i, a_j > \mid < a_i, a_j > \in POR(S)$, but $< a_j, a_i > \notin POR(S)\}$.

Clearly, for a set of given sample S, $CS(S) \cap NCS(S) = \emptyset$. If $CS(S_1) \neq CS(S_2)$ (or $NCS(S_1) \neq NCS(S_2)$), then $S_1 \neq S_2$.

Definition 2. $PS(P,s)$. $PS(P, s)$ is a function in which P is a finite set of symbols and s is a string. Each symbol s_i of s in $PS(P, s)$ is defined as follows: $\pi_s(P, s_i) = s_i$ if $s_i \in P$; otherwise $\pi_s(P, s_i) = \varepsilon$. The return value of $PS(P, s)$ is a new string s' with ε removed.

For example, let $P = \{b, c, r\}$ and $s = ebbdfc$. $s' = PS(P, s) = bbc$.

Definition 3. extended String (eS). Let Σ be a finite set of terminal symbols. An eS is a finite sequence $s_1^{c_1} s_2^{c_2} \cdots s_n^{c_n}$, where $s_i \in \Sigma$ and $c_i \in \{1, ?, +, *\}$.

Definition 4. Various CHAin Regular Expression with Interleaving (VCHARE). Let Σ be a finite alphabet. A VCHARE is a regular expression with interleaving over Σ in which each symbol occur once at most. It consists of a finite sequence of factors of two forms. One form is of $a_1^{c_1} \& a_2^{c_2} \& \cdots \& a_n^{c_n}$ where $n \geq 2$, $a_i \in \Sigma$ and $c_i \in \{1, ?, +, *\}$. The other form is of $f_1 f_2 \cdots f_m$ where $m \geq 1$. Each factor f_i is of the form of $(b_1 | b_2 | \cdots | b_n)$, $(b_1 | b_2 | \cdots | b_n)^?$, $(b_1 | b_2 | \cdots | b_n)^+$ or $(b_1 | b_2 | \cdots | b_n)^*$ where b_i has two forms: 1. terminal symbol a or a^+ with $|b_i| = 1$ for the first two forms; 2. for the last two forms, it can be an eS $s = a_1^{c_1} a_2^{c_2} \cdots a_n^{c_n}$ where $a_i \in \Sigma$ and $c_i \in \{?, *\}$ with $n \geq 1$.

Clearly, $E_1 = a^? \& b^* \& c^+$ and $E_2 = a^? (b + c^+)(c^? d^* + e^?)^+$ are both VCHAREs.

3   Inference Algorithm

In this section, we will introduce the inference algorithm $GenVCHARE$ for VCHARE. The algorithm is based on SOA and MIS.

We use the method $2T\text{-}INF$ [15] to construct a SOA for S. It was proved that $L(SOA(S))$ is inclusion-minimal of S. Finding a maximum independent set from a graph G is a well-known NP-hard problem. Therefore we use the approximation method $clique\_removal()$ [10] to find the approximative results. all_mis is the set contained all maximum independent sets iteratively obtained from G using $clique\_removal()$. $symbol(A)$ is the set of all symbols occur in A. The main procedure of $GenVCHARE$ is described as follows.

– Construct a graph $G(V, E) = SOA(S)$ using method $2T\text{-}INF$ [15].
– For each node v with a self-loop, label it with v^+ and remove the self-loop. Update the graph G.

- If G is a strongly connected component, then return the result $v_1^{c_1} \& v_2^{c_2} \& \cdots \& v_n^{c_n}$ where $v_i \in V$ and assign the repetition operator $c_i \in \{1, ?, +, *\}$ using CRX [6]. Otherwise, continue to run the following steps.
- For each non-trival strongly connected component c_i, replace it with the return value of $RepairRE()$ as one new node. All relations with any node in c_i rebuild the relations with the new node.
- Assign level numbers for the new graph and compute all skip levels.
- Nodes of each level are turned into one or more chain factors. If there are more than one non-letter nodes (label with more than one terminal symbols) with the same ln, or if ln is a skip level, then ? is appended to every chain factor on that level.

Pseudo code for GenVCHARE ALT(C) can be found on the web site: http://lcs.ios.ac.cn/~zhangxl/.

Algorithm Analysis. For graph $G(V, E) = SOA(S)$, let $n = |V|$ and $m = |E|$. It costs time $O(n)$ to find all nodes with self-loops and $O(m + n)$ to find all NTSCCs. The time complexity of $clique\_removal()$ is $O(n^2 + m)$. For each NTSCC, computation of all_mis costs time $O(n^3 + m)$ and the topological sort for each mis costs time $O(m + n)$. The number of NTSCCs in a SOA is finite. Therefore computing all_mis for all NTSCCs also costs time $O(n^3 + m)$. Assigning level numbers and computing all skip levels will be finished in time $O(m + n)$. All nodes will be converted into specific chain factors of $VCHARE$ in $O(n)$. Therefore, the time complexity of $GenVCHARE$ is $O(n^3 + m)$.

Theorem 1. *Suppose that $\alpha = GenVCHARE(SOA(S))$ where S is a set of given sample. If there exists another VCHARE β such that $S \subseteq L(\beta) \subset L(\alpha)$, then $L(\beta) = L(\alpha)$.*

All detail proofs are omitted due to limited space.

4   Experiments and Analysis

In this section, we first investigate the proportion of $VCHARE$ based on real-world data, and then analyze our inference algorithm on $DBLP$ downloaded from the Web[1]. $DBLP$ is a Computer Science Bibliography corpus, a data-centric database of information on major computer science journals and proceedings. All our experiments were conducted on a machine with Intel Core i5-5200U@2.20 GHz, 4G memory, OS: Ubuntu 16.04. All codes were written in python 3.

---

[1] http://aiweb.cs.washington.edu/research/projects/xmltk/xmldata/www/repository.html.

4.1 Usage of VCHARE in Practice

To investigate the proportion of *VCHARE* in practice, we crawled 29414 DTDs, 38554 XSDs and 4526 Relax NGs files from the Web and extracted 118242, 476804 and 509267 regular expressions from them respectively. The coverage proportions of subclasses: *VCHARE, SORE, DME, ME, Ghelli* [17], *SIRE* are shown in Fig. 1. Clearly, we can find out that the proportions of *VCHARE* are the highest for XSDs and Relax NG which are 94.95% and 95.28% respectively.

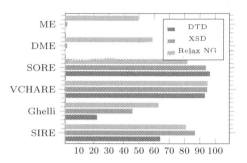

Fig. 1. Proportions of subclasses

For DTDs, the proportion (93.54% for *VCHARE*) is a little lower than *SORE* (96.69%). This is because interleaving operator is not supported in DTD. Interleaving is defined in an unlimited manner in Relax NG with any symbol in strings to interleave in any order while it is limited in XSD with only single symbols to interleave in any order. For example, $(ab^*)\&(c^+d^?)$ is not valid in XSD but it is allowed in Relax NG. Although interleaving defined in *SIRE* conforms to Relax NG, the proportion of *VCHARE* is still higher than *SIRE*. This means that in actual data, interleaving is used mostly in a quite simple and concise form. Therefore, *VCHARE* is more practical in real-world applications.

4.2 Analysis of Inference Results Compared with *GenVCHARE*

In this section, we analyze the inference results by *GenECHARE* [12] (algorithm for inferring *eSimplified CHARE*), *Soa2Chare* [13], *Original Schema, Trang,*

Table 1. Results of inference using different methods on **inproceedings**

Sample size	From	Element name	ND	$\lvert RE\rvert$
1610138	DBLP	Inproceedings		
Methods		Regular expression		
1. Original Schema	$(a_1\lvert a_2\lvert a_3\lvert a_4\lvert a_5\lvert a_6\lvert a_7\lvert a_8\lvert a_9\lvert a_{10}\lvert a_{11}\lvert a_{12}\lvert a_{13}\lvert a_{14}\lvert a_{15}$ $\lvert a_{16}\lvert a_{17}\lvert a_{18}\lvert a_{19}\lvert a_{20}\lvert a_{21}\lvert a_{22}\lvert a_{23})^*$			1	48
2. IntelliJ IDEA	$a_2^*(a_1\lvert a_3\lvert a_4\lvert a_5\lvert a_6\lvert a_{10}\lvert a_{11}\lvert a_{12}\lvert a_{13}\lvert a_{14}\lvert a_{15}\lvert a_{17}\lvert a_{18})^+$			1	31
3. Liquid Studio	$(a_1\lvert a_2\lvert a_3\lvert a_4\lvert a_5\lvert a_6\lvert a_{10}\lvert a_{11}\lvert a_{12}\lvert a_{13}\lvert a_{14}\lvert a_{15}\lvert a_{17}\lvert a_{18})^+$			1	30
4. Trang	$a_2^*(a_1\lvert a_3\lvert a_4\lvert a_5\lvert a_6\lvert a_{10}\lvert a_{11}\lvert a_{12}\lvert a_{13}\lvert a_{14}\lvert a_{15}\lvert a_{17}\lvert a_{18})^+$			1	31
5. Soa2Chare	$a_2^*(a_1\lvert a_3\lvert a_4\lvert a_5\lvert a_6\lvert a_{10}\lvert a_{11}\lvert a_{12}\lvert a_{13}\lvert a_{14}\lvert a_{15}\lvert a_{17}\lvert a_{18})^+$			1	31
6. GenEchare	$a_2^*(a_1^+\lvert a_3\lvert a_4^+\lvert a_5\lvert a_6\lvert a_{10}\lvert a_{11}\lvert a_{12}\lvert a_{13}\lvert a_{14}^+\lvert a_{15}^+\lvert u_{17}\lvert a_{18}^+)^!$			2	36
7. conMiner	$a_1^*a_{17}^?a_{14}^*\&a_2^*a_{11}^?a_4a_{10}^?\&a_3a_6\&a_5^?\&a_{12}\&a_{13}^?\&a_{18}^*a_{15}^*$			1	37
8. GenVCHARE	$a_2^*(a_1^?a_{17}^?a_{14}\lvert a_3a_{12}^?a_{15}\lvert a_4\lvert a_5\lvert a_6\lvert a_{13}^*\lvert a_{18}^*a_{11}^?a_{10}^?)^+$			2	40

conMiner [29] (algorithm for inferring SIRE), *IntelliJ IDEA* and *Liquid Studio* compared with *GenVCHARE* on *inproceedings, incollection, phdthesis, mastersthesis*. Using two indicators: *Nesting Depth* [21] and *length of regular expressions* (the number of symbols together with operators), we only give the analysis of inferred regular expressions on *inproceedings* due to limited space reason. Analysis on other elements can be found on the web site: http://lcs.ios.ac.cn/~zhangxl/.

From Table 1, we can find that a_2 must occur in the first position if it appears. However, its position is not fixed in regular expressions inferred from methods 1, 3, 7 which lead to over-generalization.

5   Conclusion and Future Work

After a detailed analysis of real-world data, we propose a new subclass *VCHARE* of restricted regular expressions considering interleaving operator. Each terminal symbol in a *VCHARE* can only occur at most once. Compared with existing subclasses, *VCHARE* can cover more real-world data. This is useful for applications such as data process and integration and so on. Further, we proposed an inference algorithm *GenVCHARE* for *VCHARE* based on *SOA* and *MIS*. It is proved that regular expressions inferred by *GenVCHARE* are descriptive generalized. Experimental results show that regular expressions inferred by *GenVCHARE* is more accurate.

One future work is to consider constructing an automaton for regular expression with interleaving which is useful for schema inference. In addition, we will also study *SORE* extended with interleaving.

References

1. Abiteboul, S., Bourhis, P., Vianu, V.: Highly expressive query languages for unordered data trees. Theor. Comput. Syst. **57**(4), 927–966 (2015)
2. Benedikt, M., Fan, W., Geerts, F.: XPath satisfiability in the presence of DTDs. J. ACM **55**(2), 1–79 (2008)
3. Bernstein, P.A.: Applying model management to classical meta data problems. In: CIDR. vol. 2003, pp. 209–220. Citeseer (2003)
4. Bex, G.J., Gelade, W., Neven, F., Vansummeren, S.: Learning deterministic regular expressions for the inference of schemas from XML data. ACM Trans. Web **4**(4), 1–32 (2010)
5. Bex, G.J., Neven, F., Bussche, J.V.D.: DTDs versus XML schema: a Practical Study. In: International Workshop on the Web and Databases, pp. 79–84 (2004)
6. Bex, G.J., Neven, F., Schwentick, T., Tuyls, K.: Inference of concise DTDs from XML data. In: International Conference on Very Large Data Bases, Seoul, Korea, pp. 115–126, September 2006
7. Bex, G.J., Neven, F., Schwentick, T., Vansummeren, S.: Inference of concise regular expressions and DTDs. ACM Trans. Database Syst. **35**(2), 1–47 (2010)
8. Bex, G.J., Neven, F., Vansummeren, S.: Inferring XML schema definitions from XML data. In: International Conference on Very Large Data Bases, University of Vienna, Austria, pp. 998–1009, September 2007

9. Boneva, I., Ciucanu, R., Staworko, S.: Simple schemas for unordered XML. In: International Workshop on the Web and Databases (2015)
10. Boppana, R., Halldrsson, M.M.: Approximating maximum independent set by excluding subgraphs. Bit Numer. Math. **32**(2), 180–196 (1992)
11. Che, D., Aberer, K., Özsu, M.T.: Query optimization in XML structured-document databases. VLDB J. **15**(3), 263–289 (2006)
12. Feng, X.Q., Zheng, L.X., Chen, H.M.: Inference algorithm for a restricted class of regular expressions. Comput. Sci. **41**, 178–183 (2014)
13. Freydenberger, D.D., Kötzing, T.: Fast learning of restricted regular expressions and DTDs. Theor. Comput. Syst. **57**(4), 1114–1158 (2015)
14. Freydenberger, D.D., Reidenbach, D.: Inferring Descriptive Generalisations of Formal Languages. Academic Press Inc., Cambridge (2013)
15. Garcia, P., Vidal, E.: Inference of k-testable languages in the strict sense and application to syntactic pattern recognition. IEEE Trans. Pattern Anal. Mach. Intell. **12**(9), 920–925 (2002)
16. Garofalakis, M., Gionis, A., Shim, K., Shim, K., Shim, K.: XTRACT: learning document type descriptors from XML document collections. Data Min. Knowl. Disc. **7**(1), 23–56 (2003)
17. Ghelli, G., Colazzo, D., Sartiani, C.: Efficient inclusion for a class of XML types with interleaving and counting. Inf. Syst. **34**(7), 643–656 (2009)
18. Gold, E.M.: Language identification in the limit. Inf. Control **10**(5), 447–474 (1967)
19. Grijzenhout, S., Marx, M.: The quality of the XML web. Web Semant. Sci. Serv. Agents World Wide Web **19**, 59–68 (2013)
20. Koch, C., Scherzinger, S., Schweikardt, N., Stegmaier, B.: Schema-based scheduling of event processors and buffer minimization for queries on structured data streams. In: Thirtieth International Conference on Very Large Data Bases, pp. 228–239 (2004)
21. Li, Y., Zhang, X., Peng, F., Chen, H.: Practical study of subclasses of regular expressions in DTD and XML schema. In: Li, F., Shim, K., Zheng, K., Liu, G. (eds.) APWeb 2016. LNCS, vol. 9932, pp. 368–382. Springer, Cham (2016). https://doi.org/10.1007/978-3-319-45817-5_29
22. Manolescu, I., Florescu, D., Kossmann, D.: Answering XML queries on heterogeneous data sources. In: International Conference on Very Large Data Bases, pp. 241–250 (2001)
23. Martens, W., Neven, F.: Typechecking top-down uniform unranked tree transducers. In: Calvanese, D., Lenzerini, M., Motwani, R. (eds.) ICDT 2003. LNCS, vol. 2572, pp. 64–78. Springer, Heidelberg (2003). https://doi.org/10.1007/3-540-36285-1_5
24. Martens, W., Neven, F.: Frontiers of tractability for typechecking simple XML transformations. In: ACM Sigmod-Sigact-Sigart Symposium on Principles of Database Systems, pp. 23–34 (2004)
25. Martens, W., Neven, F., Schwentick, T.: Complexity of decision problems for XML schemas and chain regular expressions. SIAM J. Comput. **39**(4), 1486–1530 (2013)
26. Melnik, S.: Generic Model Management: Concepts and Algorithms. Springer, Heidelberg (2004). https://doi.org/10.1007/b97859
27. Min, J.K., Ahn, J.Y., Chung, C.W.: Efficient extraction of schemas for XML documents. Inf. Process. Lett. **85**(1), 7–12 (2003)
28. Papakonstantinou, Y., Vianu, V.: DTD inference for views of XML data. In: Nineteenth ACM Sigmod-Sigact-Sigart Symposium on Principles of Database Systems, pp. 35–46 (2000)

29. Peng, F., Chen, H.: Discovering restricted regular expressions with interleaving. In: Cheng, R., Cui, B., Zhang, Z., Cai, R., Xu, J. (eds.) APWeb 2015. LNCS, vol. 9313, pp. 104–115. Springer, Cham (2015). https://doi.org/10.1007/978-3-319-25255-1_9
30. Rahm, E., Bernstein, P.A.: A survey of approaches to automatic schema matching. VLDB J. **10**, 334–350 (2001)
31. Thompson, H.S.: XML schema part 1: structures. Recommendation **6**, 291–313 (2001)

Efficient Query Reverse Engineering
for Joins and OLAP-Style Aggregations

Wei Chit Tan[(⊠)]

Singapore University of Technology and Design, Singapore, Singapore
weichit_tan@mymail.sutd.edu.sg

Abstract. Query reverse engineering is getting important in database usability since it helps users to gain technical insights about the database without any intentional knowledge such as schema and SQL. In this paper, we review some existing techniques that focus on join query discovery, and we devise our efficient algorithm to discover the SQL queries that contain both joins and OLAP-style aggregations which are substantially for querying OLAP data warehouses. We show that our algorithm is adaptable and scalable for large databases by performing an empirical study for TPC-H benchmark dataset.

1 Introduction

Since every organization may have its unique data warehouse and it is always managed and maintained by a team of technical experts, it is rather hard for ordinary users to make full use of these generated data, especially those spreadsheets from the data warehouse. For a general purpose, database users are required to learn both schema and query language, which are important for them to invoke the tuples from the relevant relations precisely. Thus, the SQL join operations are definitely important for combining the relevant columns from these tables into a common (denormalized) table. Besides, these combined data are often associated with OLAP-style aggregations (e.g., basic mathematical operators) for offering more valuable insights about the numerical data.

Figure 1 illustrates a motivating example. Figure 1(a) is an example spreadsheet table, and Fig. 1(b) shows are a pair of or even better minimal join graphs that could regenerate this spreadsheet table through different join tables, projections, and aggregations. A candidate join graph is akin to a schema graph. Each node represents a relation, and it is starred if it contains a projection column. Therefore, from the candidate join graphs, only the validated join graph would be executed for discovering other SQL classes, e.g., OLAP group-by, aggregations and selection filters.

1.1 Related Work

Instead of using a keyword query that is made up of several keywords, there are many proposals have been implemented to discover join queries by using a tabular list of tuples as the implication of keyword search in relational databases [7].

© Springer International Publishing AG, part of Springer Nature 2018
Y. Cai et al. (Eds.): APWeb-WAIM 2018, LNCS 10988, pp. 53–62, 2018.
https://doi.org/10.1007/978-3-319-96893-3_5

N_NAME	L_LINESTATUS	MAX(O_TOTALPRICE)	SUM(L_QUANTITY)
ARGENTINA	O	530604.44	2284691.00
CANADA	F	510061.60	2343191.00
CANADA	O	515531.82	2284164.00
FRANCE	F	508668.52	2343432.00
IRAN	F	522644.48	2276219.00
JAPAN	O	502742.76	2287464.00
MOZAMBIQUE	O	508047.99	2348205.00
PERU	F	544089.09	2269762.00
PERU	O	522720.61	2264220.00
RUSSIA	F	555285.16	2359354.00
UNITED STATES	O	525590.57	2316886.00
VIETNAM	F	504509.06	2301689.00

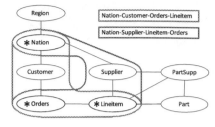

(a) An example spreadsheet table   (b) Join candidate graphs

Fig. 1. The different join queries that are possible to generate an example spreadsheet.

Most of the existing solutions (e.g. [10,11,13]) depend on schema-based approach [1,2], and the database schema is illustrated as a graph by taking the relations as nodes and the foreign key references as edges. In DISCOVER [2] and its extended works [3,7], given that a set of candidate networks discovered by a keyword query, the candidate network evaluation needs an optimized execution plan which is depicted as an operator tree in order to translate each of them into SQL. Nonetheless, the full-text search is another technique to verify candidate queries by emphasizing keyword containments as SQL predicates, which it is exceptionally useful for text attributes and built-in indexes are required in advance. Several works [5,6,8] support this full-text search feature, thus the query discovery is restricted to textual databases in lieu of the OLAP data warehouses.

Another critical factor that could optimize the join execution is the indexing techniques. In lieu of joining every projection attribute for candidate query evaluation, the implementation of join indices [12] only require those relevant primary keys to form a temporal relation so that the overhead memory cost can be avoided. The well known TALOS framework [10,11] uses the join indices to build an intermediate join relation and thus applies the decision tree classifier to classify the tuples for selection predicate generation. In addition, as indicated in [13], the unique tuple identifiers (*tids*) within each relation are used to examine each schema-based connected tree at instance-level in order to invalidate any schema trees that cannot generate a random output tuple.

Apart from that, besides those fundamental SQL classes which can determine the schema tables and attributes for query discovery, other classes such as HAVING and ORDER BY clauses have their specifications to produce the finalized SQL results. PALEO framework [4] uses the concept of ranked list of tuples to reverse engineer OLAP queries where each query contains an ORDER BY column.

1.2   Contributions

Our contributions in this paper are presented as follows:

– We provide a solution that generates the candidate join graphs through the schema and metadata exploration to characterize each distinct column of query output table.

- We improve the expressiveness of our solution by discovering SQL HAVING clause for aggregation queries.
- We prove that our algorithm is adaptable and scalable by conducting an experimental evaluation over the standard TPC-H dataset.

2   Problem Definition

A relational database $\mathcal{D}$ consists of a set of relations and every relation is linked by referential integrity constraints. The relational schema is defined as a schema graph $\mathcal{SG}(\mathcal{R}, \zeta)$, where each $\mathcal{R}$ is a table and each ζ is an fk/pk constraint. A subgraph $\mathcal{J}(\mathcal{G})$ entails a join query where a relation $\mathcal{R} \in \mathcal{SG}$ may appear more than once as a node in it. The Project-Join (PJ) queries should contain at least both projection (π) and join ($\bowtie$) operations where the projection determines the number of columns and the join determines the number of relations. A subgraph $\mathcal{J}(\mathcal{G})$ connects all the relevant relations through their fk/pk constraints while it may contain other relations as well as intermediate nodes to interconnect all the relevant relations. Hence, the schema size is directly proportional to the size of $\mathcal{J}(\mathcal{G})$. To prune the overwhelming unnecessary tuples from the outputted join table, the selection operation (σ) is used as a filter by specifying the necessary conditions for the query output table Out. The formulated queries with these three SQL operations are named as Select-Project-Join (SPJ) queries. In our work, we intend to discover more complex queries than the SPJ queries, i.e., the OLAP queries. Given the query output table Out as input, the GROUP BY operator will correspond to the number of tuples (groups) of Out. Each group will be used to produce one or multiple aggregations where each aggregation takes an aggregate operator (e.g. MAX, MIN, AVG, SUM, and COUNT) for a numeric attribute. Upon the above OLAP specifications, we define the queries as Select-Project-Join-Aggregation (SPJA) queries.

3   Join Query Discovery

In this section, we discuss how to discover the possible subgraphs based on a given query output table Out. Its columns are essential to delimit the schema size for query regeneration. Our join query discovery relies on a graph search algorithm to determine the possible candidate subgraphs. For instance, the breadth-first search algorithm in DISCOVER [2] finds the subgraphs where the nodes that contain the given keywords are taken as the leaf nodes. Apart from just considering the keywords, our problem is to find out all the possible subgraphs that can cover all columns in Out. Algorithm 1 indicates the join query discovery.

3.1   Column Mapping Table

Consider a column of Out, it is outputted by the projection operation (π) for a schema attribute A, either is operated as group-by or aggregation. An SPJA query, that aggregates the output tuples from the sets of grouped tuples; there are some columns whose aggregate values cannot be directly mapped to any schema attributes. Due to the possibility of unidentified/anonymous schema attribute(s), the column mapping details may be incomplete. To solve this problem, for each unmapped column, it can match a set of covering attributes; otherwise it has to be an integer column that can be corresponded to COUNT aggregation. These covering attributes are discovered due to different mathematical properties that are possessed by different aggregate functions. However, it is non-trivial to determine the set of covering attributes intuitively if the unmapped column tuples are far beyond any minimum/maximum values of schema attributes which the only possibility is the SUM computation that relates with both COUNT and AVG.

3.2   Candidate Subgraph Generation

By assuming the schema graph is undirected, the current (in)complete column mapping table is used to search for the (partial) candidate subgraphs. The mapped relations are set as leaves so that they must be contained in the candidate subgraph generation. A set of partial subgraphs is generated due to incomplete column mapping. Given a partial subgraph, it will be either explored or expanded to find the covering attribute(s) for the unmapped column(s) of column mapping table. In Fig. 1, the relations named Nation, Lineitem and Orders are the leaves because the schema attributes Nation.name, Lineitem.linestatus, and Orders.totalprice are mapped. When exploring the discovered partial subgraph, the attribute Lineitem.quantity can be the covering attribute for the last column of *Out*.

Partial Subgraph. Consider a set of leaves, the least connected leaf node is selected as root to connect other leaves to form a subgraph through the undirected schema graph via breadth-first search exploration. If there exists a pair of same leaf nodes, the node duplication is allowed where a node can be visited for twice. The schema size thus is determined by the total number of visited nodes. To control the schema size as well as the cost complexity, the number of intermediate nodes should be kept as fewer as possible. By heuristically, the candidates are sorted by the schema size for evaluation.

Join Table Size Estimation. Upon a partial subgraph, by doing schema exploration, the utmost task is to complete the column mapping table. Once every column in *Out* has its corresponding schema attribute(s), the partial subgraph thus becomes the complete candidate subgraph. For an unmapped column that contains aggregation results, the idea is to find the corresponding numeric attributes. Among the possible candidates, the priority is to quickly prune the inappropriate ones by inferring its join size. For a partial subgraph, its join size, Υ is determined by the total number of tuples to generate *Out*. In addition, its schema is equivalent to a set of attributes, denoted as $\mathcal{A}$. If an unmapped column λ contains only natural numbers, its total number is considered the estimated

---

Algorithm 1. Join Query Discovery

> **input** : $\mathcal{SG}$: schema graph, Out: query output table
> **output**: $\{\mathcal{J}(\mathcal{G})\}$: set of candidate subgraphs
>
> //Column Mapping
> mapping table $\phi = \emptyset$
> covering table $\bar{\phi} = \emptyset$
> **foreach** $column$ $\lambda \in Out$ **do**
> > **if** $\phi(\lambda) = schema$ $attribute$ A **then**
> > > update $\phi(\lambda) \leftarrow A$
> >
> > **else**
> > > insert λ into $\bar{\phi}$
>
> //Candidate Subgraph Generation
> **foreach** $mapping$ ϕ **do**
> > find partial subgraphs from $\mathcal{SG}$
> > **foreach** $partial$ $subgraph$ **do**
> > > **if** $\bar{\phi} \neq \emptyset$ and $\bar{\phi}(\lambda) = schema$ $attribute$ A **then**
> > > > update $\bar{\phi}(\lambda) \leftarrow A$
> > > > $\{\mathcal{J}(\mathcal{G})\} \leftarrow$ partial subgraph
> > >
> > > **else if** $\bar{\phi} \neq \emptyset$ and $!\bar{\phi}(\lambda) = schema$ $attribute$ A **then**
> > > > find set of neighbour nodes $\{R\}$
> > > > **while** $expandPartialSubgraph(R)$ **do**
> > > > > **if** $\bar{\phi}(\lambda) = schema$ $attribute$ A **then**
> > > > > > update $\bar{\phi}(\lambda) \leftarrow A$
> > > > > > $\{\mathcal{J}(\mathcal{G})\} \leftarrow$ partial subgraph
> >
> > $\{\mathcal{J}(\mathcal{G})\} \leftarrow$ partial subgraph

---

join size, $\Gamma = \sum \Lambda$, where $\Lambda \in \lambda$. The estimated Γ should be within the range of $\alpha * \Upsilon$ and Υ where α is the selectivity factor that delimits the number of tuples to generate Out as the impact of applied selection conditions. We assume the default value of α as 0.1 and all data are in normal distribution. If the statement is true, then it can be delineated as the computation of COUNT($*$). However, for the implication of $SUM(A)$, given that A is a schema attribute where $A \in \mathcal{A}$ and an unmapped numerical column λ, the estimated join size, Γ can be calculated by a simple formula as below.

$$\frac{\sum_{i=1}^{|\lambda|} \Lambda_i}{AVG(A)} = \Gamma \quad \begin{cases} \alpha * \Upsilon \leq \Gamma \leq \Upsilon & \text{true} \\ \text{otherwise} & \text{false} \end{cases}$$

An attribute $A \in \mathcal{A}$ is acceptable if the estimated Γ is between $\alpha * \Upsilon$ and Υ.

Expanding Neighbour Nodes. If it still does not have any covering attribute for any unmapped column of Out, the partial subgraph cannot establish as a candidate subgraph. An alternative approach is to expand the current partial subgraph by adding one of its neighbour nodes to form a new subgraph for schema exploration. If every column of Out is being mapped or covered, the new subgraph is added to the set of candidate subgraphs; otherwise, another new subgraph is generated by adding another selective neighbour node. This process is iterated until the set of candidate subgraphs is found.

4   Group-By Discovery, Aggregates Pruning and Filter Discovery

After determining the possible joins, the next step is to determine the group-by candidates for query discovery. According to the mapping columns, a group-by lattice is built where its nodes are the group-by candidates and its edges are the superset-subset relationships. The invalid nodes are pruned by exploring the lattice, and the remaining nodes are the possible candidates for subsequent aggregates pruning. The rule-based aggregation checking is used to generate a set of group-by key-aggregation pairs. Besides, the candidate SQL queries may contain any possible selection filters. A selection filter is illustrated as a fuzzy bounding box that can be cross-validated over a group of multi-dimensional matrices, which corresponds to a conjunction of selection predicates. The full implementation is depicted in REGAL [9].

5   Group Selection

Ideally, any constructed SPJA query $\mathcal{Q}'$ should reproduce the given output table Out, or at least $Out \subset \mathcal{Q}'(\mathcal{D})$. Since the Out itself may have been skimmed by source query $\mathcal{Q}$ for a summarized version, it contains only some groups whose corresponding aggregate values are passed a threshold. However, this threshold is considered as an additional SQL functionality, and it is less being discussed in the query reverse engineering. The SQL HAVING clause is a specific term can be used to decide whether a set of groups will be outputted in Out based on the current query result $\mathcal{Q}'(\mathcal{D})$. The HAVING clause contains a condition which involves one or two output columns. For all groups within current query result $\mathcal{Q}'(\mathcal{D})$, a satisfied HAVING condition will separate them into two distinct subsets, where one subset is similar as those groups in Out and another subset is taken as $\mathcal{Q}'(\mathcal{D}) - Out$. On the one hand, if the HAVING condition involves only one column, the current query result $\mathcal{Q}'(\mathcal{D})$ is examined by all its groups are arranged based on one of the numeric columns. On the other hand, if the HAVING condition takes two columns, where these columns are being compared so that the Out exists a specified relationship between them, such as one column whose values are always larger than those from another column. A candidate query $\mathcal{Q}'$ for the motivating example in Fig. 1 is given as:

select	*N.name, L.linestatus, max(O.totalprice), sum(L.quantity)*
from	*Nation N, Customer C, Orders O, Lineitem L*
where	*N.nationkey=C.nationkey and C.custkey=O.custkey and*
	O.orderkey=L.orderkey and L.linenumber > 1
group by	*N.name, L.linestatus*

The generated table $\mathcal{Q}'(\mathcal{D})$ contains 50 tuples (groups). However, Out contains only 12 tuples (groups), and it is a subset of $\mathcal{Q}'(\mathcal{D})$. By searching through the aggregation candidates, e.g., *max(O.totalprice)* and *sum(L.quantity)*, the twelve tuples can be discerned by formulating a HAVING condition, as *having max(O.totalprice) > 500000*.

6   Experimental Evaluation

Implementation and Dataset. We implemented our proposed algorithm in Java with MySQL server as DBMS. The experiments were conducted on an Ubuntu machine with 2.40 GHz Intel CPU and 16 GB RAM. TPC-H benchmark is the dataset that used for experiments, with a scale factor of 1 and size of 1 GB.

TPC-H Test Queries. There are a total of 22 test queries for TPC-H benchmark. Most of them include different number of joins, except for TQ1 and TQ6, with the absence of join. We neglect the complex join query discovery, i.e. the nested joins, fk/fk joins, and equijoins, which are exhibited in TQ5 (e.g., S_nationkey = C_nationkey) and TQ21 (e.g., L1_orderkey = L2_orderkey) respectively. We test for the remaining join queries and scale them based on the number of joins, i.e. from 1 to 5.

Query Output Table Generation. Given that a test query Q, we execute it over TPC-H dataset D to generate the query output table $Q(D) = Out$ which later the Out will be used as the input of our proposed algorithm to discover for such a query Q' where $Q'(D) = Out$. As we have selected those TPC-H benchmark queries, however, except for the join relations and join predicates are remained, other SQL operations are altered. We set several parameters for the experiments to control the variety of query output tables. For example, we will produce the query output table Out with the cardinality of m and the arity of n, and the test query Q contains a N-dimensional filter.

6.1   TPC-H Join Queries

For each of these test queries, we generate a query output table with moderate row size m and column size $n = 4$ where it must contain both group-by

Table 1. Effect of number of joins.

# Joins	TQ	Tables	Runtime (s) min-max	# Graphs min-max
1	4, 12	L, O	144.189–281.529	1
	13, 22	C, O		
	14, 17, 19	L, P		
	15	L, S		
2	11	PS, S, N	60.937–294.624	1
	16	P, PS, S		
	3, 18	C, O, L		
3	10	N, C, O, L	313.882	1
4	2	R, N, S, PS, P	68.648–338.248	1–2
	20	N, S, L, PS, P		
5	7	N1, S, L, O, C, N2	417.508–443.197	2
	9	N, S, L, O, PS, P		

statements and aggregations which the number of group-by columns is set at most two while the other columns are used for aggregations. Each query contains one-dimensional filter as $\mathcal{N} = 1$. Table 1 records the experimental results based on the number of joins of TPC-H benchmark test queries. First, the inferred join table size is essentially crucial as it will impact the time cost for a table scan. Some of these test queries like TQ2, TQ11, and TQ20 contain the least inferred join table size (0.8 million tuples), which take less than 70 s for discovering these queries. Second, the number of join graphs is directly proportional to the total execution time. For example, test queries like TQ7, TQ9 and TQ20, they need to evaluate two join candidates to generate the discovered queries.

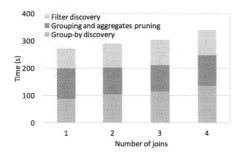 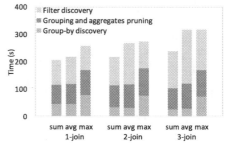

Fig. 2. Average time for query discovery against number of joins.

Fig. 3. Individual phase performances for different aggregations.

6.2 Individual Phase Performances vs. Joins

In order to further analyze the important factors that influence the total execution time besides the join operations, those TPC-H benchmark queries under similar considerations (i.e. the same inferred join table size) are examined. As there are multiple test queries for each number of joins, we run those queries individually and take their average running time. To avoid the cost of exploring all candidate queries, the execution time is taken once the least complex $\mathcal{Q}'$ is returned for a given test query $\mathcal{Q}$. The experimental results are illustrated as shown in Fig. 2. According to the experimental results, the total execution time is proportional to the number of joins. As the number of joins is increased, it indulges more schema tables/attributes for the query discovery and takes longer time for evaluation. Furthermore, all three individual phases involve table scans. In the phase of group-by discovery, the *Out* tuples are verified at instance-level to validate the group-by nodes. Second, during the phase of grouping and aggregates pruning, the inferred join instance is partitioned by the group-by nodes to find out the possible aggregations based on the derived constraint rules. Third, the schema attributes are used to construct the $\mathcal{N}$-dimensional matrices, so that the selection filter(s) can be found within these matrices.

6.3   Joins vs. Aggregations

Figure 3 shows the experimental results by comparing the selected aggregations for test queries w.r.t. different number of joins. Among five basic aggregate operators that we have discussed, three of them are chosen for this experiment, namely MAX, SUM and AVG, since MIN and MAX are symmetrical whereas COUNT is assumed as another SUM operation of a special attribute whose each of its values is set to 1. For the experiment settings, we set the parameters to output every Out with $n = 4$, where there must be one aggregation column that is selected between MAX, SUM and AVG with three group-by columns. First, by looking at each individual aggregation w.r.t. joins, as the number of joins is increased, the total running time is also increased. By comparing these aggregations, it is apparent that MAX takes the largest running time in the phase of group-by discovery regardless the number of joins if compared to both SUM and AVG aggregations. The size of group-by lattice for MAX is $2^4 = 16$ nodes whereas the size of group-by lattice for SUM or AVG is smaller, which is $2^3 = 8$ nodes. However, AVG takes more time in the phase of filter discovery as compared to SUM and MAX, since the computation for AVG is more complex than that SUM. Thus, AVG takes the second largest time for the query discovery.

7   Conclusion

In this paper, we bring these two main features together by integrating the promising approaches from both existing works with optimizations. Our empirical study has shown that our proposed solution can work in practice with the TPC-H benchmark dataset.

References

1. Agrawal, S., Chaudhuri, S., Das, G.: DBXplorer: a system for keyword-based search over relational databases. In: ICDE, pp. 5–16 (2002)
2. Hristidis, V., Papakonstantinou, Y.: DISCOVER: keyword search in relational databases. In: VLDB, pp. 670–681 (2002)
3. Markowetz, A., Yang, Y., Papadias, D.: Keyword search on relational data streams. In: SIGMOD, pp. 605–616 (2007)
4. Panev, K., Michel, S.: Reverse engineering top-k database queries with PALEO. In: EDBT, pp. 113–124 (2016)
5. Psallidas, F., Ding, B., Chakrabarti, K., Chaudhuri, S.: S4: top-k spreadsheet-style search for query discovery. In: SIGMOD, pp. 2001–2016 (2015)
6. Qian, L., Cafarella, M.J., Jagadish, H.V.: Sample-driven schema mapping. In: SIGMOD, pp. 73–84 (2012)
7. Qin, L., Yu, J.X., Chang, L.: Keyword search in databases: the power of RDBMS. In: SIGMOD, pp. 681–694 (2009)
8. Shen, Y., Chakrabarti, K., Chaudhuri, S., Ding, B., Novik, L.: Discovering queries based on example tuples. In: SIGMOD, pp. 493–504 (2014)
9. Tan, W.C., Zhang, M., Elmeleegy, H., Srivastava, D.: Reverse engineering aggregation queries. PVLDB **10**(11), 1394–1405 (2017)

10. Tran, Q.T., Chan, C.-Y., Parthasarathy, S.: Query by output. In: SIGMOD, pp. 535–548 (2009)
11. Tran, Q.T., Chan, C.Y., Parthasarathy, S.: Query reverse engineering. VLDB J. **23**(5), 721–746 (2014)
12. Valduriez, P.: Join indices. ACM Trans. Database Syst. **12**(2), 218–246 (1987)
13. Zhang, M., Elmeleegy, H., Procopiuc, C.M., Srivastava, D.: Reverse engineering complex join queries. In: SIGMOD, pp. 809–820 (2013)

DCA: The Advanced Privacy-Enhancing Schemes for Location-Based Services

Jiaxun Hua[1], Yu Liu[1], Yibin Shen[1], Xiuxia Tian[3(✉)], and Cheqing Jin[2]

[1] School of Computer Science and Software Engineering,
East China Normal University, Shanghai, China
{vichua,leoliu,ybshen}@stu.ecnu.edu.cn
[2] School of Data Science and Engineering, East China Normal University,
Shanghai, China
cqjin@dase.ecnu.edu.cn
[3] Shanghai University of Electric Power, Shanghai, China
xxtian@fudan.edu.cn

Abstract. With the popularity of Location-based Services, LBS providers have been obtaining more data, by analyzing which they may infer users' real locations and patterns of behavior. Unfortunately, most previous schemes using *k-anonymity* can hardly resist such fiercer side information-based privacy attacks. To address existing problems, we design a novel metric to accurately measure the resulted privacy level. Additionally, Dual Cloaking Anonymity (*DCA*) and *enhanced-DCA* (*enDCA*) algorithms, which are based on our metric, are also proposed. The former (*DCA*) constructs a k-anonymity set via carefully selecting *k-1* users according to *various query probabilities* of each area and correlations between users' *query preferences*. Then, *enDCA* further employs *caching* and *location blurring* to enhance the privacy preservation. Evaluations show that our proposals can significantly improve the privacy level.

Keywords: LBS privacy · k-anonymity · Confusion degree

1 Introduction

Location-based services are springing up around us, whereas leakages of users' privacy are inevitable during these services. Even worse, adversaries may analyze intercepted service data, and extract more privacy like hobbies, health and property. Hence, privacy preservation is an indispensable guarantee on LBS.

Among existing privacy preservation approaches, ones based on k-anonymity are widely researched. However, some privacy concern will be aroused if these schemes are adopted directly. For example (in Fig. 1), an area is divided into 4×4 cells, where a target user U_t issues a query "Find the nearest hotel" (his privacy profile $k = 4$). *DLS* algorithm [6] selects four blue cells to construct a cloaking set because their *gross query probabilities* are similar. Although such a set reached the maximum entropy, experienced adversaries can exclude some

© Springer International Publishing AG, part of Springer Nature 2018
Y. Cai et al. (Eds.): APWeb-WAIM 2018, LNCS 10988, pp. 63–71, 2018.
https://doi.org/10.1007/978-3-319-96893-3_6

cells if they have richer side information, such as features of each cell and users in the cells.

According to querying features of different cells and U_t's query content, adversaries may exclude cell b & d from the set. With the help of further analyses of query preferences, if adversaries learn that U_t is a businessman, they can confidently locate U_t. Thus, location privacy of U_t is invaded.

To address those defects, we propose a novel privacy metric which first takes into account the impact of richer side information on privacy. Then, DCA and $enDCA$ algorithms are designed. They both fulfill our objectives while either one has different advantages. Major contributions are summarized as follows:

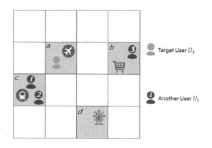

Fig. 1. An example of a cloaking set. More queries about hotels and transport occur in cell a & c, while more queries about entertainment and shopping occur in cell b & d. U_t prefers to query for hotels and conference centers via LBS. U_1 and U_2 mainly search for entertainment.

- A newly-proposed entropy-based privacy metric may measure the privacy level, and depict the impact of richer side information on privacy.
- We design DCA algorithm, which considers richer side information (query probabilities & preferences) when constructing k-anonymity sets.
- Based on DCA, *location blurring* and *caching* are introduced to $enDCA$. These techniques impede invading location privacy, promote the low bandwidth overhead and resist the disclosure of users' preference privacy.
- We adopt a novel Wi-Fi access point based Peer-to-Peer structure.

2   Related Work

Recently, many research efforts have been concentrated in LBS privacy.

Among cryptography based techniques, Ghinita et al. [2] used Computational PIR, which needs two stages to retrieve POI data. Papadopoulos et al. [10] proposed cPIR which reduces computational overhead.

Kido et al. [3] cloaked user's real location by generating $k-1$ dummy locations, but side information is ignored. *Casper* [5] provided cloaking regions according to user's privacy profile and minimum area, whereas maintaining the pyramid structure leads to high costs. Niu et al. [6,7] designed AP-based k-anonymity schemes considering query probabilities and caching. However, constructing cloaking sets and caching data need high computational and storage overhead for APs, and k-anonymity isn't effectively guaranteed due to negligence in the variety of queries.

Palanisamy et al. [9] constructed adaptive mix-zones centered at road intersections, which replace actual query time with shifted ones, to resist timing attacks. However, these schemes limit the submissions of queries in Mix-zones.

Miguel et al. [1] migrated differential privacy to LBS privacy preservation by adding Laplace noise to users' coordinates.

3 Preliminaries

3.1 Basic Concepts

Query Probabilities. We classify LBS queries into m types with respect to contents of queries. Then we define *various query probabilities* in Eq. 1. For simplicity, an m-dimensional vector $\mathcal{P}_i$ is used to represent respective probabilities of all m types of queries in $cell_i$.

$$\mathcal{P}_i = (p_i^1, p_i^2, \ldots, p_i^m), \quad p_i^j = \frac{\# \text{ of type-}j \text{ queries in } cell_i}{\# \text{ of total queries over all cells}} \quad (1)$$

Users' Query Preferences. Different users have various *query preferences*, which are closely related to their life patterns. We use a vector $\mathcal{W}_i$ to describe the query preference of user U_i (see Eq. 2). Preference vectors will be updated periodically using *Aging Algorithm*.

$$\mathcal{W}_i = (w_i^1, w_i^2, \ldots, w_i^m), \quad w_i^j = \frac{\# \text{ of } U_i's \text{ type-}j \text{ queries (over all cells)}}{\# \text{ of } U_i's \text{ total queries (over all cells)}} \quad (2)$$

Moreover, we use *standardized preference vector* $\mathcal{W}_i' = (w_i^{1'}, w_i^{2'}, \ldots, w_i^{m'})$ instead to preserve users' preference privacy (Different preference vectors may have the same standardized vector), where $w_i^{j'} = \frac{w_i^j - \mu_{\mathcal{W}_i}}{\sigma_{\mathcal{W}_i}}$ ($\mu_{\mathcal{W}_i}, \sigma_{\mathcal{W}_i}$ are the mean and the standard deviation of $\mathcal{W}_i$ respectively). Then, the correlation coefficient between arbitrary two LBS users U_x, U_y is defined in Eq. 3.

$$\rho(U_x, U_y) = \frac{covariance(\mathcal{W}_x, \mathcal{W}_y)}{\sigma_{\mathcal{W}_x} \cdot \sigma_{\mathcal{W}_y}} = covariance(\mathcal{W}_x', \mathcal{W}_y') \quad (3)$$

3.2 Adversary Model

In this paper, we resist *eavesdropping attack* performed by passive adversaries via applying SSL on communication channels. We consider LBS servers, who own global data, as active adversaries. Even worse, those untrusted servers may *collude* with malicious users to *infer* normal users' query preferences and behavior patterns by exchanging extra information and analyzing obtained data.

3.3 Privacy Metrics

In order to demonstrate the impact of *query preferences* and *various query probabilities* on privacy quantitatively, we improve the definition of entropy [6].

Supposing a user U_t issues a type-j query in $cell_t$ under the protection of a k-anonymity set. The query preference of U_t is $\mathcal{W}_t$, and the type-j query probability of $cell_t$ is p_t^j. In addition, $k-1$ other users are located in $cell_1, cell_2, \ldots, cell_{k-1}$ (type-j query probabilities of these cells are $p_1^j, p_2^j, \ldots, p_{k-1}^j$). So the *confusion degree* (ξ) of the k-anonymity set is defined in Eq. 4.

$$\xi = -\sum_{i=1}^k \rho(U_t, U_i) \cdot q_i^j \cdot \log_2 q_i^j = -\sum_{i=1}^k r_i \cdot q_i^j \cdot \log_2 q_i^j \quad (q_i^j = \frac{p_i^j}{\sum_{s=1}^k p_s^j}) \quad (4)$$

4 Our Proposed Schemes

4.1 System Model

Figure 2 shows our novel AP-based P2P structure. APs[1] are designed to undertake such *light* workloads as collecting query probabilities, forwarding data, locating users, and storing caches. Maintenance of users' query preference vectors and calculations are conducted by users locally. Besides, LBS users may communicate with APs *anonymously* (i.e. using pseudonyms) to preserve privacy against APs.

4.2 Schemes Overview

We introduce how APs work via the example in Fig. 2. Suppose that Peter issues a query Q in $cell_t$. APs construct an anonymity set by taking following steps.

(1) After an AP receives Q and Peter's real location $cell_t$ (together with $\mathcal{W}'_{Peter}$ and some other parameters), it will determine the query type of Q.
(2) If Q is a type-j query, APs will search for nearby cells with similar type-j query probabilities to $cell_t$. (subject to probability threshold β).
(3) APs forward $\mathcal{W}'_{Peter}$ to users in cells found in step (2).
(4) Any user U_x who has received $\mathcal{W}'_{Peter}$ computes the correlation coefficient $\rho(U_x, Peter)$ between his preference vector and Peter's. U_x will reply APs with the coefficient if the value is greater than the preference threshold θ.
(5) APs reply Peter with users who have similar query preferences, together with coefficient values, indexes of probability differences, and indexes of distance between Peter and them. The distance can be measured by # of hops on the grid-based map (e.g. In Fig. 1, the distance between U_t and U_3 is 2).
(6) Peter filters out $k-1$ optimal users locally according to side information above. Then, he will construct a k-anonymity set and issue the formal query.

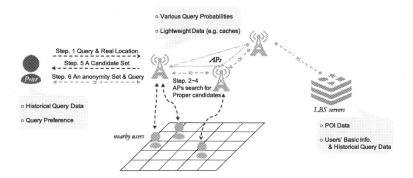

Fig. 2. Schemes overview (data owned by each role is shown in gray blocks)

---

[1] AP-based schemes [4,6–8] have been widely applied to LBS in mobile environments.

Algorithm 1. Client: DCA Sub-algorithm (issuing a query)

Input: target user U_t's standardized preference vector $\mathcal{W}_t'$, an LBS query
$Q(qtype, qdetail)$, real location $cell_t$, privacy profile k_t, distance
preference μ, # of sets ns

Output: an optimal k-anonymity set AS

1 send $(\mathcal{W}_t', Q, cell_t, k_t, \mu)$ to AP (run Algorithm 2);

2 wait until AP returns CS to it; //Alg. 2 (Line 9) shows data structure of CS

3 **for** $(i = 0; i < \min(ns, \binom{3k_t}{k_t-1}); i++)$ **do**

4 $\quad$ construct set C_i with U_t and $k_t - 1$ other users (in set CS) at random;

5 $\quad score_{C_i} = \sum_{j=1}^{k_t} (index\_prdiff_{ij} \cdot index\_dis_{ij} \cdot r_{ij});$

6 **return** $\arg\max_{C_i}(score_{C_i});$

4.3 The Dual Cloaking Anonymity Algorithm

According to the division of work, we implement our schemes in three sub-algorithms. Algorithms 1 and 3 run on clients, and Algorithm 2 runs on APs.

Algorithm 1 demonstrates *DCA* Sub-algorithm which runs on the client of target user U_t (who issues the query actually). It corresponds to Step 1, 6 in last section.

Next, we present Algorithm 2 running on APs. This process corresponds to Step 2, 3, 5 in Sect. 4.2. Index of differences in type-j query probability between the real location $cell_t$ and other cells can be achieved by $index\_prdiff = 1 - \frac{|pr - p_t^{qtype}|}{\beta}$. In addition, we use the index of distance $index\_dis = e^{-\frac{(dis-\mu)^2}{8}}$ to describe users' distance preference. If there aren't enough candidates in CS, AP will extend searching areas (Line 2).

Algorithm 3 computes correlation coefficient between query preferences.

Algorithm 2. AP: DCA Sub-algorithm (forwarding information)

Input: U_t's standardized preference vector $\mathcal{W}_t'$, an LBS query $Q(qtype, qdetail)$,
real location $cell_t$, privacy profile k_t, distance preference μ

Output: a candidate set CS

1 CS=NULL;

2 **for** $(d = 1; CS.size() < 3k_t; d++)$ **do**

3 $\quad$ searching for $cell_x$ in d-hop area around $cell_t$, s.t. $\forall x, |p_t^{qtype} - p_x^{qtype}| < \beta;$

4 $\quad$ send $\mathcal{W}_t'$ to users who are located in these found cells (run Algorithm 3);

5 $\quad$ **while** $\exists$ tuples $(\widetilde{user}, r)$ returned from users **do**

6 $\quad\quad index\_dis = e^{-\frac{(d-\mu)^2}{8}};$

7 $\quad\quad pr = \text{getPr}(user, qtype);$ //retrieve the query probability of a cell

8 $\quad\quad index\_prdiff = 1 - \frac{|pr - p_t^{qtype}|}{\beta};$

9 $\quad\quad$ add tuples $(\widetilde{user}, r, index\_dis, index\_prdiff)$ to CS;

10 **return** CS;

4.4   The Enhanced Dual Cloaking Anonymity Algorithm

We introduce more advanced techniques: *location blurring* and *caching* to *enDCA*, which may upgrade users' privacy at the expense of limited compromise in QoS.

Location Blurring. When applying k-anonymity, the real location is likely to be inferred if k is large, as all dummies are distributed around the real one.

Algorithm 3. Client: compute_corr

    Input: U_t's standardized preference vector $\mathcal{W}_t'$, other's preference vector $\mathcal{W}_a$
    Output: Pearson correlation coefficient between U_t and himself(herself)
1 standardize the vector $\mathcal{W}_a$ as $\mathcal{W}_a'$;
2 **if** $(r = covariance(\mathcal{W}_t', \mathcal{W}_a')) > \theta$ **then**
3   **return** $(\widetilde{user}, r)$;   //user's ID will be replaced by a pseudonym

---

To address that privacy issue, *location blurring* is introduced into *enDCA*. Target user's real location will be shifted to a cell which is randomly selected from the nearby ones (in the 1-hop area) with similar same-type query probabilities.

Caching. Different from previous work [7,11], we propose the idea of caching the anonymity sets. Supposing an LBS user U_a (privacy profile is k_a) issues a query $Q(qtype_a, qdetail_a)$. A cached set t can be used to preserve U_a's location privacy if Eq. 5 holds. *Caching* may relieve the workload of APs, reduce the bandwidth overhead, and preserve query preference privacy (reducing transmission of users' preferences). Cache will be maintained by APs in background.

$$\exists t \in AS, \ s.t. \ (1) \ t.qtype = qtype_a; \ (2) \ t.k \geq k_a; \ (3) \ \exists i \in [1,k], \ t.U_i = U_a. \ (5)$$

The data structure of the cached anonymity sets is as follows:
$AS(qtype, k, expire, U_1, U_2, \ldots, U_k)$, where $expire$ is the lifetime of a set.

Algorithm 4. Client: enDCA Sub-algorithm (issuing a query)

    Input: U_t's standardized preference vector $\mathcal{W}_t'$, an LBS query $Q(qtype, qdetail)$,
           real location $cell_t$, privacy profile k_t, distance preference μ, # of sets ns
    Output: an optimal k-anonymity set AS (or a cached set CAS)
1 send $(\mathcal{W}_t', Q, cell_t, k_t, \mu)$ to AP (run Algorithm 5);
2 wait until CS or CAS returned from AP ;
3 **if** $CAS \ != NULL$ **then**
4   **return** CAS or a subset of CAS according to k_t;
5 **else**
6   run Lines 3-6 in Algorithm 1 (Client: DCA Sub-Algorithm);

Algorithm 4 presents *enDCA* Sub-algorithm which runs on clients. If there exists an appropriate cached set, it'll call Algorithm 1 to construct the set (Line 6).

Algorithm 5. AP: enDCA Sub-algorithm (forwarding information)

Input: U_t's standardized preference vector $\mathcal{W}'_t$, an LBS query $Q(qtype, qdetail)$,
real location $cell_t$, privacy profile k_t, distance preference μ
Output: a candidate set CS or a cached anonymity set CAS
1 CS=NULL, T=NULL; //T stores cached anonymity set temporarily
2 **foreach** t in $cache[qtype]$ **do**
3   **if** $t.k \geq k_t$ **and** $(\exists i \in [1, t.k], t.U_i == U_t)$ **then**
4        $T = T \bigcup \{t\}$;
5 **if** $T \;!= NULL$ **then**
6   **return** $\arg\max_{t \in T}(\frac{\xi_t}{\log_2 t.k})$;   //return the set with highest confusion degree
7 run AP: DCA Sub-Algorithm($\mathcal{W}'_t$, Q, shiftLocation($cell_t$), k_t, μ); //run Algorithm 2

Algorithm 5 illustrates *enDCA* Sub-algorithm running on APs. After AP receives U_t's query, it will check in cache whether there exist appropriate anonymity sets. Otherwise, Algorithm 5 shifts U_t's real location first, and then follows ordinary steps to construct a candidate set CS (Line 7).

4.5   Security Analysis (Resistance to Colluding and Inference Attacks)

Adversaries try to infer U_t's real location in the way described in Sect. 3.2. However, the idea of maximizing confusion degree and randomization in our schemes will obstruct their conspiracies. Compared with *DCA*, *caching* in *enDCA* reduces exposure of query preferences. *Location blurring* and *standardized preference vectors* may frustrate their inference of real locations when constructing new anonymity sets.

5   Performance Evaluation

5.1   Simulation Setup

The trajectory data of taxis (From http://soda.datashanghai.gov.cn, involving about 10,000 trajectories) is used to describe the mobility patterns of LBS users in a $10 \, \text{km} \times 8 \, \text{km}$ area in downtown Shanghai. The area is divided into 8,000 cells, with the size of each being $100 \, \text{m} \times 100 \, \text{m}$. The real deployment of APs in that area will also be simulated. Query probabilities are computed as the users' density in each cell, and the query preferences of users are randomly assigned under normal distribution. Parameters used in our simulation are as follows:

Privacy profile k is set from 2 to 15. # of query types $m = 5$, # of sets $ns = 100$. Threshold $\beta = 0.0015$, $\theta = 0.2$.

We select *Random* [3] as the baseline scheme. *DLS* (*enhanced-DLS*) [6], one of state-of-the-art methods, is also chosen as a comparison.

5.2   Evaluation Results

k vs. Privacy Metrics. Figure 3(a) and (b) show the relation between k and entropy. *Gross query probability* is used in Fig. 3(a), so that all schemes except for *Random* perform well. On the contrary, *various query probability* highlights the advantages of our schemes in Fig. 3(b).

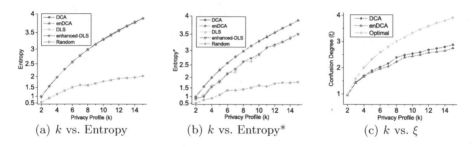

(a) k vs. Entropy   (b) k vs. Entropy*   (c) k vs. ξ

Fig. 3. Effect of k on privacy metrics

As to confusion degree (Fig. 3(c)), *DCA* edges out *enDCA*, as *enDCA* sacrifices some confusion degree to decrease bandwidth overhead. Our schemes have high but not theoretically optimal results because finding $k - 1$ nearby users having approximately the same query preferences is quite tough.

Other Performance Evaluations. Figure 4 depicts that bandwidth overhead of *enDCA* outperforms *DCA*, since *caching* can serve users' requests for anonymity sets. Figure 5 illustrates the relation among k, cache hit ratio and simulation time t. The hit ratio increases gradually with the t, and smaller k

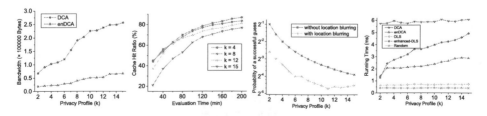

Fig. 4. Bandwidth   **Fig. 5.** Cache   **Fig. 6.** Guessing Pr.   **Fig. 7.** Efficiency

usually results in higher ratio. Figure 6 confirms that schemes without *location blurring* have the theoretical k-anonymity. *enDCA*, equipped with *location blurring*, owns significantly lower probabilities of successful guesses. Figure 7 shows the running time of all schemes. Our schemes consume moderate time to construct a k-anonymity set, and *enDCA* costs less time than *DCA* with the help of *caching*.

6 Conclusion

We propose two different LBS privacy-enhancing schemes, and a novel metric to measure the privacy level. *DCA* constructs a k-anonymity set via carefully selecting $k-1$ users according to various query probability and users' query preferences. Based on that, *caching* and *location blurring* are introduced to *enDCA*, which reduce exposure of query preferences, and decrease the bandwidth overhead. Simulations confirm the effectiveness of our schemes.

Acknowledgment. Our research is supported by the National Key Research and Development Program of China (2016YFB1000905), NSFC (61772327, 61370101, 61532021, U1501252, U1401256 and 61402180), Shanghai Knowledge Service Platform Project (No. ZF1213), Shanghai Science and Technology Committee Grant (15110500700).

References

1. Andrés, M.E., et al.: Geo-indistinguishability: differential privacy for location-based systems. In: 2013 ACM SIGSAC, pp. 901–914 (2013)
2. Ghinita, G., Kalnis, P., Khoshgozaran, A., Shahabi, C., Tan, K.L.: Private queries in location based services: anonymizers are not necessary. In: ACM SIGMOD (2008)
3. Kido, H., Yanagisawa, Y., Satoh, T.: An anonymous communication technique using dummies for location-based services. In: ICPS, pp. 88–97 (2005)
4. Luo, W., Hengartner, U.: VeriPlace: a privacy-aware location proof architecture. In: ACM SIGSPATIAL GIS, pp. 23–32 (2010)
5. Mokbel, M.F., Chow, C.Y., Aref, W.G.: The new casper: query processing for location services without compromising privacy. In: VLDB, pp. 763–774 (2006)
6. Niu, B., Li, Q., Zhu, X., Cao, G.: Achieving k-anonymity in privacy-aware location-based services. In: IEEE INFOCOM, pp. 754–762 (2014)
7. Niu, B., Li, Q., Zhu, X., Cao, G.: Enhancing privacy through caching in location-based services. In: IEEE INFOCOM, pp. 1017–1025 (2015)
8. Okamoto, M., Fujita, N., Inomae, G., Tate, H.: Wi-Fi LBS: information delivery services using Wi-Fi access point location. NTT Tech. Rev. 11(9) (2013)
9. Palanisamy, B., Liu, L.: MobiMix: protecting location privacy with mix-zones over road networks. In: IEEE ICDE, pp. 494–505 (2011)
10. Papadopoulos, S., Bakiras, S., Papadias, D.: pCloud: a distributed system for practical PIR. IEEE TDSC 9(1), 115–127 (2012)
11. Shokri, R., Theodorakopoulos, G., Papadimitratos, P., Kazemi, E.: Hiding in the mobile crowd: locationprivacy through collaboration. IEEE TDSC 11(3), 266–279 (2014)

Data Streams

Discussion on Fast and Accurate Sketches for Skewed Data Streams: A Case Study

Shuhao Sun[1] and Dagang Li[1,2(✉)]

[1] School of ECE, Peking University Shenzhen Graduate School,
Shenzhen 518055, China
shuhaosun@pku.edu.cn, dgli@pkusz.edu.cn
[2] Institute of Big Data Technologies, Peking University,
Shenzhen 518055, China

Abstract. Sketch is a probabilistic data structure designed for the estimation of item frequencies in a multiset, which is extensively used in data stream processing. The key metrics of sketches for data streams are accuracy, speed, and memory usage. There are various sketches in the literature, but most of them cannot achieve high accuracy, high speed and using limited memory at the same time for skewed datasets. Recently, two new sketches, the Pyramid sketch [1] and the OM sketch [2], have been proposed to tackle the problem. In this paper, we look closely at five different but important aspects of these two solutions and discuss the details on conditions and limits of their methods. Three of them, memory utilization, isolation and neutralization are related to accuracy; the other two: memory access and hash calculation are related to speed. We found that the new techniques proposed: automatic enlargement and hierarchy for accuracy, word acceleration and hash bit technique for speed play the central role in the improvement, but they also have limitations and side-effects. Other properties of working sketches such as deletion and generality are also discussed. Our discussions are supported by extensive experimental results, and we believe they can help in future development for better sketches.

Keywords: Sketch · Skewed data · Data structure

1 Introduction

Estimating the frequency of each item in a multiset is one of the most classic tasks in data stream applications. In many networking scenarios such as real-time IP traffic, IP phone calls, videos, sensor measurements, web clicks and crawls, massive amount of data are often generated as high-speed streams [3, 4], requiring servers to process such stream in a single-pass [5]. Calculating exact statistics (e.g., using hash tables) is often impractical, because the time and space overhead of storing the whole data stream is too high. Therefore, it is popular and widely accepted to estimate the frequencies of each item by the probabilistic data structure [6–8].

Sketches are a family of probabilistic data structure designed for the estimation of item frequencies in data streams [9, 10], which is extensively used in data stream processing. They use counters to store frequencies and have two primary operations:

© Springer International Publishing AG, part of Springer Nature 2018
Y. Cai et al. (Eds.): APWeb-WAIM 2018, LNCS 10988, pp. 75–89, 2018.
https://doi.org/10.1007/978-3-319-96893-3_7

insertion and query. By using multiple hash functions, sketches summarize massive data streams within a limited space, which means there might be two or more items sharing the same counter(s). Sketches can also be applied to other fields, such as compressed sensing [11], natural language processing [12], and data graph [13].

Conventional sketches (CM sketch [7], CU sketch [14], Count sketch [8], and Augmented sketch [6]) use a number of counters of fixed size. The size needs to be large enough to accommodate the highest frequency. However, according to the literatures [6] and confirmed by our experiments on real datasets, the items in real data streams often have unbalanced distribution, such as Zipf [15] or Power-law [16]. This means that most items have low frequency (called cold items), while a few items have high frequency (called hot items). Such data streams are often called skewed data streams. Therefore, the high-order bits in most counters of conventional sketches are wasted, as hot items are much fewer than cold items in real data streams. This kind of memory inefficiency reduces the number of counters, causing the accuracy of the conventional sketches to drop drastically. Besides, conventional sketches cannot perfectly catch up with the high speed of data streams because they need three or more hash computations and memory accesses for each insertion or query. Overall, conventional sketches fall short handling skewed data streams, and the goal of this paper is to discuss how to design better sketches for this matter.

Two novel sketches have been proposed recently, the Pyramid sketch [1] and the OM sketch [2], which can achieve both high accuracy and high speed using limited memory, especially for skewed data streams. These two sketches bring new ideas that are specifically designed for skewed data. For example, automatic enlargement and hierarchy can greatly improve the accuracy when summarizing skewed datasets, and word acceleration and hash bit technique can significantly improve the speed for each insertion or query operation. However, we found that many aspects need to be further considered when using these techniques, therefore in this paper we will discuss the strategies of automatic enlargement, the side-effect of hierarchy, the use conditions of word acceleration and hash bit technique. Furthermore, we found that there are two other aspects to improve accuracy, which are barely scratched in the original papers [1, 2]. We name these two methods as isolation and neutralization. The usage of them depends on the specific target application scenario. Moreover, when designing the sketch, other requirements and constraints brought by the target application scenario should also be considered, such as deletion and generality [17]. These are also discussed in this paper.

Our contributions can be summarized as follows.

- We sort out five important aspects to design an accurate and fast sketch for skewed data streams. Three of them, memory utilization, isolation, and neutralization are to help improve accuracy, and the other two: memory access and hash computation are important for speed. Their role in an effective and efficient solution are analyzed.
- The specific methods proposed from the latest work [1, 2] are discussed in details, including the strategies of automatic enlargement, the side-effect of hierarchy, the usages of isolation and neutralization, the use conditions of word acceleration and hash bit technique. We also discuss the deletion and generality of the sketch. These discussions will help better understanding and further utilization of these new ideas.

2 Related Work

2.1 Conventional Sketches

Typical sketches include CM sketch [7], CU sketch [14], Count sketch [8], and Augmented sketch [6]. A CM sketch consists of d arrays: $A_1...A_d$, and each array consists of w counters. There are d hash functions, $h_1...h_d$, in the CM sketch. When inserting an item e, the CM sketch first computes the d hash functions and locates the d counters: $A_1[h_1(e)]...A_d[h_d(e)]$. Then it increases all the d hashed counters. When querying an item e, the CM sketch reports the minimum of the d hashed counters as the estimated frequency of this item. The CU sketch has a slight but effective modification to the CM sketch, that is, conservative update. It only increases the smallest one(s) among the d hashed counters during insertions while the query process keeps unchanged. The Count sketch is similar to the CM sketch except that each array uses an additional hash function to smooth the accidental errors. The Augmented sketch aims to improve the accuracy by using one additional filter to dynamically capture hot items, suffering from complexities, slow insertion and query speed. Among these sketches, the CU sketch achieves the best performance in terms of both accuracy and speed. More sketches are detailed in the survey [18].

Unfortunately, the sketches above have two shortcomings for skewed data streams: (1) the accuracy is poor when using limited memory; (2) requiring multiple memory accesses and hash computations for each insertion or query thus slow the speed.

2.2 The OM Sketch

The key techniques of OM sketch are hierarchical counter-sharing, word acceleration and fingerprint check.

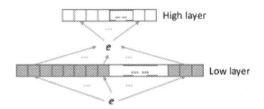

Fig. 1. Basic structure of OM sketch.

As shown in Fig. 1, the OM sketch is organized as a two-layer structure in which the high layer possesses less memory. The low layer with small counter sizes mainly records the information of cold items, while the high layer with relatively large counter sizes mainly records the information of hot items. When one or more counters overflow at the low layer, the OM sketch uses the high layer to record its number of overflows. Based on this structure, the OM sketch significantly improves the memory efficiency, thus improving accuracy. Moreover, the OM sketch constrains the hashed counters within one or several machine words by using the word acceleration technique. It also leverages the hash bit technique [19] to locate multiple hashed counters within one or several machine words at each layer through a 64-bit hash value by one hash function. Therefore, the OM sketch achieves close to one memory access and one hash computation for each insertion or query. Besides, the OM sketch records the fingerprints of the overflowed items in their corresponding machine words at the low layer in order to distinguish them from non-overflowed items during queries.

Insertion: When inserting an item, the OM sketch first computes the low layer hash function to locate the low layer hashed counters, and then increases the smallest counter(s). This method makes the low layer counters of each item always overflow concurrently. If an item overflows, the OM sketch first sets all its low layer hashed counters to zero, and then uses the fingerprint technique to distinguish it from non-overflowed items. Finally, the OM sketch computes the high layer hash function to locate the high layer hashed counters and increases the smallest counter(s).

Query: When querying an item, the OM sketch first gets the value of the smallest hashed counter(s) at the low layer, denoted by V_l. Then it checks if the item overflows. If it is, the OM sketch queries the high layer and gets the value of the smallest hashed counter(s) at the high layer, denoted by V_h. The OM sketch returns $V_l + V_h \times 2^{\delta_l}$ as the estimated size of the item, and δ_l is the counter size at the low layer.

2.3   The Pyramid Sketch

The key techniques of the Pyramid sketch are counter-pair sharing, word acceleration and Ostrich policy.

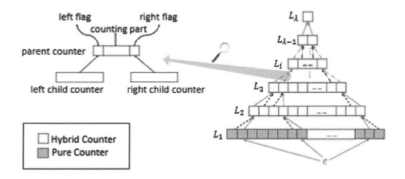

Fig. 2. Basic structure of Pyramid sketch.

As shown in Fig. 2, the Pyramid sketch employs a pyramid-shaped data structure. The ith layer L_i is associated with the $i + 1^{th}$ layer L_{i+1} in the following way: the left child counter and the right child counter at L_i are associated with the parent counter at L_{i+1}. When the child counter overflows, the Pyramid sketch uses its parent counter to record its number of overflows. In Pyramid sketch, the first layer is composed of pure counters, only used for recording frequencies. The other layers are composed of hybrid counters, which can be split into three parts: the left flag, the counting part and the right flag. The flag parts indicate whether its child counters are overflowed. Based on this counter-pair sharing technique, the Pyramid sketch dynamically assigns the appropriate number of bits for different items with different frequencies, thus improving the memory efficiency. Like OM sketch, the Pyramid sketch uses word acceleration and hash bit technique to improve its speed.

The Pyramid sketch can be applied to conventional sketches (CM, A, C and CU), and the results are denoted as Pcm, Pa, Pc and Pcu. It uses a novel strategy, Ostrich policy, to improve the insertion speed of sketches that need to know the values of the d mapped counters during each insertion. Here, we take Pcu as an example. The key idea of Ostrich policy is ignoring the second and higher layers when getting the values of the d mapped counters, only increases the smallest first layer counter(s).

Insertion: When inserting an item, the Pyramid sketch first computes the hash function to locate the hashed counters at layer L_1. Different sketches will perform different increase operations on these counters. If any of the counters overflows, the Pyramid sketch sets the counter to zero, and assigns its parent counter according to its index. Then, the left/right flag of its parent counter will be set to 1. These operations are called carryin. The Pyramid sketch repeats the carryin operation at layer L_2, and the operation will be performed layer by layer until there is no overflow.

Query: When querying an item, the Pyramid sketch first locates the hashed counters at the first layer, and then gets the values of the d mapped counters by accumulating the values of corresponding counters of each layer. Finally, the Pyramid sketch produces the query output based on the specific sketch under use.

3 Analysis and Discussion

In this section, we will discuss from five different aspects on how to design an accurate and fast sketch for skewed data streams. We use the OM sketch and the Pyramid sketch as latest examples, discussing their methods handling these important aspects. At the end of this section, we will discuss two more aspects, namely the support for deletion and generality. Depending on the target scenarios they might also become as important as the former ones.

3.1 Accuracy Improvement of Sketch for Skewed Datasets

Accuracy is one of the most important indicators of the sketch. We can try tackle the problem from three different aspects: (1) higher memory utilization, (2) isolation, and (3) neutralization to improve the accuracy of the sketch. In the following we will discuss the solutions from the literatures and our findings.

Improvement of Memory Utilization
Improvement of memory utilization means increasing the number of counters in the same memory, so as to reduce the probability of collision. Automatic enlargement and hierarchy are techniques that can be used to improve memory utilization.

Automatic Enlargement Technique

In the process of the automatic enlargement, it's unnecessary to allocate enough bits to each counter in advance. When the counter overflows, the sketch enlarges the initial counter space automatically. In view of the characteristics of skewed data streams, this technique can greatly improve memory utilization of the sketch.

When using the automatic enlargement technique, the enlargement strategy depends on the type of counter overflows. There are two types of overflows, one is called simultaneous overflow and the other is called non-simultaneous overflow. Simultaneous overflow means that all counters of an item overflow at the same time, and needs only one enlargement, while non-simultaneous overflow means that not all counters of an item overflow at the same time, so multiple enlargements are often necessary. What's more, if the initial counter space is separated from its enlarged space, during the automatic enlargement, the non-simultaneous overflow needs to establish the corresponding relationship between each initial counter space and its enlarged space, otherwise the sketch cannot be queried. However, it is unnecessary to do so for the simultaneous overflow. When querying an item, the sketch only needs to query the two spaces separately. Compared to the non-simultaneous overflow, the simultaneous overflow has fewer enlargements and relatively simpler enlargement strategy. However, realizing simultaneous overflow needs to design specific insert operations, which may require extra cost, such as some trade-offs of performances.

Both the OM sketch and the Pyramid sketch adopt automatic enlargement technique. For the OM sketch, it only increases the smallest counter(s) during insertion, thus achieving simultaneous overflow. For the Pyramid sketch, the insert operations depend on the sketch under use. Counters of an item cannot be guaranteed to overflow at the same time. Therefore, the Pyramid sketch is a non-simultaneous overflow. Besides, the initial counter space of the OM sketch and Pyramid sketch is separated from its enlarged space. The automatic enlargement strategy of OM sketch is to use another hash function to enlarge the space for all the overflowed counters. When querying an item, the OM sketch locates the counters of each space by two hash functions. The automatic enlargement strategy used in the Pyramid sketch is index calculation, which is to establish the relationship between the initial counter space and its enlarged space. When querying an item, the parent counters are located through the indices of their child counters. Specific insert operation lets the OM sketch achieve simultaneous overflow, simplifying the automatic enlargement strategy. Meanwhile, it also makes the OM sketch isolate items, which further improve the accuracy (will be described later). However, the cost is that the OM sketch cannot support deletion.

Hierarchical Structure

Considering the characteristics of skewed datasets, the higher frequency of an item is, the less proportion it occupies. Thus, we can design the sketch as a hierarchical structure. For example, lower layers have smaller size but a larger number of counters, while higher layers have larger size but a smaller number of counters. The hierarchical structure increases the number of counters, thus improving the memory utilization. The more the number of layers, the higher the accuracy of the sketch. However, hierarchy has a side-effect, which cannot be ignored. With the increase of the number of layers, the number of memory accesses will be increased, which can slow insertion and query speed. Therefore, when using hierarchy technique, the number of layers of the sketch is selected based on a tradeoff between accuracy and speed.

Both the Pyramid sketch and the OM sketch adopt the hierarchy technique. The Pyramid sketch is a multi-layer pyramid-shaped structure, while the OM sketch is a two-layer trapezoid structure. As mentioned earlier, there is a corresponding relationship between the initial counter space and its enlarged space in the Pyramid sketch. If one child counter monopolizes to one parent counter, massive memory waste will be caused because of the characteristics of skewed datasets. Therefore, the Pyramid sketch lets two child counters share one parent counter in order to improve the memory utilization. With this corresponding relationship, the Pyramid sketch gradually forms a pyramid type. Using a multi-layer structure rather than a two-layer structure like the OM sketch is to further increase the memory utilization. However, this hierarchical structure will slow the insertion and query speed of the Pyramid sketch.

To design the OM sketch as a two-layer structure is a tradeoff between accuracy and speed. For the OM sketch, the characteristics of the low layer counters conform to the characteristics of low frequency items, whose sizes are small and numbers are large, while the characteristics of the high layer counters conform to the characteristics of the intermediate and high frequency items, whose sizes are large and numbers are small. In skewed data streams, the vast majority of items are the low frequency items. Therefore, dividing into two layers can significantly improve the memory utilization of the OM sketch. The more the layers, the lower the accuracy. Thus, the OM sketch is designed as a two-layer structure.

Isolation

Improvement of memory utilization is to improve accuracy by reducing hash collisions between items. In addition to reducing hash collisions, we can limit the range of hash collisions to reduce the collisions between items of different frequency segments. For example, we can isolate the low, intermediate and high frequency items in the sketch, so that the collisions occur only within these frequency segments but not cross. This method reduces the impact of high frequency items on intermediate and low frequency items, and the impact of intermediate frequency items on low frequency items, thus improving the accuracy. Besides, we can design the sketch according to specific requirements and application scenarios. For example, in some scenarios (e.g., NLP), the accuracy of low frequency items is very important. Thus, we can design corresponding sketches to improve the accuracy of low frequency items, and the accuracy of intermediate and high frequency items can be relaxed appropriately.

As mentioned earlier, the simultaneous overflow makes the OM sketch isolate items. This is because it is unnecessary for the simultaneous overflow to establish the relationship between the initial counter space and its enlarged space. The low layer of the OM sketch plays a role of filtering all low frequency items, so only the intermediate and high frequency items can get into the high layer. Therefore, the impact of intermediate and high frequency items on low frequency items is reduced. The accuracy of low frequency items can be greatly improved. The Pyramid sketch does not isolate items. Its child counters are bound to their corresponding parent counter. The low frequency items that collide with the intermediate and high frequency items at the lower layer will also get into the higher layer. Therefore, the estimated frequencies of low frequency items are still affected by the intermediate and high frequency items.

Neutralization

The evaluation indices of accuracy can be divided into under-estimation rate, correction rate and over-estimation rate. There are different requirements for them in different application scenarios. We can improve the accuracy of the sketch by using specific application features. For example, if the application scenario allows the sketch have under-estimation error, a small amount of under-estimation error can be introduced to neutralize part of the over-estimation-error and improve the correction rate. Since the estimated frequency is already larger than the real frequency in many cases, a little under-estimation can improve the overall accuracy. However, if the target application scenario does not allow under-estimation-error, this method will not work. All sketches with under-estimation error will not be applicable to such scenarios. Therefore, this neutralization method is related to the tolerance of under-estimation-error, depending on specific application scenario.

The OM sketch has under-estimation error, which can improve a bit of accuracy. The cause of under-estimation is that when counters overflowed, the OM sketch set all these counters to zero. For the Pyramid sketch, if conventional sketch has under-estimation-error or over-estimation-error, then its corresponding Pyramid sketch will also have it, otherwise it will not. However, there is a special case, Pcu sketch. Since the Pcu sketch uses Ostrich strategy, each insertion does not necessarily increase the smallest hashed counter(s), resulting in a little under-estimation. However, this under-estimation neutralizes part of over-estimation. Thus, in the original paper [2], the experimental results show that Ostrich policy can help improve accuracy. Neither the OM sketch nor the Pcu sketch can be applied to the scenarios that do not allow under-estimation error.

3.2   Insertion and Query Speed Improvement of Sketch

Another important indicator of the sketch is speed. The speed is mainly related to the number of memory accesses and the number of hash calculations required for each insertion and query. Therefore, there are two ways to improve the speed, and in the following discussions we will reference two methods: (1) word acceleration and (2) hash bit technique as representing examples.

Reduction in the Number of Memory Accesses

For conventional sketches, the number of memory accesses for each insertion or query is the same as the number of counters assigned for each item, usually more than three. Since the counter size of conventional sketches is usually large (e.g., 16 bit), it is difficult to reduce the number of memory accesses. However, if we use certain techniques to make the counter size smaller, we can constrain the counters of one item within one or several machine word to reduce the number of memory accesses for each insertion or query. This is called word acceleration, which use condition is that the counter size should be relatively smaller. In modern CPUs, a machine word is usually 64 bits in width. In the GPU architecture, the size of a machine word is much larger. Therefore, one machine word on CPU or GPU can typically contain a reasonably large number of small counters.

The hierarchical structure of the Pyramid sketch and the OM sketch increases the number of memory accesses. However, this structure makes their counters size smaller, so that both of them can use word acceleration.

The OM sketch constrains the hashed counters of the low layer within one machine word and the hashed counters of the high layer within two machine words, and scatters these counters over these machine words evenly. As real data streams are skewed, the probability of accessing the high layer for each insertion and query is very small (e.g., 1/20). Therefore, the average number of memory accesses for each insertion and query is close to 1 (e.g., $1 + 1/20 * 2 = 1.1$). The Pyramid sketch constrains the hashed counters of each layer within one machine word. Most of the insertions only access the first layer (the Ostrich strategy also makes Pcu so). Therefore, the average number of memory accesses for each insertion is close to 1, which has been proved in the original paper [1]. However, for queries, the number of memory accesses depends on the frequency of queried item. The higher frequency of queried item, larger number of layers to be accessed and larger number of memory accesses. Therefore, although the Pyramid sketch adopts word acceleration technique, it does not make great improvement in reducing the average number of memory accesses for queries.

Reduction in the Number of Hash Computations

For conventional sketches, the number of hash calculations for each insertion or query is also the same as the number of counters assigned for each item. This is because the size of the counter address is usually large. Thus, one counter can only be positioned by one hash function. However, if we use certain techniques to make the address size smaller, we can use fewer hash functions to locate the counters. For example, if we have adopted the word acceleration to constrain the hashed counters within one or several machine words, the address size of these counters can be shortened. We can leverage the hash bit technique from the literature [19] to reduce the number of hash computations. The key idea is that split one hash value into several bit arrays to locate one or several machine words and offsets of counters in the corresponding machine words. In this way, we can use only one hash computation to handle a sketch which originally required multiple hash computations.

Both OM sketch and Pyramid sketch use hash bit technique. The OM sketch uses hash bit technique at each layer. Supposing the probability of accessing the high layer is 1/20, the average number of hash computations for each insertion or query is close to 1 (e.g., $1 + 1/20 * 1 = 1.05$). The Pyramid sketch only uses hash function and hash bit technique at the first layer. Counters of other layers are located by the index of the first layer counters. Therefore, for the Pyramid sketch, the average number of hash computations for each insertion or query is 1, achieving one hash computation.

3.3   Other Related Aspects

In addition to accuracy and speed, there are also other important properties that need to be considered in the designing of sketch. Here we will discuss a bit on two of them: the support for deletion and the support for generality, which are actually considered in [1, 2].

Deletion: In some application scenarios, the sketch is required to support deletion [18]. If the insert operation is always reversible throughout the use of the sketch, the sketch can support deletion, and the delete operation is the inverse operation of insertion. For example, the insert operation of the CM sketch is to plus 1, then its delete operation is to subtract 1. Therefore, to make the sketch support deletion, we should design a reversible insert operation. The insert operation of the OM sketch is the same as that of the CU sketch, and neither of them supports deletion.

Generality: If the goal of a sketch is to solve a common problem for all sketches, it can be applied to all sketches to improve the target performance of them. We say such a sketch has generality. If the design goal of a sketch is to enhance one or more performances and cannot be applied to all sketches, such sketch has no generality. However, sketches that have no generality are more targeted, thus may be more significant to improve the target performances. The OM sketch sacrifices generality, but brings more significant accuracy and speed.

4 Experimental Result

4.1 Metrics

Average Absolute Error (AAE): AAE is defined as $\frac{1}{|N|}\sum_{i=1}^{N}\left|f_i - \hat{f}_i\right|$ where f_i is the real frequency of the ith item, $\hat{f}_i$ is the estimated frequency of this item, and N is the total number of distinct items in the query set.

Average Relative Error (ARE): ARE is defined as $\frac{1}{|N|}\sum_{i=1}^{N}\left|f_i - \hat{f}_i\right|/f_i$.

Under-Estimation Rate (UER): UER is defined as N_{under}/N where N_{under} is the number of distinct items whose estimated frequency is less than its real frequency.

Correct Rate (CR): CR is defined as N_{acc}/N where N_{acc} is the number of distinct items whose estimated frequency equals to its real frequency.

Over-Estimation Rate (OER): OER is defined as N_{over}/N where N_{over} is the number of distinct items whose estimated frequency is larger than its real frequency.

Throughput: We simulate how sketches actually insert and query on CPU platform and calculate the throughput using mega-instructions per second (Mips).

4.2 Experimental Setup

We use the real IP trace from the main gateway at our campus. The estimation of item frequency corresponds to the estimation of the number of packets in a flow. The number of packets of the trace is 10M and the number of distinct flows is around 1M.

We implement the sketches of CM, CU, C, A, OM sketch and Pyramid sketch in C++. For the four conventional sketches, we set the counter size to 16 bits and the number of arrays to 4. Other experimental settings are the same as the original paper [1, 2]. In all our experiments, unless noted otherwise, the memory size of each sketch is 1 MB by default. We performed all the experiments on a machine with 2-core CPUs (2 threads, Pentium(R) Dual-Core CPU E5800 @3.2 GHz) and 4 GB total DRAM memory.

4.3 Performance of Different Sketches

The experiment results of Figs. 3 and 4 show that the AAE and ARE of the OM sketch and the Pyramid sketch are much smaller than those of the conventional sketches. The experiment results of Figs. 5 and 6 show that the insertion and query throughput of the OM sketch and the Pyramid sketch are much higher than those of the conventional sketches. We can see that by using automatic enlargement and hierarchy to improve accuracy and by using word acceleration and hash bit technique to improve speed, the OM sketch and the Pyramid sketch achieve a much better performance than the state-of-the-art in terms of both speed and accuracy.

From the experiment results, we find that in Pyramid versions, Pcu sketch achieves the highest accuracy and speed. Furthermore, the accuracy and speed of Pcu sketch is the closest to that of the OM sketch. Besides, the increase operation of the OM sketch is the same as the Pcu sketch and CU sketch. Therefore, in the following experiments, we use the Pcu sketch as an example of Pyramid sketch, and compare the performances of the CU sketch, Pcu sketch and OM sketch.

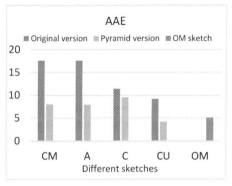

Fig. 3. AAE of sketches   **Fig. 4.** ARE of sketches

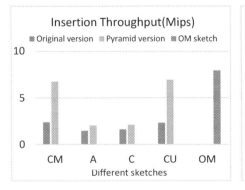

Fig. 5. Insertion throughput of sketches   **Fig. 6.** Query throughput of sketches

4.4   Performances of CU, Pcu and OM Sketch

AAE and ARE
Figures 7 and 8 plots the AAE and ARE of three different sketches on different memory sizes increasing from 0.40 MB to 2.00 MB with a step of 0.20 MB. Our experimental results show that the AAE and ARE of the Pcu sketch and the OM sketch are always lower than those of the CU sketch.

Besides, we find that the ARE of the OM sketch is always lower than that of the Pcu sketch. As mentioned in the Sect. 3.1, the OM sketch uses the isolation method of filtering all low frequency items at the low layer, improving the accuracy of low frequency items. The accuracy of low frequency items has the greatest impact on ARE. Thus, the ARE of the OM sketch can be improved. Meanwhile, the Pyramid sketch does not use the isolation method, and the estimated frequencies of low frequency items are still affected by the intermediate and high frequency items. Therefore, its ARE is always lower than that of the OM sketch. Our experimental results have proved that isolation can help improve the accuracy of the sketch.

Under-Estimation Rate, Correct Rate and Over-Estimation Rate
Figures 9, 10 and 11 plots the under-estimation rate, correct rate and over-estimation rate of three different sketches on different memory sizes increasing from 0.10 MB to 2.00 MB with a step of 0.40 MB. Our experimental results show that expect the CU sketch, both the Pcu sketch and the OM sketch have under-estimation. The under-estimation rate of the OM sketch is about 4.33 times higher than that of the Pcu sketch. The correct rate of the OM sketch is about 1.26 and 8.83 times higher than those of the Pcu and CU sketch. And the over-estimation rate of the OM sketch is about 1.27 and 1.64 times lower than those of the Pcu and CU sketch.

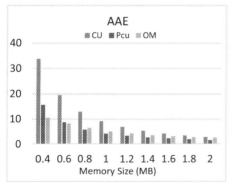

Fig. 7. AAE vs. memory sizes

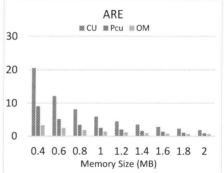

Fig. 8. ARE vs. memory sizes

The experimental results show that the higher under-estimation rate, the lower over-estimation rate and the higher correct rate. As mentioned in Sect. 3.1, a small amount of under-estimation-error can be introduced to neutralize part of the

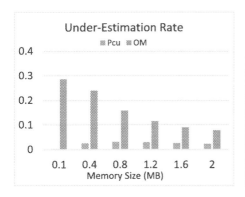

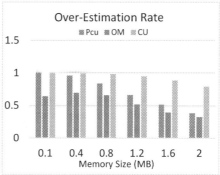

Fig. 9. UER vs. memory sizes

Fig. 10. OER vs. memory sizes

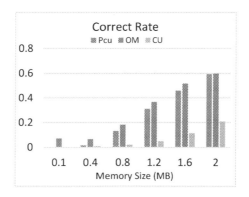

Fig. 11. CR vs. memory sizes

over-estimation-error and improve the correction rate. Our experimental results have well proved the neutralization method indeed improve the overall accuracy.

Speed

Figures 12 and 13 plots the insertion throughput and query throughput of different sketches on different memory sizes increasing from 0.40 MB to 2.00 MB with a step of 0.20 MB. Our experimental results show that the speed of the Pcu sketch and the OM sketch is always higher than that of the CU sketch. The insertion throughput of the OM sketch is about 1.18 and 3.39 times higher than those of the Pcu and CU sketch. The query throughput of the OM sketch is about 1.71 and 4.23 times higher than those of the Pcu and CU sketch.

From the experimental results, we find that the Pcu sketch can significantly improve the insertion throughput but cannot greatly improve the query throughput. As mentioned in the Sect. 3.2, the OM sketch employs two-layer structure and uses word acceleration technique, so that can achieve close to one memory access for each insertion and query. For the Pyramid sketch, although it also uses word acceleration

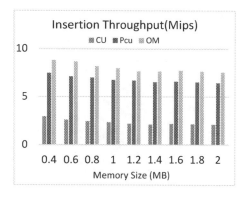 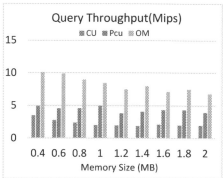

Fig. 12. Insertion throughput vs memory sizes **Fig. 13.** Query throughput vs memory sizes

technique, the hierarchical structure still highly affects the number of memory accesses for each query. Our throughput experimental results have well proved our viewpoint that the hierarchy technique can slow the insertion and query speed and is more significant to slow the query speed.

Generality

The experimental results above show that the performances of the OM sketch are better than the Pyramid sketch in terms of both speed and accuracy. As mentioned in Sect. 3.3, the sketch that does not have generality is more targeted, and may be more significant to improve the target performances. Thus, our experimental results have well proved our analysis and discussion on Generality.

5   Conclusion

Sketches have been applied to many fields. In this paper, we sort out five important aspects in improving the accuracy and speed of sketch for skewed data streams with limited memory. We provide detailed discussions on the positive and negative effects of typical and latest methods from different aspects on the performances of the sketch. Two other properties of the sketch such as deletion and generality are also discussed. Generally, although the purpose of these aspects are somehow orthogonal to each other, the methods handling them may have effects on more aspects and need more thorough considerations with their limitations and side-effects. Experimental results demonstrate the validity and extendibility of our discussions. We believe our paper can be a good help to the future study of the accurate and fast sketches.

Acknowledgements. This work was supported by Shenzhen Basic Research Program (JCYJ20160525 154348175), the Shenzhen Municipal Development and Reform Commission (Disciplinary Development Program for Data Science and Intelligent Computing) and Shenzhen Key Lab Project (ZDSYS20170303140513705).

References

1. Yang, T., Zhou, Y., Jin, H., Chen, S., Li, X.: Pyramid sketch: a sketch framework for frequency estimation of data streams. Proc. VLDB Endow. **10**(11), 1442–1453 (2017)
2. Zhou, Y., Liu, P., Jin, H., Yang, T., Dang, S., Li, X.: One memory access sketch: a more accurate and faster sketch for per-flow measurement. In: IEEE GLOBECOM (2017)
3. Manerikar, N., Palpanas, T.: Frequent items in streaming data: an experimental evaluation of the state-of-the-art. Data Knowl. Eng. **68**(4), 415–430 (2009)
4. Cormode, G., Johnson, T., Korn, F., Muthukrishnan, S., Spatscheck, O., Srivastava, D.: Holistic UDAFs at streaming speeds. In: ACM SIGMOD, pp. 35–46. ACM (2004)
5. Cormode, G., Garofalakis, M., Haas, P.J., Jermaine, C.: Synopses for massive data: samples, histograms, wavelets, sketches. Found. Trends Databases **4**(1–3), 1–294 (2012)
6. Roy, P., Khan, A., Alonso, G.: Augmented sketch: faster and more accurate stream processing. In: ACM SIGMOD, pp. 1449–1463. ACM (2016)
7. Cormode, G., Muthukrishnan, S.: An improved data stream summary: the count-min sketch and its applications. J. Algorithms **55**(1), 58–75 (2005)
8. Cormode, G., Hadjieleftheriou, M.: Finding frequent items in data streams. Proc. VLDB Endow. **1**(2), 1530–1541 (2008)
9. Chen, A., Jin, Y., Cao, J., Li, L.E.: Tracking long duration flows in network traffic. In: IEEE INFOCOM, pp. 1–5. IEEE (2010)
10. Liu, Z., Manousis, A., Vorsanger, G., Sekar, V., Braverman, V.: One sketch to rule them all: rethinking network flow monitoring with UnivMon. In: ACM SIGCOMM, pp. 101–114. ACM (2016)
11. Gilbert, A.C., Strauss, M.J., Tropp, J.A., Vershynin, R.: One sketch for all: fast algorithms for compressed sensing. In: ACM STOC, pp. 237–246. ACM (2007)
12. Durme, B.V., Lall, A.: Probabilistic counting with randomized storage. In: IJCAI, pp. 1574–1579. Morgan Kaufmann Publishers Inc. (2009)
13. Polyzotis, N., Garofalakis, M., Ioannidis, Y.: Approximate XML query answers. In: ACM SIGMOD, pp. 263–274. ACM (2004)
14. Estan, C., Varghese, G.: New directions in traffic measurement and accounting. ACM Trans. Comput. Syst. **21**(3), 270–313 (2002)
15. Powers, D.M.W.: Applications and explanations of Zipf's law. Adv. Neural. Inf. Process. Syst. **5**(4), 595–599 (1998)
16. Adamic, L.A., Huberman, B.A., Barabási, A.L., Albert, R., Jeong, H., Bianconi, G.: Power-law distribution of the World Wide Web. Science **287**(5461), 2115 (2000)
17. Yang, T., Liu, L., Yan, Y., Shahzad, M., Shen, Y., Li, X., Cui, B., Xie, G.: SF-sketch: a fast, accurate, and memory efficient data structure to store frequencies of data items. In: IEEE ICDE. IEEE (2017)
18. Graham, C.: Sketch techniques for approximate query processing. Found. Trends Databases (2011)
19. Qiao, Y., Li, T., Chen, S.: One memory access bloom filters and their generalization. Proc. IEEE INFOCOM **28**(6), 1745–1753 (2011)

Matching Consecutive Subpatterns over Streaming Time Series

Rong Kang[1,2], Chen Wang[1,2], Peng Wang[3], Yuting Ding[1,2],
and Jianmin Wang[1,2(✉)]

[1] School of Software, Tsinghua University, Beijing, China
{kr11,dingyt16}@mails.tsinghua.edu.cn,
{wang_chen,jimwang}@tsinghua.edu.cn
[2] National Engineering Laboratory for Big Data Software,
Tsinghua University, Beijing, China
[3] School of Computer Science, Fudan University, Shanghai, China
pengwang5@fudan.edu.cn

Abstract. Pattern matching of streaming time series with lower latency under limited computing resource comes to a critical problem, especially as the growth of Industry 4.0 and Industry Internet of Things. However, against traditional single pattern matching model, a pattern may contain multiple subpatterns representing different physical meanings in the real world. Hence, we formulate a new problem, called "consecutive subpatterns matching", which allows users to specify a pattern containing several consecutive subpatterns with various specified thresholds. We propose a novel representation Equal-Length Block (ELB) together with two efficient implementations, which work very well under all L_p-Norms without false dismissals. Extensive experiments are performed on synthetic and real-world datasets to illustrate that our approach outperforms the brute-force method and MSM, a multi-step filter mechanism over the multi-scaled representation by orders of magnitude.

Keywords: Pattern matching · Stream · Time series

1 Introduction

Time series are widely available in diverse application areas, such as Healthcare [21], financial data analysis [22] and sensor network monitoring [25], and they turn the interests on spanning from developing time series database [6]. In recent years, the rampant growth of Industry 4.0 and Industry Internet of Things, especially the development of intelligent control and fault prevention to complex equipment on the edge, urges more challenging demands to process and analyze streaming time series from industrial sensors with low latency under limited computing resource [24].

As a typical workload, similarity matching over streaming time series has been widely studied for fault detection, pattern identification and trend prediction, where accuracy and efficiency are the two most important measurements

© Springer International Publishing AG, part of Springer Nature 2018
Y. Cai et al. (Eds.): APWeb-WAIM 2018, LNCS 10988, pp. 90–105, 2018.
https://doi.org/10.1007/978-3-319-96893-3_8

to matching algorithms [11]. Given a single or a set of patterns and a pre-defined threshold, traditional similarity matching algorithms aim to find matched subsequences over incoming streaming time series, between which the distance is less than the threshold. However, in certain scenarios, the single threshold pattern model is not expressive enough to satisfy the similarity measurement requirements. Let us consider the following example.

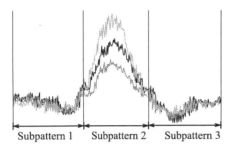

Fig. 1. Diverse patterns of Extreme Operating Gust (EOG). EOG pattern is composed of three subpatterns and users tend to specify a larger threshold for Subpattern 2 comparing with Subpattern 1 and Subpattern 3.

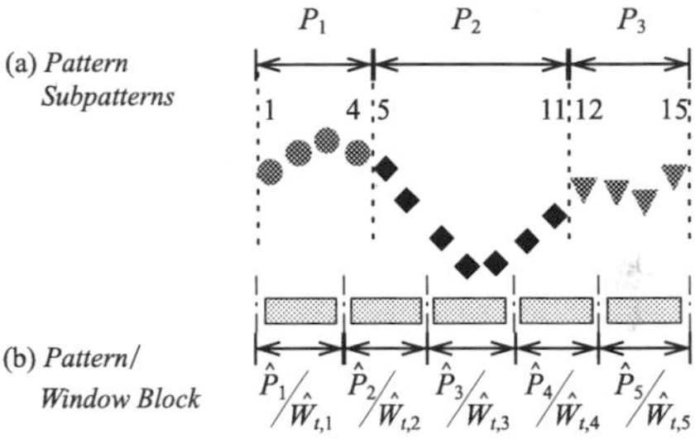

Fig. 2. (a) Pattern P is composed of three subpatterns: $P_1 = P[1 : 4]$, $P_2 = P[5 : 11]$ and $P_3 = P[12 : 15]$. (b) In ELB representation, if we set block size $w = 3$, P and W_t are divided into 5 pattern/window blocks.

In the field of wind power generation, Extreme Operating Gust (EOG) [4] is a typical gust pattern which is a phenomenon of dramatic changes of wind speed in a short period. Early detection of EOG can prevent the damage to the turbine [17]. A typical pattern of EOG has three physical phases, where its corresponding shape contains a slight decrease (Subpattern 1), followed by a steep rise, a steep drop (Subpattern 2), and a rise back to the original value (Subpattern 3). Users usually emphasize the shape feature of the second subpattern much more than its exact numeric value. In other words, users tend to specify a larger threshold of distance measurement for Subpattern 2 comparing with Subpattern 1 and Subpattern 3. For instance, all time series in Fig. 1 are regarded as correct matches of EOG, although they have diverse values in their second subpatterns.

In summary, above example shows that a complex pattern is usually composed of several subpatterns representing different physical meanings, and users may want to specify various thresholds for different parts. There are similar situations in other fields like electrocardiogram in Healthcare and technique analysis in the stock market. Therefore, we formulate a new problem, named as *consecutive subpatterns matching* over streaming time series. In this scenario, a pattern contains a list of consecutive subpatterns with different thresholds. A sliding window on stream matches the given pattern only if each of its components matches the corresponding subpattern.

Although many techniques have been proposed for time series similarity matching, they do not aim to solve the problem mentioned above. For streaming time series matching, some recent works take advantage of similarity or correlation of multiple patterns and avoid the whole matching of every single patterns [11,21]. Similarly, most of the previous approaches for subsequence similarity search explore and index the commonalities of time series in database to accelerate the query [14,20]. These approaches are not optimized for the scenario of matching consecutive subpatterns.

In this paper, we propose Equal-Length Block (ELB) representation together with the *lower bounding property*. ELB representation divides both the pattern and a sliding window into equal-length disjoint pattern/window blocks. Then ELB characterizes a pattern block as upper/lower bounds and a window block as a single value. Two ELB implementations are provided which allow us to process multiple successive windows together, so that speed up the matching process dramatically while guaranteeing no false dismissals.

In summary, this paper makes the following contributions:

- We introduce a new model, consecutive subpatterns matching, which allows us to describe pattern more expressively and process streaming time series more precisely.
- We propose a novel ELB representation which accelerate the matching process dramatically under all L_p-norms and guarantees no false dismissals.
- We illustrate the efficiency of our algorithms with sufficient experiments on real-world and synthetic datasets and a comprehensive theoretical analysis.

The rest of the paper is arranged as follows: Sect. 2 gives a brief review of the related work. Section 3 formally defines our problem. Section 4 proposes ELB representation together with its two implementations. Section 5 conducts extensive experiments. Finally, Sect. 6 concludes the paper.

2   Related Work

There are two categories of the related works, multiple patterns matching over streaming time series and subsequence similarity search.

Multiple Patterns Matching over Streaming Time Series. Traditional single pattern matching over the stream is relatively trivial, hence recent research works put more focus on optimizing the multiple pattern scenario. Atomic wedge [21] is proposed to monitor stream with a set of pre-defined patterns, which exploits the commonality among patterns. Sun et al. [18] extend atomic wedge for various length queries and tolerances. Lian et al. [11] propose a multi-scale segment mean (MSM) representation to detect static patterns over streaming time series. They discuss the batch processing optimization and the case of dynamic patterns in its following work [10]. Lim et al. [12] propose SSM-IS which divides long sequences into smaller windows. Although these techniques are proposed for streaming time series and some of them speed up the distance calculation between the pattern and the candidate, most of them focus on exploring the commonality and correlation among multiple patterns for pruning unmatched pattern candidates, which doesn't reduce the complexity brought by the problem of consecutive subpatterns matching.

Subsequence Similarity Search. FRM [5] is the first work for subsequence similarity search which maps data sequences in database into multidimensional rectangles in feature space. General Match [16] divides data sequences into generalized sliding windows and the query sequence into generalized disjoint windows, which focuses on estimating parameters to minimize the page access. Loh et al. [14] propose a subsequence matching algorithm that supports normalization transform. Lim et al. [13] address this problem by selecting the most appropriate index from multiple indexes built on different windows sizes. Kotsifakos et al. [9] propose a framework which allows gaps and variable tolerances in query and candidates. Wang et al. [20] propose DSTree which is a data adaptive and dynamic segmentation index on time series. This category of researches focuses on indexing the common features of *archived* time series, which is not optimized for pattern matching over the stream.

3   Problem Definition

Pattern P is a time series which contains n number of elements $(p_1, \cdots, p_n)$. We denote the subsequence $(p_i, \cdots, p_j)$ of P by $P[i : j]$. Logically, P could be divided into several consecutive subpatterns which have varied thresholds of matching deviation. Given a pattern P, P is divided into b number of non-overlapping subsequences in time order, represented as $P_1, P_2, \cdots, P_b$, in which the k-th subsequence P_k is defined as the k-th subpattern and associated with a specified threshold ε_k.

As shown in Fig. 2(a), for instance, pattern P is composed of three subpatterns: $P_1 = P[1 : 4]$, $P_2 = P[5 : 11]$ and $P_3 = P[12 : 15]$. These subpatterns may be specified different thresholds.

A streaming time series S is an ordered sequence of elements that arrive in time order. We denote a sliding window on S which starts with timestamp t by $W_t = (s_{t,1}, s_{t,2}, \cdots, s_{t,n})$. We denote the subsequence $(s_{t,i}, s_{t,i+1}, \cdots s_{t,j})$ in W_t by $W_t[i : j]$. According to the sub-pattern division of P, W_t is also divided into b sub-windows $W_{t,1}, W_{t,2}, \cdots, W_{t,b}$. For convenience, we refer to p_i and $s_{t,i}$ as an *element pair*.

There are many distance functions such as DTW [3], $LCSS$ [19], L_p-norm [23], etc. We choose L_p-norm distance which covers a wide range of applications [1,5,15]. Given two n-length sequences where $X = (x_1, x_2, \cdots, x_n)$ and $Y = (y_1, y_2, \cdots, y_n)$, the L_p-*Norm Distance* between X and Y is defined as follows:

$$L_p(X, Y) = \left(\sum_{i=1}^{n} |x_i - y_i|^p \right)^{\frac{1}{p}}$$

Since the L_{norm} is a distance function between two equal-length sequences, there are $|W_t| = n$ and $|W_{t,k}| = |P_k|$ for $k \in [1, b]$. In addition, we denote by $L_p[i : j]$ the normalized Euclidean distance between $P[i : j]$ and $W_t[i : j]$.

Problem Statement: Given a pattern P which contains b number of sub-patterns $P_1, P_2, \cdots, P_b$ with specified thresholds $\varepsilon_1, \varepsilon_2, \cdots, \varepsilon_b$. For a stream S, *consecutive subpatterns matching* is to find all sliding windows W_t on S, where it holds that $L_p(P_k, W_{t,k}) \leqslant \varepsilon_k$ for $k \in [1, b]$ (denoted by $W_{t,k} \prec P_k$).

4   Equal-Length Block

In this section, we first sketch a novel representation, Equal-Length Block (ELB), together with *Lower Bounding Property*, which enables us to process several successive windows together while guaranteeing no false dismissals. After that, we will introduce two ELB implementations in turn.

ELB representation is inspired by the following observation. To avoid false dismissals, a naive method is to slide the window over the stream by one element and calculates the corresponding distance, which is computationally expensive. However, one interesting observation is that in most real-world applications, the majority of adjacent subsequences of time series might be similar. This heuristic gives us the opportunity to process multiple successive windows together. Based on this hint, we propose Equal-Length Block (ELB), and the corresponding lower bounding property.

ELB divides the pattern P and the sliding window W_t into several disjoint w-length *blocks* while the last indivisible part can be safely discarded. The block division is independent of pattern subpatterns. A block may overlap with two or more adjacent subpatterns, and a subpattern may contain more than one block. The number of blocks is denoted by $N = \lfloor n/w \rfloor$. Based on the concept of block, P and W_t are split into $\hat{P} = \{\hat{P}_1, \cdots, \hat{P}_N\}$ and $\hat{W}_t = \{\hat{W}_{t,1}, \cdots, \hat{W}_{t,N}\}$ respectively, where $\hat{P}_j$(or $\hat{W}_{t,j}$) is the j-th block of P (or W_t), that is, $\hat{P}_j = \{p_{(j-1)\cdot w+1}, p_{(j-1)\cdot w+2}, \cdots, p_{j\cdot w}\}$, similarly for $\hat{W}_{t,j}$. As shown in Fig. 2(b), we set $w = 3$, thus P and W_t are divided into 5 blocks. Based on blocks, each pattern block $\hat{P}_j$ is represented by a pair of bounds, upper and lower bounds, which are denoted by $\hat{P}_j^u$ and $\hat{P}_j^l$ respectively. Each window block $\hat{W}_{t,j}$ is represented by a feature value, denoted by $\hat{W}_{t,j}^f$.

It is worth noting that the ELB representation is only an abstract format description, which doesn't specify how to compute upper and lower bounds of $\hat{P}_j$ and the feature of window $\hat{W}_{t,j}$. We can design any ELB implementation, which just needs to satisfy the following lower bounding property:

Definition 1. (Lower Bounding Property): given $\hat{P}$ and $\hat{W}_t$, if $\exists\, i \in [0, w)$, W_{t+i} is a result of consecutive subpatterns matching of P, then $\forall j \in [1, N]$, $\hat{P}_j^l \leqslant \hat{W}_{t,j}^f \leqslant \hat{P}_j^u$ (marked as $\hat{W}_{t,j} \prec \hat{P}_j$).

We first provide our matching algorithm based on ELB which satisfies lower bounding property before introducing our ELB implementation. Instead of processing sliding windows one-by-one, lower bounding property enables us to process w successive windows together in the pruning phase. Given N number of window blocks $\{\hat{W}_{t,1}, \cdots, \hat{W}_{t,N}\}$, if anyone in them (e.g. $\hat{W}_{t,j}$) doesn't match its aligned pattern block ($\hat{P}_j$ correspondingly), we could skip w consecutive windows, $W_t, W_{t+1}, \cdots, W_{t+w-1}$, together. Otherwise, the algorithm takes these w windows as candidates and calculate exact distances one by one. The lower bounding property enables us to extend the sliding step to w while guaranteeing no false dismissals. The critical challenge is how to design ELB implementation which is both computationally efficient and effective to prune sliding windows.

4.1 Element-Based ELB Representation

In this section, we present the first ELB implementation, element-based ELB, denoted by ELB_{ele}. The basic idea is as follows. According to our problem statement, if window W_t matches P, for any subpattern P_k and corresponding $W_{t,k}$, their L_p-Norm distance holds that:

$$L_p(W_{t,k}, P_k) \leqslant \varepsilon_k \tag{1}$$

It's easy to infer that any element pair p_i together with $s_{t,i}$, which falls into the k-th subpattern, satisfies that:

$$|s_{t,i} - p_i| \leqslant \varepsilon_k \tag{2}$$

In other words, if $s_{t,i}$ falls out of the range $[p_i - \varepsilon_k, p_i + \varepsilon_k]$, we know that W_t cannot match P.

Based on this observation, we construct two envelope lines for pattern P, as illustrated in Fig. 3(b). The upper line $U = \{U_1, U_2, \cdots, U_n\}$ and the lower line $L = \{L_1, L_2, \cdots, L_n\}$ are defined as follows, $1 \leqslant i \leqslant n$:

$$\begin{cases} U_i = p_i + \varepsilon_k \\ L_i = p_i - \varepsilon_k \end{cases} \tag{3}$$

The envelope guarantees that if $s_{t,i}$ falls out of $[L_i, U_i]$, we know that W_t cannot match P.

Now we consider how to construct ELB implementation satisfying the lower bounding property, i.e., how to construct upper/lower bounds of pattern block and the feature of window block so that we could prune w number of successive windows together. We show the basic idea with an example in Fig. 3(a). Assume $w = 3$ and $N = 5$. At the sliding window W_t, element $s_{t,9}$ aligns with p_9. Accordingly, in W_{t+1} (or W_{t+2}), $s_{t,9}$ aligns with p_8 (or p_7). Obviously, if $s_{t,9}$ falls out of all upper and lower envelopes of p_9, p_8 and p_7, these 3 corresponding windows can be pruned together. Note that $s_{t,9}$ is the last element of block $\hat{W}_{t,3}$, and only in this case, all three elements of P aligning with $s_{t,9}$ belong to a same pattern block $\hat{P}_3$. Based on this observation, we define $\hat{P}_j^u$, $\hat{P}_j^l$ and $\hat{W}_j^f$ as follows:

$$\begin{cases} \hat{P}_j^u = \max_{0 \leqslant i < w} (U_{j \cdot w - i}) \\ \hat{P}_j^l = \min_{0 \leqslant i < w} (L_{j \cdot w - i}) \\ \hat{W}_{t,j}^f = last(\hat{W}_{t,j}) = s_{t,j \cdot w} \end{cases} \tag{4}$$

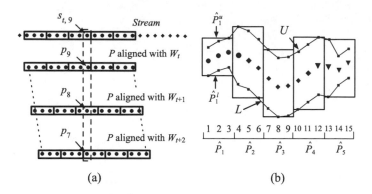

(a)   (b)

Fig. 3. (a) The element $s_{t,9}$ aligns with p_9, p_8 and p_7 at W_t, W_{t+1} and W_{t+2} respectively. (b) $\hat{P}_j^u$ and $\hat{P}_j^l$ are constructed by U and L.

As shown in Fig. 3(b), for each pattern block, its upper and lower bounds are set to the maximum and minimum of its two envelope lines respectively. It's obvious that ELB_{ele} satisfies the lower bounding property.

4.2   Subsequence-Based ELB Representation

In this section, we introduce the second ELB implementation, subsequence-based ELB, denoted by ELB_{seq}. Compared to ELB_{ele}, ELB_{seq} has a tighter bound which brings higher pruning power, although it is a little costlier on computing features of window blocks.

Different from ELB_{ele} which uses the tolerance of the whole subpattern to constrain one element pair, in ELB_{seq}, we use the same tolerance to constrain a w-length subsequence. Referring to [10], given two sequences $X = (x_1, \cdots, x_w)$ and $Y = (y_1, \cdots, y_w)$, it holds that:

$$w \left| \mu_x - \mu_y \right|^p \leqslant \sum_{i=1}^{w} |x_i - y_i|^p \tag{5}$$

where μ_x and μ_y are the mean values of X and Y. This theorem allows us to construct upper/lower envelope with the mean value of the subsequence.

Consider two w-length subsequences $P[i' : i]$ and $W_t[i' : i]$ where $i' = i - w + 1$ ($i' > 0$ so $i \geqslant w$). We first consider the case that all elements in $P[i' : i]$ (or $W_t[i' : i]$) belongs to only one subpattern (like P_k) and the corresponding subwindow (like $W_{t,k}$). If $W_{t,k}$ matches P_k, referring to Eq. 1, we know that:

$$L_p(P[i' : i], W_t[i' : i])^p = \sum_{j=i'}^{i} (p_j - s_{t,j})^p \leqslant \varepsilon_k^p \tag{6}$$

We denote by $\mu_{P[i':i]}$ and $\mu_{W_t[i':i]}$ that the mean value of $P[i' : i]$ and $W_t[i' : i]$ respectively. By combining Eqs. 5 and 6, we have:

$$\left| \mu_{P[i':i]} - \mu_{W_t[i':i]} \right| \leqslant (\frac{1}{w} \varepsilon_k^p)^{1/p} \tag{7}$$

We construct the envelope of pattern P as follows, $w \leqslant i \leqslant n$:

$$
\begin{cases}
U_i = \mu_{P[i':i]} + (\frac{1}{w}\varepsilon_k^p)^{1/p} \\
L_i = \mu_{P[i':i]} - (\frac{1}{w}\varepsilon_k^p)^{1/p}
\end{cases}
\tag{8}
$$

Now we consider the case that the interval $[i' : i]$ overlaps with more than one subpattern. Suppose $P[i' : i]$ overlaps with $P_{k_l}, P_{k_l+1}, \cdots, P_{k_r}$. Due to the additivity of the p-th power of L_p-Norm, we deduce from Eq. 6 that:

$$
L_p(P[i' : i], W_t[i' : i])^p = \sum_{j=i'}^{i}(p_j - s_{t,j})^p \leqslant \sum_{k=k_l}^{k_r} \varepsilon_k^p
\tag{9}
$$

By combining Eqs. 5 and 9, we have that:

$$
|\mu_{P[i':i]} - \mu_{W_t[i':i]}| \leqslant (\frac{1}{w}\sum_{k=k_l}^{k_r}\varepsilon_k^p)^{1/p}
\tag{10}
$$

We denoted the right term as $\theta_{seq}(i)$ and provide the general case of the pattern envelope as follows, $w \leqslant i \leqslant n$:

$$
\begin{cases}
U_i = \mu_{P[i':i]} + \theta_{seq}(i) \\
L_i = \mu_{P[i':i]} - \theta_{seq}(i)
\end{cases}
\tag{11}
$$

Note that Eq. 8 is the special case of Eq. 11.

The construction of upper and lower bounds are very similar to ELB_{ele}, while the feature of window block is adopted to the mean value. We show the basic idea with an example in Fig. 4(a). At the sliding window W_t, the subsequence $W_t[7 : 9]$ aligns with $P[7 : 9]$. Similarly, in W_{t+1} (or W_{t+2}), this subsequence aligns with $P[6 : 8]$ (or $P[5 : 7]$). According to Eq. 11, we know that if the mean value of $W_t[7 : 9]$ falls out of all upper and lower bounds of $P[7 : 9], P[6 : 8]$ and $P[5 : 7]$, these 3 corresponding windows can be pruned together. Based on this observation, we give the formal implementation of ELB_{seq} as follows:

$$
\begin{cases}
\hat{P}_j^u = \max_{0 \leqslant i < w} (U_{j \cdot w - i}) \\
\hat{P}_j^l = \min_{0 \leqslant i < w} (L_{j \cdot w - i}) \\
\hat{W}_{t,j}^f = mean(\hat{W}_{t,j}) = \mu_{W_t[(j-1) \cdot w + 1 : j \cdot w]}
\end{cases}
\tag{12}
$$

Note that, the upper and lower bounds of $\hat{P}_1$ are meaningless according to the definition of the envelope of ELB_{seq}.

Figure 4(b) provides an example of ELB_{seq} implementation. For clarity, we only illustrate the bounds of $\hat{P}_3$. The lower bound $\hat{P}_3^l$ is set to the minimum of L_7, L_8 and L_9 and covers 3 successive windows W_t, W_{t+1} and W_{t+2}.

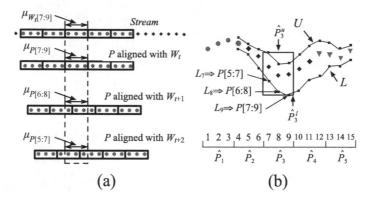

Fig. 4. (a) The subsequence $W_t[7:9]$ aligns with $P[7:9], P[6:8]$ and $P[5:7]$ at W_t, W_{t+1} and W_{t+2} respectively. (b) $\hat{P}_j^u$ and $\hat{P}_j^l$ are constructed by U and L.

4.3   Complexity Analysis

We first analyze ELB_{ele}. For each block $\hat{W}_{t,j}$, the time complexities of computing feature and determining $\hat{W}_{t,j} \prec \hat{P}_j$ are both $O(1)$. Therefore, the amortized pruning cost of ELB_{ele} is $O(1/w)$. Its space complexity is $O(N) = O(\lfloor n/w \rfloor)$. Although ELB_{ele} is very efficient, it constrains one element pair with the tolerance of the whole subpattern, which makes the envelope loose. Its pruning effectiveness is better when thresholds are relatively small, or pattern deviates from the normal stream far enough.

ELB_{seq} calculates the mean value of each window block with $O(w)$ and determining $\hat{W}_{t,j} \prec \hat{P}_j$ with $O(1)$. Considering a window block appears in several consecutive sliding windows, we store feature values in memory to avoid repeated calculation. Therefore, the amortized pruning cost of ELB_{seq} is reduced to $O(1)$. Same as ELB_{ele}, the space complexity of ELB_{seq} is $O(N)$.

5   Experimental Evaluation

In this section, we first describe datasets and experimental settings in Sect. 5.1 and then present the results of performance evaluation comparing the brute-force approach Sequential Scanning (SS), the classic method MSM [10] and our two approaches based on ELB_{ele} (ELB-ELE) and ELB_{seq} (ELB-SEQ) respectively. As presented in Sect. 2, although there are many works after MSM addressing time series similarity matching, most of them focus on utilizing the commonality among multiple patterns to build indexes, but not speeding up the problem of matching stream with a list of consecutive subpatterns.

Our goal is to:

– Demonstrate the efficiency of our approach on all L_p-Norm distance and different thresholds.

- Demonstrate the robustness of our approach on different pattern occurrence probabilities.
- Investigate the impact of block size on performance which helps to choose the appropriate parameter.

5.1   Experimental Setup

The experiments are conducted on both synthetic and real-world datasets.

Datasets. Real-world datasets are collected from a wind turbine manufacturer, where each wind turbine has hundreds of sensors generating streaming time series with sampling rate from 20 ms to 7 s. Our experimental datasets are from 3 turbines. In each turbine, we collect data of 5 sensors including wind speed, wind deviation, wind direction, generator speed and converter power. We replay the data as streams with total lengths of 10^8. For each stream, a pattern containing consecutive subpatterns with thresholds is given by domain experts.

Synthetic datasets are constructed based on UCR Archive [7]. UCR Archive is a popular time series repository, which includes a set of datasets widely used in time series mining researches [2,8,10]. To simulate patterns with various lengths, we select four datasets, Strawberry (Straw for short), Meat, NonInvasiveFa-talECG_Thorax1 (ECG for short) and MALLAT whose time series lengths are 235, 448, 750 and 1024. Referring to [10], for each selected UCR dataset, we choose the first time series of class 1 as the pattern and divide it into several subpatterns according to its shape and trend. Numbers of subpatterns of these four datasets are 5, 6, 8 and 7 respectively.

Concerning threshold of synthetic datasets, we define *threshold_ratio* as the ratio of the average threshold to the value range of this subpattern. Given a *threshold_ratio* and a subpattern P_k, the L_p-Norm threshold of P_k is defined by:

$$\varepsilon_k = |P_k|^{1/p} \times threshold\_ratio \times value\_range(P_k)$$

In practice, we observe that *threshold_ratio* being larger than 30% indicates that the average deviation from a stream element to its aligned pattern element is more than 30% of its value range. In this case, the candidate may be quite different from given pattern where similarity matching becomes meaningless. Therefore, we vary *threshold_ratio* from 5% to 30% in Sect. 5.2.

As for streaming data of synthetic datasets, referring to [2], we first generate a random walk time series S with length of 10^8 for each UCR dataset. Element s_i of S is $s_i = R + \sum_{j=1}^{i}(\mu_j - 0.5)$, where μ_j is a uniform random number in $[0, 1]$. As value ranges of the four patterns are about -3 to 3, we set R as the mean value 0. Then we randomly embed some time series of class 1 of each UCR dataset into corresponding steaming data with certain occurrence probabilities.

Algorithm. We compare our approaches to SS and MSM [10]. SS matches the sliding window one by one. For each window, SS calculates the L_p-Norm distances between all subpatterns and subwindows sequentially. In our scene, we let MSM build hierarchical grid index for each subpattern. For fair comparison, we adopt its batch version where the batch size is equal to ELB block size. We perform three schemes of MSM to choose the best one: stop the pruning phase at the first level of grid index (MSM-1), the second level (MSM-2), or never early stop the pruning phase (MSM-MAX).

Default Parameter Settings. There are three parameters for datasets: distance function, threshold and pattern occurrence probability. There is a parameter for our algorithm: block size. The default distance function is set to L_2-Norm (i.e., Euclidean distance). The default value of *threshold_ratio* and pattern occurrence probability are set to 20% and 10^{-4} respectively. We set the default value of block size to 5% of the pattern length. The impact of all above parameters will be investigated in following sections.

Performance Measurement. We regard the brute-force method SS as the baseline and measure the speedup of MSM and our algorithms. Streams and patterns are loaded into memory in advance where data loading time is excluded. To avoid the inaccuracy due to cold start and random noise, we run all algorithms over 10,000 ms and average them by their cycle numbers. All experiments are run on 4.00 GHz Intel(R) Core(TM) i7-4790K CPU, with 8 GB physical memory.

5.2 Performance Analysis

In this set of experiments, we first show our algorithms together outperform compared approaches on both synthetic and real-world datasets under different L_p-Norm functions and provide detailed analysis. After that, we perform experiments on diverse synthetic datasets by varying threshold ratio and pattern occurrence probability to demonstrate efficiency and robustness of our approaches. At last, we also evaluate the impact of block size for optimal parameter determination.

Performance Under Different L_p-Norm Distance. In this section, we report experiments of ELB-ELE and ELB-SEQ comparing to SS and MSM under different distance functions. We performed these experiments on all real-world and synthetic datasets using L_p-Norm where $p = 1, 2, 3, \infty$.

Figure 5 shows the experimental results. For real-world datasets, the results are similar among different turbines, so we only illustrate the wind turbine 1. Our algorithms show a great advantage over MSM and SS. As the distance function varies from L_1-Norm to L_∞-Norm, the advantage of our approaches over other methods gets larger.

We provide the experimental detail on a wind generator dataset in Table 1. The first two columns present the total and pruning time on each sliding window. Column *pruning power* is the percentage of pruned windows. Comparing to SS, our algorithms could prune numerous windows in the pruning phase, while

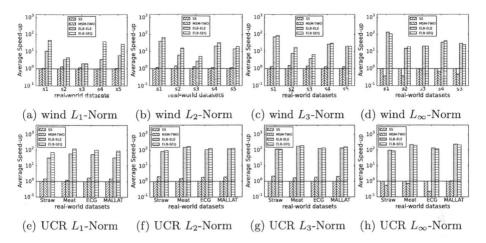

(a) wind L_1-Norm   (b) wind L_2-Norm   (c) wind L_3-Norm   (d) wind L_∞-Norm

(e) UCR L_1-Norm   (f) UCR L_2-Norm   (g) UCR L_3-Norm   (h) UCR L_∞-Norm

Fig. 5. Speedup vs. L_p-Norm. s_1: wind speed, s_2: wind deviation, s_3: wind direction, s_4: generator speed, s_5: converter power.

Table 1. The detail statistics on wind generator dataset

Algorithm	L_1-Norm			L_∞-Norm		
	Total time (ns)	Pruning time (ns)	Pruning power (%)	Total time (ns)	Pruning time (ns)	Pruning power (%)
ELB_SEQ	**13.04**	0.52	97.16	**8.78**	0.41	97.43
ELB_ELE	146.10	0.51	6.48	10.78	0.30	96.18
MSM_ONE	556.39	543.15	98.20	691.11	667.34	87.05
MSM_TWO	548.42	547.40	99.89	670.51	668.73	99.21
MSM_MAX	549.43	548.94	99.96	682.40	682.00	99.97
SS	562.84	-	-	413.34	-	-

SS has to perform exact matching for each sliding window, resulting in high time cost. Regarding MSM, its pruning power gets better from MSM-ONE to MSM-MAX (increased from 98.20% to 99.96% in L_1-Norm). Although MSM is more accurate, our pruning phase is much more efficient than MSM. Concerning ELB_SEQ and MSM_TWO (the best one among three MSM schemes) on L_1-Norm, our approach has slightly lower pruning power (97.16% vs. 99.89%), yet much more efficient pruning cost (0.52 vs. 547.40). On the whole, ELE_SEQ has an advantage of more than one order of magnitude over MSM_TWO.

Now we analyze the different performance of ELB on different L_p-Norm. From L_1-Norm to L_∞-Norm, the pruning effectiveness of ELB gets better. Although ELB-ELE spends less time on pruning phase than ELB-SEQ, its pruning power is very low at L_1-Norm (6.48%) due to its too loose bound. As p increases, its bound becomes tighter and performance gets better. In the case of L_∞-Norm, its performance has been flat with, and even outperformed ELB-SEQ on several datasets, as shown in Fig. 5(d) and (h). In contrast to ELB-ELE, ELB-SEQ is efficient under all L_p-Norms.

Impact of Distance Threshold. In this section, we compare the performance of ELB-ELE, ELB-SEQ, SS and MSM under different thresholds. We vary *threshold_ratio* from 5% to 30% on synthetic datasets, as described in Sect. 5.1.

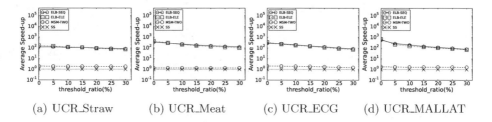

(a) UCR_Straw   (b) UCR_Meat   (c) UCR_ECG   (d) UCR_MALLAT

Fig. 6. Speedup vs. *threshold_ratio*

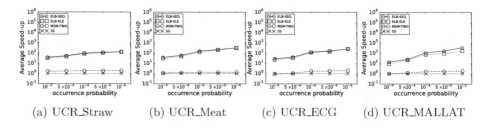

(a) UCR_Straw   (b) UCR_Meat   (c) UCR_ECG   (d) UCR_MALLAT

Fig. 7. Speedup vs. pattern occurrence probability.

The result on synthetic datasets is shown in Fig. 6. The performances of our two algorithms are very similar in synthetic datasets. Both ELB-ELE and ELB-SEQ outperforms MSM and SS by orders of magnitude. As the threshold gets larger, the speedups of ELB-ELE and ELB-SEQ decrease slightly. Nevertheless, our algorithms keep their advantage over other approaches even though *threshold_ratio* increases to 30%.

Impact of Pattern Occurrence Probability. In this section, we further examine the performance by varying the pattern occurrence probability. When the probability becomes lower, more windows are filtered out in the pruning phase. In contrast, when the probability becomes higher, more windows enter the post-processing phase. A good approach should be robust to these situations.

We perform this experiment on synthetic datasets and vary the occurrence probability over $\{10^{-3}, 5 \times 10^{-4}, 10^{-4}, 5 \times 10^{-5}, 10^{-5}\}$. The largest probability is set to 10^{-3} since in this case, the stream of MALLAT, which has largest pattern length, has been filled up by embedded UCR time series. As illustrated in Fig. 7, Our algorithms outperform MSM and SS in all examined probabilities. Furthermore, our algorithms show a larger speedup when the pattern occurrence probability becomes lower. This experiment demonstrates the robustness of our algorithms over different occurrence probabilities.

Impact of Block Size. The block size is an important parameter affecting the pruning power of our approach. In this experiment, we investigate the effect of block size by comparing ELB-ELE, ELB-SEQ and MSM on both synthetic and real-world datasets. We vary the ratio of the block size to the pattern length from 1% to 40%. A ratio being larger than 50% indicates that the entire pattern contains only one block, which makes ELE-SEQ meaningless.

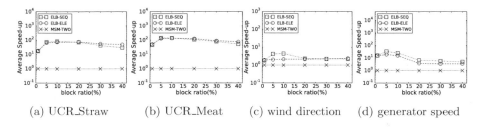

(a) UCR_Straw   (b) UCR_Meat   (c) wind direction   (d) generator speed

Fig. 8. Speedup vs. *block_ratio*.

Figure 8 shows the experimental results on some representative synthetic and real-world datasets while the rest are consistent. A too small or too large block size results in performance degradation. In detail, a smaller block size leads to a tighter bound for each block which improves the pruning effectiveness. Nevertheless, a small block size, corresponding to a small sliding step, results in more block computation and higher cost in the pruning phase. A larger block size may bring less block computation, but a looser bound meanwhile. The loose bound incurs degradation of the pruning effectiveness. In practice, our algorithms achieve the optimal performance when the block ratio is about 5% to 10%.

6 Conclusion

In this paper, we propose a new problem, called "consecutive subpatterns matching", which allows users to specify a pattern containing a list of consecutive subpatterns with different distance thresholds. We present a novel ELB representation to prune sliding windows efficiently under all L_p-Norms. We conduct extensive experiments on both synthetic and real-world datasets to illustrate that our algorithm outperforms the baseline solution and prior-arts.

References

1. Agrawal, R., Faloutsos, C., Swami, A.: Efficient similarity search in sequence databases. In: Lomet, D.B. (ed.) FODO 1993. LNCS, vol. 730, pp. 69–84. Springer, Heidelberg (1993). https://doi.org/10.1007/3-540-57301-1_5
2. Begum, N., Keogh, E.: Rare time series motif discovery from unbounded streams. PVLDB 8(2), 149–160 (2014)
3. Berndt, D.J., Clifford, J.: Using dynamic time warping to find patterns in time series. In: KDD Workshop, vol. 10, pp. 359–370 (1994)
4. Branlard, E.: Wind energy: on the statistics of gusts and their propagation through a wind farm. In: ECN-Wind-Memo 2009, vol. 5 (2009)
5. Faloutsos, C., Ranganathan, M., Manolopoulos, Y.: Fast subsequence matching in time-series databases. In: SIGMOD, pp. 419–429. ACM (1994)
6. Jensen, S.K., Pedersen, T.B., Thomsen, C.: Time series management systems: a survey. TKDE PP(99), 1 (2017)
7. Keogh, E.: Welcome to the UCR Time Series Classification/Clustering Page. www.cs.ucr.edu/~eamonn/time_series_data
8. Keogh, E.: Exact indexing of dynamic time warping. In: PVLDB, Hong Kong, China, pp. 406–417 (2002)
9. Kotsifakos, A., Papapetrou, P., Hollmén, J., Gunopulos, D.: A subsequence matching with gaps-range-tolerances framework: a query-by-humming application. PVLDB 4(11), 761–771 (2011)
10. Lian, X., Chen, L., Yu, J.X., Han, J., Ma, J.: Multiscale representations for fast pattern matching in stream time series. TKDE 21(4), 568–581 (2009)
11. Lian, X., Chen, L., Yu, J.X., Wang, G., Yu, G.: Similarity match over high speed time-series streams. In: ICDE, pp. 1086–1095. IEEE, April 2007
12. Lim, H.-S., Whang, K.-Y., Moon, Y.-S.: Similar sequence matching supporting variable-length and variable-tolerance continuous queries on time-series data stream. Inf. Sci. 178(6), 1461–1478 (2008)
13. Lim, S.-H., Park, H.-J., Kim, S.-W.: Using multiple indexes for efficient subsequence matching in time-series databases. In: Li Lee, M., Tan, K.-L., Wuwongse, V. (eds.) DASFAA 2006. LNCS, vol. 3882, pp. 65–79. Springer, Heidelberg (2006). https://doi.org/10.1007/11733836_7
14. Loh, W.-K., Kim, S.-W., Whang, K.-Y.: A subsequence matching algorithm that supports normalization transform in time-series databases. DMKD 9(1), 5–28 (2004)
15. Luo, G., Yi, K., Cheng, S.W., Li, Z., Fan, W., He, C., Mu, Y.: Piecewise linear approximation of streaming time series data with max-error guarantees. In: 2015 IEEE 31st International Conference on Data Engineering, pp. 173–184, April 2015
16. Moon, Y.-S., Whang, K.-Y., Han, W.-S.: General match: a subsequence matching method in time-series databases based on generalized windows. In: SIGMOD, pp. 382–393. ACM (2002)
17. Pace, A., Johnson, K., Wright, A.: LIDAR-based extreme event control to prevent wind turbine overspeed. In: 51st AIAA Aerospace Sciences Meeting including the New Horizons Forum and Aerospace Exposition, p. 315 (2012)
18. Sun, H., Deng, K., Meng, F., Liu, J.: Matching stream patterns of various lengths and tolerances. In: CIKM, pp. 1477–1480. ACM (2009)
19. Vlachos, M., Kollios, G., Gunopulos, D.: Discovering similar multidimensional trajectories. In: ICDE, pp. 673–684. IEEE (2002)

20. Wang, Y., Wang, P., Pei, J., Wang, W., Huang, S.: A data-adaptive and dynamic segmentation index for whole matching on time series. PVLDB **6**(10), 793–804 (2013)
21. Wei, L., Keogh, E., Van Herle, H., Mafra-Neto, A.: Atomic wedgie: efficient query filtering for streaming time series. In: ICDM, p. 8-pp. IEEE (2005)
22. Wu, H., Salzberg, B., Zhang, D.: Online event-driven subsequence matching over financial data streams. In: SIGMOD, pp. 23–34. ACM (2004)
23. Yi, B.-K., Faloutsos, C.: Fast time sequence indexing for arbitrary Lp norms. In: PVLDB, pp. 385–394. Morgan Kaufmann Publishers Inc. (2000)
24. Zhao, J., Liu, K., Wang, W., Liu, Y.: Adaptive fuzzy clustering based anomaly data detection in energy system of steel industry. Inf. Sci. **259**(Suppl C), 335–345 (2014)
25. Zhu, Y., Shasha, D.: Efficient elastic burst detection in data streams. In: SIGKDD, pp. 336–345. ACM (2003)

A Data Services Composition Approach for Continuous Query on Data Streams

Guiling Wang[1,2(✉)], Xiaojiang Zuo[1], Marc Hesenius[3], Yao Xu[2], Yanbo Han[1], and Volker Gruhn[3]

[1] Beijing Key Laboratory on Integration and Analysis of Large-Scale Stream Data, North China University of Technology, No. 5 Jinyuanzhuang Road, Shijingshan District, Beijing 100144, China
`wangguiling@ict.ac.cn`
[2] Ocean Information Technology Company, China Electronics Technology Group Corporation (CETC Ocean Corp.), No. 11 Shuangyuan Road, Badachu Hi-Tech Park, Shijingshan District, Beijing 100041, China
[3] paluno - The Ruhr Institute for Software Technology, University of Duisburg-Essen, Schützenbahn 70, 45127 Essen, Germany

Abstract. We witness a rapid increase in the number of data streams due to Cloud Computing, Big Data and IoT development. We would like to access and share data streams using a data service approach. In this paper, we propose a flexible continuous data service model and a continuous data service composition algorithm for answering queries across data streams. Service operation instance is modeled as a view defined on data streams composed of two parts: a data part and a time synchronization part. The composition algorithm extends the traditional Bucket algorithm to find the contained rewriting of user query on views satisfying the containment relationship of both data part and time synchronization part. We also present use case and experimental studies indicating that the approach is effective and efficient.

Keywords: Data streams · Query rewriting · Data services
Service composition · Continuous query

1 Introduction

Web services technology is a general medium for sharing data and functionality and enabling cross-organization collaboration for enterprise and web systems. Data services [1] or data-providing services [2] are a kind of services that allow query-like access to an organization's data sources. Although the existing data processing framework provides composition models or query languages which allow us to retrieve desired data from multiple data sources, data services provide a flexible, controlled and standardized approach to access or query an organization's data sources without exposing its databases directly [3]. Furthermore,

Y. Cai et al. (Eds.): APWeb-WAIM 2018, LNCS 10988, pp. 106–120, 2018.
https://doi.org/10.1007/978-3-319-96893-3_9

when queries require to access data sources across organizations, several services can be composed to generate a response [4–6].

To bring the benefits of data services, we would like to access and share data streams using a data service approach. However, data streams are very different from traditional data sources. This makes the problem of data service modelling and composition challenging for accessing and sharing data streams. Firstly, unlike traditional snap-shot queries over data tables, queries over data streams are continuous. A continuous query is issued once and remains active for a long time. The answer to a continuous query is constructed progressively as new input stream tuples arrive [7]. Once executed, data services for queries on data streams need to continuously return results and consider temporal constraints. Secondly, for queries over multiple data sources, traditional data providing services are often modeled as parameterized views over data schemas [3,4]. Based on the service model, services can be composed using a query rewriting approach to answer queries over multiple data sources [3,4]. Because most of the stream query language do not support views [7], how to model data services as views over data streams is not trivial. And because queries for data streams need to be updated continuously, the traditional query rewriting approach is inapplicable to rewrite query over data streams directly.

In this paper, we introduce a data service model for continuous query over data streams, and call it "continuous data services". Service operation inputs are not modeled as fixed query conditions. They are arbitrary query conditions modeled as a set of optional attributes of the underlying data model and condition predicates. "sliding window" is introduced into the service model to describe the temporal feature of services. The instance of the service operation can be modeled as a view defined on data streams. Based on the continuous data service model, we propose a continuous data service composition algorithm for answering queries across data streams. It improve the Bucket algorithm [8] for "answering queries using views" on persistent relation data to find the contained rewriting by checking the containment relationship between time synchronization part of the query and the rewriting. We describe an implementation, a use case and provide a performance evaluation of the proposed approach.

The rest of this paper is organized as follows: In Sect. 2, we motivate the need for conjunctive queries across data streams, and discuss the underlying challenges. In Sect. 3, we describe the continuous data service model. In Sect. 4, we propose the continuous data service composition algorithm. In Sect. 5, we describe our implementation and evaluate our approach. We overview related work in Sect. 6. We provide concluding remarks and future research outlook in Sect. 7.

2   Motivation

In this section, we describe a motivating scenario we use throughout the paper. Various systems for maritime freight logistics collect data like vessel trajectories, vessel basic information and so on. Among these data sources, the data stream `vesseltraj(mmsi, long, lat, speed)` records trajectory points of a vessel, where `mmsi` is the Maritime Mobile Service Identity, `long` and `lat` is the longitude and latitude of the vessel location, and `speed` is the vessel's speed. The relation data `vesselinfo(mmsi, imo, callsign, name, type, length, width, po-sitionType, eta, draught)` records static information of ships including the `mmsi`, the International Maritime Organization (imo) code, call sign, name, type, length, width, the Estimated Time of Arrival (eta), draught of the vessel. The relation data `vesseltravelinfo(mmsi, dest, source)` records the destination and the identification of the position message source.

These systems are subordinate to different management domains and won't expose full data access to their data sources directly. They provide access to the set of services with constraints described in Table 1. The underlying data streams of DS_1 are `vesselinfo` and `vesseltraj`. They have constraints that `mmsi` must be greater than 3000 and `speed` greater than 50 km with a time-based sliding window of window size 5s and slide size 1s. The underlying data streams of DS_2 is `vesselinfo` and `vesseltraj`. The time window of the stream has window size 5s and slide size 2s. The underlying data stream of DS_3 is `vesseltraj`. This data stream has constraints that the `speed` must be less than 40 km with window size 5s and slide size 2s. The underlying data streams of DS_4 are `vesseltravelinfo` and `vesseltraj`. This service has constraints that the `mmsi` must be less than 2000 with window size 5s and slide size 2s. We also express the underlying query of the services as conjunctive queries extended with time-based sliding window semantics. Note that join predicates in this notation are expressed by multiple occurrences of the same variables.

Table 1. Continuous data services in the ocean data query scenario

Service	Functionality and constraints	Formal expression of the underlying data streams
DS_1	Query on those vessels whose *mmsi* number greater than 3000 and speed greater than 50 km with a time-based sliding window of window size 5s and slide size 1s	*vesselinfo(mmsi, imo, callsign, name, type, length, width, positionType, eta, draught), vesseltraj(mmsi, long, lat, speed), mmsi > 3000, speed ≥ 50* km, *wsize(5), slide(1)*
DS_2	Query on those vessels with a time-based sliding window of window size 5s and slide size 2s	*vesselinfo(mmsi, imo, callsign, name, type, length, width, positionType, eta, draught), vesseltraj(mmsi, long, lat, speed), wsize(5), slide(2)*
DS_3	Query on those vessels whose speed is less than 40 km with a time-based sliding window of window size 5s and slide size 2s	*vesseltraj(mmsi, long, lat, speed), speed < 40* km, *wsize(5), slide(2)*
DS_4	Query on those vessels whose *mmsi* number less than 2000 with a time-based sliding window of window size 5s and slide size 2s	*vesseltravelinfo(mmsi, dest, source), vesseltraj(mmsi, long, lat, speed), mmsi < 2000, wsize(5), slide(2)*

Those services with sliding window constraints continuously push output to the service consumer once the consumer creates a connection with the service producer. The output is the query results in range of the configured window size that will be updated every slide size. So we call these services "continuous data services".

Now assume the following query asks for vessels that have outstanding speed over a defined sliding window. Note we express the query as conjunctive queries extended with time-based sliding window semantics. And note that join predicates in this notation are expressed by multiple occurrences of the same variables.

Q(mmsi, draught, dest, speed):-vesselinfo(mmsi, imo, callsign, name,

type, length, width, positionType, eta, draught), vesseltraj(mmsi, long,

lat, speed), vesseltravelinfo(mmsi, dest, source), speed $\geq$ 40 km,

wsize(5), slide(4)

Obviously service DS_3 is not useful to satisfy this query request, because DS_3 has information only on vessels whose speed is less than 40 km whereas we are interested in vessels which has speed greater than 40 km. Although DS_1 is relevant to user query, it only has mmsi information and need to retrieve destination information by invoking other service like DS_4. However, DS_1 only has information on vessels with mmsi greater than 3000, and DS_4 has information on vessels with mmsi less than 2000, meaning DS_1 and DS_4 are disjoint. So service DS_1 is also not useful to answer this user query. We are left with one possible plan to use the services to answer this query. Firstly invoke DS_2 to retrieve the list of vessels with a sliding window of window size 5s and slide size 2s. Then invoke DS_4 where mmsi is less than 2000 with a sliding window of window size 5s and slide size 2s. Results from both services are joint to answer Q. Note that the sliding window constraints of DS_2 and DS_4 is different, we also need to judge if the joint results can satisfy the query requirement. Also note that the results only vessels with mmsi less than 2000, which can satisfy the query is not equivalent with it. Note in this example, there is only one service composition plan satisfying the query, but there may be multiple plans in other examples.

3   Model of Continuous Data Service

3.1   Data Model

We use the synchronized relation model for describing the contents of data stream sources. The data model includes:

- S and $\Re(S)$. S is a tagged stream with the format of "Tag⟨Attrs⟩ts", where Tag can be either insert (+), update (u), or delete (-) and ts indicates the time at which the modification takes place. For detailed explanation of what is a tagged stream, please refer to [7]. Any tagged stream S has a corresponding time-varying relation $\Re(S)$. The relation is continuously modified by S's tuples.

- **Attrs**. **Attrs** are the attributes of the time-varying relation $\Re(S)$.
- **ts**. **ts** is the time point where the relation $\Re(S)$ is modified by the underlying S's tuples.
- **sync**. **sync** synchronized stream is a special tagged stream "+$\langle$timepoint$\rangle$ts", where **timepoint** represents a time point which is the only attribute of **sync**. Synchronized stream is a kind of tagged stream. So it also has a corresponding time-varying relation $\Re($sync$)$.
- $\Re_{sync}(S)$. $\Re_{sync}(S)$ is a synchronized relation of any arity. Figure 1 illustrates a synchronized stream of $\Re_{sync_2}($Vesseltraj$)$. For traditional persistent data (e.g. data tables in a database), the tuples are reflected at any time. Here we denote the synchronized stream associated with the traditional persistent data as $sync_0$.

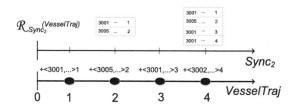

Fig. 1. A synchronized stream of $\Re_{sync_2}($Vesseltraj$)$

DataModel of $\Re_{Sync}(S)$ can be represented as a tuple: $\langle$Attrs, SyncUnits$\rangle$, where **Attrs** = {attr} is a set of attributes, **SyncUnits** is the subscript index of the synchronization stream **sync**. For example, the value of **SyncUnits** is 2 for $sync_2$, 3 for $sync_3$ and 4 for $sync_4$ etc.

3.2 Continuous Query Containment

Query containment and equivalence provide a formal framework to compare different queries in a data integration system. In relational databases, a query Q_1 is said to be contained in Q_2, denoted by $Q_1 \subseteq Q_2$, if and only if $Q_1(D) \subseteq Q_2(D)$ for any database instance D. Q_1 is *equivalent* to Q_2 if and only if $Q_1 \subseteq Q_2$ and $Q_2 \supseteq Q_1$.

In stream processing system, a continuous query over n tagged streams $S_1 \ldots S_n$ is semantically equivalent to a materialized view that is defined by a SQL expression over the time-varying relations $\Re(S_1) \ldots \Re(S_n)$ [7]. The big difference between time-varying relations and traditional relations is that the time-varying relations have arbitrary refresh conditions. The solution is to isolate the time synchronization streams out of the continuous query expression. Then the containment relationship is tested from two aspects: (1) test data containment using traditional query containment test method, and (2) test synchronization containment.

For example, if we want to check the containment relationship of a query Q and a data service instance of DS' like this:

$$Q(\texttt{mmsi}, \texttt{draught}, \texttt{dest}, \texttt{speed}) :- \Re(T), I, \texttt{TRAVEL}, \texttt{speed} \geq 40,$$
$$(\texttt{currTime-5}) < \texttt{TS} < \texttt{currTime}, \texttt{sync}_4$$

and

$$Q_{\texttt{DS}'}(\texttt{mmsi}, \texttt{speed}, \texttt{imo}) :- \Re(T), I, \texttt{speed} \geq 30,$$
$$(\texttt{currTime-5}) < \texttt{TS} \leq \texttt{currTime}, \texttt{sync}_1$$

We first test containment of time part of $Q_{\texttt{DS}'}$ and Q. The synchronization relation part of Q (i.e. $\Re(\texttt{sync}_4)$) is contained in the synchronization relation part of $Q_{\texttt{DS}'}$ (i.e. $\Re(\texttt{sync}_1)$). Because any tuples satisfied by the selection and projection conditions of Q also satisfied $Q_{\texttt{DS}'}$, the data part of Q is contained in data part of $Q_{\texttt{DS}'}$. We can conclude that Q is contained in $Q_{\texttt{DS}'}$.

3.3 Continuous Data Service

We model a continuous service as a view defined on the underlying data streams. Any service subscribes one or multiple data streams or database tables, which is defined as Subs. Any service has zero to multiple operations in which inputs, outputs, window range, window slide size should be defined. Input and output parameters are from the attributes of the underlying synchronized relations corresponding with Subs. Every service instance publishes one tagged stream on message queue.

Such service can be expressed as follows: $DS = \langle \texttt{ID}, \texttt{SubS}, \texttt{PubS}, \texttt{Ops} \rangle$, where:

- ID is the unique identity of the service.
- SubS is the stream set of the service subscribed from message queue. $\texttt{SubS} = \{\langle \texttt{DS}_{\texttt{sub}}, \texttt{DataConstrs}, \texttt{TimeConstr} \rangle\}$, where $\texttt{DS}_{\texttt{sub}}$ is a tagged stream defined in Section II. A Data model $\langle \texttt{Attrs}, \texttt{SyncUnits} \rangle$ is corresponding with a time-varying relations $\Re(\texttt{DS}_{\texttt{sub}})$. DataConstrs and TimeConstr are the constraints applied on content and time of the tagged stream.
- PubS is the stream set of the service published to message queue. $\texttt{PubS} = \{\langle \texttt{DS}_{\texttt{pub}}, \texttt{DataConstrs}, \texttt{TimeConstr} \rangle\}$, where $\texttt{DS}_{\texttt{pub}}$ is a tagged stream. It is corresponding with a time-varying relation $\Re(\texttt{DS}_{\texttt{pub}})$.
- DataConstrs = {DataConstr}, where DataConstr = $\langle \texttt{attr}, \texttt{condop}, \texttt{constant} \rangle$. attr is the attribute of $\Re(\texttt{DS}_{\texttt{sub}})$ for SubS and $\Re(\texttt{DS}_{\texttt{pub}})$ for PubS. condop can be one of the condition operator from $>, =, <, \geq, \neq, \leq$. constant is a constant value.
- TimeConstr = $\langle \texttt{range} \rangle$, where range is range size of the sliding window of synchronized relation. Note that tumbling window and hopping window are both a special form of the sliding window. For tumbling window, range size is equal to slide size. And for hopping window, range size is a multiple of slide size.

– Ops = {⟨inputs, outputs, range, slide⟩} is the service operations. inputs = {input} are a set of attributes of S_{sub}, the corresponding condition operator $>, =, <, \geq, \neq, \leq$ and constants. outputs = {output} are a set of output parameters of the service operation. range and slide are the time constraint of the service request. A *SyncSQL* expression can be generated from Ops.

The elements of the input and output set Ops are determined when a service is instantiated. PubS of a service are also determined when a service is instantiated.

Given a specific user inputs, the service has an associated instance. A service instance can also be defined as a query view on the underlying time-varying relations. We use the notation of *conjunctive queries* extended with synchronization stream to express the view. A data service DS = ⟨ID, SubS, PubS, Ops⟩ is transformed into a view:

$$DS(\bar{X}) :- \Re(S_{sub_1}), \ldots, \Re(S_{sub_n}), c_1, \ldots, c_n, tc, sync_1 \cap \ldots \cap sync_n$$

where $\bar{X}$ is all the attributes from all S_{sub} elements of SubS, $\Re(S_{sub_i})$ are the underlying time-varying relation corresponding with all the elements of SubS. Note that not all subscribed streams have data constraints applied on them. If S_{sub_i} has no data constraint, we can add a data constraint c on it: $-\infty \leq c \leq +\infty$. Thus all subscribed streams have data constraints represented as $c_1, \ldots, c_n$. tc is the intersection of all the window range size constraints applied on them. $sync_i$ is the synchronization stream applied on $\Re(S_{sub_i})$.

A service instance of S = ⟨ID, SubS, PubS, Ops⟩ can be transformed into a view like this:

$$DS(\bar{X}) inst :- \Re(S_{sub_1}), \ldots, \Re(S_{sub_n}), c_1, \ldots, c_n, c_{op_1}, \ldots, c_{op_s}, tc,$$
$$sync_1 \cap \ldots \cap sync_n, \cap sync_1 \cap \ldots \cap sync_t$$

$c_{op_1}, \ldots, c_{op_s}$ are data constraints from inputs of service operations. $sync_1 \cap \ldots \cap sync_t$ are synchronization stream from the time constraints of service operations. tc is the intersection of all the window range size constraints applied on $\Re(Sub_i)$ and from service operations.

4   Data Services Composition for Answering Continuous Query

When services and service instances are transformed into views on time-varying relations, given a conjunctive query Q, we need to find the service composition plans to answer it. The problem of answering conjunctive query using views for traditional persistent data is NP-complete [9]. Bucket algorithm or *minicon* algorithm are the approaches to drastically reduce the number of rewritings we need to consider for a query given a set of views. So we can improve the Bucket algorithm [8] or *MiniCon* algorithm [10] to find the service composition plans to answer query Q. Here we give the improved Bucket algorithm. The main idea of

Bucket algorithm is that we first consider each subgoal in the query in isolation, and determine which views may be relevant to that subgoal. Thus the number of query rewritings that need to be considered can be drastically reduced. In order to support finding relevant continuous data services or service instances, we improve the Bucket algorithm by adding the synchronization stream containment judgement and determining the service operation inputs and outputs after the relevant services are found.

The first step is shown in Algorithm 1. It constructs for each subgoal g in the query a bucket of relevant service or service instance atoms. In this algorithm, we check the containment relationship between the query sub-goal and the view transformed from the service or service instance.

Algorithm 1. Create buckets

Input: conjunctive query Q in two parts:
   data part Q^d of the form:
   $Q^d(\bar{X}):-\Re(R_1)(\bar{X}_1), \ldots, \Re(R_n)(\bar{X}_n), c_1, \ldots,$
   c_n, tc
   synchronization part $Sync_Q$ of the form:
   $Sync_Q = Sync_1 \cap \ldots Sync_n;$
   a set of views $\mathcal{V}$ transformed from services $\mathcal{S}$ and service instances $\mathcal{S}inst;$
Output: list of buckets
 1: **for** $1 \leq i \leq n$ **do**
 2:     Initialize $Bucket_i$ to $\emptyset$
 3: **end for**
 4: **for** each subgoal g_i in Q **do**
 5:     **for** each $V \in \mathcal{V}$ **do**
 6:         Let V be of the form:
         $V(\bar{Y}):-\Re(S_1)(\bar{Y}_1), \ldots, \Re(S_m)(\bar{Y}_m),$
         $d_1, \ldots, d_m, sync_1 \cap \ldots \cap sync_m \cap sync_1 \cap$
         $\ldots \cap sync_t$
 7:         **if** $\Re(Sync_V) \subseteq \Re(Sync_q)$ **then**
 8:             **if** g_i is an element of subgoals set of V **then**
 9:                 **if** each $x \in X_i$ is also an element of $\bar{Y}$ **then**
10:                     **if** the data constraints of V satisfy the data constraints of Q **then**
11:                         add V into $Bucket_i$
12:                     **end if**
13:                 **end if**
14:             **end if**
15:         **end if**
16:     **end for**
17: **end for**

The second step considers all the possible combinations of services and service instances. Each combination should include one of the service or service instance atoms from every bucket. Generate the candidate composition plans by checking if each combination is satisfied (if there exists no self-contradictory in the same combination). Keep those plans that is satisfied and delete those that is unsatisfied.

Algorithm 2. Check whether a candidate plan is equivalent

Input: candidate services and service instances composition plan p($\bar{Y}$);
    conjunctive query Q($\bar{X}$);
    a set of executable equivalent services and/or service instances composition plan eqCompPlans
Output: the updated result of eqCompPlans
 1: Let the set of subgoals of p in the form of goalsOfp, and the subgoals of each plan eqPlan in eqCompPlans in the form of goalsOfeqPlan
 2: Denote the intersection of data constraints of p and Q as D ∩ C, where D is the data constraints set of p and C is the data constraints set of Q
 3: Get all of the elements exist in set D ∩ C that don't exist in set of data constraints of p, denoted as A = D ∩ C \ D. This set is the additional data constraints that should be added on p in order to be equivalent to Q
 4: **if** Q ⊆ p **then**
 5:   **if** there exists no plan eqPlan in eqCompPlans satisfying the condition that goalsOfeqPlan ⊂ goalsOfp **then**
 6:     **if** there exists services (not service instance) in p **then**
 7:       **for** each subgoal g of p **do**
 8:         **if** g is a service **then**
 9:           **if** D ∩ C ≠ ∅ **then**
10:             A = genInstance($\bar{Y}$ ∩ $\bar{X}$, A, sync)
11:           **else**
12:             genInstance($\bar{Y}$ ∩ $\bar{X}$, ∅, sync)
13:           **end if**
14:         **end if**
15:       **end for**
16:       **if** A = ∅ **then**
17:         delete the redundant plan than p and add p into eqCompPlans
18:       **end if**
19:     **else**
20:       **if** p ⊆ Q **then**
21:         delete the redundant plan than p and add p into eqCompPlans
22:       **end if**
23:     **end if**
24:   **end if**
25: **end if**

In the example explained in Sect. 2, the returned results contain only vessels with mmsi less than 2000, which is not equivalent with the query. In fact, the service composition plans that can answer user query can be divided into two categories: the equivalent composition plans and the contained composition plans. The former is equivalent with the query and the latter is contained in the query. There exists a maximally contained composition plan among the contained composition plans. So if a continuous query can be supported by mutiple composition plans, we can choose the equivalent or maximally contained composition plan among the candidates.

The third step searches the equivalent service composition plans or the contained service composition plans. Take the equivalent service composition plan as the example, the basic idea is to consider each candidate composition plan p, check if $p \equiv Q$ when there exists no service atom in p. If there exists services and there exists data constraint atoms C and synchronization constraint atoms sync such that $Q \wedge C \equiv Q$ and they can be used as the additional constraints on service when we instantiate it. The concrete steps for considering each p are shown in Algorithm 2.

In steps 4 and 20, when we judge the containment relationship between the plan and query, time synchronization containment relationship is checked first.

In step 5, we check if the equivalent composition plan that is more concise than the current plan p already exists. If it already exists, the current plan is abandoned. In steps 10 and 12, we use the additional data constraints A to instantiate a service. A method genInstance(output, dataConstr, timeConstr) is called to determine the input and output parameters of the service operation. In this method, the output parameter value is taken as the output parameter value of the service operation. In step 10, we take additional data constraints in A as the input parameter values of the service operation. The time constraints of Q are taken as the time constraints of the service operation. In this method, we update A with the unsatisfied data constraints and returned. After the loop 7, all the services in p are instantiated. If the attributes of all the additional data constraints are also the data attributes of $\Re(g_{sub})$, it means that all the additional data constraints can be applied on the services, in other words, the services can satisfy the data constraints after instantiation. Otherwise, the services can not satisfy the data constraints and the service composition plan is abandoned.

In step 20, if $Q \subseteq p$, $Q \supseteq p$ and all atoms of p are service instances, delete the redundant plan than p (in other words, the redundant plan rePlan satisfying the condition that goalsOfrePlan $\supset$ goalsOfp) from the result set and add p into equivalent result set.

To search the contained composition plan, if $Q \supseteq p$ and all atoms of p are service instances, add p into equivalent result set directly. If Q is not contained in p and sub-goals of Q overlap with that of p, and there exist service atoms in p, we should instantiate the services. Check whether all the additional constraints can be applied on the services when instantiating them. If they can't be applied, this means that the services can not satisfy the data constraints after instantiation, in other words, the plan is not executable. We omit the pseudo code of this algorithm for searching contained composition plans due to limited space.

5   Implementation and Evaluation

In this section, we first describe an implementation of our approach. Then we provide a use case and experimental evaluation.

5.1   Implementation

The architecture of our system is shown in Fig. 2. Firstly, relational databases and data stream sources should be registed and managed. When a query is posed, the query rewriter module uses the information from service registry to decide the candidate service composition plan. The service executor module is responsible for invocation and join/compose the service execution results.

Every service is implemented as a Spark Streaming job. The underlying data streams are subscribed by the service using Kafka. And the outputs of a service are published to Kafka, which can be subscribed by later services. For those Web based clients, we expose continuous data service as REST-like API over HTTP protocol based on a Web-based push technology - Sever-Sent Events (SSE) [11]. It allows the service to push query results to clients continuously. The client sends a request to a service and opens a single long-lived HTTP connection. The service then sends data continuously to the client without further action from the client.

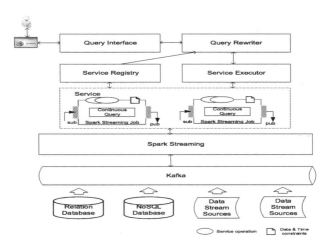

Fig. 2. Architecture of the implementation.

5.2   Case Study

In this section, we take the example introduced in Sect. 2 as the use case to introduce how our approach works.

Assume the outputs of Ops of an instance of DS_1 are {mmsi, imo} and no input parameters. range and slide are 5 and 1 separately.

This instance of DS_1 can be expressed as:

DS_1inst(mmsi, speed, imo):-T(mmsi, long, lat, speed), I(mmsi, imo,

callsign, ...), mmsi > 3000, speed $\geq 50, 5, sync_1$

In a similar way, the instance of DS_2 is:

DS_2inst(mmsi, draught, speed):-T(mmsi, long, lat, speed), I(mmsi, imo, ..., draught), 5, sync$_2$

The instance of DS_3 can be expressed as follows:

DS_3inst(mmsi, speed):-T(mmsi, long, lat, speed), speed < 40, 5, sync$_2$

Assume there is no instance for service DS_4, so it is express as:

DS_4(mmsi, speed, dest, source, long, lat):-TRAVEL(mmsi, dest, source), T(mmsi, long, lat, speed), mmsi < 2000, 5, sync$_2$

Query is expressed as Sect. 2. This query has sub-goals $\Re$(T), I and TRAVEL. According to our algorithm, the steps to answer user query are as follows:

In the first step the algorithm creates buckets for each sub-goal of Q. The contents of bucket for sub-goal $\Re$(T) are: DS_1inst,DS_2inst, and DS_4. DS_3inst is not in this bucket because the interpreted predicates of the view and the query are not mutually satisfiable. The contents of bucket for sub-goal TRAVEL are: DS_4(mmsi, speed, dest, long, ...). The contents of bucket for sub-goal I are: DS_2inst(mmsi, draught, speed).

In the second step of the algorithm, we combine elements from the buckets. The first combination, involving the first element from each bucket, yields the rewriting

Q_1(mmsi, draught, speed, dest):-DS_1inst(mmsi, speed, imo$'$), DS_4(mmsi, speed, dest$'$, long$'$), DS_2inst(mmsi, draught, speed)

However, while both DS_1inst and DS_4 are relevant to the query in isolation, their combination is guaranteed to be empty because they cover disjoint sets of vessel identifiers.

Consider the second elements in the left bucket yields the rewriting

Q_2(mmsi, draught, speed, dest):-DS_2inst(mmsi, draught, speed), DS_4(mmsi, speed, dest$'$, lo-ng$'$, ...), DS_2inst(mmsi, draught, speed)

Then we remove the first sub-goal, which is redundant, and generate service instance with the additional data constraints speed ≥ 40. The output parameters of DS_4 instance operation are set to be variables from attributes of the underlying data stream which are also in the head of Q, which is mmsi, dest, speed. The inputs parameters are speed ≥ 40.

So we would obtain Q_2, which is the only contained composition plan the algorithm finds.

5.3   Experimental Evaluation

In this section, we give an experimental evaluation of our approach. The goal of the experimental evaluation is to analyze the factors that affect the performance of the service composition algorithm.

The service composition algorithm experiments were run on a computer with Intel(R) Core(TM) i5-2400 CPU 3.10 GHz and 8 GB memory. In order to experimentally evaluate our approach, we generated a set of continuous data services and service instances. We use three representative queries including the query example shown in Sect. 2. According to 80/20 rule (also known as Pareto principle), The method guarantees that the number of services and service instances that are related to user queries are about 20% of the total services and service instances generated. For each query, we generated various number of data services and data service instances from 100, 200, ... to 500. Figure 3 plots the total and average time to generate all composition plans for each query against the number of data sources. We can observe that the average generation time per composition plan is within 10ms, which is acceptable in real application.

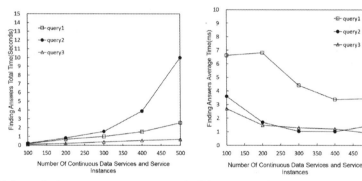

(a) total time to generate composition plans

(b) average time to generate comosition plans

Fig. 3. Total and average time to generate compostion plans.

6   Related Work

Most of the research work on web service composition focus on traditional *Effect-Providing* services or *application-logic* services instead of *Data-Providing* services or data services. The traditional *application-logic* service composition algorithms are inapplicable and inefficient to data services that all share the same business function (i.e. data query) and have no side-effects [4].

Data integration approach is often adopted for the purpose of data services composition. Some use the query rewriting techniques as the composition algorithm [2–5,12]. Others use visual mashup languages or constructs as composition approach [13,14]. However, the data services model and composition algorithm in these work are inapplicable to data stream sources and data stream integration.

There are some related research work from data integration area such as Info-Master [15] and Information Manifold [8]. Our work differs with these works in many ways. First, these works target toward resolving specific queries given a set of data sources, whereas in our work the focus is on constructing a composition of services that is independent of a particular input value. The composite service can be reused to answer a set of queries instead of a specific queries. Second, compared to previous query rewriting algorithms [10,16] that were proposed for the traditional static relational data model, our composition algorithm is based on data stream model. As far as we know, our continuous data service model is the first service model to support data stream query and our algorithm is the first to address the problem of composing continuous data services to support data stream integration.

There are some related research work on service modeling for data streams such as [17,18], however, the work cannot be used to solve the problem of query across various data sources directly. Some work has addressed the problem of supporting views in data stream management systems [7], however, the work is limited only to answering specific queries based on a set of data sources. Our work propose a continuous data service model which provides a flexible, controlled and standardized approach to access or query data stream. We address data stream integration problem by providing service composition approach. The composite service can access a set of conditions as input instead of limiting to answering specific queries.

7 Conclusion

In this paper, we presented an approach for conjunctive query on data streams by composing continuous data services. We introduce a flexible continuous data service model with continuous query as service operation. Service operation instance is modeled as a view defined on data streams in which the data part and time synchronization part are separated from each other. A continuous data service composition algorithm is introduced for answering queries across data streams. An experimental study is provided to evaluate the performance of our approach. As a future work, we plan to address location concerns when composing continuous data services.

Acknowledgments. This work is supported by Beijing Natural Science Foundation No. 4172018, National Natural Science Foundation of China No. 61672042, and University Cooperation Projects Foundation of CETC Ocean Corp.

References

1. Carey, M.J., Onose, N., Petropoulos, M.: Data services. Commun. ACM **55**(6), 86–97 (2012)
2. Vaculín, R., Chen, H., Neruda, R., Sycara, K.: Modeling and discovery of data providing services. In: 2008 IEEE International Conference on Web Services, pp. 54–61, September 2008

3. Barhamgi, M., Benslimane, D., Ouksel, A.M.: Composing and optimizing data providing web services. In: Proceedings of the 17th International Conference on World Wide Web, pp. 1141–1142. ACM (2008)
4. Barhamgi, M., Benslimane, D., Medjahed, B.: A query rewriting approach for web service composition. IEEE Trans. Serv. Comput. **3**(3), 206–222 (2010)
5. Zhou, L., Chen, H., Yu, T., Ma, J., Wu, Z.: Ontology-based scientific data service composition: a query rewriting-based approach. In: AAAI Spring Symposium: Semantic Scientific Knowledge Integration, pp. 116–121 (2008)
6. Zhang, F., Wang, G., Han, Y.: Automatic generation of service composition plans for correlated queries. In: 2013 10th Web Information System and Application Conference, pp. 143–149, November 2013
7. Ghanem, T.M., Elmagarmid, A.K., Larson, P.Å., Aref, W.G.: Supporting views in data stream management systems. ACM Trans. Database Syst. **35**(1), 1–47
8. Levy, A.Y., Rajaraman, A., Ordille, J.J.: The world wide web as a collection of views: query processing in the information manifold. In: VIEWS, pp. 43–55 (1996)
9. Doan, A., Halevy, A., Ives, Z.: Principles of Data Integration, 1st edn. Morgan Kaufmann Publishers Inc., San Francisco (2012)
10. Pottinger, R., Halevy, A.: MiniCon: a scalable algorithm for answering queries using views. Int. J. Very Large Data Bases **10**(2–3), 182–198 (2001)
11. Hickson, I.: Server-sent events. https://www.w3.org/TR/eventsource/. Accessed 25 October 2015
12. Zhao, W., Liu, C., Chen, J.: Automatic composition of information-providing web services based on query rewriting. Sci. China Inf. Sci. **55**(11), 2428–2444 (2012)
13. Wang, G., Yang, S., Han, Y.: Mashroom: end-user mashup programming using nested tables. In: Proceedings of the 18th International Conference on World Wide Web, pp. 861–870. ACM (2009)
14. Han, Y., Wang, G., Ji, G., Zhang, P.: Situational data integration with data services and nested table. Serv. Oriented Comput. Appl. **7**(2), 129–150 (2013)
15. Genesereth, M.R., Keller, A.M., Duschka, O.M.: Infomaster: an information integration system. SIGMOD Rec. **26**(2), 539–542 (1997)
16. Levy, A.Y., Rajaraman, A., Ordille, J.J.: Querying heterogeneous information sources using source descriptions. In: Proceedings of the 22th International Conference on Very Large Data Bases. In: VLDB 1996, pp. 251–262. Morgan Kaufmann Publishers Inc., San Francisco (1996)
17. Han, Y., Liu, C., Su, S., Zhu, M., Zhang, Z., Zhang, S.: A proactive service model facilitating stream data fusion and correlation. Int. J. Web Serv. Res. (IJWSR) **14**(3), 1–16 (2017)
18. Gil, D., Ferrández, A., Mora-Mora, H., Peral, J.: Internet of things: a review of surveys based on context aware intelligent services. Sensors **16**(7), 1069 (2016)

Discovering Multiple Time Lags of Temporal Dependencies from Fluctuating Events

Wentao Wang$^{(\boxtimes)}$, Chunqiu Zeng, and Tao Li

School of Computing and Information Sciences, Florida International University,
Miami, FL 33199, USA
{wwang041,czeng001,taoli}@cs.fiu.edu

Abstract. As one of the key features of temporal dependency, time lag plays an important role in analyzing sequential data and predicting the developing trend. Huge number of temporal mining approaches have been successfully applied in many applications, like finance, environmental science and health-care. However, these approaches cannot effectively deal with a more realistic scenario, where more than one types of time lags are existed in sequences and all of them are fluctuating due to the inevitable noise. In this paper, we study the problem of discovering multiple time lags of temporal dependencies from event sequences considering the randomness property of the hidden time lags. We first present a parametric model as well as an EM-based solution for solving this problem. Then two approximate approaches are proposed for efficiently finding diverse types of time lags without significant loss of accuracy. Extensive empirical studies on both synthetic and real datasets demonstrate the efficiency and effectiveness of our proposed approaches.

Keywords: Time lag · Temporal dependency · Event mining

1 Introduction

In the past several decades, temporal data mining has been widely applied in many domains, such as finance [8], computer science [19], environmental science [2]. The goal of temporal data mining is to discover hidden temporal dependencies, unexpected trends or other subtle relationships in sequential data [15,27]. As an important task in temporal data mining, temporal dependency discovery has been extensively studied for identifying hidden interactions and mining useful information from sequential data. Specifically, suppose A and B are two types of items, a temporal dependency for A and B, written as $A \rightarrow B$, could be discovered when the occurrence of B depends on the occurrence of A.

Traditional temporal mining methods either utilize some statistical techniques [18] or employ a predefined window [6] to discover temporal dependencies. The main drawback of these previous methods is that they cannot discover interleaved dependencies, since all of these methods are based on an assumption that

© Springer International Publishing AG, part of Springer Nature 2018
Y. Cai et al. (Eds.): APWeb-WAIM 2018, LNCS 10988, pp. 121–137, 2018.
https://doi.org/10.1007/978-3-319-96893-3_10

every item A only has a dependency relation with its first following B. However, interleaved dependencies are very common in real application scenarios, where an item A could have a dependency relation with any following B. For example, as shown in Fig. 1, an event *High CPU Utilization Alert* can be triggered by an event *Abnormal Process* in system management domain. Since sometimes one abnormal process may be solved very quickly after it appeared, it would not trigger any *High CPU Utilization Alert*. Hence, this abnormal process (at time point 38) does not have the corresponding *High CPU Utilization Alert*. Two well-designed algorithms are presented in [28] for mining interleaved temporal dependencies from sequential data with satisfactory time cost and space cost.

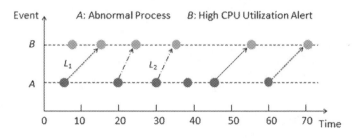

Fig. 1. Two types of time lags for temporal dependency *Abnormal Process* → *High CPU Utilization Alert*.

Time lag, one of the key features of temporal dependency, plays an essential role in interpreting the cause of discovered temporal dependencies and predicting the evolving trends for future data. Existing work related to time lag discovery suppose the time lag between two correlated events is constant and fluctuations can be ignored [9,28]. However, because of fluctuation, noise and missing data, there are more than one types of time lags are existed in event sequences in the real application scenarios, and each of them involves randomness property.

As summarized in [30], taking randomness of time lag into consideration in temporal dependencies discovery is a big challenge, since (1) the number of time lag candidates in large sequential datasets are tremendous, and (2) the hidden time lags may oscillate with noise formed in data collection process. The model proposed in [30] assumes the distribution of time lags follows a normal distribution $\mathcal{N}(\mu, \sigma^2)$. Nevertheless, in practice, we find that the interleaved time lags often follow multiple distributions in one temporal dependency rather than a single normal distribution. As shown in Fig. 1, there are two types of time lags, i.e., L_1 and L_2, between event *Abnormal Process* and *High CPU Utilization Alert*. The reason caused this situation is that event *Abnormal Process* represents many different kinds of abnormal processes, and each of them may have different effects on the system with respect to CPU utilization.

In order to overcome the limitations of existing approaches and deal with real application scenarios better, in this paper, we study the problem of mining multiple time lags with randomness property for temporal dependencies. The contribution of this paper is summarized as follows:

- Investigates the problem of discovering multiple types of fluctuating inter-leaved time lags, and proposes a parametric model to formulate the random-ness of time lags for temporal dependencies between pairwise events.
- Presents an EM-based solution for mining multiple types of time lags. More-over, for efficiently mining diverse time lags from large event sequences, two approximate algorithms are proposed with satisfactory performance.
- Conducts extensive experiments on different synthetic datasets and several real datasets. The experimental results demonstrate that all our proposed algorithms could find multiple types of time lags effectively.

The rest of the paper is organized as follows. Section 2 summarizes the exist-ing work for temporal data mining. We formulate the problem for discovering multiple time lags from fluctuating events in Sect. 3. A parametric model as well as an EM-based solution are presented in Sect. 4. Section 5 presents two approximate algorithms with better efficiency. Extensive experimental results are reported in Sect. 6. Finally, we conclude this paper in Sect. 7.

2   Related Work

Temporal dependency discovery approaches have been extensively applied in numerous real applications with various dataset types. Transaction data, com-monly known as market basket transactions [27], is a collection of transactions, in which each transaction contains a set of items. Transaction data arises in many business related applications, including marketing promotions, advertise-ments and recommendation systems. Discovering temporal dependencies from transaction data is equivalent to finding frequent itemsets satisfying some pre-defined thresholds. Many algorithms are proposed for efficiently mining frequent itemsets from transaction datasets, such as GSP [26], FreeSpan [10], and Pre-fixSpan [22].

Mining temporal dependencies from time series data has been recognized as one of the key tasks in time series analysis [29]. For time series data, each record is a series of measurements taken over time. A temporal dependency, often called causal relationship, among time series can be seen as a correlation on multiple time series, which states one time series is significantly helpful to predict the future trend of another time series [7,20]. In particular, if time series A causes time series B, then the prediction of future value of B can be improved by utilizing A and B together. In recent years, the problem of identifying causal relationship between various time series has attracted widespread attention, and two effective frameworks has became very popular in temporal dependency infer-ence, i.e., Dynamic Bayesian Network [13,25] and Granger Causality [3,4].

Event data, converted from textual logs which generated by modern comput-ing systems, has been widely used in system and network management related applications [11]. Differing from time series where the value of data item is con-tinuous, event data denotes the discrete data item values [16]. An event sequence is an ordered finite sequence, in which each element is a tuple consisting of one instance of some event and its corresponding timestamp. A lot of research on event mining are proposed for discovering relationships of events [21,23]. Our work also focuses on event data, where only the timestamps of items are available and no other information can be utilized to find temporal dependencies.

Unlike previous work which can only discover fixed time lags, in this paper, we proposed a parametric model to extract the probability distributions of time lags. Considering probability distributions could precisely depict the randomness property of time lags, our method provides more flexibility and usability than fixed ones.

3   Problem Definition and Formulation

3.1   Problem Definition

Let Ω be the event space comprises all possible events. An event sequence $\mathbf{S}$ over Ω is a finite ordered list with the form $\mathbf{S} = e_1 e_2 \ldots e_u$. Every element $e_i \in \mathbf{S}$ is a tuple $e_i = (E_i, t_i)$ indicating an instance of event $E_i \in \Omega$ occurred at time t_i.

Assume A be a type of event coming from event space Ω, $\mathbf{S_A}$ be a subsequence of $\mathbf{S}$ which only consists of instances of event A. Because all elements in $\mathbf{S_A}$ belong to the same type of event, we simplify $\mathbf{S_A}$ as a sequence of timestamp, i.e., $\mathbf{S_A} = a_1 a_2 \ldots a_m$, where a_i is the ith timestamp of event A's instances. Similarly, for another type of event B, we denote $\mathbf{S_B} = b_1 b_2 \ldots b_n$.

If there is a temporal dependency $A \rightarrow_L B$, for any associated timestamp pair a_i and b_j, there always exists a relation $b_j = a_i + L$ indicating an event A occurred at a_i is followed by an event B occurred at b_j after a time lag L.

Theoretically, the time lag L should be a constant. However, the noise is inevitable during data collection process because of various factors, such as missing records, incorrect values and recording delay. In order to discover underlying temporal dependencies effectively, in this paper, the time lag L is defined as

$$L = \mu + \epsilon \tag{1}$$

where μ is a constant representing the true time lag and ϵ is a random variable indicating the noise. Hence, time lag L is a random variable.

In our practice, we find that the time lag L often follows a more complicated distribution rather than a single normal distribution supposed in previous work [30]. Without loss of generality, we assume this complicated distribution is consisted of K different probability distributions, and we want to discover multiple time lags following various probability distributions from datasets. Definition 1 provides the formal description of the problem we studied in this paper.

Definition 1. *For two event types A and B, suppose there are K different temporal dependencies $A \rightarrow_{L_1} B, A \rightarrow_{L_2} B, \ldots, A \rightarrow_{L_K} B$ existed in a given dataset, our goal is to learn a time lag L, which is consisted of K types of time lags $L_1, L_2, \ldots, L_K$ following K different probability distributions, respectively.*

3.2   Problem Formulation

Given two timestamp sequences $\mathbf{S_A}$ and $\mathbf{S_B}$, suppose the distribution of time lag L is determined by the parameters Θ, which are independent from the occurrences of event A. Therefore, the problem of discovering the temporal dependency $A \rightarrow_L B$ is equivalent to learning the distribution of time lag L through the maximum likelihood parameters Θ defined by

$$\hat{\Theta} = \arg\max_{\Theta} P(\Theta|\mathbf{S_A}, \mathbf{S_B}). \tag{2}$$

Applying Bayes' theorem to Eq. (2), we have

$$\ln P(\Theta|\mathbf{S_A}, \mathbf{S_B}) = \ln P(\mathbf{S_B}|\mathbf{S_A}, \Theta) + \ln P(\mathbf{S_A}) + \ln P(\Theta) - \ln P(\mathbf{S_A}, \mathbf{S_B}). \tag{3}$$

In Eq. (3), only $\ln P(\mathbf{S_B}|\mathbf{S_A}, \Theta)$ and $\ln P(\Theta)$ are related to the parameters Θ. Therefore, the problem of learning time lag L can be simplified into the problem of solving the following equation

$$\hat{\Theta} = \arg\max_{\Theta} \ln P(\mathbf{S_B}|\mathbf{S_A}, \Theta). \tag{4}$$

4   Modeling and Solution

4.1   Time Lag Modeling

For a temporal dependency $A \rightarrow_L B$, we assume that every occurrence of event B is only determined by event A and time lag L, i.e., every occurrence of event B is mutually independent with each other. Therefore,

$$P(\mathbf{S_B}|\mathbf{S_A}, \Theta) = \prod_{j=1}^{n} P(b_j|\mathbf{S_A}, \Theta). \tag{5}$$

For every timestamp b_j, a latent variable z_{ijk} is introduced to model the relation between b_j and one timestamp of event A, denoted as a_i. Specifically,

$$z_{ijk} = \begin{cases} 1, & \text{the } i^{th} \text{ event } A \text{ implies the } j^{th} \text{ event } B \text{ following } k^{th} \text{ distribution;} \\ 0, & \text{otherwise.} \end{cases} \tag{6}$$

Hence, the relation between b_j and sequence $\mathbf{S_A}$ can be represented by a latent matrix $\mathbf{Z_j} = \{z_{ijk}\}_{m \times K}$, in which only one element equals to 1 and all other elements are 0. If $z_{ijk} = 1$, then cell (i, k) in $\mathbf{Z_j}$ equals 1.

Based on the definition of latent variable z_{ijk}, the distribution of latent matrix $\mathbf{Z_j}$ and the conditional distribution of b_j given $\mathbf{Z_j}$ are shown as:

$$P(\mathbf{Z_j}) = \prod_{i=1}^{m} \prod_{k=1}^{K} P(z_{ijk} = 1)^{z_{ijk}}. \tag{7}$$

$$P(b_j|\mathbf{Z_j},\mathbf{S_A},\mathbf{\Theta}) = \prod_{i=1}^{m}\prod_{k=1}^{K} P(b_j|z_{ijk}=1,a_i,\mathbf{\Theta})^{z_{ijk}}. \tag{8}$$

Therefore, the joint distribution $P(b_j, \mathbf{Z_j}|\mathbf{S_A}, \mathbf{\Theta})$ can be written as follow:

$$P(b_j, \mathbf{Z_j}|\mathbf{S_A},\mathbf{\Theta}) = \prod_{i=1}^{m}\prod_{k=1}^{K} \{P(b_j|z_{ijk}=1,a_i,\mathbf{\Theta}) \times P(z_{ijk}=1)\}^{z_{ijk}}. \tag{9}$$

So the marginal distribution of b_j is obtained by

$$P(b_j|\mathbf{S_A},\mathbf{\Theta}) = \sum_{\mathbf{Z_j}} P(b_j, \mathbf{Z_j}|\mathbf{S_A},\mathbf{\Theta}) = \sum_{i=1}^{m}\sum_{k=1}^{K} P(b_j|z_{ijk}=1,a_i,\mathbf{\Theta}) \times P(z_{ijk}=1). \tag{10}$$

Combining Eqs. (4), (5) and (10) together, the log-likelihood function can be rewritten as:

$$\ln P(\mathbf{S_B}|\mathbf{S_A},\mathbf{\Theta}) = \sum_{j=1}^{n}\ln\sum_{i=1}^{m}\sum_{k=1}^{K} P(b_j|z_{ijk}=1,a_i,\mathbf{\Theta}) \times P(z_{ijk}=1). \tag{11}$$

For simplicity, let $\pi_{ijk} = P(z_{ijk}=1)$, where $0 \le \pi_{ijk} \le 1$, $\sum_{i=1}^{m}\sum_{k=1}^{K}\pi_{ijk}=1$.

Time lag L is consisted of K different time lags $L_1, L_2, \ldots, L_K$ with K different probability distributions. For any one temporal dependency $A \to_{L_t} B$ $(1 \le t \le K)$, we have $L_t = \mu_t + \epsilon_t$. Based on the Central Limit Theorem, we assume that noise ϵ follows the normal distribution with zero-mean value, i.e., $\epsilon_t \sim \mathcal{N}(0, \sigma_t^2)$, where σ_t^2 represents the variance of current distribution. Since μ_t is a constant, the distribution of L_t can be expressed as $L_t \sim \mathcal{N}(\mu_t, \sigma_t^2)$. Therefore, time lag L can be regard as a mixture of K different normal distributions with various μ and σ^2. Hence, if $z_{ijk} = 1$, then

$$P(b_j|z_{ijk}=1, a_i, \mathbf{\Theta}) = P(b_j|a_i, \mu_k, \sigma_k^2) = \mathcal{N}(b_j - a_i|\mu_k, \sigma_k^2). \tag{12}$$

Consequently, Eq. (11) can be expressed as follow:

$$\ln P(\mathbf{S_B}|\mathbf{S_A},\mathbf{\Theta}) = \sum_{j=1}^{n}\ln\sum_{i=1}^{m}\sum_{k=1}^{K} \pi_{ijk} \times \mathcal{N}(b_j - a_i|\mu_k, \sigma_k^2). \tag{13}$$

Based on Eq. (13), the problem described in Eq. (4) is equivalent to the following equation

$$(\hat{\mu}_k, \hat{\sigma}_k^2) = \mathop{\arg\max}_{\substack{\mu_k, \sigma_k^2 \\ k\in\{1,\ldots,K\}}} \sum_{j=1}^{n}\ln\sum_{i=1}^{m}\sum_{k=1}^{K} \pi_{ijk} \times \mathcal{N}(b_j - a_i|\mu_k, \sigma_k^2)$$
$$\textbf{s.t. } \textit{for every } j \in [1,n], \ \sum_{i=1}^{m}\sum_{k=1}^{K}\pi_{ijk} = 1. \tag{14}$$

4.2   Maximization

Equation (14) can be solved by expectation-maximization (EM) algorithm [5], since it is one kind of mixture model. For applying EM algorithm, consider the expected log likelihood function of complete data $\{\mathbf{S_B}, \boldsymbol{\Theta}\}$ at first. Suppose parameters $\boldsymbol{\Theta}$ is already known and $\mathbf{Z} = \{z_{ijk}\}_{m \times n \times K}$ is a latent matrix. Then, based on Eqs. (7) and (8), we have

$$P(\mathbf{S_B}, \mathbf{Z}|\mathbf{S_A}, \boldsymbol{\Theta}) = \prod_{j=1}^{n} P(b_j|\mathbf{Z_j}, \mathbf{S_A}, \boldsymbol{\Theta}) \times P(\mathbf{Z_j}) = \prod_{i=1}^{m} \prod_{j=1}^{n} \prod_{k=1}^{K} \left\{ \mathcal{N}(b_j - a_i|\mu_k, \sigma_k^2) \times \pi_{ijk} \right\}^{z_{ijk}}.$$

$$(15)$$

Therefore, the expectation can be expressed as

$$\mathcal{Q}(\boldsymbol{\Theta}, \boldsymbol{\Theta}') \triangleq \mathbb{E}[\ln P(\mathbf{S_B}, \mathbf{Z}|\mathbf{S_A}, \boldsymbol{\Theta})] = \sum_{i=1}^{m} \sum_{j=1}^{n} \sum_{k=1}^{K} \mathbb{E}[z_{ijk}] \times \left\{ \ln \mathcal{N}(b_j - a_i|\mu_k, \sigma_k^2) + \ln \pi_{ijk} \right\}.$$

$$(16)$$

where $\boldsymbol{\Theta}'$ is the parameters estimated on the previous iteration. Using r_{ijk} to denote $\mathbb{E}[z_{ijk}]$, i.e.,

$$r_{ijk} \triangleq \mathbb{E}[z_{ijk}] = P(z_{ijk} = 1|\mathbf{S_A}, \mathbf{S_B}, \boldsymbol{\Theta}') = \frac{\pi'_{ijk} \times \mathcal{N}(b_j - a_i|\mu'_k, \sigma_k'^2)}{\sum_{i=1}^{m} \sum_{k=1}^{K} \pi'_{ijk} \times \mathcal{N}(b_j - a_i|\mu'_k, \sigma_k'^2)}.$$

$$(17)$$

Then Eq. (16) can be rewritten as

$$\mathcal{Q}(\boldsymbol{\Theta}, \boldsymbol{\Theta}') = \sum_{i=1}^{m} \sum_{j=1}^{n} \sum_{k=1}^{K} r_{ijk} \times \left\{ \ln \mathcal{N}(b_j - a_i|\mu_k, \sigma_k^2) + \ln \pi_{ijk} \right\}. \qquad (18)$$

The parameters μ_k, σ_k^2 and π_{ijk} can be learned by maximizing $\mathcal{Q}(\boldsymbol{\Theta}, \boldsymbol{\Theta}')$.

$$\mu_k = \frac{1}{N_k} \sum_{i=1}^{m} \sum_{j=1}^{n} r_{ijk}(b_j - a_i) \qquad (19)$$

$$\sigma_k^2 = \frac{1}{N_k} \sum_{i=1}^{m} \sum_{j=1}^{n} r_{ijk}(b_j - a_i - \mu_k)^2 \qquad (20)$$

$$\pi_{ijk} = \frac{1}{n} \sum_{j=1}^{n} r_{ijk} \qquad (21)$$

where $N_k = \sum_{i=1}^{m} \sum_{j=1}^{n} r_{ijk}$.

Using this EM-based algorithm, called *EMLag* algorithm, we can find the maximum likelihood estimates of parameters $\boldsymbol{\Theta}$. Algorithm 1 states the pseudo-code of *EMLag* algorithm with the time complexity $\mathcal{O}(rmnK)$, where m and n are the number of timestamps of event A and B, respectively, K is the number of distributions and r indicates iteration number.

Algorithm 1. The EMLag algorithm

1: **procedure** EMLAG($\mathbf{S_A}$, $\mathbf{S_B}$)   ▷ $|\mathbf{S_A}|=m$, $|\mathbf{S_B}|=n$
2:     Initialize $r'_{ijk} = \frac{1}{mK}$, choose μ'_k and σ'^2_k randomly.   ▷ Initialization
3:     **while** TRUE **do**
4:         Evaluate r_{ijk} by Eq. (17).   ▷ Expectation
5:         Update μ_k and σ^2_k by Eqs. (19) and (20), respectively.   ▷ Maximization
6:         **if** parameters converge **then**   ▷ Convergence test
7:             **return** μ_k and σ^2_k ▷ $k = 1, 2, \ldots, K$
8:         **end if**
9:     **end while**
10: **end procedure**

---

5   Time Lag Discovery

Based on *EMLag* algorithm, we design two approximate algorithms for mining multiple time lags from large datasets more efficiently. Both of these two algorithms could achieve good efficiency without significant loss of accuracy.

5.1   winEMLag Algorithm

Intuitively, suppose a temporal dependency $A \rightarrow B$ is exist, timestamp b_j is more likely to be implied by timestamp a_i if the index i is close to the index j rather than far from j. Hence, for mining various temporal dependencies from large event sequences efficiently, for every b_j, we only select a subset of event A's timestamps whose index is close to j for calculation instead of all of them.

Algorithm 2. The winEMLag algorithm

1: **procedure** EXPECTATION($\mathbf{S_A}$, $\mathbf{S_B}$, λ)   ▷ λ is predefined, $0 < \lambda \leq 1$
2:     $l = \lambda \times m$ ▷ l is the window length, $|\mathbf{S_A}| = m$
3:     $left = 0, right = 0$ ▷ Index bound
4:     **for** each b_j **do**
5:         **if** $j - l/2 \geq 0$ **and** $j + l/2 \leq m - 2$ **then**
6:             $left = j - l/2$, $right = left + l - 1$
7:         **else**
8:             **if** $j - l/2 \geq 0$ **then**
9:                 $right = m - 1, left = m - l$
10:            **else**
11:                $left = 0$, $right = l - 1$
12:            **end if**
13:        **end if**
14:        Select a_i into w_j where $i \in [left, right]$.
15:        Evaluate r_{ijk} utilizing set w_j.
16:    **end for**
17: **end procedure**

Let w_j be a subset of A's timestamps used to estimate the relation between event A and b_j. Our goal is to fill in w_j so that, compared with remainders, the indexes of A's timestamps in w_j are much closer to j. Inspired by Sliding Window Model [1], we design an approximate algorithm $winEMLag$ for speeding up the mining process of $EMLag$ algorithm. Algorithm 2 shows the Expectation procedure of $winEMLag$ algorithm. In each iteration, the update operations in Maximization procedure of $winEMLag$ algorithm will also utilize each subset w_j.

In $winEMLag$ algorithm, parameter λ is a user-specified parameter representing the ratio between the length of window l and the length of event sequence. Therefore, the length of window l can be calculated by $l = \lambda \times |\mathbf{S_A}|$. Since the size of each w_j in $winEMLag$ algorithm is much smaller than the size of sequence $\mathbf{S_A}$ in $EMLag$ algorithm, $winEMLag$ algorithm could achieve better efficiency.

5.2   appEMLag Algorithm

During each iteration of $EMLag$ algorithm, we find that, for every specific distribution k, the responsibility r_{ijk} describing the likelihood that the ith event A implies the jth event B following k^{th} normal distribution, becomes smaller with the deviation of $b_j - a_i$ from the estimated time lag μ_k increasing. In other words, r_{ijk} will close to 0 as $|b_j - a_i - \mu_k|$ becomes larger. Based on this observation, we design an approximate algorithm for efficiently estimating parameters μ_k and σ_k^2 by ignoring those $r_{ijk}(b_j - a_i)$ and $r_{ijk}(b_j - a_i - \mu_k)^2$ with small r_{ijk} in both Eqs. (19) and (20). Since in real application scenarios, the time spans of given event sequences are very long, and most r_{ijk} are very small, the loss of accuracy of this approximate method is acceptable.

We introduce two parameters ϵ and δ to help distinguishing retained part and neglected part of r_{ijk}. Given b_j, let ϵ_j be the sum of the responsibility r_{ijk} which will be neglected, i.e., $\epsilon_j = \sum_{k=1}^{K} \sum_{\{i|a_i \ is \ neglected\}} r_{ijk}$, and ϵ be the largest one among all the ϵ_j, that is, $\epsilon = \max_{1 \le j \le n} \{\epsilon_j\}$. In practice, parameter ϵ can be predefined by users based on the application scenario.

Recall that in Eqs. (19) and (20), each pair of μ_k and σ_k are calculated by their corresponding r_{ijk}. Therefore, for every b_j, we suppose set C_{jk} includes all retained r_{ijk} which will be used to estimate μ_k and σ_k. Since all timestamps of event A are consecutive in ascending order, the index i for timestamps of event A in set C_{jk} are also consecutive. Let δ_{jk} be the ratio of the sum of retained r_{ijk} in set C_{jk} to the sum of all retained r_{ijk} for the given b_j, and hence δ is a $n \times K$ matrix filled in by all δ_{jk}. In each iteration, given b_j, δ_{jk} can be updated by the following equations:

$$AVG_{jk} \triangleq \frac{\sum_{\{r_{ijk} \in C_{jk}\}} r_{ijk}}{|C_{jk}|}, \quad \delta_{jk} = \frac{AVG_{jk}}{\sum_{k=1}^{K} AVG_{jk}} \qquad (22)$$

To guarantee the sum of neglected r_{ijk} is less than ϵ, for every b_j, the sum of retained r_{ijk} should be greater than $1 - \epsilon$, i.e., $\sum_{k=1}^{K} \delta_{jk} \ge 1 - \epsilon$. In order to minimize the size of C_{jk}, we adopt a greedy way to select timestamps a_i from all timestamps of event A. Specifically, given b_j and distribution k, we add a_i

into C_{jk} with its corresponding r_{ijk} in decreasing order until the summation of r_{ijk} in C_{jk} is greater than δ_{jk}. Algorithm 3 describes how to find the minimum and maximum indexes of a_i in C_{jk}.

Algorithm 3. The greedyBound algorithm

1: **procedure** GREEDYBOUND($\mathbf{S_A}$, b_j, μ'_k, δ'_{jk}, ϵ)
2:　　$t = b_j - \mu'_k$
3:　　Locate the closest a_i to t using binary search.
4:　　$min_{jk} = i$, $max_{jk} = i$
5:　　$prob = 0.0$
6:　　**while** $prob < \delta'_{jk} \times (1 - \epsilon)$ **do**
7:　　　　**if** $r_{(min_j-1)jk} \geq r_{(max_j+1)jk}$ **then**
8:　　　　　　$i = min_{jk} - 1$
9:　　　　　　$min_{jk} = i$
10:　　　　**else**
11:　　　　　　$i = max_{jk} + 1$
12:　　　　　　$max_{jk} = i$
13:　　　　**end if**
14:　　　　Add a_i to C_{jk}.
15:　　　　$prob = prob + r_{ijk}$
16:　　**end while**
17:　　**return** min_{jk} and max_{jk}
18: **end procedure**

Based on *greedyBound* algorithm, we present an approximate algorithm, called *appEMLag* algorithm, to efficiently estimate parameters μ_k and σ_k^2 without significant loss of accuracy. As described in Algorithm 4, the time cost of

---

Algorithm 4. The appEMLag algorithm

1: **procedure** APPEMLAG($\mathbf{S_A}$, $\mathbf{S_B}$, ϵ)　　　　　　　▷ ϵ is predefined, $0 < \epsilon \leq 1$
2:　　Initialize $\delta'_{jk} = \frac{1}{K}$, choose μ'_k and σ'^2_k randomly.　　　▷ Initialization
3:　　**while** TRUE **do**
4:　　　　**for** each b_j **do**
5:　　　　　　**for** $k \leftarrow 1, K$ **do**
6:　　　　　　　　Get min_{jk}, max_{jk} by *greedyBound*.　　▷ Find the index bound of a_i
7:　　　　　　**end for**
8:　　　　　　Evaluate r_{ijk} utilizing sets $C_{j1}, \ldots, C_{jK}$.　　　　▷ Expectation
9:　　　　**end for**
10:　　　　Update μ_k and σ_k^2 by Eqs. (19) and (20) within the　　▷ Maximization
　　　　　　bound, respectively, and update δ_{jk} by Eq. (22).
11:　　　　**if** parameters converge **then**　　　　　　　　▷ Convergence test
12:　　　　　　**return** μ_k and σ_k^2　　　　　　　　　　▷ $k = 1, 2, \ldots, K$
13:　　　　**end if**
14:　　**end while**
15: **end procedure**

appEMLag algorithm is $\mathcal{O}(rnK(\log m + t))$, where m and n are the number of timestamps of event A and B, respectively, r indicates iteration number, K is the number of distributions, and t is the average size of all C_{jk}. Since $t \ll m$ and $\log m \ll m$, *appEMLag* algorithm is much faster than *EMLag* algorithm.

6 Empirical Study

This section presents empirical studies of our proposed algorithms on both synthetic datasets and real datasets with respect to effectiveness and efficiency. To demonstrate the performance of our proposed algorithms, we implement all of them using Java 1.7, and execute them on a computer with Linux 2.6.32. This computer is equipped with Intel(R) Xeon(R) CPU with 24 cores running at speed of 2.50 GHz, and the total memory of it is 126G.

6.1 Synthetic Data

Synthetic Data Generation. In our experiments, we execute our proposed algorithms on five different synthetic datasets. Parameters shown in Table 1 are utilized to generate synthetic datasets. Each dataset consists of two event sequences $\mathbf{S_A}$ and $\mathbf{S_B}$ with same length, and we assume there are two types of temporal dependency exist in each dataset. That is to say, $K = 2$. Moreover, we use the exponential distribution to simulate the inter-arrival time between two adjacent events [17]. The way of generating $\mathbf{S_A}$ and $\mathbf{S_B}$ is shown below.

Table 1. Parameters used for generating synthetic data

Name	Description
N	The number of events in one synthetic event sequence
K	The number of types of time lag
β_{min}	The minimum value for the average inter-arrival time β
β_{max}	The maximum value for the average inter-arrival time β
μ_{min}	The minimum value for the true time lag μ
μ_{max}	The maximum value for the true time lag μ
σ^2_{min}	The minimum value for the variance of time lag
σ^2_{max}	The maximum value for the variance of time lag

1. Randomly choose parameters β from $[\beta_{min}, \beta_{max}]$, μ_1 and μ_2 from $[\mu_{min}, \mu_{max}]$ and σ^2_1 and σ^2_2 from $[\sigma^2_{min}, \sigma^2_{max}]$, respectively.
2. Generate $N/2$ timestamps for event A, where the inter-arrival time between two neighbors follows the exponential distribution with parameter β.
3. For each timestamp a_i of event A, the time lag is randomly generated according to normal distribution with parameters μ_1 and σ^2_1.

4. Combine all the timestamps associated with their types to form two event sequence $\mathbf{S_{A_1}}$ and $\mathbf{S_{B_1}}$.
5. Repeat Steps 2–4 to generate another two event sequences $\mathbf{S_{A_2}}$ and $\mathbf{S_{B_2}}$ with parameters μ_2 and σ_2^2 chosen in Step 1.
6. Merge $\mathbf{S_{A_1}}$ and $\mathbf{S_{A_2}}$ to form the sequence $\mathbf{S_A}$ with timestamps in ascending order, then merge $\mathbf{S_{B_1}}$ and $\mathbf{S_{B_2}}$ to form the sequence $\mathbf{S_B}$ based on the indexes of their corresponding a_i.

In our experiments, we set $\beta_{min} = 5$, $\beta_{max} = 50$, $\mu_{min} = 25$, $\mu_{max} = 100$, $\sigma_{min}^2 = 5$ and $\sigma_{max}^2 = 400$ to generate five synthetic datasets with different parameter N. The number of events in sequence $\mathbf{S_A}$ in these five synthetic datasets are $0.5k$, $1k$, $2k$, $10k$, $20k$, respectively. Note that there are only two types of events we simulated in synthetic datasets. In practice, a real dataset typically includes more than hundreds of events types. Therefore, we believe $20k$ events of two types is enough to represent the real application scenarios in miniature.

Synthetic Data Evaluation. Since all of our proposed algorithms are based on the EM algorithm, which cannot guarantee the global optimum [5], we define a batch operation to avoid this problem as much as possible. Specifically, every 10 rounds execution of the algorithm with different initial parameters chosen at random is regarded as a batch. For each batch, we choose the output with the maximum likelihood among 10 rounds as the result of a batch. For each synthetic dataset, we conduct five such batches on it, and calculated the average values of the results of five batches as the final result. Table 2 shows the outcome of experiments running *EMLag*, *winEMLag*, and *appEMLag* on such five synthetic datasets with different parameters settings, respectively.

Each algorithm terminates execution when it satisfies one of the following conditions: (1) it converges; (2) the number of iterations exceeds 500; or (3) the differences of all learned parameters between two adjacent iterations are less than 10^{-5}. *winEMLag* algorithm takes one more parameter λ as its input, where λ determines the length of windows used in the algorithm. In our experiments, we set λ to 0.002, 0.02 and 0.2. Similarly, *appEMLag* algorithm has a predefined parameter ϵ, which is used to calculate the proportion of the neglected part during the parameter estimation of each iteration. In order to evaluate the performance of *appEMLag* algorithm sufficiently, ϵ is set to 0.001, 0.01, and 0.1. As shown in Table 2, we find that parameters μs learned by *EMLag*, *winEMLag*, and *appEMLag* are quite close to the ground truth.

In addition, for evaluating the difference between the distributions of time lags given by the ground truth and learned by our proposed algorithms shown in Table 2, we introduce Kullback-Leibler (KL) divergence [14]. Figure 2 shows the evaluation results with various sizes of dataset from 0.5k to 10k, respectively. Here we can see, all proposed algorithms could effectively discovery the distributions of time lags from event sequences. Moreover, compared with *winEMLag*, *appEMLag* algorithm performs better in terms of KL divergence.

Table 2. The experimental results for various synthetic datasets with sizes from $0.5k$ to $20k$. The values of μ and σ^2 for "ground truth" are given in advance; $\overline{\mu}$ and $\overline{\sigma}^2$ represent the average values of μ and σ^2; LL_{opt} is the maximum log-likelihood obtained by running the algorithm; Entries with "N/A" are not available since they take more than 7 days to get corresponding results.

Dataset	Ground Truth			EMLag			winEMLag $\lambda = 0.002$			winEMLag $\lambda = 0.02$		
	(μ_1, σ_1^2)	(μ_2, σ_2^2)	LL_{opt}	$(\overline{\mu}_1, \overline{\sigma}_1^2)$	$(\overline{\mu}_2, \overline{\sigma}_2^2)$	LL_{opt}	$(\overline{\mu}_1, \overline{\sigma}_1^2)$	$(\overline{\mu}_2, \overline{\sigma}_2^2)$	LL_{opt}	$(\overline{\mu}_1, \overline{\sigma}_1^2)$	$(\overline{\mu}_2, \overline{\sigma}_2^2)$	LL_{opt}
$N = 0.5k$	(28.41, 17.97)	(59.16, 18.94)	-	(28.14, 10.02)	(58.49, 14.65)	-4222.01	(28.52, 19.06)	(59.04, 18.48)	-1442.53	(28.11, 10.04)	(58.50, 14.59)	-2267.04
$N = 1k$	(35.67, 24.89)	(65.93, 33.12)	-	(35.72, 10.55)	(66.32, 25.03)	-9188.64	(35.16, 22.31)	(64.60, 24.45)	-3402.28	(35.71, 10.62)	(66.24, 24.91)	-5275.52
$N = 2k$	(35.68, 14.34)	(73.70, 28.30)	-	(35.59, 6.97)	(72.68, 17.75)	-19353.02	(35.82, 11.61)	(72.47, 21.53)	-7367.57	(35.58, 6.94)	(72.59, 18.06)	-11526.52
$N = 10k$	(46.83, 34.41)	(69.96, 29.62)	-	N/A	N/A	N/A	(47.20, 24.30)	(70.13, 12.40)	-53044.11	(47.03, 22.17)	(69.99, 13.64)	-76051.98
$N = 20k$	(55.46, 32.51)	(75.66, 29.42)	-	N/A	N/A	N/A	(55.62, 14.58)	(75.19, 15.79)	-117563.57	(55.39, 13.01)	(75.15, 16.91)	-163582.36

Dataset	winEMLag $\lambda = 0.2$			appEMLag $\epsilon = 0.001$			appEMLag $\epsilon = 0.01$			appEMLag $\epsilon = 0.1$		
	$(\overline{\mu}_1, \overline{\sigma}_1^2)$	$(\overline{\mu}_2, \overline{\sigma}_2^2)$	LL_{opt}	$(\overline{\mu}_1, \overline{\sigma}_1^2)$	$(\overline{\mu}_2, \overline{\sigma}_2^2)$	LL_{opt}	$(\overline{\mu}_1, \overline{\sigma}_1^2)$	$(\overline{\mu}_2, \overline{\sigma}_2^2)$	LL_{opt}	$(\overline{\mu}_1, \overline{\sigma}_1^2)$	$(\overline{\mu}_2, \overline{\sigma}_2^2)$	LL_{opt}
$N = 0.5k$	(28.12, 10.03)	(58.44, 14.64)	-3408.06	(28.98, 11.51)	(59.31, 13.84)	-1581.37	(28.85, 11.56)	(59.27, 13.56)	-1573.18	(28.83, 11.57)	(59.03, 13.14)	-1561.07
$N = 1k$	(35.70, 10.59)	(66.25, 24.92)	-7561.03	(36.15, 13.51)	(67.03, 21.41)	-3318.73	(36.06, 13.21)	(66.93, 21.53)	-3310.18	(36.05, 12.99)	(66.82, 21.25)	-3290.54
$N = 2k$	(35.55, 6.92)	(72.45, 18.77)	-16102.33	(36.11, 9.52)	(74.27, 14.52)	-6229.23	(36.04, 9.59)	(74.25, 14.45)	-6211.82	(35.92, 9.45)	(74.01, 14.02)	-6183.27
$N = 10k$	(47.08, 23.21)	(70.09, 12.87)	-98914.56	(47.56, 19.82)	(71.03, 17.51)	-33681.02	(47.46, 19.42)	(70.87, 17.78)	-33585.87	(47.42, 19.43)	(70.69, 17.38)	-33416.45
$N = 20k$	N/A	N/A	N/A	(56.97, 17.23)	(76.55, 14.40)	-65431.36	(56.59, 16.22)	(76.35, 14.84)	-65266.24	(56.67, 16.43)	(76.35, 14.57)	-64848.79

(a) Size = 0.5k   (b) Size = 1k   (c) Size = 2k   (d) Size = 10k

Fig. 2. The KL distance between the ground truth and the one learned by each algorithm over different datasets.

The efficiency comparison between all proposed algorithms is measured by the CPU running time. As shown in Fig. 3, the time cost of *winEMLag* and *appEMLag* are much less than the *EMLag* algorithm. Since both parameter ϵ and λ could effectively decrease the number of events needed to be considered in each iteration, the efficiency of these two algorithms are satisfactory.

In summary, based on the extensively comparative experiments on synthetic data, all of our proposed algorithms have the capabilities for finding time lags from fluctuating events effectively. Two approximate algorithms *winEMLag* and *appEMLag* could achieve a good balance in terms of accuracy and efficiency.

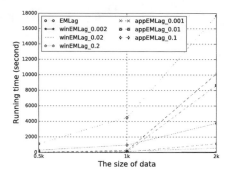

Fig. 3. Time cost comparison

6.2 Real Data

We employ two real datasets collected from several IT outsourcing centers by IBM Tivoli monitoring system [12] to verify the performance of our proposed algorithms in real application scenarios. Each dataset is a collection of system events generated by the automatic monitoring system running on servers. Most of these events are system alerts triggered by some monitoring situations, for example, the disk capacity is almost full. Table 3 lists the statistical information of these two real event datasets.

Table 3. Real event datasets

Dataset	Time span	# of events	# of types
Dataset1	32 days	100k	104
Dataset2	54 days	1,000k	136

In order to discover the time lags of temporal dependencies more effectively and efficiently, we choose *appEMLag* algorithm with parameter setting $\epsilon = 0.001$ to deal with these two real event datasets. For increasing the probability of acquiring the global optimal value, we run *appEMLag* in a batch of 30 rounds with randomly initialize parameters every round. Table 4 provides a snippet of some discovered temporal dependencies with multiple time lags from two real datasets. We employ the metric signal-to-noise ratio [24], a concept in signal processing domain, to evaluate the impact of noise relative to the expected time lag. Signal-to-noise ratio (SNR) can be calculated as the ratio of the expect time lag μ to the standard deviation σ. Here, we use the average value of SNR for two discovered distributions as the measure.

The time lags of temporal dependency *AIX_HW_Error* $\rightarrow_L$ *NV390MSG_ MVS* discovered from dataset1 follow two types of normal distribution, one is $\mu = 55.41$ and $\sigma^2 = 0.39$, and the other one is $\mu = 93.09$ and $\sigma^2 = 0.32$. Compared with algorithms proposed in [30], which can only find one time lag

Table 4. Snippet of discovered temporal dependencies with multiple time lags

Dataset	Temporal dependency	(μ_1, σ_1^2)	(μ_2, σ_2^2)	Ave. SNR
Dataset1	$AIX\_HW\_Error \rightarrow_L NV390MSG\_MVS$	(55.41, 0.39)	(93.09, 0.32)	126.64
	$generic\_postemsg \rightarrow_L NV390MSG\_AO\_Platform\_Server$	(63.55, 9.34)	(18.23, 16.37)	12.65
	$generic\_postemsg \rightarrow_L Sentry2\_0\_diskusedpct$	(219.38, 10.18)	(146.27, 10.41)	57.05
	$NV390MSG\_AO\_Platform\_Server \rightarrow_L$ $Info\_set\_ticket\_number\_using\_eventid$	(2.62, 4.49)	(18.09, 2981.51)	0.78
	$MQ\_CONN\_NOT\_AUTHORIZED \rightarrow_L ITM\_NT\_Services$	(1912.78, 47.63)	(88.16, 34.00)	146.14
	$Ticket\_Retry \rightarrow_L TEC\_Notice$	(360.59, 41.20)	(269.17, 44.76)	48.21
	$TEC\_Error \rightarrow_L ITM\_KGB\_AVAILABILITY$	(468.66, 76.77)	(434.34, 36.88)	62.50
Dataset2	$Generic\_Source\_Event \rightarrow_L Candle\_Universal\_Messages$	(383.49, 104.02)	(43.17, 24.80)	23.13
	$ITM\_Process \rightarrow_L PATROL\_APP$	(461.84, 0.93)	(168.13, 2433.03)	241.16
	$ITM\_Process \rightarrow_L OV\_IF\_Down$	(36.66, 256.39)	(244.63, 261.73)	8.71

with the expected time lag $\mu = 33.89$ and the variance $\sigma^2 = 1.95$, our method discover one more type of time lag. Moreover, since the variances of these two discovered time lags are quit small, these two time lags are very close to the true time lags.

The temporal dependency $generic\_postemsg \rightarrow_L Sentry2\_0\_diskusedpct$ has two time lags with different expected time lags μ and similar variances σ^2. Event $Sentry2\_0\_diskusedpct$ appears 2.5 or 3.5 min later after $generic\_postemsg$ occurs. Conversely, the expected time lags between $ITM\_KGB\_AVAILABILITY$ and TEC_Error are similar, while the variances are different. Because the expected time lags are very similar with each other, it is not trivial to capture two normal distributions from large datasets. Previous temporal dependency mining methods only return one constant time lag as the result, due to they ignore the existence of the noise and are not able to distinguish two very similar time lags.

The variances of the time lags between $Info\_set\_ticket\_number\_using\_eventid$ and $NV390MSG\_AO\_Platform\_Server$ in dataset1 are quite large relative to the expected time lags, since the average SNR is less than 1. For every single time lag, the variance is still relative large. Hence, we think this is a week dependency between these two events due to the discovered time lags contain too much noise.

We find the time lags of temporal dependency $ITM\_Process \rightarrow_L PATROL\_APP$ in dataset2 are quite different with other dependencies. Specifically, one time lag has large value of the expect time lag μ and small value of the variance σ^2, and the other one has small value of μ and large value of σ^2. Both of them are very difficult for previous inter-arrival pattern mining methods to discover, where the inter-arrival time lags are small time lags. In our methods, we use the expected time lag μ and its variance σ^2 to find multiple interleaved time lags.

7   Conclusions

In this paper, we study the problem of discovering multiple time lags of temporal dependencies over event sequences, where the time lags between two pairwise events are fluctuating since the existence of the noise. To solve this problem, an EM-based algorithm is proposed to capture the distribution of time lags. We also propose two approximate algorithms for speeding up the time lag mining process. Extensive empirical studies on both synthetic and real datasets demonstrate the efficiency and effectiveness of our proposed algorithms.

In future work, we plan to implement distributed versions of our algorithms for handling applications with massive data. Furthermore, mining dependencies among multiple events other than pairwise events is also attractive to us.

References

1. Aggarwal, C.C.: Data Streams: Models and Algorithms, vol. 31. Springer, Heidelberg (2007). https://doi.org/10.1007/978-0-387-47534-9
2. Ailamaki, A., Faloutos, C., Fischbeck, P.S., Small, M.J., VanBriesen, J.: An environmental sensor network to determine drinking water quality and security. ACM SIGMOD Rec. **32**(4), 47–52 (2003)
3. Arnold, A., Liu, Y., Abe, N.: Temporal causal modeling with graphical granger methods. In: Proceedings of the 13th ACM SIGKDD International Conference on Knowledge Discovery and Data Mining, pp. 66–75. ACM (2007)
4. Bahadori, M.T., Liu, Y.: An examination of practical granger causality inference. In: Proceedings of the 2013 SIAM International Conference on Data Mining, pp. 467–475. SIAM (2013)
5. Bishop, C.M.: Pattern Recognition and Machine Learning. Springer, Heidelberg (2006)
6. Bouandas, K., Osmani, A.: Mining association rules in temporal sequences. In: IEEE Symposium on Computational Intelligence and Data Mining CIDM 2007, pp. 610–615. IEEE (2007)
7. Dhurandhar, A.: Learning maximum lag for grouped graphical granger models. In: 2010 IEEE International Conference on Data Mining Workshops (ICDMW), pp. 217–224. IEEE (2010)
8. Du, X., Jin, R., Ding, L., Lee, V.E., Thornton Jr, J.H.: Migration motif: a spatial-temporal pattern mining approach for financial markets. In: Proceedings of the 15th ACM SIGKDD International Conference on Knowledge Discovery and Data Mining, pp. 1135–1144. ACM (2009)
9. Golab, L., Karloff, H., Korn, F., Saha, A., Srivastava, D.: Sequential dependencies. Proc. VLDB Endow. **2**(1), 574–585 (2009)
10. Han, J., Pei, J., Mortazavi-Asl, B., Chen, Q., Dayal, U., Hsu, M.C.: FreeSpan: frequent pattern-projected sequential pattern mining. In: Proceedings of the Sixth ACM SIGKDD International Conference on Knowledge Discovery and Data Mining, pp. 355–359. ACM (2000)
11. Hellerstein, J.L., Ma, S., Perng, C.S.: Discovering actionable patterns in event data. IBM Syst. J. **41**(3), 475–493 (2002)
12. IBM Tivoli Monitoring. http://www-01.ibm.com/software/tivoli/
13. Jansen, R., Yu, H., Greenbaum, D., Kluger, Y., Krogan, N.J., Chung, S., Emili, A., Snyder, M., Greenblatt, J.F., Gerstein, M.: A bayesian networks approach for predicting protein-protein interactions from genomic data. Science **302**(5644), 449–453 (2003)
14. Kullback, S., Leibler, R.A.: On information and sufficiency. Ann. Math. Stat. **22**(1), 79–86 (1951)

15. Laxman, S., Sastry, P.S.: A survey of temporal data mining. Sadhana **31**(2), 173–198 (2006)
16. Li, T.: Event Mining. Chapman and Hall/CRC, Boca Raton (2015)
17. Li, T., Liang, F., Ma, S., Peng, W.: An integrated framework on mining logs files for computing system management. In: Proceedings of the Eleventh ACM SIGKDD International Conference on Knowledge Discovery in Data Mining, pp. 776–781. ACM (2005)
18. Li, T., Ma, S.: Mining temporal patterns without predefined time windows. In: Fourth IEEE International Conference on Data Mining ICDM 2004, pp. 451–454. IEEE (2004)
19. Li, T., Peng, W., Perng, C., Ma, S., Wang, H.: An integrated data-driven framework for computing system management. IEEE Trans. Syst. Man Cybern.-Part A: Syst. Hum. **40**(1), 90–99 (2010)
20. Lozano, A.C., Abe, N., Liu, Y., Rosset, S.: Grouped graphical granger modeling methods for temporal causal modeling. In: Proceedings of the 15th ACM SIGKDD International Conference on Knowledge Discovery and Data Mining, pp. 577–586. ACM (2009)
21. Ma, S., Hellerstein, J.L.: Mining partially periodic event patterns with unknown periods. In: Proceedings 17th International Conference on Data Engineering 2001, pp. 205–214. IEEE (2001)
22. Pei, J., Han, J., Mortazavi-Asl, B., Pinto, H., Chen, Q., Dayal, U., Hsu, M.: Prefixspan: mining sequential patterns by prefix-projected growth. In: Proceedings of the 17th International Conference on Data Engineering, pp. 215–224. IEEE Computer Society, Washington, DC, USA (2001)
23. Peng, W., Perng, C., Li, T., Wang, H.: Event summarization for system management. In: Proceedings of the 13th ACM SIGKDD International Conference on Knowledge Discovery and Data Mining, pp. 1028–1032. ACM (2007)
24. Schroeder, D.J.: Astronomical Optics. Academic press, Cambridge (1999)
25. Song, L., Kolar, M., Xing, E.P.: Time-varying dynamic Bayesian networks. In: Advances in Neural Information Processing Systems, pp. 1732–1740 (2009)
26. Srikant, R., Agrawal, R.: Mining sequential patterns: generalizations and performance improvements. In: Apers, P., Bouzeghoub, M., Gardarin, G. (eds.) EDBT 1996. LNCS, vol. 1057, pp. 1–17. Springer, Heidelberg (1996). https://doi.org/10.1007/BFb0014140
27. Tan, P.N., et al.: Introduction to Data Mining. Pearson Education India, London (2006)
28. Tang, L., Li, T., Shwartz, L.: Discovering lag intervals for temporal dependencies. In: Proceedings of the 18th ACM SIGKDD International Conference on Knowledge Discovery and Data Mining, pp. 633–641. ACM (2012)
29. Zeng, C., Wang, Q., Wang, W., Li, T., Shwartz, L.: Online inference for time-varying temporal dependency discovery from time series. In: 2016 IEEE International Conference on Big Data (Big Data), pp. 1281–1290 (2016)
30. Zeng, C., Tang, L., Li, T., Shwartz, L., Grabarnik, G.Y.: Mining temporal lag from fluctuating events for correlation and root cause analysis. In: 2014 10th International Conference on Network and Service Management (CNSM), pp. 19–27. IEEE (2014)

A Combined Model for Time Series Prediction in Financial Markets

Hongbo Sun[ID], Chenkai Guo$^{(\boxtimes)}$[ID], Jing Xu[ID], Jingwen Zhu[ID],
and Chao Zhang[ID]

College of Computer and Control of Engineering, Nankai University, Tianjin, China
luodidegesunhb@126.com, chenkai.guo@163.com, xujing@nankai.edu.cn,
zhujingwenNK@163.com, chaos_zc@163.com

Abstract. Time series prediction is not easy to achieve high accuracy. non-linear and unstable characteristics make the time series prediction difficult. The variety of dataset make the prediction result debatable. In order to solve this problem, in this paper we propose a deep learning prediction method based on decomposition, reconstruction and combination, which combines ways of communication field. The model is decomposed by Empirical Mode Decomposition, Principal Component Analysis and Long Short-Term Memory networks (EPL below). And also, the proposed interval EPL (IEPL below) improve and consummate the EPL model. The EPL and IEPL experiment results will bring average 5% higher accuracy than that of existing research.

Keywords: Time series prediction · Deep learning · EPL · IEPL

1 Introduction

Time series prediction with high accuracy benefits the investors. In recent years, the emergence of relevant approaches make the time series prediction for financial market no longer out of reach. However, the prediction of financial market is still a challenging subject because of the linkage between the global financial markets and the particularity of the prediction time span. Generally, most of the data in financial market are non-linear, non-stationary and multi-scale interval time series with many noisy components. The difficulty is self-evident when choosing the effective information set from complex intervals with numerous noises and setting up the predicting mathematical model. Modern research approaches are divided into linear and non-linear methods to deal with the above problems. For linear ones, the ARIMA model is the most classical method, where non-stationary time series are transformed into stationary time series by differential operations. Yet, using differential data instead of actual data will reduce the prediction accuracy. Overall, the above method is somewhat farfetched as well as other linear approaches. For non-linear ones, the methods are able to better deal with the data, such as Artificial Neural Network, Support Vector Machine

Y. Cai et al. (Eds.): APWeb-WAIM 2018, LNCS 10988, pp. 138–147, 2018.
https://doi.org/10.1007/978-3-319-96893-3_11

and so on. Comparing with linear models, non-linear models are more prevalent in financial markets prediction. In order to find a more accurate prediction model of financial market time series, scholars have done a lot of efforts to build model algorithms and a large number of neural networks based on different algorithms are studied. Although results show that the "neural network-based" models have made some achievements, they are still difficult to cover the global minimum, which hardly do well in prediction. Therefore, this paper tries to find a new way to combine the idea of decomposition, reconstruction and combination, where the innovation devotes to improve the input and output of neural networks. The ideal goal is to create a mathematical model that can effectively predict the short-term trend of the time series of financial markets. At the same time, the interval EPL (IEPL) combined model is further introduced. So as to effectively make comparison between the combined model proposed in this paper and the existing research models, ARIMA, GARCH and other traditional models are used as contrast models in this paper. The selected dataset concludes S&P500 series, FTSE index, US dollar index and BDI index in the last 6 years. A variety of financial derivatives are chosen to test the effectiveness of the method.

The reminder of the rest contests is organized as follows: Sect. 2 is related work; Sect. 3 is the construction of the model; Sect. 4 is the experimental results and analysis; Sect. 5 is the summary.

2   Related Work

There are many ways to make predictions. Usually, the original approach is the "linear-based" method, which is suitable for stock index prediction, effective market test and asset pricing model establishment. Some traditional models like ARIMA or GARCHs provide a theoretical basis. However, most of these methods are linear stationary models and need to be based on efficient market hypothesis. Financial market time series itself is non-linear and non-stationary time data, so that the results predicted by the above methods are more or less biased.

Gradually, "non-linear-based" methods domain the prediction field. non-linear models have higher prediction accuracy than the "linear-based" methods, which have been already verified by related scholars. Hu et al. [1] proposed deep belief network used for stock prediction, in the shallow layer, the Bayesian network is used, and the Restricted Boltzmann Machine is used in depth. Cao et al. [2] put forward to Autoencoder self-coder, which can be used to quantify transactions of financial time series. Sezer et al. [3] proposed a Convolutional Neural Network for visual pattern recognition, which was mainly applied to automatic feature selection of time series. Pérez-Ortiz et al. [7] set up long and short memory units and modified the memory function that was realized by the switch of the door, so as to prevent the gradient from disappearing. Although the "non-linear-based" models can alleviate the high degree of non-convexity of multilayer neural networks to some extent, a large number of local advantages might cause severe overfitting in practical applications. With the continuous progress of technology, the prediction accuracy of a single model is also questioned. Combined models explored a new way for thinking. Lahmiri [5] used wavelet neural networks to quantify the high frequency trading financial calendar, and the empirical results showed that neural network can also effectively depict the trend of financial calendar. Ismail et al. [6] optimized the initial input weights of BP neural network, and established the PSO-BP neural network prediction model. The deviation between predicted and actual results was reduced to negligible. The research results of the above scholars show the combined model with combined structure in an effective method. Paralleling to combined model's research route, in some steps, the decomposition of time series in signal processing provides us with another method for constructing input and output of deep neural network model. For example, Empirical Mode Decomposition (EMD below) has a technical similarity to the establishment of multi-scale or multi-level time series models. Furlaneto et al. [4] recently proposed a prediction model based on EMD. The model first decomposed and transformed the original data of time series, and thus multiple levels of decomposition sequences were obtained. This is an interdisciplinary approach and is currently less involved in the prediction of time series in financial markets. The method of separating time series into stable term and influencing factor is worthy of reference. Based on above steps, a new combined model prediction scheme is proposed in this paper.

3   Model Construction

3.1   EMD for Financial Time Series

It is assumed that the time series of financial markets can be expressed as $X(t) = (X_O(t),\ X_H(t),\ X_L(t),\ X_C(t),\ X_V(t))$. The above five parameters represent the opening price time series $X_O(t)$, maximum price time series $X_H(t)$, lowest price time series $X_L(t)$, closing price time series $X_C(t)$ and trading volume time series $X_V(t)$. The EMD method can be decomposed by the above series dimensions. Because of the non-linear and non-stationary characteristics of financial time series, the EMD is different from the traditional time series. The traditional scheme directly transforms the original time series without progressive stripping, but the EMD is indirect. Usually, time series is decomposed into Intrinsic Mode Function (IMF below) sequences and a trend term. By corresponding with the Hilbert Spectrum, multi-scale oscillation characteristics of time series will be obtained. The IMF component explains the oscillation characteristics reflecting different time scales of the original time series, while the trend term reflects the long-term trend of the original time series, in which series are monotonic and smooth. However, the above IMF component contains two constraints: One is that the number of zero crossing and extreme points need to be the same, the other is that the IMF signals should be about the zero mean. Suppose $X(t)$ is the original signal, with the upper envelope and the lower envelope. The upper envelope is connected with all local maxima by the three-order spline interpolation curve, and the lower envelope is obtained by connecting three local spline interpolation curves to all local minima. The whole concluded maxima and minima are between the upper and lower envelopes. It is assumed the mean line of the upper and lower envelopes is $m_1(t)$. The $h_1(t)$ is the difference between the original time series $X(t)$ and the mean line $m_1(t)$:

$$h_1(t) = X(t) - m_1(t) \tag{1}$$

If $h_1(t)$ satisfies the two conditions of the IMF at one point, then $h_1(t)$ is a filtered IMF component, and vice versa. The signal characteristics of the IMF must be satisfied that the absolute value for the total number of zeros and maxima of the signal are not more than one in the whole-time series. The average between the upper and lower envelopes is zero mean the whole time. That is, the upper and lower envelopes are locally symmetric at the zero axis. In general, there are many clutters at the crest or trough of a filter, which cannot be completely screened. Therefore, the screening process needs to be repeated times. The screening process is supposed to reach two goals, one is that the decomposed time series is transformed from maximum value to zero mean value, then from zero mean to minimum value, and then the repeated waveform is used to eliminate the clutters through the above shock process. The other is to make the local peaks and valleys symmetric. At the next screening, $h_1(t)$ becomes the original sequence. Let $h_{11}(t)$ be the difference between the new original sequence $h_1(t)$ and the new mean line $m_{11}(t)$:

$$h_{11}(t) = h_1(t) - m_{11}(t) \tag{2}$$

Repeatedly iterating until k times, if the results meet the two characteristics of the IMF sequence. The first component of input characteristics is C_1:

$$C_1 = h_{1k} = h_{1(k-1)} - m_{1k} \tag{3}$$

When the screening process stops, a precision requirement is needed. In this paper, the ST coefficient is used as the threshold of the screening process, and the formula is expressed as the threshold value of the screening process:

$$ST = \sum_{n=1}^{T} \frac{|h_{1(k-1)}(t) - h_{1k}(t)|^2}{h_{1(k-1)}^2(t)}, k = 1, 2, 3... \tag{4}$$

The ST value is determined by the empirical value. The magnitude of the signal sequence elements is different, which depends on the value of ST. In the process of repeated experiments, if the order of time series elements is 10, 0.3–0.5 is more appropriate. As for the order of magnitude is 100, 1 3 is more appropriate. When the ST value of the screening results is within or below the threshold in five consecutive times, the screening can be stopped and the $C_1(t)$ takes as an IMF component. If the first IMF component is selected, the $C_1(t)$ is deleted from the original sequence, and the remaining sequence is r_1:

$$r_1 = X(t) - C_1 \tag{5}$$

According to the above operations, the IMF component in "n-term" results can be selected as:

$$r_2 = r_1 - C_2, \cdots, \cdots, r_n = r_{(n-1)} - C_n \tag{6}$$

A series of finite IMF components are obtained through the screening process. When the trend item r_n becomes a monotonic function, the IMF component cannot be sifted out from the new sequence, and the whole EMD empirical mode decomposition process finishes. Screening the IMFs can be considered as features in the limited time scale from signal which are from high to low frequency process. The below is the sum of the expression of EMD:

$$ST = \sum_{i=1}^{n+1} C_i = \sum_{i=1}^{n} C_i + r_n \tag{7}$$

where $C_{n+1} = r_n$.

3.2　Interval EMD Application

The general EMD decomposition process uses the three-spline interpolation function to calculate the upper and lower envelope. But for a financial series $X(t) = (X_O(t), X_H(t), X_L(t), X_C(t), X_V(t))$, the information is not used in all. This paper will make extra use of these loss information, and proposes interval EMD (IEMD below) algorithm on the basis of EMD algorithm.

As the Algorithm 1 below, for arbitrary time series $X(t)$, all the local maxima of the highest price $X_H(t)$ and all the local minima of the lowest price $X_L(t)$ should be recognized. The three-spline interpolation method is used to calculate the maximum envelope $U(t)$ and the lower envelope $L(t)$ of the lowest price.

The difference between EMD algorithm and IEMD algorithm is that when the IEMD algorithm calculates the mean of the upper and lower envelope lines, the upper envelope is calculated by the maximum value, and the lower envelope is calculated by the minimum value. However, the EMD algorithm uses the upper and lower envelopes of the closing price. The difference lies in the chosen of the envelope.

Algorithm 1. Interval EMD decomposition algorithm.

Begin:

  Step1: Make **c(t)=x(t)**;

  Step2: Caculate the U(t) and L(t), U(t)=Upper envelope curves of c(t), L(t)=Lower envelope curves of c(t);

  Step3: Caculate the m(t), **m(t)=(U(t)+L(t))/2**;

  Step4: Make iteration of c(t), update the **c(t)=c(t)-m(t)**;

  Step5: Make a judgment whether the IMF condition is satisfied, if so, go to *Step6* ; otherwise, go back to *Step2* and put iter=iter+1;

  Step6: Output c(t) and update the **x(t)=x(t)-c(t)**;

  Step7: Make a judgment whether the IMF decomposition is completed(comp=M), if so, end the program; otherwise, go to *Step1* and put comp=comp+1.

3.3　PCA Dimensionality Reduction and LSTM

In the EPL, the signal will be extracted from the sliding window; while the IMF sequence will be obtained by EMD decomposition. Each time in the EPL, the sequence expresses as a matrix. The first k principal component is selected as the training sample data for input into LSTM after the PCA. According to the theoretical experience, the cumulative contribution of the variance of the k principal components should be more than 95%, and the KMO coefficient determines the principal component:

$$KMO = \frac{\sum_{i=1}\sum_{j=1,j\neq i} r_{ij}^2}{\sum_{i=1}\sum_{j=1,j\neq i} r_{ij}^2 + \sum_{i=1}\sum_{j=1,j\neq i} p_{ij}^2} \tag{8}$$

r_{ij} is the correlation coefficient of row i in column j, and p_{ij} is the partial correlation coefficient in column i of row j. Generally, the values of the KMO series are between zero and one. When the value of KMO is more than 0.9, the correlation degree of the original variables is higher; and when the value is more than 0.8, the correlation is high; the value of 0.7 is normal; the correlation degree is weak about 0.6; the weakest correlation degree is below 0.5 and 0.5.

Different from the conventional LSTM, in this paper the structure adds up the peephole connections proposed by Pérez-Ortiz et al. [7]. The memory cell state is transmitted to the t time as input to the input gate and the output gate, where memory states are not processed by output gates at $t - 1$ time.

3.4   Model Evaluation Criteria

In order to verify the prediction result of the new combined EPL and IEPL model proposed in this paper, we use the following 3 indicators for performance evaluation: Mean absolute percentage error (MAPE below), Root mean square error (RMSE below), DS rate (DS below). According to their definitions and usages, $MAPE$ and $RMSE$ are used to assess the prediction error, and the lower the prediction error is the smaller the two values are. The DS index is used to measure the direction of the development of the time series. It is expressed in the form of percentage, which describes the rise and fall rather than the accurate prediction. The calculation formula of the three indexes is as follows:

$$MAPE : MAPE = \frac{1}{n}\sum_{i=1}^{n}|\frac{T_i - A_i}{T_i}| * 100\% \tag{9}$$

$$RMSE : RMSE = \sqrt{\frac{1}{n}\sum_{i=1}^{n}(T_i - A_i)^2} \tag{10}$$

$$DS : DS = \frac{100}{n} * \sum_{i=1}^{n} c_i, c_i = \begin{cases} 1 & (T_i\text{-}A_i) * (A_i\text{-}A_{i-1}) \geq 0 \\ 0 & \text{otherwise} \end{cases} \tag{11}$$

A_i and T_i represent real and predicted values respectively, and n represents the number of samples.

4   Experiment Results

4.1   Data Set Selection

This paper selects four kinds of time series data for experiment. Each group follows a reference standard in the field. The data sources come from Yahoo financial database. In this paper, the selection time of empirical data sets is from January 4, 2011 to December 30, 2016, and the total is 1510. The data is divided into 80% data training sets, 10% data validation sets and 10% data test sets. In the EMD, three-spline interpolation algorithms are needed to ensure that each IMF sequence is orthogonal to each other. Here is an example of the S&P500 data set. The results of the specific IMF decomposition are as follows: It can be seen from the Fig. 1 that each IMF component gradually presents a regular state with the number of decomposition increasing, which reflects the fluctuation characteristics of external shocks at different frequencies. High frequency IMF1 and IMF2 reflect the impact of a short period of internal and external microcosmic factors on the index, while low frequency IMF8 and IMF9 reflect the macro impact of a long period of time, such as the policy economy.

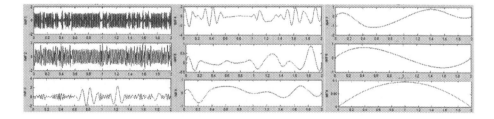

Fig. 1. IMF decomposition results.

4.2   Comparison and Analysis of Empirical Results Between EPL and IEPL

In this paper, a new combined prediction model EPL used in the time series of financial markets is proposed. Referring to the research results of relevant scholars, this paper will adopt the following single or combined models for the horizontal comparison. The contrast models used are ARIMA, GARCH, BP-NN, LSTM, SVM-LSTM, WD-LSTM. The ARIMA and GARCH models belong to linear models, while the remaining ones are non-linear models. All the data are averaged in 100 times experiment. The following is the evaluation of the performance between the models: As shown in Table 1. The results show that the EPL/IEPL combined model with LSTM in this paper is more forward-looking.

Table 1. Comparison of different model in prediction results.

Model	MAPE	RMSE	DS rate(%)
ARIMA	2.0358	239.3565	54.69
GARCH	2.1536	262.5644	56.25
BP-NN	2.0865	218.6302	59.30
RNN (LSTM)	1.7734	193.2633	65.39
SVM-LSTM	1.7587	182.2698	68.57
WD-LSTM	1.7087	178.2698	70.57
EPL	1.3598	148.3983	72.53
IEPL	1.3368	146.9801	73.96

4.3   Performance of Combined Models on Different Data Sets

In addition to the S&P 500 data set, this paper also compares the other three time series, FTSE index, US dollar index and BDI index. The purpose is to verify whether the same combined model can achieve similar results when the random selection of sequence data sets is selected. As shown in Fig. 2, the EPL combined

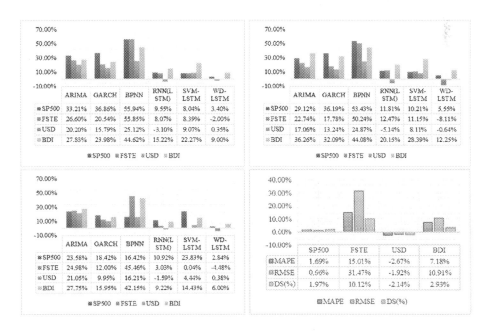

Fig. 2. The first three figures are comparison of different models in MAPE, RMSE and DS rate observation index. The last figure is the increasing rate of each evaluation criteria.

model has poor prediction results on some models except a few indicators, most of the predictions are outstanding. In particular, when using EMD to split the original data, the prediction results are better than the original LSTM or data processing used by SVM and WD methods. Therefore, the combined model proposed in this paper can be regarded as an improvement to the original LSTM or improvements to the other combined models. As for the IEPL model, it's more excellent than EPL in most cases. We can regard the IEPL model as a more microscopic prediction on the basis of the EPL model.

5   Conclusion

In this paper, a new combined model, EPL, is proposed. Combined with the EMD empirical mode decomposition method of signal processing, the prediction performance of neural network model can be improved. To some extent, the EPL model proposed in this paper is superior to the non-linear models and most of the combined prediction models. On this basis, this paper puts forward the interval EPL model, that is, IEPL. The difference between IEPL and EPL is in the envelope of EMD decomposition selection of different calculation methods. Experimental results show that the experimental results of IEPL is superior to EPL model. To sum up, this paper draws the following conclusions:

The EMD algorithm makes effective sense in helping the neural network to improve the prediction performance. The EMD algorithm works by means of decomposing the time series into IMF sequences with multiple frequencies from high to low, and IMF sequences are regarded as signals of external shocks, which can explain the inherent law of time series. The PCA method improves the prediction performance of the model as an optimization tool, and its mechanism devotes to remove the redundancy of data. The IEPL model will help optimizing the EPL model to some extent.

Acknowledgement. This research work here is supported by the Science and Technology Planning Project of Tianjin (Grant No. 17JCZDJC30700 and 17YFZCG X00610).

References

1. Hu, J., Zhang, L., Cai, Z., et al.: An intelligent fault diagnosis system for process plant using a functional HAZOP and DBN integrated methodology. Proc. Eng. Appl. Artif. Intell. **45**, 119–135 (2015)
2. Cao, L., Huang, W., Sun, F.: Building feature space of extreme learning machine with sparse denoising stacked-autoencoder. Neurocomputing **174**(A), 60–71 (2016)
3. Sezer, O.B., Ozbayoglu, M., Dogdu, E.: A Deep neural-network based stock trading system based on evolutionary optimized technical analysis parameters. Proc. Comput. Sci. **114**, 473–480 (2017)

4. Furlaneto, D.C., Oliveira, L.S., Menotti, D., et al.: Bias effect on predicting market trends with EMD. Expert Syst. Appl. **82**, 16–29 (2017)
5. Lahmiri, S.: Wavelet low- and high-frequency components as features for predicting stock prices with backpropagation neural networks. J. King Saud Univ. Comput. Inf. Sci. **26**(2), 218–227 (2014)
6. Ismail, A., Jeng, D.-S., Zhang, L.L.: An optimized product-unit neural network with a novel PSO-BP hybrid training algorithm: applications to load-deformation analysis of axially loaded piles. Eng. Appl. Artif. Intell. **26**(10), 2305–2314 (2013)
7. Pérez-Ortiz, J., Gers, F.A., Eck, D., et al.: Kalman filters improve LSTM network performance in problems unsolvable by traditional recurrent nets. Neural Netw. **16**(2), 241–250 (2003)

Data Minging and Application

Location Prediction in Social Networks

Rong Liu[1], Guanglin Cong[1], Bolong Zheng[2,3], Kai Zheng[1], and Han Su[1(✉)]

[1] Big Data Reaserch Center,
University of Electronic Science and Technology of China, Chengdu, China
{kaizheng,hansu}@uestc.edu.cn, lrong0913@gmail.com, nofloat@163.com
[2] School of Data and Computer Science, Sun Yat-sen University, Guangzhou, China
zblchris@gmail.com
[3] Department of Computer Science, Aalborg University, Aalborg, Denmark

Abstract. User locations in social networks are needed in many applications which utilize location information to recommend local news and places of interest to users, as well as detect and alert emergencies around users. However, considering individual privacy, only a small portion users share their location on social networks. Thus, to predict the fine-grained locations of user tweets, we present a joint model containing three sub models: content-based model, social relationship based model and behavior habit based model. In the content-based model, we filter out those location-independent tweets and use deep learning algorithm to mine the relationship between semantics and locations. User trajectory similarity measure is used to build a social graph for users, and historical check-ins is used to provide users' daily activity habits. We conduct experiments using tweets collected from Shanghai during one year. The result shows that our joint model perform well, especially the content-based model. We find that our approach improves accuracy compared to the state-of-the-art location prediction algorithm.

1 Introduction

Since the on-line social media grows, Twitter, Facebook and Sina Weibo have accumulated a large number of users up to now. In China, Sina Weibo, a form of unstructured short texts, has become one of the most popular social networking tools. In Sina Weibo, people post tweets about their daily routines, emergencies they meet, and comments to news. They also attach to their tweets with current locations, a.k.a. check-ins. Check-ins play an important role in location based recommendation and emergency detection/alert, which are utilized by a large number of business organizations. For example, when a user comes to a place and posts her location, she can get recommendation about local news and places of interest around, and also get alerts of unexpected events nearby. But recently, due to concerns about data privacy, weibo users have been increasingly avoiding sharing location information while posting tweets. According to a recently statistical analysis in [9] over 1 billion tweets spanning three months, only 0.58% tweets have location tags. It is becoming harder and harder for business organizations to extract user locations, hindering recommendation and detection.

© Springer International Publishing AG, part of Springer Nature 2018
Y. Cai et al. (Eds.): APWeb-WAIM 2018, LNCS 10988, pp. 151–165, 2018.
https://doi.org/10.1007/978-3-319-96893-3_12

In this paper, we present a novel approach which combines three features including textual content, social relationships and user behavior habits to predict user's current locations for tweets without any location tags. Recent works only consider the first two features, while ignore users' behavior habits. However, based on our study in our work, users' behavior habits play an important role in tweets location prediction. Actually, a Weibo user with a regular everyday life, will have similar daily activities. That is to say, his or her trajectories are similar. In content based model, we leverage Convolutional Neural Network (CNN) to mine location information in tweets. And we also mine another social relationship called user similarity in social relationship based model which cluster users with similar trajectories. Based on behavior similarities, we build a similarity graph which cluster similar users together to help find users with similar daily behavior for a specific user. Through a probabilistic model, we predict a user's location from his or her similar users.

The remainder of this paper is organized as follows. Section 2 overviews related works. The location prediction model is introduced in Sect. 3. Section 4 describes the experiments conducted to verify the accuracy of our model. Finally, Sect. 5 draws the conclusion of the paper.

2   Related Work

With the wide use of social networks, mining user location information from them and apply this to many occasions is significant, such as location based recommendation, emergency detection and alert. Thus, many related works utilized different features and approaches to roughly predict where the users were when they post tweets in their personal devices. The features used in previous works can be categorized into two types: content based and social relationship based.

Content Based. User's tweets content often contains many features, such as textual content, photos, videos and user URLs. [15] generates probabilistic language model based on the photo tags posted by users, and then estimate the location of each photo rely on the language model and Bayesian inference. Comparing with photos, textual content often contains more location clues, since users may mention location names or location related words when posting tweets.

Location prediction approaches based on text are classified to two basic types as well. One is identifying the related geographic terms from textual content, the other is building a probabilistic language model to predict tweets locations. For the reason that a small number users post exact geographic terms in their tweets, most recent works prefer to construct probabilistic models for location prediction based on the statistical linguistic features in textual content. In [1], author uses a variation of probabilistic framework in [2] which adds the feature of relationship between tweets and related reply-tweets, in order to enhance accuracy by estimating the geographic location of the user. [16] proposes a probabilistic model leveraging the Maximum Likelihood Estimation to infer users resident locations, which mines the relationship between locations and words.

Some related works are dissatisfied with such coarse-grained location prediction. In [9], Lee et al. utilize external location sharing services platforms, and the user 'Check-In' information to study the mobility characteristics of the users. This work builds language model for each PoI (Place of Interest) which is the basis of location prediction. However, the cost of building language models is huge, and they just predict site located in a part of city. Our work keeps the idea of mining the relationship between the semantics of tweet content and the locations, but builds language model for all locations through a novel approach.

Graph Based. Social relationship is one of the most essential part of on-line networks. Friendship as a kind of social relationship, always provides pivotal clues for predicting user locations. The way of building user friendship network is usually utilizing users profile, response and dialogues. Backstrom et al. propose a probabilistic model representing the likelihood of relationship between any two users in [3]. Based on this model, user locations can be inferred when given the geographic distribution of the locations. [8] presents several extensions to the model shown in [3], which adding weighting strategies that user friends have different influence on user. In [14], Sadilek et al. add up the time overlaps two users spend at their respective locations and scale each overlap by distance between the locations. Thus, distance value can be used for detecting friendship between users and representing the tightness of this relationship. Friends may stay in the same city in most situations, however, they may not always stay at the same site or regions in the city all times. So we define another social relationship called user similarity to solve this problem. Thus, user tweets can be regarded as a trajectory with timestamps and coordinates. And we leverage trajectory similarity algorithm shown in [4,5,13] to calculate the user similarity.

3   Problem Formulation

User location information in social media plays an important role in many applications, however, only a small portion users share their location for protecting privacy. Our goal is to estimate the location based on features that are minded from tweets which are lack of check-ins. Sina Weibo, one of social applications with a large number of users, provides users a lot of choices that they can post tweets containing textual data, photos and videos, interact with other users or record their lives. Tweets content, especially textual content, often contains location names or location related words, which can extract location related information directly. Trajectory similar users can be clustered through user similarity graph, they probably share the same location at most circumstances. Besides, when users record their daily lives, we can extract their behavior habits through these records. So except check-in data, we can mine user location information through other methods as well. In this work, users tweets content, social relationships and user behavior habits are used to get location information.

Location Estimation Problem: Given a set of tweets $T_{tweets}(u)$ posted by a Weibo user u, estimate a user's probability of being located at a site, such that the location with maximum probability $l_{cur}(u)$ is the user's actual location $l_{act}(u)$.

With the definition of the problem solved in this paper, we list the Notations in Table 1 used throughout the remainder of this work.

Table 1. Notations used in the paper

Notation	Explanation	Notation	Explanation
U	User set of u_i	$e(u_i, u_j)$	Relationship between u_i and u_j
E_i	Relationship set of u_i	A	Tweet matrix constructed with word vector
$T_{tweets}(u)$	Tweet set of u	X_i	The i-th word vector
$S_{words}(u)$	Words set of u's tweets	$\theta_{similar}$	Trajectory similarity threshold
T	A trajectory	$s(T_i, T_j)$	Similarity between two trajectories T_i and T_j
p	A trajectory point	l_{pre}	Location of previous tweet
V	Similar user set	l_{cur}	Predictive location of tweet
Vct	Region vector	l_{act}	Actual location of tweet
N	Region number of city	M	User location transition matrix
L	Candidate location set	P	Location transition probability matrix

As we noted that location estimation is a difficult and challenging problem. The check-in data in users' tweets is always sparse, and the frequency of user posting tweets is not high. So we divide the map of the city into square grids of different degrees to overcome the sparsity of locations, which are described in Sect. 4.2.

3.1   System Architecture

In this work, we propose a joint probabilistic model which contains the three sub models. Figure 1 provides a sketch of our system architecture for predicting the city area which the tweet belongs to. In this architecture, we define three channels to mine location information:

(1) **Textual Content**: Since users may post their locations or location related words in their tweets, the textual part of tweets becomes the most important clue of location prediction. We extract the textual part of tweets in dataset, filter out tweets without any location clues and train CNN model to predict tweet location based on these textual data.

(2) **Social Relationship**: On-line social relationship has different definitions in this work, we define it as user similarity for the reason that it has stronger connection on location than friendship mentioned in related works. User historical data can be regarded as a trajectory and used to calculate the similarity between users.

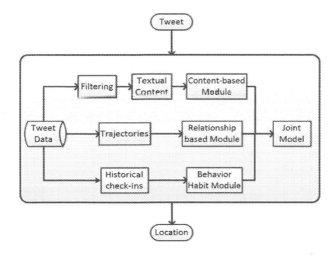

Fig. 1. Framework architecture

(3) **Behavior Habit**: Behavior habit of users provides clues of tweet location for the fact that users prefer to take their own familiar routes. User behavior habits are extracted from historical data as well. From these habits, we know how a user is moving between his or her resident locations.

Based on these features which we considered as location dependent information, we also present corresponding models to predict users' current locations. Specifically, CNN model is used to mine location information from $n - gram$ words of a tweet, user similarity cluster model finds similar users of a specific user and then follows them to where the user is, Markov Chain and Transition Probability Matrix build a customary trajectory for each user based their historical check-in data.

3.2   Content-Based Model

Textual content is the most frequently-used feature, since users may mention location names or location related words while posting tweets. However most words are distributed consistently with the population across different locations, meaning that most words provide little power at distinguishing the location of a user. For example, any user may post tweets like "I'm eating dinner", so tweets like this are called location-independent tweets. Without filter, many noise tweets in dataset increase the difficulty of extracting and distinguishing the location feature for our model. So we utilize $tfidf$ Value to evaluate whether a word is related to a location and filter out location-independent tweets without these location related words firstly. Besides, we use grid-based neighborhood smoothing approach which clusters locations into grids according to their coordinates, to overcome the sparsity of location across tweets in dataset. Thus we divide the entire city into equal-sized grid cells, which applies in the whole work.

Under this circumstance, a novel approach, Convolutional Neural Network (CNN), is used to mine the relationship between textual data and locations avoiding complex feature extractions and data reconstruction process comparing prior works. To get better training effect, traditional content-based models always need multiple parameter adjustments, while we just need pre-training word vectors in CNN model whose parameters are adjusted through backpropagation algorithm. In addition, CNN can extract information from different n-gram words sequence at the same time and is more suitable for large-scale data processing. Thus, we present a CNN architecture for tweets location prediction based on the model in [6] with a slight variance.

Firstly, when given a text portion with several sentences of a tweet, we segment these sentences into tokens which then are converted to a $TweetMatirx$ $A \in \mathbb{R}^{n \times d}$. Suppose that X_i is the $d-$dimensional word vector for the $i-$th word in n words tweet and the text portion of any user's tweet can be also descried as a matrix $X_{1:n}$. Then we define $X_{i:i+j}$ as the cascade of words $X_i, X_{i+1}, \cdots, X_{i+j}$ in tweets. To extract feature in tweet, a $filter$ $W \in \mathbb{R}^{h \times d}$ applied to a window of h words is used in convolution operation. For example, given a tweet with n words, a fixed $filter$ W and filter width h, a new feature c_i generated from sub-matrix $X_{i:i+h-1}$ by Eq. 1. Here $b \in \mathbb{R}$ is the bias term and f is an activation function such as the hyperbolic tangent (tanh) or Rectified Linear Units (ReLu).

$$c_i = f(W \cdot X_{i:i+h-1} + b) \tag{1}$$

Then, a feature map $c = [c_1, c_2, \cdots, c_{n-h+1}]$, with $c \in \mathbb{R}^{n-h+1}$, are produced by each possible sub-matrix of the given tweet $\{X_{1:h}, X_{2:h+1}, \cdots, X_{n-h+1:n}\}$. A maximum value $\hat{c} = max\{c\}$ is taken as the feature corresponding to this particular filter after applying this max-pooling operation. This representation is then fed through a softmax function to generate the final classification. During the training process, the purpose is minimizing categorical cross-entropy loss, and optimizing the parameters including weight vectors for filtering and biases in activation function.

3.3   Social Relationship Model

Although location information can be extracted from textual data of tweets in most cases, users may not always post tweets containing hints about locations. For example, the location-independent tweets like "I'm eating dinner" are not suitable for the CNN model. By mining the social relationship of users, we use the location of the neighbors in users similarity graph to predict their current locations. Related works define user social relationship as friendship through user interactions in social applications. However, not all users may share the closely similar trajectory with their on-line friends even they are off-line friends at all times. We can just estimate which city a user locate in through friendship, but not her real-time locations in the city. Thus, we define another on-line social relationship as similar users who have similar trajectories. For instance, some users work in a same area such as same office buildings may have similar daily activities but not friends whether in real life or on the Internet. Thus, we can predict a user's current location according to her similar users' locations instead of her friends' locations. The core components include the similar user clustering model and location prediction model that are described in the next part.

Similar User Clustering Model. In our work, we present an novel approach to cluster similar users based on the fact that they share nearly the same trajectories. Next, we define the notations used in this part and the operations on them.

Definition 1 (Trajectory Point). *A trajectory point is a pair:* (p, t)*, where p is a location in d-dimensional space, and t is the timestamp at which p is observed.*

In this model, users move in a two-dimensional space, that is, p is a 2-dimensional vector, and the time attribute is discrete.

Definition 2 (Trajectory). *Trajectory T is a sequence of trajectory points, extracted from user's check-ins and ordered by timestamps t. Trajectory T is represented as a sequence of trajectory sample points. Therefore, $T = [(p_1, t_1), (p_2, t_2), ...,(p_n, t_n)]$, where $(t_1 < t_2 < ... < t_n)$.*

With the definition of user trajectory, we define some operations on the point p and trajectory T.

1. $s(T_1, T_2)$: $s(T_1, T_2)$ represents the similarity rate of two users' trajectories T_1 and T_2.
2. $Head(T)$: For trajectory $T = [p_1, p_2, \cdots, p_n]$, $Head(T)$ is to get the first point of the trajectory, that is $Head(T) = p_1$.
3. $Time(p_1, p_2)$: $Time(p_1, p_2)$ represents the time difference of points p_1 and p_2.
4. $Rest(T)$: For trajectory $T = [p_1, p_2, \cdots, p_n]$, $Rest(T)$ is to get the tail points of the trajectory except the first point. Therefore, $Rest(T) = p_2, p_3, \cdots, p_n$.

In order to calculate the similarity of user trajectories, we use Spatial-Temporal Longest Common Subsequence Similarity (STLCSS) measure. This measure involved two constants:

1. δ: a real number which controls how far in time we can go in order to match a given point from one trajectory to a point in another trajectory.
2. ε: a real number that is the matching threshold. Only when the distance between two points is less than ε, can these two points be regarded as the same point.

$$s_{\delta,\varepsilon}(T_1, T_2) = \begin{cases} 0 & \text{if } T_1 \text{ or } T_2 \text{ is empty} \\ 1 + s_{\delta,\varepsilon}(Rest(T_1), Rest(T_2)) & \text{if } |Head(T_1) - Head(T_2)| < \varepsilon \text{ and} \\ & |Time(Head(T_1), Head(T_2))| \leq \delta \\ \max(s_{\delta,\varepsilon}(Rest(T_1), T_2), s_{\delta,\varepsilon}(T_1, Rest(T_2))) & \text{otherwise} \end{cases}$$

$$(2)$$

Given δ and ε, we define the similarity measure $s(T_1, T_2)$ between two trajectories T_1 and T_2, shown as follows:

$$s(T_1, T_2) = \frac{s_{\delta,\varepsilon}(T_1, T_2)}{\min(n, m)} \tag{3}$$

It is observable that the larger the value of $s(T_1, T_2)$ is, the more similar two trajectories are according to the Eq. 3. So if the similarity of two user trajectories is high, they are deemed to be similar users. Based on this conclusion, we can infer each other's positions when no location clues in user's tweets, which is shown in following part. Here we define a parameter $\theta_{similar}$ as a threshold to judge whether the two trajectories are similar. That is to say, if the similarity $s(T_i, T_j)$ between trajectories of u_i and u_j exceeds $\theta_{similar}$, they are similar users to each other. So the influence $e(u_i, u_j)$ between two users is described by their trajectory similarity $s(T_i, T_j)$.

Location Prediction Model. The goal of Location prediction Model is to infer the most likely location of user u while posting a tweet. Based on Similar User Clustering Model, we get any user u_i's personal similarity graph E_i which contains the influences on her similar users. For example, the influence of user u_i on u_j as well as u_j on u_i is described as $e(u_i, u_j)$ that can also be regarded as the weights of user similarity. Since the higher the similarity of user trajectories is, the greater the influence they have on each other.

We firstly define similar user list of u as $V_u = \{v_1, v_2, \cdots, v_m\}$, the location set of user u's similar users as $L = \{l_1, l_2, \cdots, l_m\}$ and the influence on u as $E_u = \{e(v_1, u), e(v_2, u), \cdots, e(v_m, u)\}$. Then we define $d(l_i, l_j)$ as the Euclidean distance between location l_i and l_j, and $t(v_i, u)$ as the time difference between tweets posted by u and v_i. In this model, user u's current location l_{cur} can be estimated through u's previous location l_{pre} and location list L of his similar users during this period, and the weights list E_u of u.

$$p(l_i|u) = \begin{cases} [1 - \frac{d(l_i, l_{pre})}{\sum_{j=1}^{m} d(l_j, l_{pre})}] \times e(v_i, u) & \text{if } d(l_i, l_{pre}) \leqslant \varepsilon \text{ and } t(v_i, u) \leqslant \delta \\ 0 & \text{otherwise} \end{cases} \tag{4}$$

Equation 4 shows the probability of user u appearing at i-th similar user's location l_i. Since user's moving distance during a set time period is limited, the closer the distance between u and his similar user v_i, the higher the probability that u is at l_i. There are two kinds of measures to finally estimate user's actual location: one is obtaining location with the top one probability, another is gaining the locations whose probability rank in the top $k(k<m)$.

$$l_{cur} = \underset{l_i \in L}{argmax}\, p(l_i|u) \tag{5}$$

Here we choose the first measure which deems the site l_{act} with the highest probability among the candidate locations collected from similar users as user's current location.

3.4   Behavior Habit Model

Although most users' current location can be predicted through tweets content and their similar users, there are still part of users who have few similar users and dislike to post textual content. For example, a user may have relatively stationary trajectory. Therefore, we use user behavior habits based on the supposition that users have their own daily behavior habits to predict their locations. Thus, a location point can be regarded as a state of user. When given a list of states, we can use Markov chain introduced in [12] to predict the next state on the basis of Markov property that the current state is only related to the previous state but not to the earlier states. Then we construct two matrices extended from [10] to depict user behavior habits. For overcoming the sparsity of tweets across location, the city is divided into equal-sized grid cells which represent the regions of city in this subsection.

Definition 3 (Location Transition Matrix). *The location transition matrix* $M_i \in \mathbb{R}^{N \times N}$ *of user* v_i, *where* N *refers to the number of regions divided from the city. Any element of this matrix* $M_i(r, c)$ *represents the frequency that* u_i *transferred from region* r *to region* c.

Definition 4 (Region Vector). *The region vector* Vct_i *of user* v_i *is a* N-*dimension vector. Element* $Vct_i(r)$ *refers to the number of trajectories that* v_i *transfers from region* r *to other regions.*

Definition 5 (Location Transition Probability Matrix). $P_i(r, c)$ *in this matrix* $P_i \in \mathbb{R}^{N \times N}$ *is the probability of transferring to region* c *when the current position of user* v_i *is region* r.

According to the Location Transition Matrix M_i and Region Vector Vct_i, we can get the Location Transition Probability Matrix rely on the Eq. 6.

$$P_i(r, c) = M_i(r, c)/Vct_i(r) \tag{6}$$

Then when v_i's previous location belongs to region c, her current location is calculated as:

$$l_{cur} = \underset{j \in N}{argmax}\, P_i(r_j, c) \tag{7}$$

3.5   Joint Model

The framework architecture and sub models are introduced specifically in the previous subsections. These models leverage textual content, user social relationship and individual behavior habit to mine location information. For the purpose of building a fruitful content-based location prediction model, we use $tfidf$ Value to measure the influence of each word on the locations and define location related words whose $tfidf$ values exceed the threshold θ_{tfidf}. Depending on whether the textual content contains location related words, we determine if the location of a given tweet can be predicted using content-based model, since almost no location features can be extracted from location-independent tweets. Then the locations of tweets which were filtered out can be predicted by a linear combination model that combines social relationship model and behavior habit model.

4 Experiments

In this section, we evaluate the location prediction framework presented in above section through a set of experiments. Our framework is built by three sub models: content-based model, relationship-based model and habit-based model. We report the dataset used in our work and define the general setup of models. Then we design a set of experiments to illustrate the prediction accuracy of different models. In addition, there are several thresholds in these models, such as ε and $\theta_{similar}$. So we also test how these parameters influence the result.

4.1 Datasets

It is not a difficult task to predict users resident cities for the reason that lots of city information can be extracted according to related works. So we can naturally suppose that we know which city the user belongs to. In this paper we gathered data from shanghai, China, since shanghai is one of the most densely populated cities of world whose population is more than 10 million and coverage area is nearly 6340 km^2. We collected about 90 million tweets from nearly 60 thousand users, but only 9 million tweets as well as 10% of the initial data are tagged. While some of them post few tweets or just post links or advertisements of other applications, which do few favor of predicting location. After removing these users and tweets, there are only 1036386 tweets from 10 thousand users left in the dataset.

4.2 Experiment Setup

To predict the fine-grained location of a tweet, we divide the entire city into equal-sized grid cells and each cell is labeled by its diagonal latitude/longitude coordinates. And those locations whose coordinates are falling into the same cell cluster into one category. In this part, we use a turning parameter $cellsize$ to control the granularity of city area division which also is regarded as the prediction error distance. And then we vary the parameter $cellsize$ form 1 to 15 km with the step of 5 km.

In order to evaluate the capability of our model, we calculate the accuracy (ACC) of prediction by:

$$ACC = \frac{|\{l_{cur}|l_{cur} = l_{act}\}|}{|l_{cur}|} \tag{8}$$

Here l_{cur} represents the location of a tweet predicted by our model, and l_{act} is the actual location of this tweet. Throughout our work, we set threshold θ_{tfidf} to

0.1 and filter out tweets without words whose $tfidf$ values exceed θ_{tfidf}. Thus, on the basis of whether mining location information from contents, we can divide the dataset into two parts: one is location correlation dataset and the other is location-independent dataset.

4.3 Capability of Content-Based Model

Data filter using $tfidf$ Value in our work is significant, because the more noise or location-independent tweets, the worse the effect of CNN model will be. So we filter out location dependent tweets utilizing the threshold θ_{tfidf}. After the data filter, we obtain the correlation dataset whose tweets directly or indirectly contain location information and use about 80% as the training set, the rest as the testing set. To build CNN model for location prediction, we firstly turn the check-ins of tweets to the corresponding grid cell, a classification label. And then we segment the textual part of tweets into tokens, remove stop words and punctuations in these tokens at the same time. After these operations, we get a set of words of each tweet and turn these words into word vectors based on the *word2vector* model trained by Wikipedia Chinese corpus and incrementally trained by weibo corpus extracted from dataset. Thus, each tweet turns to an matrix whose rows represent word vectors with 60 dimension and the matrix composed by word vectors can be used as a word embedding in this model. What's more, to solve the problem of different length of tweet, we specify a maximum input tweet length and fill the part whose length is not enough with zero.

To illustrate the significance of data screening, we design experiments to compare the accuracy of this model using raw dataset and filtered dataset respectively. Meanwhile, to better illustrate the validity of our model, we compare other two approaches in related works [2,7] with our method. In [2], a probability model, Content-Based User Location Estimation (CBULE), based on maximum likelihood estimation, builds the probability distribution over regions in the city for each word in the dataset. [7] uses external location-specific data source Foursquare to train language model for each region and then uses $tfidf$ Value approach to predict user's current locations. Here we use the training set to build language model for regions. The prediction results on testing set are shown as Fig. 2.

Fig. 2(a) shows the prediction accuracy of our content-based model with and without data filter. The accuracy of model using filtered dataset is much higher than using row dataset, for data filter greatly reduces the ratio of location-independent tweets in dataset. The results of different content-based approaches are shown in Fig. 2(b). Using CNN model has better performance than using $tfidf$ Value and CBULE model. The prediction accuracy of CNN model is 40.63% within 1 km. What's more, when the error distance is increased by 5 km, the accuracy is raised by nearly 10%. Content-Based User Location Estimation also perform well within different error distances, but $tfidf$ Value measure can just reach 36.92% at maximum error distance. The probable reason of this result is that location related words in training set used to build language model (LM)

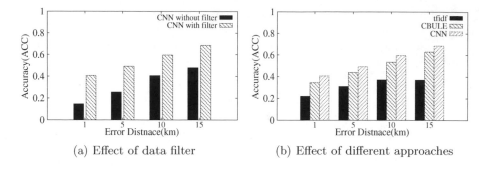

(a) Effect of data filter   (b) Effect of different approaches

Fig. 2. The capability of content based model

are sparse across locations. Overall, CNN model can handle the data with sparse distribution of location related words better than other two models.

4.4   Capability of Relation Based Model

In this section, we predict a user's current location based on the location of his or her similar users' locations. For each user u, the first step is finding users who have similar trajectories with u. Here, we use Spatial-Temporal Longest Common Subsequence Similarity (STLCSS) to calculate the trajectory similarity between users according to Eq. 3 shown in Sect. 3.3. After this operation, the similarity graph among all users is built. For any user u in the dataset, his or her similarity graph can be consisted of $V_u = \{v_1, v_2, \cdots, v_m\}$ and the corresponding relationships are described as $E_u = \{e(v_1, u), e(v_2, u), \cdots, e(v_m, u)\}$, where $e(v_i, u)$ is equivalent to $s(T_i, T_u)$ that exceeds $\theta_{similar}$. To explain the influence of the similarity of users on the accuracy of location prediction, we define different threshold values: $\theta_{similar}$ which is set from 0.3 to 0.5 at the interval of 0.05. As with content-based model, we also test the influence of different error distance on prediction accuracy. The result is shown in Fig. 3.

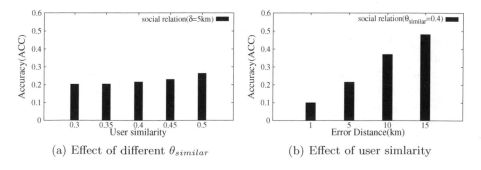

(a) Effect of different $\theta_{similar}$ (b) Effect of user simlarity

Fig. 3. The capability of relation based model

Figure 3(a) shows that the accuracy of prediction is increasing gently with the interval of 0.05 of $\theta_{similar}$ when the error distance is 5 km. Although the accuracy rate shows the trend of overall rise, the increase is small. The reason is that there is a trade-off between the accuracy and the threshold $\theta_{similar}$. Namely, when $\theta_{similar}$ increases, a smaller number of similar users are selected for the prediction. So we fix the threshold $\theta_{similar}$ on 0.4 when testing the effect of error distance based on the result shown in Fig. 3(a). From Fig. 3(b), the prediction accuracy leveraging user similarity continuously increases with the raise of error distance. Because of the low frequency of user posting tweets and small scale of similar user sets in current dataset, the accuracy of this model is low at a very fine granularity.

4.5   Capability of Behavior Habit Based Model

User behavior habit is another important feature for predicting user location, since users always lead regular everyday lives that they have similar daily activities. So we split the whole dataset into two parts, the prior part as the training set and the rest part as testing set. In the training set, we obtain all users' historical data which contains time stamps and latitude/longitude coordinates. For the reason that users are more likely to move within certain regions, we cluster the user locations into regions according to the approach mentioned in Sect. 4.2. Then we will show the effect of different error distance leveraging behavior habit model.

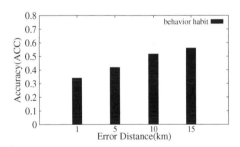

Fig. 4. The capability of behavior habit model

Figure 4 shows that the accuracy of this location prediction model increases with the raise of error distance and locates about 41% of predicted tweets within 5 km from their actual locations. Because users daily activity habits are always repeat nearly everyday. In addition, the larger error distance which also represents the grid cell size, the more locations clustered into a region and the higher accuracy of prediction is.

4.6 Capability of Linear Combination Model

In order to evaluate the effects of last two models, we build a linear combination model to combine these two features and predict tweets locations. Next, we test the effect of different user similarity threshold $\theta_{similar}$ and error distance on behavior habit based model and the combination model of this and social relation based model.

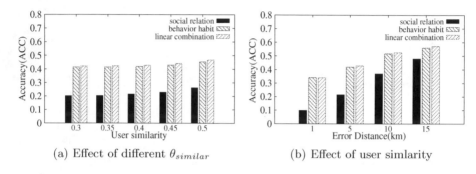

(a) Effect of different $\theta_{similar}$ (b) Effect of user simlarity

Fig. 5. The capability of behavior habit based model

Figure 5(a) shows that user behavior habit performs better than user social relationship in location prediction, and combining these two models can nearly double the accuracy of prediction leveraging the social relationship based model. Because almost all users' daily behavior habits are nearly settled, we mine a lot of location information from users' historical data. As with the reason mentioned in Sect. 4.4, the accuracy of combination model is a little higher than behavior habit based model separately used. In Fig. 5(b), the accuracy of behavior habit model increases with the raising of error distance. We also find that comparing the social relationship based model, behavior habit based model has obvious advantages in predicting fine-grained locations of tweets, while the advantages are gradually weakened in larger error distance. The probable reason is that the number of similar users increases as well as candidate locations in coarse-grained prediction.

5 Conclusion

We present a joint model for tweets location prediction which contains three sub models based on different features mining from tweets data. These models utilize textual content, social relationship and user behavior habit to extract location information, and obtain high prediction precision. From the experimental results, we conclude that content-based model is more suitable for tweets containing location related words, while other tweets can use the combination model to predict current locations.

Moreover, this work can be extended in user social relationships that takes into account user interaction information for building more sophisticated user social graphs. We would like to further reduce the prediction error to get a more granular predictive location of a given tweet as well.

References

1. Chandra, S., Khan, L., Muhaya, F.B.: Estimating Twitter user location using social interactions-a content based approach. In: IEEE Third International Conference on Privacy, Security, Risk and Trust, pp. 838–843 (2012)
2. Cheng, Z., Caverlee, J., Lee, K.: You are where you tweet: a content-based approach to geo-locating Twitter users, vol. 19, no. 4, pp. 759–768 (2010)
3. Crandall, D.J., Backstrom, L., Huttenlocher, D., Kleinberg, J.: Mapping the world's photos. In: International Conference on World Wide Web, pp. 761–770 (2009)
4. Ichiye, T., Karplus, M.: Collective motions in proteins: a covariance analysis of atomic fluctuations in molecular dynamics and normal mode simulations. Proteins Struct. Funct. Bioinf. **11**(3), 205 (1991)
5. Kearney, J.K., Hansen, S.: Stream editing for animation (1990)
6. Kim,Y.: Convolutional neural networks for sentence classification. CoRR, abs/1408.5882 (2014)
7. Kim, Y., Chiu, Y., Hanaki, K., Hegde, D., Petrov, S.: Temporal analysis of language through neural language models. CoRR, abs/1405.3515 (2014)
8. Kong, L., Liu, Z., Huang, Y.: SPOT: locating social media users based on social network context. VLDB Endow. **7**, 1681–1684 (2014)
9. Lee, K., Ganti, R.K., Srivatsa, M., Liu, L.: When Twitter meets foursquare: tweet location prediction using foursquare. In: International Conference on Mobile and Ubiquitous Systems: Computing, Networking and Services, pp. 198–207 (2014)
10. Lin, S.K., Sheng-Zhi, L.I., Qiao, J.Z., Yang, D.: Markov location prediction based on user mobile behavior similarity clustering. J. Northeast. Univ. (2016)
11. Backstrom, L., Sun, E., Marlow, C.: Find me if you can: improving geographical prediction with social and spatial proximity. In: International Conference on World Wide Web, pp. 61–70 (2010)
12. Gasparini, M.: Markov chain Monte Carlo in practice. Technometrics **39**(3), 338 (1997)
13. Robinson, M.: The temporal development of collision cascades in the binary collision approximation. Nucl. Inst. Methods Phys. Res. B **48**(1–4), 408–413 (1990)
14. Sadilek, A., Kautz, H., Bigham, J. P.: Finding your friends and following them to where you are. In: ACM International Conference on Web Search and Data Mining, pp. 723–732 (2012)
15. Serdyukov, P., Murdock, V., Zwol, R.V.: Placing flickr photos on a map, pp. 484–491 (2009)
16. Xu, D., Yang, S.: Location prediction in social media based on contents and graphs. In: International Conference on Communication Systems Network Technologies, pp. 1177–1181 (2014)

Efficient Longest Streak Discovery in Multidimensional Sequence Data

Wentao Wang[1,2], Bo Tang[1], and Min Zhu[2(✉)]

[1] Shenzhen Key Laboratory of Computational Intelligence,
Department of Computer Science and Engineering,
Southern University of Science and Technology, Shenzhen, China
wangwt@mail.sustc.edu.cn, tangb3@sustc.edu.cn
[2] College of Computer Science, Sichuan University, Chengdu, China
zhumin@scu.edu.cn

Abstract. This paper studies the problem of discovering longest streak in multidimensional sequence dataset. Given a multidimensional sequence dataset, the contextual longest streak is the longest consecutive tuples in a context subspace which match with a specific measure constraint. It has various applications in social network analysis, computational journalism, etc. The challenges of the longest streak discovery problem are (i) huge search space, and (ii) non-monotonicity property of streak lengths. In this paper, we propose a novel computation framework with a suite of optimization techniques for it. Our solutions outperform the baseline solution by two orders of magnitude in both real and synthetic datasets. In addition, we validate the effectiveness of our proposal by a real-world case study.

Keywords: Computational journalism
Multidimensional sequence data

1 Introduction

Computation Journalism [2,3] has been emerged as an interdisciplinary field in recent years. In traditional news media organizations, reporters always manually bring out attention-seizing factual statements backed by data, which may lead to news stories and investigation. However, the qualities of findings are hard to guarantee. In contrast, *Computation Journalism* brings together experts in traditional news organization, social science and computer science, and advances journalism by innovations in computational techniques. One of the goals of *Computation Journalism* is to discover newsworthy facts efficiently and effectively. In data engineering community, unstructured data analytical techniques become more sophisticated, which push the database and data mining researchers be the frontiers of the *Computation Journalism* filed. E.g., discovering newsworthy themes [4], finding prominent streaks [6,13], exploring situational facts [8], etc.

B. Tang—is co-first author.

© Springer International Publishing AG, part of Springer Nature 2018
Y. Cai et al. (Eds.): APWeb-WAIM 2018, LNCS 10988, pp. 166–181, 2018.
https://doi.org/10.1007/978-3-319-96893-3_13

Contextual longest streak is a set of maximal contiguous tuples with measure values satisfying a threshold in a context (i.e., a subset of multidimensional sequence data), which has found applications in computational journalism. Consider NBA dataset as an example, the following two news articles are user cases of contextual longest streaks:

- (I) **Longest** 100+ **Streak at Home:** *March 10, 2017, Golden State Warriors' 100+ Points Streak at Home Ends at 56 Games.*[1]
- (II) **Longest** 100+ **Streak at the History of Cleveland Cavaliers:** *December 3, 2017, Cleveland Cavaliers' 16-game streak of 100+ points ties longest in franchise history.*[2]

In news (I) the contextual longest streak is a 56 consecutive Home games of Golden State Warriors with the score values are 100 or above, and in news (II) the contextual longest streak is a subset of 16 consecutive game records of Cleveland Cavaliers with the score values are 100 or above.

Given data table $R(O, D, M)$, where O is a(n) sequential/ordinal dimension, D is a set of dimensional attributes, and M is a set of measure attributes. A context subspace S is a subset of R, resulting from a conjunctive constraints on a one or more dimension attributes in D. The measure constraint C was defined by specific values in M. The longest streak in a context subspace (i.e.,contextual longest streak) is the longest consecutive tuples/records in context subspace S, which matches with the given measure constraint C. For example, in news (I), the context subspace S is {Court = Home, Team = GoldenStateWarriors} and the measure constraint C is {Score $\geq$ 100}, the length of its corresponding longest streak is 56. In this paper, we study how to discover longest streak of a specified measure constraint C in an append-only multidimensional sequence data table R, i.e., it discovers the largest contextual longest streak by exploring every possible context subspace in $R(O, D, M)$. The technical challenges of the longest streak discovery problem are:

- **Exponential subspaces in R:** The number of subspaces is exponential to the number of dimension attributes. It also is polynomial to the domain sizes of dimensions. We will analyze it shortly.
- **Non-monotonicity of the longest streak length:** As we will explain in Sect. 4, the length of streaks is not monotonic. For example, with measure constraint C = {Score $\geq$ 100}, the length of longest streak at subspace {Team = ClevelandCavaliers} is 16, it does not guarantee that the length of longest streak of its descendant subspaces is no larger than 16.

While the longest streak discovery problem is challenging, we develop a suite of optimization techniques to address it. The contributions of this paper are:

---

[1] http://blcacherreport.com/articles/2697055-golden-state-warriors-100-point-streak-at-home-ends-at-56-games.

[2] https://cavsnation.com/cavs-news-clevelands-16-game-streak-of-100-points-ties-longest-in-franchise-history/.

1. We study the novel problem of discovering longest streak in multidimensional sequence data.
2. We devise efficient algorithms by exploiting the upper bound of context subspace and segments among tuple sets.
3. We evaluate the efficiency and effectiveness of our proposal in synthetic and real datasets, respectively.

The rest of the paper is organized as follows. Section 2 reviews related works. Section 3 gives the formal problem definition. Section 4 develops the computation framework. Section 5 devises a suite of optimization techniques to reduce the computation cost. Section 6 presents the case studies and performance evaluations. Section 7 concludes the paper.

2   Related Work

Computation Journalism: It has emerged recently as a interdisciplinary filed, which exploits the advances in computer science to assist the experts in traditional news organization [2,3] . We summarize the related works as follows.

One-of-the-k problem was investigated in [12]. For a given object o, it reports the non-empty subset of measure attributes (a.k.a., measure attributes in OLAP) if o is dominated by fewer than k objects. Fan et al. [4] proposed a novel k-Sketch query that aims to find k streaks to best summarize a given subject with scoring function. Finding a set of maximal contiguous subsequences with value all above (or below) a certain threshold (a.k.a., prominent streaks) was proposed in [6]. In its extension version [13], the authors extend the techniques in [6] for discovering general top-k, multi-sequence, and prominent streaks in multiple measures dataset. Both assume there is single dimension (i.e., ordinal dimension) in the dataset. In contrast, our proposal take an OLAP table with is a set of dimensions (i.e., D). Discovering a set of records (i.e., skyline) with regard to a context and several measures (a.k.a., prominent situational facts) was studied in [8]. It finds situational facts pertinent to new tuple in an append-only database. In our proposal, user are focusing on whole dataset, instead of the new append tuple. We summarize the major differences between these works and our proposal in Table 1.

Table 1. Comparison with related works

Related work	User input	Problem output	Sequence data?	Multi-dimensions?	Multi-measures?
Wu et al. [12]	Object o, k	Measure subset	No	No	Yes
Jiang et al. [6,13]	Nil	Streaks	Yes	No	Yes
Sultana et al. [8]	New tuple t	Dimension subset and measure subset	No	Yes	Yes
Fan et al. [4]	Subject	Top-k streaks	Yes	No	No
This work	**Measure subset**	**Longest streak**	**Yes**	**Yes**	**Yes**

Longest Subsequence Discovery: In addition, discovering longest subsequences is a well-known research problem computer science [1,5,7]. However, these existing techniques cannot be adapted our problem as we consider all possible contexts (i.e., subspaces in OLAP) in the dataset.

Exploratory Analysis: The data engineering community has developed subspace mining techniques to discover interesting subspaces from a dataset [9–11]. Our problem differs from them in two ways. First, we consider the multidimensional sequence dataset, while the others does not requires that. Second, our findings (i.e., longest streak) have not been studied yet.

3   Problem Statement

In this section, we give the formal definitions of the concepts in our problem, and formulate our longest streak discovery problem.

Given a multidimensional sequence dataset $R(\mathsf{O}, \mathsf{D}, \mathsf{M})$, where O is a(n) sequence/ordinal attribute, $\mathsf{D} = \{D_1, \cdots, D_d\}$ is a set of dimension attributes, and $\mathsf{M} = \{M_1, \cdots, M_m\}$ is set of measure attributes. For example, Table 2 is a NBA dataset, the ordinal attribute of NBA dataset is $\mathsf{O} = \mathsf{ID}$, the dimension attributes are $\mathsf{D} = \{\mathsf{Player}, \mathsf{Oppteam}, \mathsf{Season}\}$, and the measure attributes are $\mathsf{M} = \{\mathsf{Points}, \mathsf{Rebounds}\}$. We denote the domain size of dimension D_i as $|dom(D_i)|$, e.g., $|dom(\mathsf{Player})| = 3$ in Table 2, there are three players in dimension Player.

Table 2. NBA dataset

ID	Player	Oppteam	Season	Points	Rebounds
t_1	Russell Westbrook	Los Angeles Clippers	2015–16	35	6
t_2	Kevin Love	Golden State Warriors	2015–16	18	15
t_3	James Harden	Indiana Pacers	2015–16	17	8
t_4	Kevin Love	Golden State Warriors	2015–16	17	16
t_5	Russell Westbrook	Cleveland Cavaliers	2015–16	31	11
t_6	Kevin Love	Phoenix Suns	2016–17	23	8
t_7	James Harden	Sacramento Kings	2016–17	22	13
t_8	James Harden	Los Angeles Clippers	2016 17	18	9
t_9	Russell Westbrook	Los Angeles Clippers	2016–17	35	6
t_{10}	Kevin Love	Golden State Warriors	2016–17	18	15
t_{11}	James Harden	Indiana Pacers	2016–17	21	14
t_{12}	Russell Westbrook	Phoenix Suns	2016–17	31	9

Definition 1 (Context Subspace). *A context subspace is an array $S = \langle a_1, \cdots, a_d \rangle$, where $a_i \in dom(D_i) \cup \{*\}$ and $*$ refers to all values. We denote the tuples in context subspace S as T_S. A context subspace $S_1 = \langle a_1, a_2, \cdots, a_d \rangle$,*

which is called a child subspace of $S_2 = \{b_1, b_2, \cdots, b_d\}$, *iff there exists a j such that* $a_j \neq * \wedge b_j = *$ *and* $a_i = b_i$ *for any* $i \neq j$. *Conversely, S_2 is a parent subspace of S_1.*

Example. Given a context subspace $S = \langle \text{Kevin Love}, *, * \rangle$ in Table 2, $T_S = \{t_2, t_4, t_6, t_{10}\}$.

We shall find longest streak in a multidimensional sequence dataset with regard to the user specific measure constraint. We then introduce *Measure Constraint* and *(Longest) Streak*.

Definition 2 (Measure Constraint). *A measure constraint is a measure record* $\mathsf{C} = \langle v_1, \cdots, v_m \rangle$. *A tuple t in R matches with constraint* $\mathsf{C} = \langle v_1, \cdots, v_m \rangle$ *iff* $t.M_i \geq v_i, \forall i \in [1, m]$.

For the sake of presentation, we assume the constraints in C are "greater than" operators (i.e., $t.M_i \geq v_i$). Nevertheless, our solution is extensible to other kinds of operators, e.g., "less than", "range from", etc. We will discuss the extensions in Sect. 4.4.

Example. In Table 2, given a constraint $\mathsf{C} = \langle 10, 10 \rangle$, it is *double-double*[3] on combination of Points and Rebounds. t_2 is qualified tuple w.r.t. C as $t_2.\text{Points} = 18 \geq 10$ and $t_2.\text{Rebounds} = 15 \geq 10$, and t_6 is not qualified as $t_6.\text{Rebounds} = 8 < 10$.

The definitions of *Streak* and *Longest Streak* are as follows.

Definition 3 (Streak in a Context Subspace). *Given a constraint* C, *a streak in a context subspace S is a set of consecutive tuples* ST_S, *where* $\mathsf{ST}_S \subseteq T_S$, *and* $\forall t_i \in \mathsf{ST}_S$ *matches with* C. *The length of the streak is* $|\mathsf{ST}_S|$.

Example. Given constraint $\mathsf{C} = \langle 10, 10 \rangle$, context subspace $S = \langle \text{Kevin Love}, *, * \rangle$. The tuple set of S is $T_S = \{t_2, t_4, t_6, t_{10}\}$, a streak in S with constraint C is $\mathsf{ST}_S = \{t_2, t_4\}$. Its length is $|\mathsf{ST}_S| = 2$, i.e., Kevin Love got *double-double* in two consecutive games.

For a given context subspace S and constraint C, there are probably multiple streaks in it. Take Fig. 1(a) as an example, $S = \langle *, *, * \rangle$ with $\mathsf{C} = \langle 20, 0 \rangle$. It has four streaks, they are $\mathsf{ST}_S^1 = \{t_1\}$, $\mathsf{ST}_S^2 = \{t_5, t_6, t_7\}$, $\mathsf{ST}_S^3 = \{t_9\}$, and $\mathsf{ST}_S^4 = \{t_{11}, t_{12}\}$, respectively.

Definition 4 (Longest Streak in a Context Subspace). *Given a constraint* C, *the longest streak in context subspace S is* LST_S, *such that* $\forall \mathsf{ST}_S^i \in T_S$, *we have* $|\mathsf{LST}_S| \geq |\mathsf{ST}_S^i|$.

As shown in Fig. 1(a), (b), (c) and (d), the longest streak in these four context subspaces are $\{t_5, t_6, t_7\}$, $\{t_1, t_5, t_9, t_{12}\}$, $\{t_6\}$ and $\{t_7\}$ (or $\{t_{11}\}$), respectively. The lengths of the corresponding longest streaks are 3, 4, 1, and 1.

Given a constraint C, Definition 5 defines the longest streaks in a multidimensional sequence dataset $R(\mathsf{O}, \mathsf{D}, \mathsf{M})$.

---

[3] https://en.wikipedia.org/wiki/Double_(basketball).

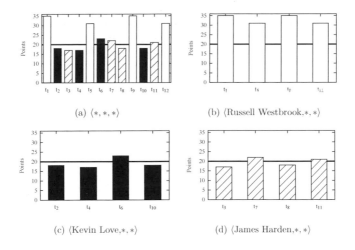

Fig. 1. (Longest) streaks in different context subspaces

Definition 5 (Longest Streak in $R(\mathsf{O}, \mathsf{D}, \mathsf{M})$). *Given a constraint* C, *the longest streaks in* $R(\mathsf{O}, \mathsf{D}, \mathsf{M})$ *is* $\mathcal{LS} = \{(S, \mathsf{LST}_S) : \forall\ S' \subseteq R, |\mathcal{LS}.\mathsf{LST}_S| \geq |\mathsf{LST}_{S'}|\}.$

Given NBA dataset in Table 2 with constraint $\mathsf{C} = \langle 20, 0 \rangle$, the longest streak $\mathcal{LS}$ in it is $(\langle\text{Rusell Westbrook}, *, *\rangle, \{t_1, t_5, t_9, t_{12}\})$.

Problem 1 (Longest Streak Discovery). Given a multi-dimensional dataset $R(\mathsf{O}, \mathsf{D}, \mathsf{M})$ and a constraint $\mathsf{C} = \langle v_1, \cdots, v_m \rangle$. The longest streak discovery problem is finding $\mathcal{LS}$, which consists of context subspace and its corresponding longest streaks pairs, such that context subspaces have the longest streaks length in the whole search space.

Example. Consider records in Table 2 with constraint $\mathsf{C} = \langle 10, 10 \rangle$, the longest streak $\mathcal{LS} = \{(\langle\text{Kevin Love, Golden State Warriors}, *\rangle, \{t_2, t_4, t_{10}\})\}$. In computational journalism, it shows the fact that Kevin Love achieved three consecutive *double-double* against Golden State Warriors.

Problem Complexity Analysis: Before presenting our solutions, we first analyze the complexity of our problem from the following two aspects.

Search space: The search space of Problem 1 is equivalent to the total number of context subspaces in $R(\mathsf{O}, \mathsf{D}, \mathsf{M})$, it is as follows. Please note that the search space is exponential to the number of dimension attributes $|\mathsf{D}|$.

Lemma 1 (Problem search space). *The number of context subspaces of data table* $R(\mathsf{O}, \mathsf{D}, \mathsf{M})$ *is* $\prod_{i=1}^{d}(|dom(D_i)| + 1).$

Proof. For each $D_i \in S$, the number of distinct values is $dom(D_i) + 1$, i.e., the domain size and the "all" value. Thus, the total number of context subspace in R is $\prod_{i=1}^{d}(|dom(D_i)| + 1).$

Non-monotonicity property: Typical sequence mining algorithms exploit the monotone property to devise efficient pruning techniques. To make the matter worse, in Problem 1, it does not hold the monotone property. Formally:

Lemma 2 (Non-monotonicity).   *Given two context subspaces S and S', where S is the parent of S'. It holds that $|\mathsf{LST}_S|$ is neither monotone increasing nor monotone decreasing with respect to $|\mathsf{LST}_{S'}|$.*

We provide a counter example to demonstrate the non-monotonicity as follows. Consider constraint $\mathsf{C} = \langle 10, 10 \rangle$ in Table 2. Context subspace $S = \langle \text{Kevin Love}, *, * \rangle$, $\mathsf{LST}_S = \{t_2, t_4\}$ with length 2. However, the longest streaks of its child subspaces $S_1 = \langle \text{Kevin Love, Phoenix Suns}, * \rangle$ and $S_2 = \langle \text{Kevin Love, Golden State Warriors}, * \rangle$ are $\mathsf{LST}_{S_1} = \emptyset$ with length 0, and $\mathsf{LST}_{S_2} = \{t_2, t_4, t_{10}\}$ with length 3, respectively.

4   Discovering Longest Streak

In this section, we propose a novel computation framework to discover longest streak in multidimensional sequence dataset (i.e., data table $R(\mathsf{O}, \mathsf{D}, \mathsf{M})$).

4.1   Computation Framework

Algorithm 1 is the pseudo-code of our computation framework. It employs a set $\mathcal{LS}$ to store the longest streaks found so far and use an integer $maxlen$ to denote the length of longest streak (Line 1–2). The algorithm traverses all

---

Algorithm 1. LongestStreaks(dataset $R(\mathsf{O}, \mathsf{D}, \mathsf{M})$, constraint C)

1: Initialize result $\mathcal{LS} \leftarrow \emptyset$, longest streak length $maxlen \leftarrow 1$ ▷ Initialization
2: Initialize root $S \leftarrow \langle *, *, \cdots, * \rangle$ and its corresponding tuple set $\mathsf{T}_S \leftarrow R$
3: $\mathsf{LST}_S \leftarrow \text{FindLS}(\mathsf{T}_S, \mathsf{C})$ ▷ Find longest streak in context subspace S
4: **if** $|\mathsf{LST}_S| \geq maxlen$ **then**
5:       Update $\mathcal{LS}$ by (S, LST_S), $maxlen \leftarrow |\mathsf{LST}_S|$
6: **for** $i \leftarrow 1$ **to** d **do**   ▷ Enumerate context subspace by instantiating D_i
7:       EnumerateSubspace(S, T_S, D_i)
8: **return** $\mathcal{LS}$

Function: EnumerateSubspace(S, T_S, D_i)
9: **foreach** $v \in dom(D_i)$ **do**
10:       $S' \leftarrow S, S'_i \leftarrow v$
11:       Obtain $\mathsf{T}_{S'}$ from T_S
12:       $\mathsf{LST}_{S'} \leftarrow \text{FindLS}(\mathsf{T}_{S'}, \mathsf{C})$ ▷ Find longest streak in context subspace S'
13:       **if** $|\mathsf{LST}_{S'}| \geq maxlen$ **then**
14:             Update $\mathcal{LS}$ by $(S', \mathsf{LST}_{S'})$, $maxlen \leftarrow |\mathsf{LST}_{S'}|$
15:       **for** $j \leftarrow i + 1$ **to** d **do**   ▷ Enumerate context subspace at dimension D_j
16:             EnumerateSubspace($S', \mathsf{T}_{S'}, D_j$)

---

possible context subspaces from the root subspace $\langle *, *, \cdots, * \rangle$. For each context subspace S, it discovers the longest streak LST_S in it by calling function FindLS, which will be discussed shortly in Sect. 4.2, and updates $\mathcal{LS}$ accordingly.

The context subspaces are traversed by recursive function Enumerate Subspace. All child context subspaces of S are enumerated by instantiating the value on dimension D_i (Line 10). For each child S', its tuple set $\mathsf{T}_{S'}$ was derived from S's tuple set T_S (Line 11). Then FindLS was called to discover the longest streak in S' and update result set $\mathcal{LS}$ (Line 12–14). Finally, we further traverse the context subspaces by instating the values in dimension D_j where $j > i$ at Line 15–16.

4.2 Discover Longest Streak in a Context Subspace S

Discovering the longest streak in a context subspace S, i.e., Function FindLS in Algorithm 1, is a core subroutine of Problem 1. In this section, we first present the straight forward solution of it. We then propose two simple but effective optimizations to improve its performance.

Algorithm 2. FindLS(context subspace tuple set T_S, constraint C)

1: Initialize result $\text{LST}_S \leftarrow \emptyset$
2: Initialize streak $\text{ST} \leftarrow \emptyset$
3: **for** $i \leftarrow 1$ **to** $|\mathsf{T}_S|$ **do**
4:      $j \leftarrow i, \text{ST} \leftarrow \emptyset$
5:      **while** $\mathsf{T}_S[j]$ matches with C **do**
6:           $\text{ST} \leftarrow \text{ST} \cup \mathsf{T}_S[j]; \quad j \leftarrow j + 1$
7:      **if** $|\text{ST}| \geq |\text{LST}_S|$ **then**
8:           $\text{LST}_S \leftarrow \text{ST}$
     return LST_S

Reduce Tuple Testing: In Algorithm 2, we verify whether tuple $\mathsf{T}_S[j]$ matches with constraint C at Line 5. It incurs expensive computation cost as the measure values of tuple $\mathsf{T}_S[j]$ are compared $|\mathsf{T}_S| - j + 1$ times. In order to avoid it, we compare the measure values of each tuple in R with constraint C once, and store the result in a bit set B. The corresponding bit set of T_S is denoted as B_S. As illustrated in Fig. 2, given measure constraint $\mathsf{C} = \langle 10, 10 \rangle$, we can convert each tuple in Table 2 to a bitset B_S. Thus, Line 5 can be replaced by a cheaper boolean testing $B_S[j] ==$ True.

Fig. 2. Convert T_S to a bitset B_S

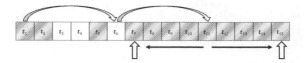

Fig. 3. Skip unqualified candidates

Skip Unqualified Candidates: Unlike discovering the streak with different starting position one by one in Line 3, Algorithm 2. We propose to set step size as *maxlen* based on the observation below.

Observation 1. *Given context subspace S, its corresponding tuple set T_S and bitset B_S, and the longest streak length found so far is maxlen. Consider a streak with start position j in it, if $B_S[j + maxlen]$ is False, then all streaks with starting position from j to $j + maxlen$ can be pruned.*

Consider the example illustrated in Fig. 3, it includes 15 tuples, $maxlen = 5$. The gray cells are the tuples which match with constraint C, and the white cells are not. Suppose the streak with starting tuple t_1, we found that $t_{1+5} = t_6$ does not match with C, then we can conclude the streaks with starting tuple $t_1, \cdots, t_6$ are not the longest streak in S. We next process the case $B_S[j + maxlen]$ is True, e.g., t_{11} in Fig. 3. Since t_{11} is true, there is a streak which includes t_{11}. In order to identify the starting and ending tuple of the above streak, we search from both directions of t_{11}. Then we found t_7 and t_{15} are starting and ending tuple, respectively. The length of the streak is $15-7+1 = 9$, it is large than $maxlen = 5$, LST_S is updated accordingly.

Through this, the time complexity of FindLS can be reduced from $O(|\mathsf{T}_S|^2)$ to $O(|\mathsf{T}_S|)$, where $|\mathsf{T}_S|$ is the tuple size of context subspace S.

4.3   Time Complexity Analysis of Algorithm 1

The time complexity of Algorithm 1 is the number of context subspaces and the cost of discovering the longest streak in each subspace. the number of context subspaces in data table R is $\prod_{i=1}^{d}(|dom(D_i)| + 1)$ (cf. Lemma 1). The cost of discovering the longest streak in a given subspace is $O(|\mathsf{T}_S|) \leq O(n)$, where n is the total tuples in R (cf. Sect. 4.2). Thus, the time complexity of Algorithm 1 is $O(n \cdot \prod_{i=1}^{d}(|dom(D_i)| + 1))$.

4.4   Extensions

We provide a computation framework for longest streak discovery problem. We suggest the constraint comparison operator, i.e., "greater than". Nevertheless, our framework is extensible.

Comparison Operators: Since we convert constraints verification to a preprocessed bitset (cf. Sect. 4.2), then our framework can adapt to any other operators, e.g., $\geq, >, \leq, <$. We also can put different comparison constraints in different measure attributes. For example, our system could discover the longest streak with constraint $C = \{\text{Points} \geq 25, \text{Rebound} \geq 10, \text{Turnover} \leq 5\}$. In practice, it is finding the longest streak of the player who achieves 25+ points, 10+ rebounds and 5− turnovers.

Top-k Longest Streaks: In this work, we only focus on the longest streak in the dataset, however, our framework can extend to discover top-k longest streaks. The adoption is straight forward by maintaining the top-k longest streaks in a min-heap in Algorithm 1.

Customize Search Space: In this work, we search the whole space to identify the longest streak. However, user also can specify the search space to discover the longest streak with her interests.

5   Optimizations

In this section, we devise several optimization techniques to reduce the computation cost of our computation framework in Algorithm 1.

5.1   Context Subspace Pruning

As Sect. 4.3 shown, the dominate part is the number of context subspaces, i.e., $\prod_{i=1}^{d}(|dom(D_i)| + 1)$. In order to reduce the total computation cost, for a given context subspace we propose an upper bound of its longest streak length.

Given a constraint C and data table $R(O, D, M)$, the upper bound of the longest streak length in context subspace S is the total number of tuples in T_S which match with C,

$$UB_S = \{|CT| : CT = \cup_{t_i \in T_S}(B_S[i] = \text{True})\}.$$

According to the definition of longest streak length upper bound, we have Lemma 3.

Lemma 3 (The upper bound of longest streak length in S). *The longest streak in a context subspace S is LST_S, we have $UB_S \geq |\text{LST}_S|$. We denote S' is the child subspace of S, we have $UB_S \geq |\text{LST}_{S'}|$.*

Proof. For the first part $UB_S \geq |\text{LST}_S|$, the proof is trivial as the tuples in the longest streak must match with constraint C, i.e., $\text{LST}_S \subseteq CT$. For the second part, suppose there exists a child context subspace S' in S that $|\text{LST}_{S'}| > UB_S$. Since $UB_{S'} \geq |\text{LST}_{S'}|$, we can conclude $UB_{S'} > UB_S$. It is contradicted with S' is a child subspace of S.   □

Example: In Table 2, given constraint $C = \langle 10, 10 \rangle$ and context subspace $S = \langle \text{Kevin Love}, *, * \rangle$. $|\text{LST}_S| = 2$ and its length upper bound is $UB_S = 3$. $S' = \langle \text{Kevin Love}, \text{Golden State Warriors}, * \rangle$ is a child subspace of S, $|\text{LST}_{S'}| = 3 \leq UB_S$.

In addition, with the upper bound property we sort S's child context subspaces by its upper bound in descending order, it will turn to a local best first search strategy. We then incorporate these two optimization techniques to the computation framework and show it in Algorithm 3.

Algorithm 3. LongestStreaksI(dataset $R(\mathsf{O}, \mathsf{D}, \mathsf{M})$, constraint C)

1: Same with Lines 1–8 in Algorithm 1

    Function: EnumerateSubspace(S, T_S, D_i)
2: Same with Line 12 in Algorithm 1
3: Derive S's children context subspaces in D_i upper bounds from T_S
4: Sort S's children context subspaces by the upper bounds of D_i
5: **foreach** $v \in dom(D_i)$ **do**
6:     $S' \leftarrow S, S'_i \leftarrow v$
7:     **if** $UB_{S'} \geq maxlen$ **then**
8:         Same with Lines 11–16 in Algorithm 1

Analysis: Compare with Algorithms 1 and 3 computes the upper bounds of each children context subspace S'. However, these bounds can be derived on-the-fly as in Line 11, Algorithm 1, it also need obtain its corresponding tuple set $\mathsf{T}_{S'}$. Thus, in worst case, the time complexity of Algorithm 3 is the same as Algorithm 1 in terms of Big-O notation. However, Algorithm 3 performs extremely better than Algorithm 1, we will show it in experimental section, as it equips upper bound pruning techniques and local best first search strategy.

5.2   Segmentation Techniques

In our previous discussion, for a given context subspace S, we process the tuples in its corresponding tuple set T_S one by one. In this section, we propose to process these records one segment by one segment, which provide benefits in two aspects: (i) we can use a divide-and-conquer methodology to find the longest streak in the context subspace, and (ii) we could prune the unqualified records during the context subspace traversal progress. We first define the segment of tuples in T_S.

Definition 6 (Segment of Tuples).   *Given a constraint* C *on dataset* $R(\mathsf{O}, \mathsf{D}, \mathsf{M})$, *consider a context subspace* S *with tuple set* T_S, *a segment in* T_S *is an interval* $[i, j]$, *where (1)* $\forall k \in [i, j]$, $\mathsf{T}_S[k]$ *matches with* C *(or does not match with* C), *and (2)* $\mathsf{T}_S[i-1], \mathsf{T}_S[j+1]$ *do not match with* C *(or match with* C).

According to Definition 6, we transform the tuple set T_S in Fig. 4(a) to a segment set SG_S in Fig. 4(b).

Suppose $\mathsf{SG}[i]$ records the starting position of the tuples in this segment, as the arrows shows in Fig. 4.

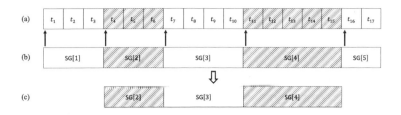

Fig. 4. Tuple set T_S to segment set SG_S

Observation 2. *We obtain the following observations.*

- *(1) If* $\mathsf{SG}[i]$ *matches with constraints* C*, then* $\mathsf{SG}[i-1]$ *and* $\mathsf{SG}[i+1]$ *either do not match with* C *or do not exist (i.e.,* $\mathsf{SG}[i]$ *is the first or last segment). For example,* $\mathsf{SG}[2]$ *matches with* C*, and* $\mathsf{SG}[1]$*,* $\mathsf{SG}[3]$ *do not match with* C*.*
- *(2) The length of* $\mathsf{SG}[i]$ *is* $\mathsf{SG}[i+1] - \mathsf{SG}[i]$*. For example, the length of* $\mathsf{SG}[2]$ *is* $\mathsf{SG}[3] - \mathsf{SG}[2] = 7 - 4 = 3$*.*
- *(3) If the first segment (or last segment) does not matches with* C*, it can be removed as it will not affect the length of the longest streak in it or its descendant context subspaces.*

Pruning Unqualified Tuples/Segments: Tuples in $\mathsf{SG}[1]$ and $\mathsf{SG}[5]$ could be removed from that context subspaces S (cf. Observation (3)), none of S's descendant context subspaces will consider them as the two segments will not affect the length of longest streaks in them. After pruned the unqualified tuples (or segments), the segment set turns to Fig. 4(c).

Divide-and-Conquer Strategy: Unlike previous FindLS function, here we commence the FindLS find at the middle of segment set SG. Suppose the current $maxlen = 4$, take Fig. 4(c) as an example, we first check $\mathsf{SG}[3]$, it skipped as it does not match with C. Then it incurs two FindLS function calls, the first one is for the left side of $\mathsf{SG}[3]$ (i.e., $\mathsf{SG}[2]$), the second one is for the right side of $\mathsf{SG}[3]$ (i.e., $\mathsf{SG}[4]$). Since the length of its left-side is 3, it is less than $maxlen = 4$, so we can skip it directly. Finally, we discovered the longest streak of the context subspace S is $\mathsf{SG}[4]$ with length 5. In general, we could use the divide-and-conquer strategy with the additional pruning (cf. the pruning progress of the left side of $\mathsf{SG}[3]$) recursively.

Generating Segment Set of S's Children: The idea of generating the segment set of S's children context subspaces by instantiating the values in dimension D_i is straight forward, We only need to maintain $|dom(D_i)|$ segment sets and process the tuples in T_S one by one. The upper bound of each context subspace S' also can be derived with the above progress.

Put All it Together: We incorporate the above ideas to Algorithm 3, we omit the details of the algorithm due to space limitation. The time complexity of the algorithm is the same as Algorithm 3, however it improves the performance as shown in Sect. 6.

6   Experimental Evaluation

In this section, we evaluate the effectiveness of longest streaks in real dataset and the efficiency of our proposed techniques. We first introduce the experiment setting in Sect. 6.1. We then conduct the case studies in Sect. 6.2, and finally, we evaluate the superiority of our proposal in Sect. 6.3.

6.1   Experiment Setting

We describe the details of the datasets, implementation of our proposed algorithms and hardware configurations in this section.

Dataset: We used two real datasets **NBA** and **Weather**, and one synthetic dataset **Synthetic** in our experiments.

NBA.[4] It contains 795,149 tuples of NBA box scores from 1985–2016 regular seasons. There are 8 dimension attributes and 5 measure attributes in each tuple. The dimension attributes are: Player, Team, Oppteam, RegularOrPlayoff, Court, Week, Month, Season. The measure attributes includes: Points, Rebounds, Assists, Steals and Blocks.

Weather.[5] There are 28,661,310 tuples about site specific forecasts for over 6,000 sites in the UK from 2012 to 2014. Each tuple has 8 dimension attributes and 6 measure attributes. The dimension attributes includes site name, wind direction, visibility, etc. The measure attributes are: Wind Speed Day, Wind Speed Night, Relative Humidity Day, Relative Humidity Night, Wind Gust Day and Wind Gust Night.

Synthetic. We generate 50,000,000 tuples by using uniform distribution. Each tuple has 8 dimension attributes and 6 measure attributes. The domain size of the 8 dimensions are range from 13 to 3000.

Compared Algorithms: For brevity, we denote Algorithm 1 with optimizations in Sect. 4 as **LS**, Algorithm 3 in Sect. 5.1 as **LSI** and Algorithm 3 in Sect. 5.1 with the segmentation techniques in Sect. 5.2 as **LSII**.

All algorithms are written in C++ and compiled using g++ 6.4.0 with optimization on level 3 in Ubuntu. All experiments were conducted on a machine with 3.4 GHz i7-6700 processors, 8 GB of memory with single thread. We study the performance of our methods for various parameters. The default parameter setting is: the number of tuples $n = 25,000,000$ (for Weather dataset)/ 30,000,000 (for Synthetic dataset), the number of dimensions $d = 6$, the number of measures $m = 3$. We randomly generated 100 measure constraints for each dataset, and measure the average response time in each dataset.

6.2   Case Study

In this section, we demonstrate the effectiveness of longest streak discovered by our algorithm in NBA dataset. In summary, (I) our method could find the

---

[4] https://www.basketball-reference.com/leagues/.
[5] https://www.metoffice.gov.uk/datapoint/product/uk-3hourly-site-specific-forecast.

well-known news facts, i.e., these also were discovered by journalists and published in the news website. E.g., "Kevin Love obtained double-doubles in 53 consecutive games during 2010–11 season"[6], "Russell Westbrook's Jordan-esque triple-double streak ends at 7"[7]. (II) our method could explore some interesting facts, which are not found by the journalists. Journalist found that Kobe Bryant obtained 40+ points in 9 consecutive games in 2002–03 season which tied the record of Michael Jordan in 1986-87 season, as shown in https://goo.gl/vpp55R. However, they omitted a fact that Michael Jordan got 40+ points in 9 consecutive road games from Nov.28th, 1986 to Dec.18th, 1986, which found by our algorithm. (III) our algorithm could explore these news facts in milliseconds in a commodity computer by a freshman. However, it probably could not be found without the domain expert in several minutes or hours.

6.3   Performance Evaluation

We start by comparing the overall performance of our solutions in **Weather** dataset with default parameters setting. As shown in Fig. 5(a), both of **LSI** and **LSII** performs better than **LS** by two orders of magnitude. Even though **LSII** is only slightly better than **LSI** in terms of response time. However, the number of tuple visited times of **LSII** is only 10% of **LSI** as shown in Fig. 5(b), it implies **LSII** has better scalability than **LSI**.

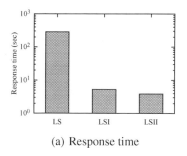
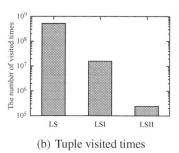

(a) Response time   (b) Tuple visited times

Fig. 5. Overall performance on **Weather** dataset

Effect of n: We evaluate the effect of the number of tuples n in **Weather** and **Synthetic** datasets, the results shows in Fig. 6(a) and (d), respectively. **LSI** and **LSII** outperform **LS** by at most two orders of magnitude on both datasets. Interestingly, the performance gap between **LSI** and **LSII** widens as n increases. It also verified the efficiency of segmentation techniques in **LSII**.

---

[6] http://www.cleveland.com/ohio-sports-blog/index.ssf/2011/03/kevin_loves_double-double_stre.html.

[7] https://www.cbssports.com/nba/news/russell-westbrooks-jordan-esque-triple-double-streak-ends-at-7/.

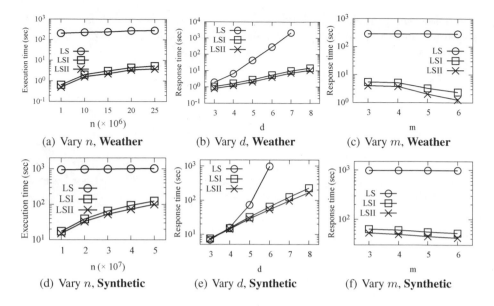

Fig. 6. Performance evaluation by varying different parameters

Effect of d: We then verify the effect of d in both datasets. We vary the number of dimension attributes from 3 to 9. We omit the response time of **LS** when $d \geq 8$ on Weather dataset and $d \geq 7$ on Synthetic dataset, as it is too slow. The response time of all solutions rise exponentially with d in Fig. 6(b) and (e), as the time complexity analysis shown in the paper. However, the increasing speed of **LSII** is the smallest as it employed all optimization techniques.

Effect of m: As illustrated in Fig. 6(c) and (f), we vary the number of measure attributes m from 3 to 6. As we discussed in Sect. 4.2, we convert the measure constraints comparison to bitset testing. Thus, the number of measure attributes will not effect the execution time of all algorithms. However, in **LSI** and **LSII**, the more measure constraints, the tighter upper bound we can derive from context subspace S. Thus, the response time slightly goes down with the growing of m.

7   Conclusion

We proposed a novel computation framework with a suite of performance optimization techniques for the longest streak discovery problem in multidimensional sequence data in this work. The case study demonstrated the effectiveness of the discovered longest streaks. Our best solution outperforms the baseline solution by two orders of magnitude in both real and synthetic dataset. We plan to investigate automatic news generation techniques in computation journalism area in the future.

Acknowledgement. This work was supported by the Science and Technology Innovation Committee Foundation of Shenzhen (Grant No. ZDSYS201703031748284).

References

1. Aldous, D., Diaconis, P.: Longest increasing subsequences: from patience sorting to the Baik-Deift-Johansson theorem. Bull. Am. Math. Soc. **36**(4), 413–432 (1999)
2. Cohen, S., Hamilton, J.T., Turner, F.: Computational journalism. Commun. ACM **54**(10), 66–71 (2011)
3. Cohen, S., Li, C., Yang, J., Yu, C.: Computational journalism: a call to arms to database researchers. In: CIDR, vol. 2011, pp. 148–151 (2011)
4. Fan, Q., Li, Y., Zhang, D., Tan, K.-L.: Discovering newsworthy themes from sequenced data: a step towards computational journalism. IEEE Trans. Knowl. Data Eng. **29**, 1398–1411 (2017)
5. Hirschberg, D.S.: Algorithms for the longest common subsequence problem. J. ACM (JACM) **24**(4), 664–675 (1977)
6. Jiang, X., Li, C., Luo, P., Wang, M., Yu, Y.: Prominent streak discovery in sequence data. In: Proceedings of the 17th ACM SIGKDD International Conference on Knowledge Discovery and Data Mining, pp. 1280–1288. ACM (2011)
7. Li, Y., Zou, L., Zhang, H., Zhao, D.: Computing longest increasing subsequences over sequential data streams. Proc. VLDB Endowment **10**(3), 181–192 (2016)
8. Sultana, A., Hassan, N., Li, C., Yang, J., Yu, C.: Incremental discovery of prominent situational facts. In: 2014 IEEE 30th International Conference on Data Engineering (ICDE), pp. 112–123. IEEE (2014)
9. Tang, B., Han, S., Yiu, M.L., Ding, R., Zhang, D.: Extracting top-k insights from multi-dimensional data. In: Proceedings of the 2017 ACM International Conference on Management of Data, pp. 1509–1524. ACM (2017)
10. Wu, T., Xin, D., Han, J.: Arcube: supporting ranking aggregate queries in partially materialized data cubes. In: Proceedings of the 2008 ACM SIGMOD International Conference on Management of Data, pp. 79–92. ACM (2008)
11. Wu, T., Xin, D., Mei, Q., Han, J.: Promotion analysis in multi-dimensional space. Proc. VLDB Endowment **2**(1), 109–120 (2009)
12. Wu, Y., Agarwal, P.K., Li, C., Yang, J., Yu, C.: On one of the few objects. In: Proceedings of the 18th ACM SIGKDD International Conference on Knowledge Discovery and Data Mining, pp. 1487–1495. ACM (2012)
13. Zhang, G., Jiang, X., Luo, P., Wang, M., Li, C.: Discovering general prominent streaks in sequence data. ACM Trans. Knowl. Discov. Data (TKDD) **8**(2), 9 (2014)

Map Matching Algorithms:
An Experimental Evaluation

Na Ta[1(✉)], Jiuqi Wang[2], and Guoliang Li[3]

[1] School of Journalism and Communication, Renmin University of China,
Beijing 100872, China
`tanayun@ruc.edu.cn`
[2] College of Software, Beihang University, Beijing 100191, China
`wangjiuqi1@163.com`
[3] Department of Computer Science, Tsinghua University, Beijing 100084, China
`liguoliang@tsinghua.edu.cn`

Abstract. Map matching is an important operation of location-based services, which matches raw GPS trajectories onto real road networks, and facilitates tasks of urban computing, such as intelligent traffic systems, etc. More than ten algorithms have been proposed to address this problem in the recent decade. However, existing algorithms have not been thoroughly compared under the same experimental framework. For example, some algorithms are tested only on specific datasets. This makes it rather difficult for practitioners to decide which algorithms should be used for various scenarios. To address this problem, in this paper we provide a survey on a wide spectrum of existing map matching algorithms, classify them into different categories based on their main techniques, and compare them through extensive experiments on a variety of real-world and synthetic datasets with different characteristics. We also report comprehensive findings obtained from the experiments and provide new insights about the strengths and weaknesses of existing map matching algorithms which can guide practitioners to select appropriate algorithms for various scenarios.

1 Introduction

Given a set of raw GPS trajectories generated by vehicles on an urban road network, the map matching algorithm is to align each raw trajectory onto underlying road network, where a raw trajectory is a sequence of sampling points of discrete locations at each sampling time, and a road network is a graph of vertices and edges modeling an urban traffic network. The need of such algorithms arises because: (1) the GPS devices have measurement errors, which may incorrectly report the actual location of a vehicle, and (2) sampling rates are not always set to high frequency due to transmission, storage and other costs, making it hard to tell the exact route. Therefore, map matching is an important operation for applications utilizing trajectory data, such as data management for traffic analysis [6], frequent path finder [15], taxi pick-up recommending system [20],

© Springer International Publishing AG, part of Springer Nature 2018
Y. Cai et al. (Eds.): APWeb-WAIM 2018, LNCS 10988, pp. 182–198, 2018.
https://doi.org/10.1007/978-3-319-96893-3_14

discovery of functional urban zones [28], location-aware publish/subscribe framework in digital content communication [13], etc. The basic idea of map matching is to align each sampling point to a "proper" location along some real road, to recover the actual route travelled by the vehicle. To this end, a number of map matching algorithms have been proposed in the past two decades [1,3–12,14–29].

A typical map matching framework includes two steps (after proper pre-processing such as data cleaning and indexing): (1) candidate selection step: road segments or road vertices are selected as candidates of actual locations according to certain measurements, and (2) actual route construction step: the route with the highest matching score is selected as the actual route of the vehicle reporting that particular raw trajectory. The candidate selection algorithms are crucial in terms of the map matching quality, and they vary as different strategies are taken. For example, early algorithms use the closest road segment of each sampling point as their candidate road and connect all candidate segments as the actual route, while later algorithms would employ more sophisticated models (such as the Hidden Markov Model) to address the candidate selection step.

Existing map matching algorithms can be categorized by different perspectives. Algorithms in [18,26] can be used for off-line map matching tasks, and algorithms in [9,22,23] are proper for on-line map matching. Algorithms in [1,14,27] are designed for low sampling rate (no more than one sample point within a minute), while most algorithms can work better on higher sampling rate data sets. According to sampling points used in the candidate selection step, there are incremental [5,8,10,25] and global [5,14,26,29] map matching algorithms. Besides, map matching algorithms can also be classified into geometry-based [11], topology-based [4,5,22,26], probability-based [3,17,19,21], and advanced algorithms such as [16] utilizing the Hidden Markov Model.

However these algorithms have not been thoroughly compared under the same experimental framework. For example, most algorithms are tested only on specific datasets, and there is no uniform quality metrics to demonstrate qualities of these algorithms. This makes it rather difficult for practitioners to decide which algorithms should be used for various scenarios.

To address this problem, in this paper we thoroughly compare existing map matching algorithms on the same experimental framework. We make the following contributions. (1) We provide a comprehensive survey on a wide spectrum of existing map matching algorithms and classify them into different categories based on their techniques. (2) We compare existing algorithms through extensive experiments on a variety of real-world and synthetic datasets with different characteristics. (3) We report comprehensive findings obtained from the experiments and provide new insights about the strengths and weaknesses of existing algorithms which can guide practitioners to select appropriate algorithms for various scenarios.

2 Preliminaries

We introduce following concepts before we formally define the map matching problem.

Definition 1 (Trajectory).[1] *A trajectory T is a sequence of sample points, $T = \{p_1, p_2, \cdots, p_{|T|}\}$, where p_k is a sample point (i.e., a geo-location with a sampling timestamp), and $|T|$ is the number of sample points in T.*

Definition 2 (Road Network). *A road network is a directed graph $G(V, E)$, where $V = \{v_i(x_i, y_i)\}$ is the set of vertices, a vertex v_i is represented by a pair of latitude (x_i) and longitude (y_i); and $E = \{e_j(v_k, v_m)\}$ is the set of edges which are road segments directly connected by vertices in V.*

Thus an actual road is composed by one or more road segments sequentially connected by road vertices.

Definition 3 (Route). *Given two road vertices v_i and v_j, a route R is a sequence of connected road segments starting from v_i and ending at v_j.*

Therefore, the problem of map matching is to align a raw trajectory T to the underlying road network and find a **matching route** R of the highest matching quality to T, where matching quality can be measured by some *matching metrics*. We can broadly classify existing matching quality metrics into several categories: geometry-based, topology-based, probability-based and statistical metrics.

Geometry-Based Metrics. These metrics quantify the matching quality based on the similarity of geometry characteristics between a trajectory and a route, such as distance, angle between the two curves formed by the trajectory and the route on the digital map. These metrics are fit for high-sampling-rate trajectories with low measurement error. For low-sampling-rate trajectories, the connectivity between sampling points can not be measured properly. In early incremental map matching algorithms (e.g. [11]), nearest road vertices to each trajectory points of T are selected to compose the route of T, and minimal distance between sample points and road vertices are used as the matching metric.

Topology-Based Metrics. The topology information employed by this kind of metrics include connectivity, adjacency, bounding relationship, etc., between curves or polygons. For example, in the global map matching algorithm [5] the Fréchet distance [2] is used to measure the matching quality between a trajectory and a route. The topology-based metrics consider not only the distance between sample points and the potential matching route, but also the topology connectivity inside the route itself, therefore, they are better metrics for noisy low-sampling-rate trajectories than the geometry-based metrics.

Probability-Based Metrics. These metrics use the probability that a trajectory may actually go through a certain route to measure the matching quality. Due to measurement precision, the actual location of each GPS sample point is restricted to an ellipse confidence area, thus the probability that a sample point goes through certain route can be calculated according to the relationship between the point and the part of route within the confidence area [3, 17, 19, 21].

---

[1] In this paper we use 'trajectory' to represent any raw GPS trajecotry for simplicity.

Fig. 1. Matching route example (Color figure online)

In [9], the Hidden Markov Model is used to measure the possibility that a trajectory may actually go through a certain route, and the route with the highest possibility is selected as the matching route.

Accuracy Metrics. These metrics use statistical knowledge to measure how accurate a route matches a trajectory. For example, given a route, the accuracy metric [14] uses the ratio of correctly matched road segments over total number of segments in the trajectory, to evaluate the accuracy and thus the quality of the matching route. The problem with such metrics is that ground truth has to be provided when evaluating matching quality, while the actual routes are unknown for most raw trajectories, thus restricting the usability of accuracy metrics.

Problem Formulation. Now, we can formalize the problem of map matching.

Definition 4 (Map Matching). *Given a raw trajectory T, a road network G and a matching metric M, the map matching of T onto G is to find a route R_{best} in G, so that M is maximized:*

$$R_{best} = arg \ max_{R_k} M(G, T, R_k) \tag{1}$$

In the matching process, more than one possible routes can be generated, they are referred to as **candidate routes**. For example, Fig. 1 visualizes a raw trajectory T (the blue curve) and two candidate routes R_1 (the red dot-curve) and R_2 (the green curve) using Google Map. Suppose the vehicle actually went through route R_1, then the accuracy of R_1 is higher than R_2 because R_1 has more correctly matched road segments for T than R_2 does.

3 ST Methods for the Map-Matching Problem

For the map matching problem, a classic method called ST-Matching [14] considers geometric and topological structures of the road network, as well as the temporal/speed constraints of the trajectories. Based on spatial-temporal analysis, a candidate route is concluded, from which the best matching score matrix is identified. The ST-Matching algorithm could be divide into two major parts: *Candidate Filtering*, and *Spatial and Temporal Analysis*. We review how to filter the candidate points in Sect. 3.1 and discuss its spatial and temporal analysis in Sect. 3.2. After that we focus on two improved versions of the ST-Matching algorithm in Sect. 3.3.

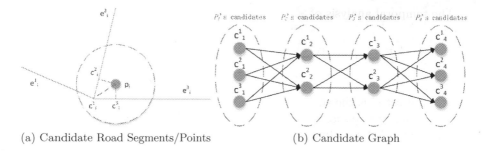

(a) Candidate Road Segments/Points   (b) Candidate Graph

Fig. 2. Candidate road segments/points & candidate graph

3.1 Candidate Filtering

Given trajectory $T = \{p_1, p_2, \cdots, p_{|T|}\}$, ST-Matching first obtains a set of candidate road segments within radius r of each unmatched trajectory point p_i ($1 \leq i \leq |T|$). As illustrated in Fig. 2(a), within the circle of radius r, c_i^1, c_i^2, and c_i^3 are candidate points for trajectory point p_i; and e_i^1, e_i^2, and e_i^3 are candidate road edges for p_i.

Once the candidate point sets are proposed for all points in trajectory T, the problem becomes how to choose one candidate from each set in order to make $c_1^{j_1} \rightarrow c_2^{j_2} \rightarrow \cdots \rightarrow c_n^{j_n}$ best matches T.

3.2 Spatial and Temporal Analysis

The spatial analysis function measures the similarity of the unmatched part between two trajectory points and the link with the shortest path between the two corresponding candidate points. First, a candidate graph is constructed (Fig. 2(b)). The distribution of the GPS measurement error is assumed to take the Gaussian distribution $N(\mu, \sigma^2)$. For each candidate point in the candidate point set, its observation probability to p_i is:

$$N(c_i^j) = \frac{1}{\sqrt{2\pi}\sigma} e^{-\frac{(x_i^j - \mu)^2}{2\sigma^2}} \tag{2}$$

where x_i^j is the Euclidean distance from candidate c_i^j to unmatched point p_i.

From candidate point c_{i-1}^t to c_i^s, the spatial analysis is defined as:

$$F_s(c_{i-1}^t \rightarrow c_i^s) = N(c_i^s) * V(c_{i-1}^t \rightarrow c_i^s), 2 \leq i \leq n. \tag{3}$$

where $V(c_{i-1}^t \rightarrow c_i^s)$ is the transition probability:

$$V(c_{i-1}^t \rightarrow c_i^s) = \frac{d(i-1, i)}{w(c_{i-1}^t, c_i^s)}. \tag{4}$$

where $d(i-1, i)$ is the Euclidian distance from p_{i-1} to p_i, and $w(c_{i-1}^t, c_i^s)$ is the length of the shortest path between c_{i-1}^t to c_i^s.

The temporal analysis of ST-Matching considers the speed information:

$$F_t(c_{i-1}^t \rightarrow c_i^s) = \frac{\sum_{u=1}^k (e_u' v * \overline{v}_{c_{i-1}^t \rightarrow c_i^s})}{\sqrt{\sum_{u=1}^k (e_u' v)^2} * \sqrt{\sum_{u=1}^k \overline{v}_{c_{i-1}^t \rightarrow c_i^s}^2}}. \tag{5}$$

where point set e' is the shortest path connecting c_{i-1}^t and c_i^s.

Combining the spatial and temporal analysis, the ST-Matching function to score the route between two candidate points can be achieved:

$$F(c_{i-1}^t \rightarrow c_i^s) = F_s(c_{i-1}^t \rightarrow c_i^s) * F_t(c_{i-1}^t \rightarrow c_i^s). \tag{6}$$

Therefore, for trajectory T, the route with the best score from one candidate point of the starting point of T to one candidate point of the end point of T is identified as the matching route for T. However, ST-Matching is based on an assumption that a driver always chooses the shortest route, which may not be consistent with the real world.

3.3   Improvements of ST-Matching

The GridST [7] tries to improve the first part of ST-Matching, and the IVMM [27] algorithm aims to improve the second part of ST-Matching.

(1) **The ST-Matching Based on the Locality of Road Networks**

The GridST algorithm ameliorates the candidate filtering of ST-Matching. The error circle radius and the maximum number of selected candidate points are dynamically adjusted according to the locality of the road network. Subsequently, the number of shortest path computations is reduced, shortening the overall running time. In order to generate the locality of road network, GridST splits the road network graph into grids. Before the map-matching process, all grids' information are calculated and organized to ensure the running time of this algorithm. If a grid has a higher density of road segments, the candidate filtering will have higher possibilities to select enough number of candidate point in a smaller error circle, vice versa. Therefore, GridST reduces the number of shortest path computations and the overall running time of map-matching process.

(2) **The Interactive Voting-Based Map-Matching algorithm (IVMM)**

This algorithm utilizes a voting process among all sampling points to reflect their interactive influence after spatial and temporal analysis of candidate points. For each sampling point, IVMM will repeatedly select an optimal route which passes through it. Every candidate point will get one vote when the optimal path includes this candidate point. Then the global optimal route will be chosen according to the vote result.

Given the spatial and temporal result of ST-Matching: $F(c_{i-1}^t \rightarrow c_i^s) = F_s(c_{i-1}^t \rightarrow c_i^s) * F_t(c_{i-1}^t \rightarrow c_i^s)$, a Static Score Matrix $M = diag(M^1, M^2, \ldots, M^n)$ is built, $M^i = (F(c_{i-1}^t \rightarrow c_i^s))_{a_{i-1} \times a_i}$. Each item in this matrix represents the possibility of a candidate point to be a correct match point. However, this possibility only considers the information of two adjacent points.

To model the weighted influence of candidate points, a $(n-1)$-dimension Matrix $\mathbf{W_i}$ is created for each sampling point p_i. And these matrix only have items in diagonal line: $w_i^j = f(dist(p_i, p_j)), (j = 1, 2, \ldots, n)$, where j is the sequence number of diagonal line in $\mathbf{W_i}$. And $dist()$ is the Euclidean distance. $f(x) = \mathrm{e}^{-\frac{x^2}{\beta^2}}$, where β is a parameter related to the road network. Then M is recalculated with the weighted influence in $\mathbf{W_i}$, and every item in M is multiplied by their weighted score. Matrix M becomes weighted score matrix Φ.

Next, voting based on the interaction of candidate points starts. For each candidate point c_i^k, IVMM attempts to find an optimal route using the weighted score matrix Φ. If a candidate point c_i^k is included in an optimal route, this candidate point c_i^k gains one vote. Then the candidate point with the largest number of votes for each sampling point p_i is identified. Finally, the best route which passes through every corresponding candidate point is selected as the matching route.

4   Other Algorithms

4.1   The Fuzzylogic Algorithm

The FuzzyLogic algorithm [12] is different from afore-mentioned algorithms: it exploits fuzzy logic to construct the degree of similarity between a matched route and a raw trajectory. The matching route is selected based on its possibility to achieve the best similarity.

(1) Candidate Filtering

In FuzzyLogic, it first plots the candidate area of an ellipse around the current trajectory point whose radius is the GPS positioning error. FuzzyLogic checks all roads in the candidate area and connects them with the already-matched road. If the candidate area can not satisfy the conditions, then FuzzyLogic directly gives up this matching. Otherwise, each sampling point has a candidate set including all candidate roads within the candidate area.

(2) Fuzzy Analysis

FuzzyLogic uses the fuzzy comprehensive judgement and constructs the set of fuzzy factors $F = \{F_x, F_y, F_z\}$, representing three aspects: car running direction, the distance between candidate road and sampling point and comparability of unmatched trajectory with candidate roads.

(2.1) The Membership Factor of Direction

Let $\theta'_{(j,k)}$ denote the direction angle between the jth sampling point and the kth candidate point for each sampling point, θ_j denote the direction angle of each sampling point. Their difference $\Delta\theta_j$ denotes direction angle factor set F_x, and it represents the angle between the vehicle's running direction and the candidate road direction (Fig. 3). Five classes of degree are identified: "very small", "small", "medium", "big" and "very big" for fuzzy reasoning.

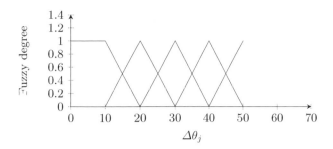

Fig. 3. Membership function of direction

(2.2) The Membership Factor of Distance

Let $\Delta d_{(j,k)}$ represent the projection distance from the jth sampling point to the kth candidate road. $\Delta d_{(j,k)}$ is regarded as the distance factor set: F_y. This distance can be classified to five fuzzy degrees: "very small", "small", "medium", "big" and "very big", as shown in Fig. 4.

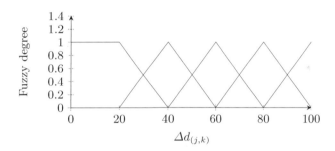

Fig. 4. Membership function of distance

(2.3) The Membership Factor of Comparability of Positioning Trajectory

The handling for this factor is similar to the previous two factors. Candidate roads are resembled to an assumption point with the computing rules used by foregone sampling point and use the distance between the assumption point and sampling point as the third factor in this fuzzy model.

With all three fuzzy factors, `FuzzyLogic` performs fuzzy transform. The fuzzy vector Q would be the result set aimed at $F = \{F_x, F_y, F_z\}$, where each element within Q denotes the possibility degree of candidate road for each sampling point. The candidate road with the largest matching degree is the matched road [12].

4.2 The Statistic Algorithm

The `Statistic` algorithm [24] is based on multiple hypothesis technique. For one unmatched trajectory, `Statistic` first selects all nodes within the radius r around the sampling point. After that, it adds all roads in the network which connect to at least one of these selected nodes to the candidate road set. For each road candidate, the sampling point is assigned to the road, and the matching score is calculated and stored in the list of current road candidates. The matching score is calculated by combining the heading of sampling point compared to the heading of road, the current speed of sampling point and the free-flow speed on the road, which is shown below:

$$Score_{road} = d(p_i, l_j) + ((v(p_i) - v_{ff}(l_j))^2 \theta_{ij}) \tag{7}$$

where p_i represents the ith sampling point and l_j is the jth candidate road for p_i, $v(p_i)$ stands for current speed of p_i and $v_{ff}(l_j)$ represents the free-flow speed of l_j road. The parameter θ equals 1 if $v(p_i) > v_{ff}(l_j)$, and 0 otherwise.

When all candidates road have been processed for the current sampling point, this algorithm selects the road which got the highest score to be the matched road. But if the number of candidate roads is not enough, it repeatedly increase radius r until there are enough candidates for each sampling point.

5   Experimental Study

We experimentally compare existing map matching algorithms. Our experimental goal is to evaluate the matching quality, running time, and impacts of parameters to the performance of different algorithms. The matching quality is measured by accuracy-based metrics to reveal how close the matching results are to the actual routes. The running time is the total time to match a given set of raw trajectories. The parameters in question are (1) number of candidate points, and (2) sampling rate, as these two are crucial to the algorithms' performance.

5.1   Experimental Settings

Algorithms. We compare the following algorithms: ST-Matching [14], IVMM [27], GridST [7], FuzzyLogic [12] and Statistic [24].

Data Sets

Road Network. We use the road network of Beijing which has 1,285,215 vertices and 2,690,296 edges.

Real Trajectory Data. We use two real datasets: Taxi (www.datatang.com/data/45888) and UCar (www.10101111.com/). Taxi contains trajectories generated by more than 8,000 public taxicabs in Beijing of one month; UCar contains trajectories of nearly 2,000 cars registered in the platform of ShenZhou Zhuanche(like Uber) within one week in Beijing.

Synthetic Trajectory Data. We implement a simulator to generate synthetic data as follows. First, a starting point v_s and a destination point v_d are randomly selected from the vertex set of the road network. Then, a connected path from v_s to v_d is generated (this path does not have to be the shortest path between v_s and v_d). Next, assuming the vehicle is moving at some fixed speed (e.g.,

Table 1. Trajectory data sets

Data set	Num. of traj.	Avg point num.	Max point num.	Min point num.	Avg sample rate
Taxi	200,000	27	50	5	x
UCar	120,000	16	20	3	x
Syn	10,000	388.6	1333	10	20 s

60 km/h), the simulator selects a set of sample points along the path for a given sampling rate (e.g., 1 min), and randomly deviates the sampling point (which is originally on the road) to a location within an error range of latitude and longitude. Our default settings are: the vehicular speed is 45 km/h, the sampling rate is 20 s, and the latitude and longitude deviations are both ±0.0002°.

Table 1 shows the statistics of the three datasets.

Ground Truth. As stated before, our synthetic trajectories are generated by first selecting a route from a starting location to a destination, and then adding some noises to simulate real trajectories. Therefore, the correct routes are known and can be used as ground truth. In addition to our synthetic data, we also provide a set of 30 real trajectories manually labeled as ground truth, denoted as HL-30. These trajectories are selected from Taxi and UCar datasets and manually labelled with the true routes. Trajectory lengths varies from 5.090 km to 23.933 km, averaging at 10.568 km; number of points set to 30.

Settings. All the algorithms are implemented by C++, compiled by Visual C++. All the experiments are conducted on a Windows Server 2012 with an Intel Xeon E52682 CPU (two cores, 2.5 GHz) and 4 GB memory.

5.2 Evaluating Accuracy

Accuracy Metrics. Given a trajectory T whose ground truth is denoted as $\overline{T}$, we measure the matching quality of a route R to T as follows:

$$N_{Acc} = \frac{\text{num. of road segments in } R \bigcap \overline{T}}{\text{num. of all road segments in } R} \qquad (8)$$

$$L_{Acc} = \frac{\sum \text{length of road segments in } R \bigcap \overline{T}}{\text{length of } R} \qquad (9)$$

Parameter Selection and Default Values. For ST-Matching algorithm, we set $k = 5$, $r=100$ m, $\mu = 0$, and $\delta=20$ m. For IVMM algorithm, we set $k = 5$, $r = 100$ m, $\mu = 0$, and $\delta=20$ m. For GridST algorithm, we set $\mu = 0$, and $\delta=20$ m. These settings are used as default values through out our experiments.

Figures 5 and 6 show the results on HL-30 and Syn datasets respectively. We have the following observations.

First, for HL-30 dataset (sampling rate $\geq$ 1 min), the IVMM algorithm and ST-Matching algorithm achieve top N_{Acc} and L_{Acc} accuracy. For IVMM, this is because the voting step strengthens scores of candidate points which have higher possibility to be on the real route. For ST-Matching, the spatial and temporal analysis can return high quality candidates especially for low-sampling-rate trajectories, explaining the top accuracy achieved by ST-Matching. GridST achieves the third accuracy, demonstrating that the policy to divide the road network into grids and adjust candidate numbers dynamically can not beat the original ST-Matching algorithm in case of low-sampling-rate. For FuzzyLogic, it chooses the best route based on the similarity between the trajectory and the route, when the sampling rate decreases, the similarity between the trajectory and the route is discounted, resulting in the fourth accuracy among all the algorithms tested. The Statistic algorithm is inferior to other algorithms on accuracy because its scoring model for candidate route sometimes could not select the "right"candidate.

Second, for synthetic dataset Syn (sampling rate 20 s), the IVMM algorithm and FuzzyLogic algorithm achieve top N_{Acc} and L_{Acc} accuracy. For IVMM, the voting step provides stable functionality despite the sampling rate as just analyzed. For FuzzyLogic, this is because that it chooses the best route based on the similarity between the trajectory and the route, the geometry and topology factors can filter high quality candidate in case of high-sampling-rate trajectories. The ST-Matching and GridST algorithms can also achieve 80%+ N_{Acc} and L_{Acc} accuracy because the spatial and temporal analysis can return high quality candidates. The Statistic algorithm is inferior to other algorithms on accuracy because its scoring model for candidate route sometimes could not select the "right" candidate, despite the sampling rate.

Third, for a given dataset, all five algorithms have similar ranking for both accuracy metrics. Although N_{Acc} focuses on the number of correctly matched road segments, and L_{Acc} focuses on the length of correctly matched road segments, on average, the number of road segments in a trajectory is proportional to the length of a trajectory, because lengths of road segments vary within a limited range (e.g., 20 m–50 m).

Fourth, for the two datasets, the ST-Matching and IVMM algorithms report similar accuracy, demonstrating stable matching quality on different sampling rates. The FuzzyLogic, GridST and Statistic algorithms work better for high-sampling-rate trajectories.

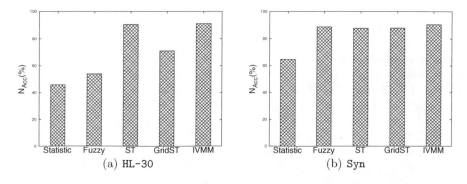

(a) HL-30   (b) Syn

Fig. 5. Evaluating accuracy: N_{Acc}

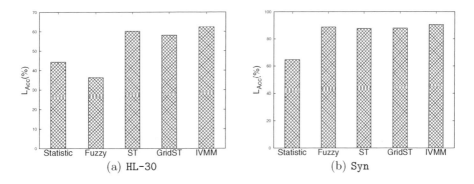

Fig. 6. Evaluating accuracy: L_{Acc}

5.3 Evaluating Running Time

We compare the running time of all the algorithms by varying the number of trajectories tested from the Taxi, UCar and Syn datasets. Figure 7(a), (b) and (c) show the respective results. We have the following observations.

First, the runtime efficiencies of all five algorithms present similar trends on all of the three datasets tested, demonstrating stability despite the underlying trajectory data.

Second, the Statistic and FuzzyLogic algorithms have top runtime efficiency, which is linear to the dataset size. This is because the logic of these two algorithms does not involve time-consuming matrix calculation as in the ST-Matching and GridST algorithms.

Third, the ST-Matching and GridST algorithms also present linear runtime efficiency in terms of dataset size. The reason that these two algorithms is less efficient than the Statistic and FuzzyLogic algorithms, as just stated, is because their logic involves matrix calculation which is time-consuming.

Fourth, the IVMM algorithm does not scale well as the data size increases. As indicated in [27], this algorithm has to be parallel-programming in order to achieve satisfying efficiency, which is non-trivial work.

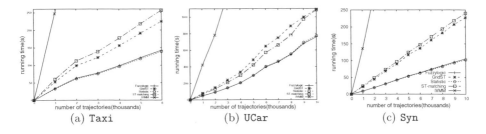

Fig. 7. Evaluating running time

5.4   Evaluating Impact on Accuracy and Running Time by Candidate Point Numbers

Both the ST-Matching and IVMM algorithms have an explicit parameter in terms of the maximum candidate points for each trajectory point. For the FuzzyLogic and Statistic algorithms, they also select a number of possible matching road points in early state of each algorithm, therefore we treat this parameter as maximum candidate points in this section as well. We evaluate the impact on accuracy by varying this parameter from 1 to 5. Figures 8, 9 and 10 show the corresponding results of N_{Acc}, L_{Acc} and runtime efficiency on HL-30 dataset and a 1000-trajectory Syn dataset. We have the following observations.

First, on each dataset, the algorithms compared exhibit similar matching quality variation and runtime efficiency trends as the number of candidate points increases, demonstrating stability despite the underlying trajectory data.

Second, for the ST-Matching and IVMM algorithm, the accuracy improvement over the number of candidate points is more obvious on the HL-30 dataset than on the Syn dataset. This is because HL-30 contains low-sampling trajectories, as the number of candidate points increases, the possibility that the "right" road segments are taken into consideration is increased, therefore, increasing accuracy.

Third, for the FuzzyLogic and Statistic algorithms, their matching quality is not comparable to the ST-Matching and IVMM algorithm. But since the Syn dataset contains high-sampling-rate trajectories, when the number of candidate points is big enough (e.g., 4 and above), the matching accuracy can be improved. Note that for FuzzyLogic, it is almost as good as ST-Matching and IVMM when the number of candidate points is 5.

Fourth, the Statistic algorithm has the best runtime efficiency, the FuzzyLogic algorithm has comparable efficiency, and the ST-Matching consumes more time as the number of candidate points increases. This indicates that the ST-Matching is not suitable for more than 5 candidate points. Besides, the runtime efficiency of IVMM is not plotted in Fig. 10 since it explodes as the number of candidate points increases, proving again that it can not scale well unless parallel programming is used.

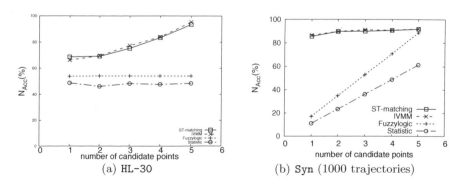

(a) HL-30   (b) Syn (1000 trajectories)

Fig. 8. Evaluating impact by max. candidate point num. on accuracy: N_{Acc}

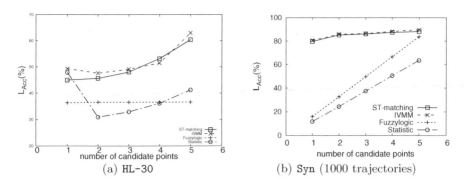

(a) HL-30   (b) Syn (1000 trajectories)

Fig. 9. Evaluating impact by max. candidate point num. on accuracy: L_{Acc}

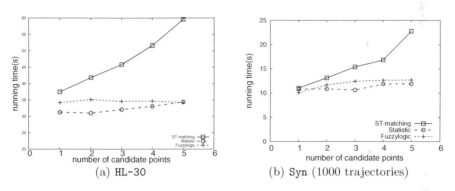

(a) HL-30   (b) Syn (1000 trajectories)

Fig. 10. Evaluating impact by max. candidate point num. on running time

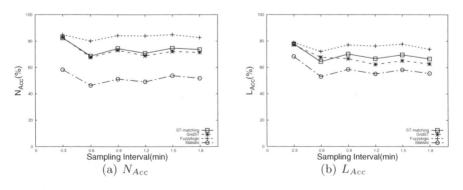

(a) N_{Acc} (b) L_{Acc}

Fig. 11. Evaluating impact by sampling rate on accuracy: N_{Acc} and L_{Acc} (Syn: 1000 trajectories for each sampling interval)

5.5   Evaluating Impact on Accuracy by Sampling Rates

In this section, we compare the matching quality in terms of accuracy with respect to the sampling rate on our synthetic dataset Syn. Figure 11(a) and (b) show the corresponding results on N_{Acc} and L_{Acc}. The result of IVMM is not reported because the running time is more than one order of magnitude to that of the other algorithms. We have the following observations.

First, the four tested algorithms exhibit similar matching quality as measured by both N_{Acc} and L_{Acc}. Second, the FuzzyLogic algorithm is the most insensitive to the sampling rate variation, and has the best matching quality. Third, the ST-Matching and GridST algorithms have similar matching quality, as their basic logic consent. Last, the Statistic has the worst matching quality, as demonstrated in above experiments.

6   Conclusion

This paper provides an experimental survey on existing map matching algorithms, including, ST-Matching, GridST, IVMM, FuzzyLogic, and Statistic, and compares them through extensive experiments on both real-world and synthetic datasets with different characteristics. We provide the following experimental findings.

(1) For better matching quality (measured by N_{Acc} and L_{Acc} accuracy), the ST-Matching and IVMM algorithms are the best choice on low-sampling-rate trajectory datasets as they outperform other algorithms; and the FuzzyLogic algorithm is also a good choice on high-sampling-rate trajectory datasets.
(2) The FuzzyLogic and Statistic algorithms always achieve better efficiency on both high-sampling-rate and low-sampling-rate trajectory datasets.
(3) Generally speaking, as the sampling rate increases, the matching quality of all tested algorithms increases.
(4) Among all tested algorithms, the Statistic algorithm reports the worst matching quality.

Acknowledgement. This research is supported in part by the Key Grant Project on Humanities and Social Sciences of MOE of China (16JJD860008), the 2018 RUC Special Fund for First-Class Universities (Majors) of Central Universites, and RUC Start-up Fund (2018030119).

References

1. Aly, H., Youssef, M.: SemMatch: road semantics-based accurate map matching for challenging positioning data. In: Proceedings of the 23rd SIGSPATIAL International Conference on Advances in Geographic Information Systems, Bellevue, WA, USA, 3–6 November 2015, pp. 5:1–5:10 (2015)
2. Alt, H., Godau, M.: Computing the fréchet distance between two polygonal curves. Int. J. Comput. Geometry Appl. **5**, 75–91 (1995)
3. Bierlaire, M., Chen, J., Newman, J.: A probabilistic map matching method for smartphone GPS data. Transp. Res. Part C-emerg. Technol. **26**, 78–98 (2013)
4. Blazquez, C., Vonderohe, A.P.: Simple map-matching algorithm applied to intelligent winter maintenance vehicle data. Transp. Res. Rec. **1935**, 68–76 (2005)

5. Brakatsoulas, S., Pfoser, D., Salas, R., Wenk, C.: On map-matching vehicle tracking data. In: Proceedings of the 31st International Conference on Very Large Data Bases, Trondheim, Norway, 30 August–2 September 2005, pp. 853–864 (2005)
6. Brakatsoulas, S., Pfoser, D., Tryfona, N.: Practical data management techniques for vehicle tracking data. In: Proceedings of the 21st International Conference on Data Engineering, ICDE 2005, 5–8 April 2005, Tokyo, Japan, pp. 324–325 (2005)
7. Chandio, A.A., Tziritas, N., Zhang, F., Xu, C.-Z.: An approach for map-matching strategy of GPS-trajectories based on the locality of road networks. In: Hsu, C.-H., Xia, F., Liu, X., Wang, S. (eds.) IOV 2015. LNCS, vol. 9502, pp. 234–246. Springer, Cham (2015). https://doi.org/10.1007/978-3-319-27293-1_21
8. Chawathe, S.S.: Segment-based map matching. In: Intelligent Vehicles Symposium, pp. 1190–1197 (2007)
9. Goh, C.Y., Dauwels, J., Mitrovic, N., Asif, M.T., Oran, A., Jaillet, P.: Online map-matching based on hidden markov model for real-time traffic sensing applications. In: 15th International IEEE Conference on Intelligent Transportation Systems, ITSC 2012, Anchorage, AK, USA, 16–19 September 2012, pp. 776–781 (2012)
10. Gonzalez, H., Han, J., Li, X., Myslinska, M., Sondag, J.P.: Adaptive fastest path computation on a road network: a traffic mining approach. In: Proceedings of the 33rd International Conference on Very Large Data Bases, University of Vienna, Austria, 23–27 September 2007, pp. 794–805 (2007)
11. Greenfeld, J.: Matching GPS observations to locations on a digital map. In: Proceedings of TRB (2002)
12. Haibin, S., Jiansheng, T., Chaozhen, H.: A integrated map matching algorithm based on fuzzy theory for vehicle navigation system, vol. 1, pp. 916–919 (2006)
13. Hu, H., Liu, Y., Li, G., Feng, J., Tan, K.: A location-aware publish/subscribe framework for parameterized spatio-textual subscriptions. In: 31st IEEE International Conference on Data Engineering, ICDE 2015, Seoul, South Korea, 13–17 April 2015, pp. 711–722 (2015)
14. Lou, Y., Zhang, C., Zheng, Y., Xie, X., Wang, W., Huang, Y.: Map-matching for low-sampling-rate GPS trajectories. In: Proceedings of 17th ACM SIGSPATIAL International Symposium on Advances in Geographic Information Systems, ACM-GIS 2009, 4–6 November 2009, Seattle, Washington, USA, pp. 352–361 (2009)
15. Luo, W., Tan, H., Chen, L., Ni, L.M.: Finding time period-based most frequent path in big trajectory data. In: Proceedings of the ACM SIGMOD International Conference on Management of Data, SIGMOD 2013, New York, NY, USA, 22–27 June 2013, pp. 713–724 (2013)
16. Newson, P., Krumm, J.: Hidden Markov Map matching through noise and sparseness. In: Proceedings of 17th ACM SIGSPATIAL International Symposium on Advances in Geographic Information Systems, ACM-GIS 2009, 4–6 November 2009, Seattle, Washington, USA, pp. 336–343 (2009)
17. Ochieng, W.Y., Quddus, M.A., Noland, R.B.: Map-matching in complex urban road networks. Revista Brasileira de Cartografia 2(55), 1–14 (2003)
18. Pereira, F.C., Costa, H., Pereira, N.M.: An off-line map-matching algorithm for incomplete map databases. Eur. Transp. Res. Rev. 1(3), 107–124 (2009)
19. Pink, O., Hummel, B.: A statistical approach to map matching using road network geometry, topology and vehicular motion constraints, pp. 862–867 (2008)
20. Qu, M., Zhu, H., Liu, J., Liu, G., Xiong, H.: A cost-effective recommender system for taxi drivers. In: The 20th ACM SIGKDD International Conference on Knowledge Discovery and Data Mining, KDD 2014, New York, NY, USA, 24–27 August 2014, pp. 45–54 (2014)

21. Quddus, M.A., Noland, R.B., Ochieng, W.Y.: A high accuracy fuzzy logic based map matching algorithm for road transport. J. Intell. Transp. Syst. **10**(3), 103–115 (2006)

22. Quddus, M.A., Ochieng, W.Y., Zhao, L., Noland, R.B.: A general map matching algorithm for transport telematics applications. GPS Solut. **7**(3), 157–167 (2003)

23. Rohani, M., Gingras, D., Gruyer, D.: A novel approach for improved vehicular positioning using cooperative map matching and dynamic base station DGPS concept. IEEE Trans. Intell. Transp. Syst. **17**(1), 230–239 (2016)

24. Schuessler, N., Axhausen, K., Zurich, E.: Map-matching of GPS traces on high-resolution navigation networks using the multiple hypothesis technique (MHT), vol. 01 (2009)

25. Wenk, C., Salas, R., Pfoser, D.: Addressing the need for map-matching speed: localizing globalb curve-matching algorithms. In: Proceedings 18th International Conference on Scientific and Statistical Database Management, SSDBM 2006, 3–5 July 2006, Vienna, Austria, pp. 379–388 (2006)

26. Yin, H., Wolfson, O.: A weight-based map matching method in moving objects databases. In: Proceedings of the 16th International Conference on Scientific and Statistical Database Management (SSDBM 2004), 21–23 June 2004, Santorini Island, Greece, pp. 437–438 (2004)

27. Yuan, J., Zheng, Y., Zhang, C., Xie, X., Sun, G.: An interactive-voting based map matching algorithm. In: Eleventh International Conference on Mobile Data Management, MDM 2010, Kanas City, Missouri, USA, 23–26 May 2010, pp. 43–52 (2010)

28. Yuan, N.J., Zheng, Y., Xie, X., Wang, Y., Zheng, K., Xiong, H.: Discovering urban functional zones using latent activity trajectories. IEEE Trans. Knowl. Data Eng. **27**(3), 712–725 (2015)

29. Zheng, K., Zheng, Y., Xie, X., Zhou, X.: Reducing uncertainty of low-sampling-rate trajectories. In: IEEE 28th International Conference on Data Engineering (ICDE 2012), Washington, DC, USA (Arlington, Virginia), 1–5 April 2012, pp. 1144–1155 (2012)

Predicting Passenger's Public Transportation Travel Route Using Smart Card Data

Chen Yang[1], Wei Chen[1], Bolong Zheng[2,3], Tieke He[4],
Kai Zheng[1], and Han Su[1(✉)]

[1] Big Data Research Center,
University of Electronic Science and Technology of China, Chengdu, China
{chenyang,weichen,kaizheng,hansu}@uestc.edu.cn, suan.sue@gmail.com
[2] School of Data and Computer Science, Sun Yat-sen University, Guangzhou, China
[3] Department of Computer Science, Aalborg University, Aalborg, Denmark
zblchris@gmail.com
[4] State Key Laboratory for Novel Software Technology,
Nanjing University, Nanjing, China
hetieke@gmail.com

Abstract. Transit prediction is a important task for public transport institutions and urban planners to provide better transit scheduling and urban planning. In recent years, there are a lot of research on traffic prediction, but the existing works focus predicting the monolithic traffic trend, and few works focus on passenger's public transportation travel route. In this paper, we study the passenger's travel route and duration prediction. We propose a prediction model based on LSTM neural network to predict passenger's travel route and duration. Specifically, we leverage multimodal embedding to extract passenger's features which are highly related to passenger's travel route and then use a LSTM-based model to improve the prediction accuracy. To verify the effectiveness of our model, we conduct extensive experiments using a real dataset which is collected from Brisbane in Australia for four months. The experimental results show that the accuracy of our model is better than baseline models.

Keywords: Transit prediction · Multimodal embedding · Smart card

1 Introduction

With the growing awareness of environmental protection, people are more and more like to take public transport. Public transportation access and corridors are natural focal points for economic and social activities. These activities help create strong neighborhood centres that are economically stable, safe, and productive. A number of studies have shown that the ability to travel conveniently in an area without a car is an important component of a community's livability. Public transportation provides opportunity, access, choice, and freedom, all of which contribute to an improved quality of life.

Under the theme of intelligent city and intelligent transportation, more and more people use smart card to take buses. Therefore, large scale passengers travel data can be collected. The smart card system can automatically and efficiently record passenger's travel routes and transactions without any additional equipment [2]. It is very important for the research and development of urban computing [11]. Through the processing of the passenger's historical data, we can predict the passenger's public transportation travel route and duration. On the one hand, according to the prediction result of passenger's public transportation travel route and duration, we can predict traffic peak time of city and formulate corresponding bus scheduling policy to alleviate the current bus scheduling imbalance and reduce passenger's waiting time. Further more, by analysing passenger's public transportation travel route and duration, we can depict the connection between the urban areas and plan out more reasonable bus routes, making people transfer less times.

First, we propose two baseline prediction models, which are based on Bayesian and Random Forest(RF) respectively. One model uses naive Bayes which makes conditional independence assumption and another model uses RF which treat information gain ratio as the criteria of attribute division. According to the historical dataset, models predict the passenger's current travel route. However, the prediction models based on Bayesian and RF have the following shortcomings: (1) the models do not take into account the impact of passenger's travel route at different time periods, and they only make an independent prediction of passenger's travel route at a certain time period. (2) the models do not predict the passenger's travel duration.

To address these challenges, this paper propose Long Short-Term Memory (LSTM)-based prediction model that enables accurately predict the passenger's travel route and duration from the passenger's historical dataset. The model is based on the LSTM [7], which maps the passenger's travel duration to different time periods. Through the interaction of each neuron, the model predicts the passenger's travel route at each time period. Then the passenger's travel route and duration are predicted under the given condition of features.

Built upon the LSTM, the prediction model uses multimodal embedding to achieve higher prediction accuracy. Before the training model, the features and labels are preprocessed by multimodal embedding [17]. The multimodal embedding learner maps all the passenger, time, week, stop and route units into the same space with their correlation preserved. If two units are highly correlated, then the distance is very close between their distributed representations of vectors. The multimodal embedding not only allows us to capture the similarity between subtle semantic units, but also provides us with background information, which reveals the relationship among passenger, time, week, stop and route.

In summary, we make the following major contributions in this paper:

1. We propose two kinds of baseline models, including Bayesian-based prediction model and RF-based prediction model. But there are a lot of defects in two models. So we design the prediction model based on LSTM. The model can not only predict the passenger's travel route, but also predict the passenger's travel duration. Compared with the baseline algorithm, the model has higher accuracy.
2. We employs a multimodal embedding learner that jointly maps the passenger, time, week, stop and route into a latent space with their correlation preserved. Such multimodal embedding not only make us to capture the subtle semantics of travel records, but also serve as background knowledge to extract features for travel records.
3. We conduct massive experiments using smart card dataset by 83515 travel records over a period of four months. The experimental results show that the proposed algorithm is very effective and outperforms baselines significantly. It can predict the passenger's travel route and duration very accurately.

The remainder of the paper is organised as follows: Sect. 2 introduces preliminary concepts and the work-flow of the proposed model. Section 3 introduces the baseline models. Our proposed prediction model is presented in Sect. 4. The experimental results are presents in Sect. 5, followed by a brief review of related work in Sect. 6. Section 7 concludes the paper.

2   Problem Statement

In this section, we introduce preliminary concepts and formally define the problem. We summarize the major notations used in the rest of the paper in Table 1.

Table 1. Summary of notations

Notation	Definition
r	A bus route
u	A passenger
t	The tth time of the day
s	The sth stop in all station
w	Denote weekday or weekend
w_i	An input vector in multimodal embedding
w_o	An output vector in multimodal embedding
S	Feature set in dataset
x^t	The input vector of the tth neuron in the prediction model
y^t	The label vector of the tth neuron in the prediction model
$\tilde{y}^{(t)}$	The output vector of the tth neuron in the prediction model

2.1 Preliminary Concepts

Definition 1 *TIME PERIOD. Time period is the discrete number that identifies the time period in a day. We divide each k minute into a time period from 0:00 to 24:00 in a day.*

Definition 2 *STOP. A stop s is a fixed location in the space where a passenger u get on or get off the bus in public transportation.*

Definition 3 *ROUTE. A route r is a bus route, which is consisted of a set stops, i.e., $r = [s_1, s_2, \ldots, s_m]$.*

Definition 4 *TRAVEL ROUTE. Travel route is a route r which a passenger takes at time period t at stop s.*

Definition 5 *TRAVEL DURATION. Travel duration is the time duration which a passenger spends on a route r. When a passenger u gets on the bus at t_1 and gets off the bus at t_i, his travel duration is about $(t_i - t_1) \cdot k$.*

2.2 Problem Description

Given a passenger u, his location s and the current time t, predict the public transportation travel route which he will take and how long he will stay on the public transportation travel route.

2.3 System Overview

In this section, we present the work-flow of generating prediction result.

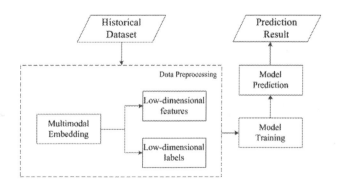

Fig. 1. System overview of model

Figure 1. Shows the system overview of prediction model. The input of the model is historical dataset, and the output is prediction result. The framework contains three main component steps: data preprocessing, model training, and model prediction. The data preprocessing uses the embedding learner including skip-gram and embedding that maps the passenger, time, week, stop and route units into the same low-dimensional space using extensive data from historical dataset. Then the model is trained and predicted based on the preprocessed data, and finally the prediction result is generated.

3   Baseline Models

We propose two baseline models. The first prediction model is based on Bayesian which makes conditional independence assumption and the second prediction model is based on RF which treat information gain ratio as the criteria of attribute division.

3.1   Prediction Model Based on Bayesian

It is very difficult to directly predict the passenger's travel route under given conditions. In order to solve the posterior probability problem, we adopt the Naive Bayes decision algorithm. The algorithm makes a strong hypothesis-attribute conditional independence assumption. It transforms the posterior probability (which is very difficult to be solved) to the prior probability(which is easy to solve).

We define the prediction model to predict passenger's travel route under certain conditions. Different models are generated for different passengers using Eq. (1):

$$r = \arg \max_r p(r|t, s, w) \tag{1}$$

where t is travel time, s is a stop by stopID, and w is a tag indicating whether it is a weekday or weekend(if w is weekend, then $w=0$, otherwise $w=1$), $p(r|t, s, w)$ is the probability of the passenger choosing to take route r under the conditions of t, s and w.

It is very natural to calculate posterior probability $p(r|t, s, w)$ with Naive Bayesian classification algorithm. The formula is shown as following:

$$p(r|t, s, w) = \frac{p(t, s, w|r)p(r)}{p(t, s, w)} \tag{2}$$

where $p(t, s, w|r)$ is conditional probability(or likelihood). Because the Naive Bayesian model assumes that the conditions are independent and identically distributed, the likelihood can be evaluated by Eq. (3). $p(r)$ is the prior probability calculated by Eq. (4). $p(t, s, w)$ is the probability of known condition and the value is constant C.

$$p(t, s, w|r) = P(t|r)p(s|r)p(w|r) = \frac{|D_{t,r}|}{|D_r|}\frac{|D_{s,r}|}{|D_r|}\frac{|D_{w,r}|}{|D_r|} \tag{3}$$

$$p(r) = \frac{|D_r|}{|D|} \tag{4}$$

D is the number of training dataset samples; $|D_r|$ is the number of records in D that the passenger takes route r; $|D_{t,r}|$ is the number of records in D that the passenger takes route r at the time period t; $|D_{s,r}|$ is the number of records in D that the passenger takes route r in the stop s; $|D_{w,r}|$ is the number of records in D that the passenger takes route r under the condition of w. After calculating likelihood probability by Eq. (3) and the prior probability by Eq. (4), we can calculate the posterior probability of passenger choosing to take route r under the conditions of t, s, and w by Eq. (5):

$$p(r|t, s, w) = p(t|r)p(s|r)p(w|r)p(r) = \frac{|D_{t,r}||D_{s,r}||D_{w,r}|}{|D_r||D_r||D|C} \tag{5}$$

Then, we enumerate all routes passing stop s and evaluate all the probabilities of a passenger to travel each route. At last, we choose the route with the greatest probability as the result.

3.2   Prediction Model Based on RF

Random Forest(RF) adopts the idea of ensemble learning. It takes decision tree as a base learner, then votes the classification results of all the decision trees, and finally selects the classification result with the largest number of votes as the result of prediction. RF makes use of sample disturbance and attribute disturbance to make the prediction model to achieve high generalization ability.

For RF-based prediction model, we select the features, i.e., passenger, stop, time, week, and use the information gain ratio as criterion. The decision tree is then constructed by m samples of bootstrap sampling and selection of a feature. In this way, we construct M decision trees to form a random forest. Finally, the relative majority voting method is adopted to select the final forecast result as shown in Eq. (6):

$$H(x) = y_{\arg\max_{j} \sum_{i=1}^{M} h_i^j(x)} \tag{6}$$

where M is number of decision trees in random forest, y_j is sample label and h_i is a decision tree in random forest. $h_i^j(x)$ is the output of h_i on y_j (0 or 1).

4   LSTM-based Prediction Model

In the baseline prediction models, passenger's travel route are regarded as discrete to predict, that is, the travel route is predicted respectively between time periods t and $t + 1$. In reality, the passenger's travel route at the time period t may have an significant impact on the travel route at the time period $t + 1$. Therefore, the prediction accuracy can be improved by considering the impact of passenger's travel route at different time periods. At the same time, the baseline prediction models do not consider getting off time and can not predict the passenger's travel duration.

The Recurrent Neural Network(RNN) model solves the above problems. The neurons in the hidden layer connect each other, and the state of neurons at $t+1$ is affected by the state of neurons at t, thus RNN can consider the influence of passenger's travel route at different time periods. Meanwhile, the RNN can make the information persistent transfer so as to conduct the serialized prediction Thus, The RNN model not only predicts the passenger's travel route at the time period t, but also can predicts the travel duration.

Thus, we propose to maximize the probability of correct choice given the passenger's features using the following formulation:

$$\theta^* = \operatorname{argmax} \sum_{(X,R)} \log p(R|X;\theta) \tag{7}$$

where X is the input feature, R is the passenger's travel route and θ denotes user-defined parameters of our models. It is common to apply the chain rule to model the joint probability over $R^{(1)}, \ldots, R^{(T)}$, where T is number of hidden layer neurons. $logp(R|X)$ is measured as following:

$$\log p(R|X) = \sum_{t=1}^{T} \log p(R^{(t)}|X, R^{(1)}, \ldots, R^{(t-1)}) \tag{8}$$

We can optimize the sum of the probabilities as described in (8) by using stochastic gradient descent(SGD).

We model $p(R^{(t)}|X, R^{(1)}, \ldots, R^{(t-1)})$ using RNN, where the state of t is determined by the state of $t-1$ and the input of t. The current state is updated by using nonlinear function f:

$$h^{(t)} = f(h^{(t-1)}, x^{(t)}) \tag{9}$$

4.1   LSTM Model

In the training process of most neural networks, there are problems of gradient vanishing and explosion [7]. In order to solve these challenges, RNN evolves the particular form,called LSTM [7]. LSTM is widely used in natural language processing [1,13], picture and sound capture [3] and sequence prediction [4], and has achieved great success.

The key to LSTM is the cell state, as shown in Fig. 2. The horizontal line running through the top is the cell state. The cell state runs directly throughout the LSTM, and it allow the state of all neurons to be easily transmitted across the entire neural network through a small number of linear operations. The core components of the LSTM structure are the forget gate, the input gate and the output gate. The blue box means forget gate, the green box means input gate and the red box means output gate. The forget gate determines what information is discarded from the cell state. The input gate determines which information is stored in the cell state. The output gate determines which information will be output. The gate value is the number between 0 and 1. If the gate value is 1, all the information will be reserved. If the gate value is 0, all the information will be discarded. The output h at time $t-1$ and the input x at time t together determine the output h at time t through three gates. Cell state c at time $t-1$ is altered by forget gate. Output h at time t is calculated by softmax function, and finally the prediction results are obtained.

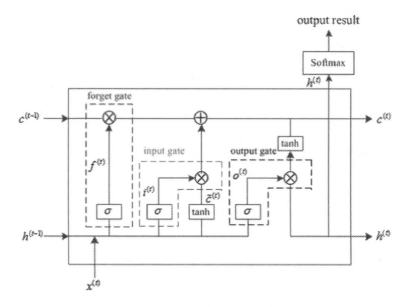

Fig. 2. LSTM cell structure (Color figure online)

The definition of gates and cell update and output are as follows:

$$i^{(t)} = \sigma(W_{ix}x^{(t)} + W_{ih}h^{(t-1)} + b_i)$$
$$f^{(t)} = \sigma(W_{fx}x^{(t)} + W_{fh}h^{(t-1)} + b_f)$$
$$\tilde{c}^{(t)} = \tanh(W_{ch}h^{(t-1)} + W_{cx}x^{(t)} + b_c)$$
$$c^{(t)} = f^{(t)} \otimes c^{(t-1)} + i^{(t)} \otimes \tilde{c}^{(t)} \tag{10}$$
$$o^{(t)} = \sigma(W_{ox}x^{(t)} + W_{oh}h^{(t-1)} + b_o)$$
$$h^{(t)} = o^{(t)} \otimes \tanh(c^{(t)})$$
$$\tilde{y}^{(t)} = soft\max(h^{(t)})$$

where $\otimes$ represents the product with a gate value, W is respectively input weights, output weights, and forget weights, b is respectively the input bias, output bias and forget bias in LSTM networks, $\sigma(\cdot)$ nonlinear sigmoid function and $tanh(\cdot)$ is hyperbolic tangent function. The last equation $\tilde{y}$ is a probability distribution over taking all routes by $h^{(t)}$ feed to a softmax. In the LSTM, multiplicative gates make it possible to deal well with exploding and vanishing gradients.

4.2   Multimodal Embedding

In LSTM-based prediction model, if one-hot encoded passenger, time, week and stop are used as input vector directly, the feature vector will be particularly discrete and the weight matrix will be sparse, which is unfriendly to the training of neural network. At the same time, the dimension of matrix and feature vector is very large, which will cause extremely long training time and prediction time. So we transform one-hot into distributed representation. We employ the method of multimodal embedding [17] which maps all the passenger, time, week, stop, and route units into the same low-dimensional space with their correlations preserved. If two units often appear together, their similarity is very large and their embedding tend to be close in latent space(for example, passenger A often takes a route at B stop, the distance between A and B is very close in latent space). When passenger, week, stop, and route are all natural and discrete units, we can directly use them as embedded units. However, time is continuous and there is no natural embedding units. To address this problem, we divide every k minute into a time period in a day and consider each time period as a basic time unit.

Our embedding algorithm is inspired by the Skip-gram model [10] that predicts the surrounding context by one unit. Here, we regard each element (passenger, time, week, stop, and route) in a travel record as a unit. Given a travel record d, We calculate the similarity between the two units, defined as

$$s(w_i, w_o) = V_{w_i}^T V_{w_o} \tag{11}$$

where V_{w_i} is the embedding of unit w_i, V_{w_o} is the embedding of unit w_o. We model the likelihood using softmax function as follows:

$$p(w_o|w_i) = \exp(s(w_i, w_o)) / \sum_{w \in d} \exp(s(w, w_i)) \tag{12}$$

where w_i is the training feature and w_o is the target feature. $s(w_i, w_o)$ is the similarity score between w_i and w_o.

For all passengers' records dataset S, the objective of the multimodal embedding is to predict all the units in S. We define the loss function as follows:

$$J_S = - \sum_{d \in S} \sum_{w_i \in d} \sum_{-m \leq j \leq m} \log p(w_{i+j}|w_i) \tag{13}$$

where m is the size of the training context. In order to minimize the above loss function, we use the method of stochastic gradient descent(SGD) and negative sampling [10] to update the weight value. We use Noise Contrastive Estimation(NCE) [5] which makes the training time shorter and improves the accuracy of the representation of feature vector. At each time we randomly select a record d from $S(d \in S)$ and randomly select a unit i from $d(i \in d)$. Then we randomly select K negative units that have the same type with i from S (not appear in d). We define negative samples(NEG) by the following objection function.

$$J_d = -\log \sigma(s(w_i, w_o)) - \sum_{k=1}^{K} \log \sigma(-s(w_k, w_i)) \tag{14}$$

where $\sigma(\cdot)$ is the sigmoid function. The updating rules of variable updates can easily be obtained by the above objective function and using SGD.

4.3   Model Description

LSTM predicts the passenger's travel route using Eq. (8). First, multimodal embedding is used to project features and labels into a same z-dimensional space. Here, using v_u, v_t, v_w, v_s and v_r to represent the features(passenger, time, week, stop) and labels(route). Then we use the mean value $\bar{v} = (v_u + v_t + v_w + v_s)/4$ of the features' distributed representation as the input of the first LSTM cell, and the input of the next LSTM cell in turn are the output of the LSTM cell in the previous time. When the output of a certain time is a special value, which does not represent any route, we stop the prediction. Figure 3 shows the expansion form of the prediction sequence. If the maximum travel time in the dataset is l, the longest length of the LSTM prediction sequence is $T = \lceil l/k \rceil$.

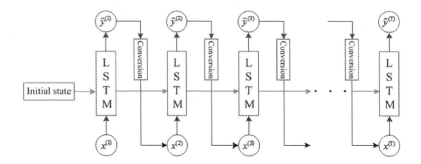

Fig. 3. LSTM-Based prediction model

And as shown in Fig. 3, $x^{(t)}$ is the input vector at the time period $t(1 \leq t \leq T)$. $\tilde{y}^{(t)}$ is the output vector of the model at the time period t, which is represented by a vector, and each element in the vector is the probability of the corresponding prediction result. From the output vector, we choose the route with the maximum probability as the prediction result of the model. *Conversion* converts the one-hot form of the prediction result into a distributed representation. When $t = 1$, $x^{(t)} = \bar{v}$ and when $t > 1$, $x^{(t)}$ is a distributed representation of prediction results through *Conversion*.

However, a passenger will leave the route at a certain time. To characterize this particular state, we add an additional dimension to the one-hot encoding vector of the route to indicate whether a passenger has left the current route. For example, $(0, \ldots, 0, 1)$ indicate that passenger isn't in a travel route, while $(0, \ldots, 1, \ldots, 0)$ indicate that passenger is in a travel route. The label vector is as follows:

$$y^{(t)} = (r_1, r_2, \ldots, r_{n+1}), \; s.t. \; r_i = 0 \; or \; 1(i = 1, 2, \ldots, n+1) \qquad (15)$$

where n is the number of routes.

4.4   Model Training

In the model training phase, we still use the mean of the features' distributed representation as the input vector of the first LSTM cell like the prediction phase. However, unlike the prediction phase, starting from the second LSTM cell, we take the distributed representation corresponding to the true travel route as the input rather than the prediction at last time step. After constructing the model, we use cross entropy as the loss function. The SGD and the back-propagation through time(BPTT) algorithm [12] are used to train the parameters in the LSTM-based model.

5   Experiment

In this section, we present our experimental results to evaluate the performance of proposed prediction model. We conduct the experiments on a computer with Intel Xeon E5-2620 2.10GHz CPU, Titan X GPU, and 128G memory.

5.1   Experimental Setup

DataSet. In our experiments, we use the passenger data in Brisbane, Australia from Translink. We use dataset from January 2013 to April 2013. The information contained in the dataset is shown in Table 2. We choose the 'Inbound' direction in dataset. According to the actual situation, most passengers do not have to travel by more than five hours in a record, so we assume that the maximum travel time is five hours. After removing the noise data. There are 1000 passengers, 3189 stops and 532 routes with a total of 83515 records in dataset. We divide the dataset into a training set and a test set. The training set contains 61646 records, and the test set contains 21869 records.

Table 2. Meaning of each field

Field	Meaning
Smartcard ID	Encrypted unique id of passenger
Direction	Inbound/Outbound
Route	Route number of the bus
Boarding time	Date/Time touch on a card
Alighting time	Date/Time touch off a card
Boarding stop	Boarding stop (ID & Description)
Alighting stop	Alighting stop (ID & Description)

Metrics. To evaluate the performance of all the models, we use the following metrics:

(1) Accuracy. The prediction accuracy is the main factor to measure the performance of the model, $p = N_{true}/N_{total}$, where N_{true} is the number of correctly predicted samples and N_{total} is the number of all the test samples.
(2) Running Time. The time that a model takes to predict the samples is also an effective indicator in measuring the performance of a model. We calculate time from beginning to ending of model predict.

5.2   Performance Evaluation

In this section, we evaluate the performance of the proposed model by conducting both objective and subjective experiments. In Effectiveness Evaluation and Efficiency Evaluation, we set the length of time period $k = 15$ min.

Effectiveness Evaluation. We use the same training set and test set to get the effectiveness of the baseline models and the LSTM-based model we have proposed. We use different number of samples (30 days, 60 days and 90 days) as training sets. The prediction accuracy of different models is shown in Fig. 4. It shows the relationship between the scale of training set and the prediction accuracies of the Bayesian-based, RF-based and LSTM-based respectively. The accuraies in corresponding partitions are referred as a_1, a_2, a_3. From Fig. 4, we have the following two main observations: (1) both a_1 and a_2 are much smaller than a_3, it is because that with multimodal embedding, distributed representation can represent the correlation between features and the interaction between LSTM cells improves the prediction accuracy; (2) as the number of samples increases, both a_1 and a_2 have small increases, while a_3 has a significant change. The reason is that using more training samples means models can better capture the passenger's diverse lifestyle. And for LSTM-based model, each weight matrix can be trained more robustly.

Efficiency Evaluation. We proceed to report the accumulated prediction time of different methods. Table 3 shows the change of running time with the different sample sizes in prediction. From Table 3, we can discover that both Bayesian-based and RF-based models can complete the forecast task in a very short time, while the LSTM-based model needs to take a few seconds due to the matrix operations between the different time steps.

Effects of Parameter. k is the length of time period and can be used to control the number of LSTM cells in LSTM-based model. As a result, k will affect all models' prediction accuracies, and it also has an impact on the prediction time for LSTM-based model. Figure 5(a) demonstrates the change of prediction time for predicting 30000 samples with the increase of the length of time period. This is because that, for a fixed length of time intend when the passenger is

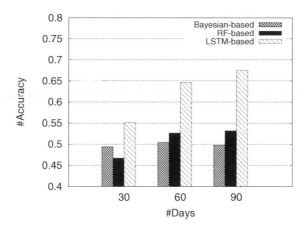

Fig. 4. Prediction accuracy. The total number of samples in different days v.s. accuracy.

Table 3. Prediction time of different models

Model	10000	20000	30000	40000	50000
Bayesian-based	0.0030	0.0055	0.0090	0.0130	0.0281
RF-based	0.2552	0.4702	0.7239	0.9430	1.1761
LSTM-based	**2.3155**	**3.0906**	**4.6520**	**6.1690**	**7.7438**

on the route, LSTM-based model only needs a few of time steps to predict the travel duration with a large k. Figure 5(b) demonstrates the relationship between the length of time period and prediction accuracy. From Fig. 5(b), we can observe that the fluctuation of accuracy is very little for Bayes-based and RF-based models, while it is obvious for LSTM-based model. But in general, the fluctuation range is in 0.05. For LSTM-based model, when the length of time period is 23, the prediction accuracy is maximum. But when the length of the time period is increased, the subtle state of the passenger can not be predicted. Therefore, setting k to 15 can better take all aspects of impact into account.

6   Related Work

We study the related work in this section. Most of the previous works are mainly divided into three parts: (1) the study of the urban structure; (2) the recommendation system; (3) passenger's travel destination prediction. However, none of these problems is same with ours.

Urban Structure Discovery. Ma *et al.* [9] put forward the use of spatial clustering and multi criteria analysis to study urban structure. Jiang *et al.* [8] measure spatial and temporal structure of cities by defining space activities with time information.

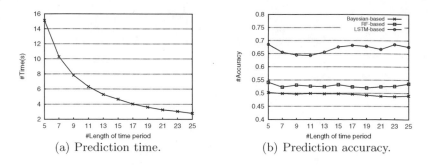

(a) Prediction time.   (b) Prediction accuracy.

Fig. 5. Efficiency Evaluation

Friends Recommend. Xiao *et al.* [15] proposed an algorithm for measuring the similarity of different users based on the historical track of semantics and then recommend friends to a user. Yu *et al.* [16] proposed a three-step statistical recommendation approach to build a heterogeneous information network.

Destination Prediction. Wang *et al.* [14] proposed a method which insteads of searching similar trajectories in sparse dataset to predict the destination. He *et al.* [6] proposed a model based on kernel density estimation to predict destination by using smart card dataset, and the model achieves a great improvement in prediction accuracy.

7   Conclusions

In this paper we have proposed a LSTM-based model to study passenger's travel route and duration in public transportation using smart card data. As far as we know, we are the first to mention passenger's travel route in urban computing. With the multimodal embedding of the passenger, time, week, stop and route, we extract the features which reserve the correlation in a low-dimension space. We conduct extensive experiments in a real dataset. The experimental results show that the performance of our model is better than the baseline models. In the future, we are interested in extending the method for passenger transfer during a journey, and have a better prediction for the details of individual transit.

References

1. Cho, K., van Merrienboer, B., Gülçehre, Ç., Bahdanau, D., Bougares, F., Schwenk, H., Bengio, Y.: Learning phrase representations using RNN encoder-decoder for statistical machine translation. In: Proceedings of the 2014 Conference on Empirical Methods in Natural Language Processing, EMNLP 2014, Doha, Qatar, 25–29 October 2014, A Meeting of SIGDAT, a Special Interest Group of the ACL, pp. 1724–1734 (2014)

2. Eagle, N., Pentland, A.S., Lazer, D.: Inferring friendship network structure by using mobile phone data. Proc. Nat. Acad. Sci. U.S.A. **106**(36), 15274–15278 (2009)
3. Gao, L., Guo, Z., Zhang, H., Xu, X., Shen, H.T.: Video captioning with attention-based LSTM and semantic consistency. IEEE Trans. Multimed. **19**(9), 2045–2055 (2017)
4. Graves, A.: Generating sequences with recurrent neural networks. CoRR abs/1308.0850 (2013)
5. Gutmann, M., Hyvärinen, A.: Noise-contrastive estimation of unnormalized statistical models, with applications to natural image statistics. J. Mach. Learn. Res. **13**, 307–361 (2012)
6. He, L., Trépanier, M.: Estimating the destination of unlinked trips in transit smart card fare data. Transp. Res. Rec. J. Transp. Res. Board **2535**(2535), 97–104 (2015)
7. Hochreiter, S., Schmidhuber, J.: Long short-term memory. Neural Comput. **9**(8), 1735–1780 (1997)
8. Jiang, S., Ferreira Jr. J., González, M.C.: Discovering urban spatial-temporal structure from human activity patterns. In: Proceedings of the ACM SIGKDD International Workshop on Urban Computing, UrbComp@KDD 2012, Beijing, China, 12 August 2012, pp. 95–102 (2012)
9. Ma, X., Liu, C., Wen, H., Wang, Y., Wu, Y.J.: Understanding commuting patterns using transit smart card data. J. Transp. Geogr. **58**, 135–145 (2017)
10. Mikolov, T., Sutskever, I., Chen, K., Corrado, G.S., Dean, J.: Distributed representations of words and phrases and their compositionality. In: Advances in Neural Information Processing Systems 26: 27th Annual Conference on Neural Information Processing Systems 2013, Proceedings of a Meeting held 5–8 December 2013, Lake Tahoe, Nevada, United States, pp. 3111–3119 (2013)
11. Paulos, E., Goodman, E.: The familiar stranger: anxiety, comfort, and play in public places. In: Proceedings of the 2004 Conference on Human Factors in Computing Systems, CHI 2004, Vienna, Austria, 24–29 April 2004, pp. 223–230 (2004)
12. Pearlmutter, B.A.: Gradient calculations for dynamic recurrent neural networks: a survey. IEEE Trans. Neural Netw. **6**(5), 1212–1228 (1995)
13. Sutskever, I., Vinyals, O., Le, Q.V.: Sequence to sequence learning with neural networks. In: Advances in Neural Information Processing Systems 27: Annual Conference on Neural Information Processing Systems 2014, Montreal, Quebec, Canada, 8–13 December 2014, pp. 3104–3112 (2014)
14. Wang, L., Yu, Z., Guo, B., Ku, T., Yi, F.: Moving destination prediction using sparse dataset: a mobility gradient descent approach. TKDD **11**(3), 37:1–37:33 (2017)
15. Xiao, X., Zheng, Y., Luo, Q., Xie, X.: Inferring social ties between users with human location history. J. Ambient Intell. Humaniz. Comput. **5**(1), 3–19 (2014)
16. Yu, X., Pan, A., Tang, L.A., Li, Z., Han, J.: Geo-friends recommendation in GPS-based cyber-physical social network. In: International Conference on Advances in Social Networks Analysis and Mining, pp. 361–368 (2011)
17. Zhang, C., Zhang, K., Yuan, Q., Peng, H., Zheng, Y., Hanratty, T., Wang, S., Han, J.: Regions, periods, activities: uncovering urban dynamics via cross-modal representation learning. In: Proceedings of the 26th International Conference on World Wide Web, WWW 2017, Perth, Australia, 3–7 April 2017, pp. 361–370 (2017)

Detecting Taxi Speeding from Sparse and Low-Sampled Trajectory Data

Xibo Zhou[1,2,3,4]($\boxtimes$), Qiong Luo[1], Dian Zhang[4], and Lionel M. Ni[5]

[1] Department of Computer Science and Engineering,
The Hong Kong University of Science and Technology, Clear Water Bay, Hong Kong
{xzhouaa,luo}@cse.ust.hk
[2] Guangzhou HKUST Fok Ying Tung Research Institute,
The Hong Kong University of Science and Technology, Clear Water Bay, Hong Kong
[3] Guangdong Key Laboratory of Popular High Performance Computers,
Shenzhen, China
[4] Shenzhen Key Laboratory of Service Computing and Applications, Shenzhen, China
zhangd@szu.edu.cn
[5] University of Macau, Macau, China
ni@umac.mo

Abstract. Taxis are a major means of public transportation in large cities, and speeding is a common problem among motor vehicles, including taxis. Unless caught by sensors or patrol officers, many speeding incidents go unnoticed, which pose potential threat to road safety. In this paper, we propose to detect speeding behaviors of individual taxis from taxi trajectory data. Such detection results are useful for driver risk analysis and road safety management. However, the taxi trajectory data are geographically sparse and the sample rate is low. Furthermore, existing methods mainly deal with the estimation of collective road speeds whereas we focus on the speeds of individual vehicles. As such, we propose to use a two-fold collective matrix factorization (CMF)-based model to estimate the individual vehicle speed. We have evaluated our method on real-world datasets, and the results show the effectiveness of our method in detecting taxi speeding behaviors.

Keywords: Speeding · Collective matrix factorization · Trajectory

1 Introduction

In many large cities, with the popularity of private cars and taxis traveling around, the incidence of traffic accidents has been rapidly increasing, which often causes damage to personal properties and public facilities, and even leads to traffic congestions. One of the most common inducements of these accidents is speeding. Taxi speeding is the most common violation among taxi drivers [4], which reduces the quality of road safety in modern cities.

In order to solve the major problem of speeding, authorities nowadays have paid a lot of attention and effort by distributing sensors (such as loops and

© Springer International Publishing AG, part of Springer Nature 2018
Y. Cai et al. (Eds.): APWeb-WAIM 2018, LNCS 10988, pp. 214–222, 2018.
https://doi.org/10.1007/978-3-319-96893-3_16

cameras) along roadways to monitor the real-time driving speeds. Due to the nontrivial cost, most of these sensors are limited in covering freeways and arterial roads. Unfortunately, collector roads, referring to secondary main roads that connect arterial roads in cities, are often sparsely covered by these sensors. As a result, the driving speed information collected in this way is not complete. On the other hand, the ubiquitous taxi trajectory data provides alternative opportunities to estimate the driving speeds and detect speeding. Most of these trajectory data contains the instant speed information recorded by speed meters embedded on taxis. Unfortunately, due to the low sampling rate, these information are usually too sparse to cover the entire travel paths. Several approaches [2,3,5–8] have been trying to predict traffic conditions in terms of road speeds by utilizing taxi trajectory data. However, these approaches are not able to monitor the driving speeds of individual vehicles, thus cannot be applied to detect taxi speeding behaviors.

In this paper, we propose a prediction system to detect individual taxi speeding behaviors by utilizing taxi trajectory data. Different from previous works, we try to estimate the driving speed for each individual taxi along the road it traveled. We first propose a two-fold collective matrix factorization (CMF)-based model to predict the individual driving speed, capturing the spatial and temporal patterns of traffic conditions, and predict individual speeding based on the estimation result. We evaluate our system on real-world taxi trajectory data, and the results show that our system is effective to detect speeding. Moreover, we conduct an empirical study on the occurrence of taxi speeding.

2   Overview

2.1   Preliminary

Definition 1 (Road Segment). *A road segment e is a directed polyline between two road intersections v_i and v_j, and there is no other road intersection on e. We denote $v_i \in e$ and $v_j \in e$.*

Definition 2 (Road Network). *A road network is a weighted directed graph $G = (V, E)$, where V is a set of road intersections (or vertices), and E is a set of road segments (or edges). The weight of a road segment is represented by its properties.*

Definition 3 (Tracing Record). *A tracing record r of a taxi is denoted as a tuple $r = \langle id, t, p, v \rangle$, where $r(id)$ is the taxi id, $r(t)$ is the record time, $r(p)$ is the location point of the taxi at $r(t)$ represented by its latitude and longitude, and $r(v)$ is the instant speed of the taxi at $r(t)$.*

Definition 4 (Trajectory). *A trajectory T of a taxi with id tid is a sequence of tracing records denoted as $T = (r_1, r_2, \cdots, r_n)$, where $r_i(id) = tid$ for $i = 1, \cdots, n$. We denote $r_i \in T$ for $i = 1, \cdots, n$ and $|T| = n$.*

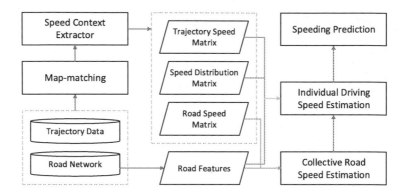

Fig. 1. The workflow of taxi speeding prediction.

Definition 5 (Path). *A path* $P = (e_1, e_2, \cdots, e_n)$ *is a sequence of road segments where* e_i *and* e_{i+1} *are connected for* $i = 1, 2, \cdots, n-1$. *Two road segments* e_i *and* e_j *are connected if there exists some intersection* v *such that* $v \in e_i$ *and* $v \in e_j$.

Definition 6 (Trajectory Speed). *Given a trajectory* T *and its map-matched path* $P = (e_1, e_2, \cdots, e_{n_e})$, *the trajectory speed of* T *is denoted as* $v(T) = (v_1(T), v_2(T), \cdots, v_{n_e}(T))$, *where* $v_i(T)$ *is the driving speed of* T *on road segment* $e_i \in P$.

Definition 7 (Speeding). *Given a trajectory* T *of a taxi and its map-matched path* $P = (e_1, e_2, \cdots, e_{n_e})$, *we say the taxi is speeding on road segment* e_i *if* $v_i(T) > v_{max}$, *where* v_{max} *is the speed limit of road segment* e_i.

In this paper, we will solve the problem of finding all the speeding behaviors from a trajectory dataset.

2.2 Framework

Figure 1 shows the process of our taxi speeding prediction system. We first map raw trajectories into connected paths constrained to the road network, and then extract the instant speed information in terms of three matrices, namely collective road speed matrix, driving speed distribution matrix, and individual trajectory speed matrix. Next, we implement a two-fold CMF-based model to estimate the driving speed of individual trajectories. The first fold of our CMF-based model is built to predict the missing values of the collective road speed matrix by utilizing the road features extracted from the road network data, and the second fold is built to predict the missing values of the individual trajectory speed matrix by utilizing the other two matrices as well as the road features. Finally, we use the completed matrices to predict speeding based on the threshold of road speed limits.

3   Methodology

3.1   Matrix Construction

In order to extract spatial features from raw trajectories, we use the Hidden Markov Model (HMM)-based map-matching algorithm [10] to convert raw location points into travel paths along the road network. In order to distinguish the traffic conditions the road network and different time slot within a day, we extract the collective road speeds from map-matched trajectories by constructing 2D matrices with the two dimensions standing for road segments and time slots. Given a trajectory dataset and the road network $G = (V, E)$, suppose we split the time in one day into a number of slots μ_t, the collective road speed v_{ij} is the average instant speed of taxis recorded on road segment $e_i \in E$ during the jth time slot. We assume that the instant speeds of all the taxis passing a certain road segment e_i during the jth time slot follows a Gaussian distribution $\mathcal{N}(\mu_{ij}, \sigma_{ij}^2)$ [7]. Thus, the speed distribution matrix can be constructed similar to the collective road speed matrix. Note that v_{ij} and $\mathcal{N}(\mu_{ij})$ is missing if no taxi is traveling on e_i during the jth time slot. Thus, the collective road speed matrix $V_c \in \mathbb{R}^{|E| \times \mu_t}$ with a large percentage of values missing due to data sparsity. In our dataset, if the time slot is set with an interval of 30 min, only 1.8% entries of V_c have values. With such a low sparsity, it is difficult to predict the missing values only using its own non-zero values.

In order to solve this problem, we build another dense matrix by extracting road features from the road network data, and use it for supplementing the speed estimation. Besides the spatial and topological information, the road network data consists of road contexts including road level, road direction, road width, road type (indicating the road is one-way or bi-directional), road length, and curvature. Each categorical feature is flattened into a vector with 0 and 1, and each numerical feature is normalized into $(0, 1)$.

3.2   Collective Road Speed Estimation

We implement a two-fold CMF-based model to estimate the driving speed of individual trajectories. The first fold of our CMF-based model is constructed to predict the missing values of the collective road speed matrix. Given the collective road speed matrix V_c, suppose the dimensionality of latent feature vectors for matrix factorization is k, V_c can be factorized into two latent feature matrices $V_c \approx W \times X = \hat{V}_c$. The dimension of W and X are $|E| \times k$ and $k \times \mu_t$. The loss function between V_c and $\hat{V}_c$ is denoted as:

$$\mathcal{J}_v = \sum_{v_{ij} \neq null} (v_{ij} - \sum_{s=1}^{k} w_{is} x_{sj})^2 \tag{1}$$

where $v_{ij} \neq null$ means that v_{ij} is not missing, w_{is} and x_{sj} represent the corresponding element in W and X, respectively.

Similarly, the road feature matrix F can be factorized into two latent feature matrices $F \approx W \times Y = \hat{F}$. Note that F and V_c shares the latent feature matrix W, and the dimension of Y is $k \times h$, where h is the number of road features. Since F is a dense matrix, roads with similar features will generate similar latent feature vectors, which will be propagated into the factorization of V_c, and thus reduce the sparsity problem of factorizing V_c.

The loss function between F and $\hat{F}$ is denoted as:

$$\mathscr{J}_f = \sum_{f_{ij} \in F} (f_{ij} - \sum_{s=1}^{k} w_{is} y_{sj})^2 \qquad (2)$$

where w_{is} and y_{sj} represent the corresponding element in W and Y, respectively. The inference process is to minimize the loss functions for both of the two matrices. Thus, the objective function is denoted as:

$$O(V_c, F) = \alpha_v \times \mathscr{J}_v + \alpha_f \times \mathscr{J}_f + R(V_c, F) \qquad (3)$$

where α_v is the weight of relative importance of V_c, α_f is the weight of relative importance of F, and $R(V_c, F)$ is the L2 regularization term, denoted as $R(V_c, F) = \lambda \times \sum_{i=1}^{k} \theta_i^2$, where λ is the weight of the regularization term, and θ_i is the ith value of latent factors.

We implement the Newton-Raphson method [9] to find W, X, Y that minimize the objective function $O(V_c, F)$. After that, the missing values of V_c can be predicted by calculating $W \times X$.

3.3   Individual Driving Speed Estimation

The second fold of our model is constructed for predicting the driving speed of individual taxis. Similar with the collective road speed estimation, for an individual taxi with id tid, we extract the instant speeds from its map-matched trajectories by constructing 2D matrices with the two dimensions standing for road segments and time slots. Since the sparsity of the speed matrix extracted from a single taxi trajectory is much lower than that of collective road speed matrix, instead of predicting the driving speed, we use standard z-score to estimate the deviation of driving speed $v(tid)$ on e_i during the jth time slot from the distribution, denoted as:

$$z(tid, i, j) = \frac{v(tid) - \mu_{ij}}{\sigma_{ij}} \qquad (4)$$

Moreover, since a single taxi tends to travel around limited area within a city, we reduce the dimensionality of its speed matrix by pruning the road segments that has never been traveled along. Based on our observation, for a taxi driver, the deviation of his driving speed from the distribution tends to be more stable than the driving speed itself. Hence, even if the testing trajectory reaches the pruned road segments, the error of predicting its speed deviation is acceptable. Thus, given the trajectories of each taxi, we can construct a $n_v \times \mu_t$

matrix Z of its travel speed deviations, where n_v is the number of road segments traversed by the taxi. Suppose the dimensionality of latent feature vectors for matrix factorization is k', Z can be factorized into two latent feature matrices $Z \approx W' \times X' = \hat{Z}'$. Similar with the collective road speed estimation, we implement the Newton-Raphson method to find W', X' that minimize the objective function. After predicting the missing values of the speed deviation matrix, we can estimate the driving speed of taxi tid passing on road segment e_i during the jth time slot as follows:

$$v(tid, i, j) = z(tid, i, j) \times \sigma_{ij} + \mu_{ij} \tag{5}$$

Finally, given the speed limit v_{max} of each road segment e_i, it is straight forward to predict the speeding behavior of a taxi, by utilizing its driving speed matrix.

4   Evaluation

4.1   Experiment Setup

The experiments are conducted on a Linux server with a CPU of Intel Core i5-4590 and 8 GB memory. The operating system is Ubuntu 14.04, and the code is written in Python 2.7.6. We use a dataset collected from a large city in China. The dataset contains 90 million taxi tracking records of 12,000 taxis for 30 days. The road network consists of 74,184 intersections and 54,723 road segments. There are five levels of road segments in our road network data, and the speed limits of each level are 30 km/h, 50 km/h, 60 km/h, 80 km/h, 90 km/h, respectively.

We use three metrics to evaluate the performance of our proposed model, namely the Mean Absolute Error (MAE), the Root Mean Square Error (RMSE), and the normalized Root Mean Square Error (NRMSE) [1]. A smaller MAE, RMSE and NRMSE indicates that the predicted values are closer to the ground truth, which means a better performance.

In order to evaluate the performance of our individual speed estimation model, we compare our framework with the following baselines: (1) the average value of the instant speeds of the two adjacent trajectory records (AAS); (2) the average speed of the sub-path between each two consecutive trajectory records (APS); (3) the collective road speed of the nearest road segment that is available during the corresponding time slot (NRS); (4) the collective road speed of the corresponding road segment and time slot predicted by our first-fold CMF-based model (CRS); and (5) the driving speed predicted by a straight forward implementation of our first-fold CMF-based model on the individual taxi speed matrix (SCMF).

4.2   Experiment Results

In our experiments, we build the collective road speed matrix and the individual driving speed matrix using the instant speed information contained in the

Table 1. Performance of speed estimation model

Method	RMSE(km/h)	NRMSE(%)	MAE(km/h)
AAS	26.279	55.465	15.793
APS	18.474	34.940	10.951
NRS	12.677	18.718	7.248
CRS	8.469	10.080	5.185
SCMF	23.584	48.720	12.604
Two-fold CMF	5.988	6.301	4.068

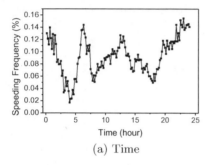
(a) Time

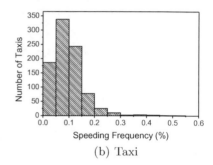
(b) Taxi

Fig. 2. Frequency of speeding occurrence among different time slots and taxis.

trajectory dataset. After building the matrices, we use 10-fold cross validation to evaluate the performance of our model. The default values of each parameter described in Sect. 3 are: $k = 5$, $k' = 15$, $\alpha_v = 0.5$, $\alpha_f = 0.8$, $\alpha_z = 0.8$, $\alpha_{f'} = 0.3$, $\lambda = 0.5$, $\lambda' = 0.2$.

We evaluate the overall performance of our model compared with the baseline listed above. The results are shown in Table 1. The first three baselines (AAS, APS and NRS) are calculated from the geographic information or statistics extracted from the dataset without matrix completion, and the performances of these methods are poor. If we directly use the collective road speed calculated by the first-fold CMF-based model to predict the individual trajectory speed, the performance is better but still not satisfactory. Meanwhile, since the individual speed matrix is too sparse, the performance is even worse if we directly implement the CMF-based model on it. Finally, it is observed that our proposed two-fold CMF-based model achieves the best performance compared to all the baselines, whose prediction error (in terms of normalized root mean square error) is only 6.301%, which is satisfactory for individual trajectory speed estimation.

4.3   Empirical Study

We evaluate the occurrence of taxi speeding detected by our proposed model among different time periods and taxis. Figure 2a demonstrates the frequency of taxi speeding occurrence on different time periods within a day. As we can see, the frequency of taxi speeding occurrence is low before dawn (around 4–6 a.m.) because most of the taxi drivers are off work, and also limited during rush hours (around 8–10 a.m. and 5–8 p.m.) because the traffic is usually congested. On the other hand, the frequency of taxi speeding occurrence is relatively high in the morning before rush hours (around 7 a.m.), after lunchtime (around 1–2 p.m.), and at midnight (from 23 p.m. to 2 a.m.). There are various possible reasons for this phenomenon, such as less congested traffic conditions, requirements of quick deliveries to the work places, or less effective monitoring (at midnight). Last but not least, Fig. 2b shows the frequency of taxi speeding occurrence among different taxi drivers. According to the statistics, most of the taxi drivers do not often cross the speed limits (with the frequency below 15%). On the contrary, we observe that a small amount of taxi drivers (around 5%) have high frequency of speeding (over 20%). Therefore, our speeding prediction system is helpful in finding out these drivers with bad habits of speeding, providing guidance to authorities.

5   Conclusion

In this paper, we propose a prediction system to detect individual taxi speeding behaviors by utilizing sparse and low-sampled trajectory data. Most of the existing approaches are designed to predict the collective speed of roads or paths, considering spatial and temporal dynamics and patterns. However, they cannot estimate the driving speed of an individual vehicle from the trajectory datasets. We implement a two-fold (CMF)-based model to predict the individual driving speed, and use the completed speed matrix to predict taxi speeding. We conduct intensive experiments on real trajectory data. The results show that our proposed system achieves a satisfactory performance.

Acknowledgments. This work is supported in part by the Guangdong Pre-national project 2014GKXM054 and the Guangdong Province Key Laboratory of Popular High Performance Computers 2017B030314073.

References

1. Chai, T., Draxler, R.R.: Root mean square error (RMSE) or mean absolute error (MAE)?-arguments against avoiding RMSE in the literature. Geosci. Model Dev. **7**(3), 1247–1250 (2014)
2. Jenelius, E., Koutsopoulos, H.N.: Travel time estimation for urban road networks using low frequency probe vehicle data. Transp. Res. Part B: Methodol. **53**, 64–81 (2013)
3. Liu, Y., Li, Z.: A novel algorithm of low sampling rate GPS trajectories on map-matching. EURASIP J. Wirel. Commun. Netw. **2017**(1), 30 (2017)
4. Tseng, C.-M.: Operating styles, working time and daily driving distance in relation to a taxi driver's speeding offenses in Taiwan. Accid. Anal. Prev. **52**, 1–8 (2013)

5. Wang, Y., Zheng, Y., Xue, Y.: Travel time estimation of a path using sparse trajectories. In: Proceedings of the 20th ACM SIGKDD International Conference on Knowledge Discovery and Data Mining, pp. 25–34. ACM (2014)
6. Wang, Z., Li, M., Wang, L., Liu, X.: Estimation trajectory of the low-frequency floating car considering the traffic control. Math. Prob. Eng. **2013**, 11 (2013)
7. Xin, X., Lu, C., Wang, Y., Huang, H.: Forecasting collector road speeds under high percentage of missing data. In: AAAI, pp. 1917–1923 (2015)
8. Xu, J., Deng, D., Demiryurek, U., Shahabi, C., van der Schaar, M.: Mining the situation: spatiotemporal traffic prediction with big data. IEEE J. Sel. Top. Sig. Process. **9**(4), 702–715 (2015)
9. Ypma, T.J.: Historical development of the Newton-Raphson method. SIAM Rev. **37**(4), 531–551 (1995)
10. Zhou, X., Ding, Y., Tan, H., Luo, Q., Ni, L.M.: HIMM: an HMM-based interactive map-matching system. In: Candan, S., Chen, L., Pedersen, T.B., Chang, L., Hua, W. (eds.) DASFAA 2017. LNCS, vol. 10178, pp. 3–18. Springer, Cham (2017). https://doi.org/10.1007/978-3-319-55699-4_1

Cloned Vehicle Behavior Analysis Framework

Minxi Li[1,2], Jiali Mao[1,2(✉)], Xiaodong Qi[1,2], Peisen Yuan[1,2], and Cheqing Jin[1,2]

[1] School of Data Science and Engineering, East China Normal University,
Shanghai, China
{minxli,jlmao1231,xdqi}@stu.ecnu.edu.cn, cqjin@dase.ecnu.edu.cn
[2] College of Information Science and Technology, Nanjing Agricultural University,
Nanjing, China
peiseny@njau.edu.cn

Abstract. Cloned vehicles brought tremendous harm to transportation management and public safety, which necessitates an efficient detection mechanism to discern the behaviors of cloned vehicles. The ubiquitous inspection spots deployed in the city have been collecting moving information of passing vehicles. Thus the positional sequences of inspection spots that vehicles passed by could form into their travelling traces. This provides us unprecedented opportunity to detect cloned vehicles. In this paper, we first propose a framework to discern the behaviors of cloned vehicles, called *CVAF*. It consists of three parts, including cloned vehicle detection, trajectory differentiation using *matching degree-based* clustering, and behavior pattern extraction. The experimental results on the real-world data show that our *CVAF* framework can identify cloned vehicle and discern their behavior patterns effectively. Our proposal can assist traffic control and public security department to solve the crime of cloned vehicle.

Keywords: Cloned vehicle · Object identification
Behavior pattern mining

1 Introduction

In recent years, more and more lawbreakers in many countries have stolen the vehicle identification number(*VIN*) of the legitimately-owned vehicle, then put it on a theft or salvaged car to gain illicit benefits. Such case is called Car cloning, which brought enormous harm to the society safety. This necessitates a high-efficiency detection mechanism to discern the behaviors of cloned vehicles. With the widespread use of the video surveillance technology, the inspection spots equipped with camera deployed in city traffic crossroads have been gathering the information of the passing vehicles. Accordingly, the positional sequences of inspection spots that vehicles visited could form into vehicles' trajectories. Analyzing the trajectories of vehicles can support the identification of cloned *VIN*.

© Springer International Publishing AG, part of Springer Nature 2018
Y. Cai et al. (Eds.): APWeb-WAIM 2018, LNCS 10988, pp. 223–231, 2018.
https://doi.org/10.1007/978-3-319-96893-3_17

Based upon *ANPR* data, [6] presented *FP-Detector* method to detect cloned *VIN* using an unified speed threshold. In actual applications, the traffic conditions behave differently across the regions and change over time, thus, *FP-Detector* attains lower precision. In addition, the detection results are not enough to help the authority to differentiate the trajectories of various objects with the same *VIN*. Further investigation reveals that the cloned vehicle may manifest with distinct behavior patterns, in which the different crime motives are hidden. Hence, it is vital to model the spatial-temporal behavior patterns of cloned vehicles.

To differentiate the distinct behaviors of cloned vehicles, the following challenges shall be addressed: (i) several vehicles with the same *VIN* may travel on the roads at the same time, accordingly, the traces of them would mix together, which increases the difficulty of vehicles' trajectories identification; (ii) different cloned vehicles have distinct spatial-temporal behavior patterns. It is hard to extract different behavior patterns using traditional time-series pattern mining algorithm with the unified parameters.

To tackle the above challenges, we first propose a Cloned Vehicle behavior Analysis Framework, called *CVAF*. It consists of three parts: (i) detect the cloned vehicle according to behavior modeling upon historical trajectory data; (ii) differentiate the traces of various cars with the same *VIN*; (iii) discern the spatial-temporal behavior patterns of different cars with the same *VIN*. For trajectory differentiation, [2] proposed *TPA* algorithm to judge the rationality of adjacency between two points in cloned vehicle trajectory. *TPA* algorithm uses the unified speed threshold and groups the points which satisfy the threshold into the same class. However, as it does not take the real-time traffic condition into account, *TPA* algorithm attains low precision in differentiating trajectory. To solve this issue, combined the spatial-temporal feature with traffic conditions, we design *matching degree* measurement to evaluate the rationality of adjacency between two points. In addition, the existing works on frequent behavior patterns mining [1,3,7] could obtain high effective results. Based on them, we design a mining method to extract different behavior pattern of cloned vehicles. Specifically, we make the following contributions: (1) we first bring forward a framework, called *CVAF*, for differentiating the behavior patterns of cloned vehicles. (2) we present *matching degree* measurement to assess the rationality of adjacency between two points, and then propose a clustering method to discern the trajectories of various objects. (3) we evaluate *CVAF* on the large scale real inspection spot data. Experimental results manifest the effectiveness of our proposal.

2   Preliminary and Problem Definition

Definition 1 (Inspection Spot). *The inspection spot $I(ID, lon, lat)$ refers to the place with the camera, which monitors the passing vehicles, here lon and lat stand for the longitude and latitude of I respectively.*

Definition 2 (Trajectory of Vehicle). *The positional sequence of inspection spots that an vehicle v_i traverses is viewed as its trajectory, denoted as $TR_{v_i} = \{p_0, p_1, \ldots p_n\}$, here p_j is the location of inspection spot that v_i visits at timestamp t_j, i.e., $p_j = (I_j, t_j)$.*

It is observed that several vehicles with the same *VIN* may travel on the road at the same period, and the traces of them would mix together. To discern the trajectories of different objects, it is available to cluster the points into classes which belongs to different objects. In addition, according to the historical behavior of the vehicle, the possibility of the next inspection spot it would visit can be calculated statistically.

Definition 3 (Transition Probability). *Given a pair of inspection spots and a time period T, a transition probability $Pr_{I_i,I_j}^{(T)}$ indicates the probability of the vehicle passing through I_i to I_j in T.*

$$Pr_{I_i,I_j}^{(T)} = n_{I_i,I_j}^{(T)} / \sum_k n_{I_i,I_k}^{(T)}$$

where $n_{I_i,I_j}^{(T)}$ denotes the number of vehicle passing from I_i to I_j in T, and $\sum_k n_{I_i,I_k}^{(T)}$ denotes the number of vehicle passing through I_i in T.

For two consecutive points in a normal trajectory, they shall have the high temporal closeness and spatial proximity. Hence, combining with the transition probability between the inspection spots, the likelihood estimation that two points are generated by the same object is defined as below.

Definition 4 (Matching Degree).

$$\varphi_{p_i,p_j} = (\lambda \times S(\frac{1}{\Delta T_{p_i,p_j}}) + (1 - \lambda) \times S(\frac{1}{\Delta D_{p_i,p_j}})) \times Pr_{I_i,I_j}^T$$

where $\Delta T_{p_i,p_j}$ and $\Delta D_{p_i,p_j}$ denote the time gap and road network distance between p_i and p_j respectively. $\lambda(0 < \lambda < 1)$ is the user-specified weight for time, $S(x)$ represents the Sigmoid function.

In general, a higher matching degree means high possibility of two points are adjacent and generated by the same vehicle. Thus the trajectories that mixed together could be grouped into the clusters which belong to different objects according to matching degree measurement.

Definition 5 (Behavior Pattern). *A behavior pattern of the vehicle v_i is defined as a spatial-temporal pair, i.e., $TP_{v_i} = (S, A)$, where $S = <I_0, I_1, \ldots, I_k>$ is a positional sequence of inspection spots that v_i visits, and $A = <\alpha_1, \ldots, \alpha_k>$ is a sequence of temporal annotations. Behavior pattern has the following form:*

$$TP_{v_i} = I_0 \xrightarrow{\alpha_1} I_1 \xrightarrow{\alpha_2} \ldots \xrightarrow{\alpha_k} I_k$$

Problem Definition: Given the trajectories of vehicles, our task is to discern the trajectories of different vehicles that used the same cloned *VIN*, and further extract their distinct spatial-temporal behavior patterns.

3   Overview

In this section, we present a framework to discern the spatial-temporal behavior patterns of various objects that use the same *VIN*, called Cloned Vehicle behavior Analysis Framework (or *CVAF*, for short). *CVAF* is comprised of three parts: (i) cloned vehicle detection, (ii) trajectory differentiation, and (iii) behavior pattern extraction. In our previous work [5], we have illustrated the detection process in detail. We leverage the historical trajectory data to establish the speed distribution among the inspection spot pair, and the shortest travel time between the inspection spot pair can be calculated by that distribution. If the traveling time of the vehicle between an inspection spot pair is less than its corresponding shortest travel time during several periods, it is identified as a cloned vehicle.

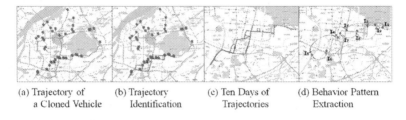

(a) Trajectory of   (b) Trajectory   (c) Ten Days of   (d) Behavior Pattern
a Cloned Vehicle   Identification   Trajectories   Extraction

Fig. 1. An example of discerning the behavior patterns of cloned vehicle (Color figure online)

After cloned vehicle detection, it is imperative to identify the trajectories of different objects that used the same *VIN*, and differentiate the distinct spatial-temporal behavior patterns of them. Figure 1 shows an example of analyzing the behaviors of cloned vehicles. The trajectory of a cloned vehicle is shown in Fig. 1(a). Through shortest travel time rationality verification and *matching degree-based* clustering, the trajectories of two objects are identified, as shown in Fig. 1(b) (highlighted in blue and yellow respectively). Figure 1(c) illustrates ten days of trajectories (in blue) of an object, and its behavior patterns in different periods of a day can be extracted, as shown in Fig. 1(d). Next, we proceed to describe the processes of trajectory identification and behavior pattern mining.

Trajectory Identification of Individual Object. When several cars with the same *VIN* drive on the roads at same time, their respective position data would mix together. To differentiate the trajectories of different objects, it is required to group the mixed position data into classes which belong to various objects. First of all, each position point pair needs to be verified the rationality according to the shortest travel time. If the time gap between any two points is shorter than the corresponding shortest travel time, they are viewed as unreachable in such short time slot and cannot be generated by the same object. After *shortest travel time*-based rationality verification, *matching degree*-based clustering method is leveraged to discern the trajectories of different objects. The matching degree and the shortest travel time can help us to judge whether these points are generated by the same vehicle.

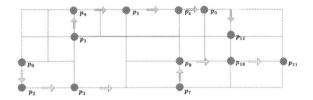

Fig. 2. An example of identifying the objects' trajectories

Algorithm 1 illustrates the detailed process of identifying the trajectory of each object. We combine the example in Fig. 2 to explain our algorithm. Suppose the first point p_0 in the cloned vehicle trajectory TR_{cv} is generated by the vehicle v_0, we put p_0 into TR_{v_0} (at lines 2–3). Then, we need to judge whether p_1 is generated by v_0 or not, we attempt to compare the time gap between p_0 and p_1 with the corresponding shortest travel time. Suppose the time gap between them is less than the shortest travel time, we cannot find the existing trajectory for p_1 (at line 6), p_1 is supposed to be generated by the other vehicle v_1. Thus, we put p_1 into TR_{v_1} (at lines 6–9). So far, there are two trajectories which are generated by v_0 and v_1, respectively. Then we need to judge the rationality, and calculate the matching degree of (p_0, p_2) and (p_1, p_2), respectively (at line 5). If the rationality of both point pairs are false, p_2 is supposed to be generated by another vehicle v_2. Otherwise, p_2 is absorbed into the existing vehicle's trajectory with the highest matching degree (at line 11). In the same way, the rest points in the cloned vehicle trajectory are processed and two trajectories are derived, as shown in Fig. 2. We

---

Algorithm 1. Differentiating the trajectories of Cloned Vehicle

Input: Cloned Vehicle Trajectory TR_{cv};
Output: Trajectory set $Traj\_set$;
1 $Traj\_set = \varnothing$;
2 $TR_{v_0} = \{p_0\}$;
3 $Traj\_set = Traj\_set \bigcup \{TR_{v_0}\}$;
4 **for** $p_i = p_1$ **to** $p_{TR_{cv}.size()-1}$ **do**
5     $TR_{match} = find\_best\_match(p_i, Traj\_set)$;
6   **if** $TR_{match} == NULL$ **then**
7         $num = Traj\_set.size()$;
8         $TR_{num} = \{p_i\}$;
9         $Traj\_set = Traj\_set \bigcup \{TR_{num}\}$;
10   **else**
11         $append(TR_{match}, p_i)$;
12 **return** $Traj\_set$;

can see that one trajectory is denoted as $TR_{v_0} = \{p_0, p_2, p_3, p_7, p_8, p_{10}, p_{11}\}$, and the other is represented as $TR_{v_1} = \{p_1, p_4, p_5, p_6, p_9, p_{12}\}$.

Behavior Pattern Extraction. The behavior pattern of the cloned vehicle can be depicted as the temporal sequence of inspection spots that an vehicle visits frequently. Through observing the trajectories of cloned vehicles for a month, the behavior patterns can be roughly divided into two categories. In the first case, vehicles have a fixed place of residence but with the random position sequence of daily activities. In the second case, vehicles visit the same places within the same time period, and almost keep the same travel trace every day. For example, on weekday, a white collar usually leaves home at 8:00 AM, arrives at the company at 9:00 AM and departures from the company at 5:00 PM. The example of this pattern can be represented as: $TP = I_i \xrightarrow{1 \text{ hour}} I_j \xrightarrow{8 \text{ hour}} I_k \xrightarrow{1 \text{ hour}} I_l$, where I_i, I_l denote the inspection spots near the home, and I_j, I_k denote the inspection spots near the company. Algorithm 2 details the process of mining the behavior pattern. Firstly, we attempt to identify the residence place of a driver. The start point and the end point of one trajectory in a day may be viewed as inhabited or adjacent to the area of residence. Therefore, we put the first k points and last k points of each trajectory into the set FLP (at lines 2). The area extracted from the points in FLP whose quantity satisfies the minimum support δ_p (at line 3) is regarded as the residence place of vehicle. Secondly (at lines 4–8), the behavior patterns of the object are mined. Vehicle may behave differently during various periods, the behavior patterns in $\mathcal{G}$ periods shall be derived respectively. Initially, each trajectory in $Traj\_Set$ is partitioned into sub-trajectories according to the period, and the sub-trajectory set contains all the sub-trajectories in the period T (at line 5). We use trajectory pattern mining algorithm in [4] to extract the behavior patterns H from each $SubTraj\_Set^T$. Given *minimum support* δ_f, the longest frequent sequence TP which satisfies the condition is regarded as the typical behavior pattern of the vehicle in this period (at lines 6–8).

Algorithm 2. Mining the Behavior Pattern of the Individual Object

---

    Input: A set of trajectories of continuous days $Traj\_set$, δ_p, δ_f, $\mathcal{G}$;
    Output: Residence places $\mathcal{R}$, Frequent Pattern set $\mathcal{F}$;
1   $\mathcal{R} = \varnothing$, $\mathcal{F} = \varnothing$, $FLP = \varnothing$;
2   Initialize FLP;
3   $\mathcal{R} = find\_residence\_place(FLP, \delta_p)$;
4   **for** $T = 1$ **to** $\mathcal{G}$ **do**
5      $SubTraj\_Set^T$: sub-trajectories in T;
6      $H = Dynamic\_I\_Pattern (SubTraj\_Set^T, \delta_f)$;
7   TP : the longest sequence in H;
8      $\mathcal{F} = \mathcal{F} \bigcup \{TP\}$;
9   **return** *the pattern* $\mathcal{R}$ *and* $\mathcal{F}$;

The daily behavior pattern and the residence place of the vehicles can be used to predict the future movement trend of cloned vehicles, and further improve the accuracy of hunting for suspect. For an vehicle v_i with the regular behavior, according to the pattern $TP_{v_i} = I_0 \xrightarrow{\alpha_1} I_1 ... \xrightarrow{\alpha_k} I_k$, we can predict that v_i may appear in one or more inspection spots of the sequence. If v_i is detected at the inspection spot I_j, we can infer the next inspection spot where v_i would most likely visit in the sequence after I_j.

4   Experiments

In this section, we conduct extensive experiments to validate the effectiveness of *CVAF*. Our framework is implemented in Java. All the experiments are run on a computer with Intel Core i7-6700 CPU (3.40 GHz) and 8 GB memory.

Experimental Setup. We use a real-world ITS surveillance data set that derived from 535 inspection spots between Sep.1 and Sep.30 in Nanjing. Each record contains 16 properties, including *VIN* of the passing vehicle, lane number, timestamp, direction, etc. Through integrating the data recorded by inspection spots, we get more than 80 million trajectories of vehicles with 2.8 million *VIN*, and we detect 101 cloned *VIN* from them using the detection method in our previous work [5]. For effectiveness validation purpose, we choose 30 most frequently recorded *VIN*, and let volunteers to verify them manually. The values of parameters are set based on our multiple experimental tuning, the weight of time in matching degree is set to 0.3, and the minimum support in pattern extraction is set to 0.6.

Effectiveness. Partial results of trajectory identification and behavior pattern extraction are visualized in Fig. 3. We can see that three days of *VIN* (No. 8867)'s trajectories (highlighted in red, black and blue respectively) in Fig. 3(a). After trajectory identification phase, we discern the trajectories of two objects (highlighted in solid and dotted lines respectively) using the same *VIN*, as illustrated in Fig. 3(b). We can observe that the car represented by the dotted lines has the stable behavior pattern, i.e., it visits the same fixed places with the same

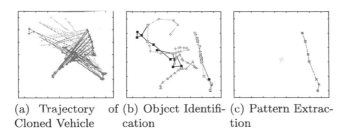

(a) Trajectory of (b) Object Identifi- (c) Pattern Extrac-
Cloned Vehicle   cation   tion

Fig. 3. Object identification and behavior pattern extraction (Color figure online)

chronological order every day. Conversely, the car represented by the solid lines does not visit the same fixed places. Noted that, it often starts from and returns back within the same area every day, on the basis of which we infer that area should be the living region of the car owner. The behavior pattern of each vehicle is shown in Fig. 3(c), among of which a star represents the living region of the vehicle without regular behavior, and black line sequence represents the pattern of the vehicle with regular behavior.

In a bid to verify the effectiveness of our proposed trajectory identification method, we conduct the contrast experiment with *TPA* algorithm. We first examine the impact of the number of vehicles (denoted as n) using the same *VIN* on trajectory identification process, here the value of n is set to $2, 3, 4$ respectively. As shown in Fig. 4, we can see that our trajectory identification algorithm performs better than *TPA* algorithm, it is due to that *TPA* algorithm dose not take the effect of traffic condition into account. With the increment of value of n, there is no significant change about the accuracy rate, indicating that our trajectory identification algorithm is not sensitive to the number of vehicles. Further, we use trajectories between Sep. 11 and Sep. 20 to mine the behavior pattern of vehicles, and leverage trajectories between Sep. 20 to Sep. 25 to verify the mining result. The effect of the minimum support in different period on pattern mining is illustrated in Fig. 5. We can see that, with the increase of the minimum support, our pattern mining algorithm always achieves high accuracy.

Efficiency. Figure 6 shows the execution overhead of the trajectory identification process with respect to the number of trajectories. We observe that the execution time increases linearly with the number of trajectories, which validates that the time complexity is proportional with the number of trajectories. The execution time increases with the increment of the number of vehicles (denoted as n), this is due to that it takes more time to compute the matching degree. Figure 7 shows the execution time of the behavior pattern extraction process with respect to the minimum support. We can see that the execution time decreases with the increment of the minimum support. As the value of the minimum support decreases, it needs to consume more time to extract the behavior patterns. The above experiments establish that our proposal can differentiate the behavior pattern of cloned vehicle in a promising efficiency.

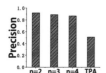

Fig. 4. Evaluation of object identification   **Fig. 5.** Evaluation of pattern extraction   **Fig. 6.** Time Cost in object identification   **Fig. 7.** Time cost in pattern extraction

5 Conclusion

In this paper, we first propose a framework to differentiate the behaviors of cloned vehicles, called *CVAF*. It consists of cloned vehicle detection, trajectory identification and behavior patterns extraction. Especially, we design matching degree measurement to evaluate the probability that two points are adjacent and generated by the same vehicle. Comparison experiment on real data shows that our proposed trajectory identification method based on *matching degree* measurement could attain better validity. Moreover, experimental results demonstrate that *CVAF* could discern different vehicles with the same cloned *VIN*, which is valuable in assisting the management to solve cloned vehicle crime.

Acknowledgements. Our research is supported by the National Key Research and Development Program of China (2016YFB1000905), NSFC (Nos. 61702423, 61370101, 61532021, U1501252, U1401256 and 61402180), Shanghai Knowledge Service Platform Project (No. ZF1213).

References

1. Cao, H., Mamoulis, N., Cheung, D.W.: Mining frequent spatio-temporal sequential patterns. In: ICDM, pp. 82–89 (2005)
2. Dai, C.: Data Analysis to the traffic checkpoint based on cloud computing. Ph.D. thesis, South China University of Technology (2016)
3. Giannotti, F., Nanni, M., Pedreschi, D.: Efficient mining of temporally annotated sequences. In: SIAM, pp. 348–359 (2006)
4. Giannotti, F., Nanni, M., Pinelli, F., Pedreschi, D.: Trajectory pattern mining. In: KDD, pp. 330–339 (2007)
5. Li, M., Mao, J., Yuan, P., Jin, C.: Detection of fake plate vehicles based on traffic data stream. J. East Chin. Normal Univ. **2**, 63–76 (2018)
6. Li, Y., Liu, C.: An approach to instantly detecting fake plates based on large-scale ANPR data. In: WISA, pp. 287–292 (2015)
7. Pei, J., Han, J., Dayal, U., Hsu, M.: Prefixspan: mining sequential patterns by prefix-projected growth. In: ICDE, pp. 215–224 (2001)

An Event Correlation Based Approach to Predictive Maintenance

Meiling Zhu[1,2,3(✉)], Chen Liu[2,3], and Yanbo Han[2,3]

[1] School of Computer Science and Technology, Tianjin University,
Tianjin 300350, China
`meilingzhu2006@126.com`
[2] Beijing Key Laboratory on Integration and Analysis of Large-Scale Stream
Data, North China University of Technology, Beijing 100144, China
`{liuchen,hanyanbo}@ncut.edu.cn`
[3] Cloud Computing Research Center, North China University of Technology,
Beijing 100144, China

Abstract. Predictive maintenance aims at enabling proactive scheduling of maintenance, and thus prevent unexpected equipment failures. Most approaches focus on predicting failures occurring within individual sensors. However, a failure is not always isolated. It probably formed by propagation of trivial anomalies, which are widely regarded as events, among sensors and devices. In this paper, we propose an event correlation discovery algorithm to capture correlations among anomalies/failures. Such correlations can show us lots of clues to the propagation paths. We also extend our previous service hyperlink model to encapsulate such correlations and propose a service-based predictive maintenance approach. Moreover, we have made extensive experiments to verify the effectiveness of our approach.

Keywords: Event correlation · Sensor data · Predictive maintenance
Proactive data service · Service hyperlink

1 Introduction

Stable and reliable operation is critical to industrial enterprises. Scheduled maintenance is widely used to prevent unexpected breakdowns and downtimes. Such maintenance is always performed separately for every component at fixed intervals. However, it is extremely costly and inefficient beyond fixed schedule.

In contrast, predictive maintenance grasps the evolvement and development of equipment running status. It aims at enabling proactive scheduling of maintenance, and thus prevent unexpected equipment failures. Nowadays, predictive maintenance is well performed when values deviate from normal behavior within individual sensors [1, 2].

However, a failure is not always isolated. Owing to the obscure physical interactions, trivial anomalies will propagate among different sensors and devices, and gradually deteriorate into a severe one in some device [3]. Mining such propagation paths is an effective means of predictive maintenance.

We will examine a scenario with the example of anomaly propagation in a coal power plant. In a coal power plant, there are hundreds of machines running continuously and thousands of sensors have been deployed. From individual sensors, anomalies, i.e., values deviating from normal behaviors, can be regarded as events [4]. As Sect. 2 shows, such events correlate with each other in multiple ways across sensors and devices. These event correlations uncover possible propagation paths of anomalies among different devices. It is very helpful to explain the root cause of an observable device anomaly/failure and perform predictive maintenance proactively.

Recently, event correlation discovery problem has received notable attentions [5–10]. The techniques can be widely applied in process discovery [5, 6], anomaly detection [7, 8], healthcare monitoring [9, 10], and so on. Motivated by our scenario, in this paper, we put our focus on a new kind of relationship between two sets of event types. It is that events of a type set are probably followed by events of another type set within a predefined time period. Such correlations provide clues for us to find possible anomaly propagation paths so as to perform predictive maintenance.

In our previous work [11, 12], we tried to map physical sensors into a software-defined abstraction, called proactive data service. A proactive data service takes values from physical sensors or events from other services as inputs and generates new event streams based on user-defined operations. We also defined the invocation relationships between two proactive data services as service hyperlinks. In [11], we used Pearson coefficient to measure such relationship. In [12], we represented original values as symbols and used a time-constrained frequent sequence to measure the correlation. In this paper, we concentrate on event correlations between two sets of event types. The main contributions include: (1) We propose an algorithm, called *EventCorrelator*, to discover such event correlations among values from different sensors. To reach this goal, we detect events from values and transform the correlation discovery problem into a time-constrained frequent co-occurrence pattern mining problem. A frequent co-occurrence pattern is a set of objects, which occurs frequently in any order within a predefined time period. We refine the concept by introducing a customized time constraint to fit our problem. (2) The discovered event correlations are encapsulated into service hyperlinks to improve our previous work. In a real application, we apply our service-based solution to make predictive maintenance in a coal power plant with refined hyperlinks. (3) Furthermore, a lot of experiments are done to show the effectiveness of our approach based on a real dataset from a coal power plant.

2   A Case Study

Figure 1 shows a real case about how trivial anomalies propagate into a severe failure. This case involves five sensors on two devices: coal feeder D device (CF-D device) and coal mill D device (CM-D device). As two maintenance record shows, a severe coal blockage failure happens twice at the CM-D device (id: 933) from 2014-10-03 09:00:00 to 2014-10-03 12:00:00, and from 2014-10-04 15:00:00 to 2014-10-04 17:30:00 respectively. The failure description is "#1D over high inlet air pressure and differential pressure of grinding bowl: coal blockage failure on coal mill D device." As the two maintenance records show, over high inlet air pressure and over high differential pressure of grinding bowl is the signs of the coal blockage failure. But they are not the root cause of the coal blockage failure. Actually, it is caused by three trivial anomalies, including over high coal feed, over high hot air valve degree and over low coal hopper level.

As Fig. 1 shows, the three causal anomalies occur in different orders in the two failures. So do the signs. The two signs occur at the same time in the first failure. But in another failure, over high differential pressure is 45 min delayed to the over high differential pressure. However, the three causal anomalies are always followed by the two signs within a short time period. It indicates that this type of failure can be predicted when the three causal anomalies happen. In other words, when the three causal anomalies occur in any order, we can perform predictive maintenance of CM-D device for the coal blockage failure proactively.

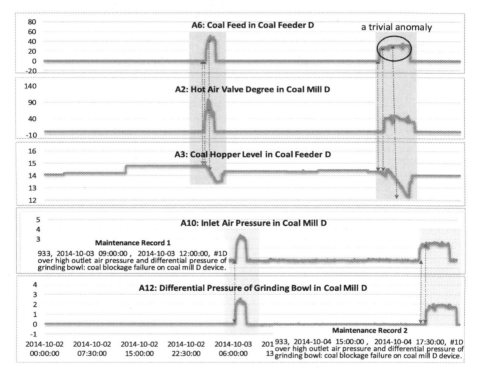

Fig. 1. Partial propagation path of trivial anomalies evolving into a server failure in a coal mill device: a real case.

This case illustrates how anomalies propagate among different devices and finally evolve into a severe failure in a specific device. However, mining a complete propagation path is a challenging problem as we cannot clarify and depict complicated physical interactions among these devices. Fortunately, the correlations among anomalies/failures show us lots of clues. If we can find such correlations, we have chances to splice them to form a complete propagation path. With these paths, we can make predictive maintenance of a specific device before trivial anomalies evolves into a severe failure of that device. Furthermore, we can also explain an anomaly/failure and find its root causes.

3   Problem Definition

Definition 1 (Sensor Record). A sensor record is a 3-tuple: $r = (timestamp, sensorid, value)$, in which *timestamp* is the generation time of r; *sensorid* represents the sensor generates r; *value* is the value of r.

Example (Sensor Record): A sensor record $r = (2014\text{-}10\text{-}02\ 00:00:00, A6, 0.0156)$ represents that the coal feed sensor (id: A6) generates a coal feed value of 0.0156 at 2014-10-02 00:00:00.

Definition 2 (Sensor Sequence). Given a sensor s_i, a sensor sequence is a time-ordered list of sensor records $R_i = \langle r_{i,1}, r_{i,2}, \ldots, r_{i,n} \rangle$, where $r_{i,k}(k = 1, \ldots, n)$ is generated from sensor s_i.

Example (Sensor Sequence): Fig. 1 shows five examples of sensor sequence, such as a coal feed sequence, and a hot air valve degree sequence.

In the above case, we observe that values deviating from normal behaviors are more valuable, because they may propagate and evolve into a severe failure. Such values are regarded as trivial anomalies and are widely described by event [4].

Definition 3 (Event). An event, which also refers to an event instance or an instance, is a 3-tuple: $e = (timestamp, eventid, type)$, in which *timestamp* is the generation time of e; *eventid* is the unique identifier of e; *type* is the type of e.

Example (Event): Fig. 1 shows several trivial anomalies. The circled trivial anomaly can be expressed as an event $e = (2014\text{-}10\text{-}04\ 05:18:00, 12986, H\text{-}CF)$. It indicates that an over high coal feed event (id: 12986) occurs at 2014-10-04 05:18:00.

Definition 4 (Event Sequence). Given a sensor s_i, an event sequence is a time-ordered list of events $E_i = \langle e_{i,1}, e_{i,2}, \ldots, e_{i,m} \rangle$, where $e_{i,k}(k = 1, \ldots, m)$ comes from s_i.

Example (Event Sequence): We can get an event sequence from the coal feed sequence in Fig. 1 as follows: $E = \langle$ (2014-10-03 01:42:00, 12985, H-CF), (2014-10-04 05:18:00, 12986, H-CF) $\rangle$.

Definition 5 (Event Correlation). Let $\mathcal{E} = \{E_1, \ldots, E_k\}$ be k event sequences from k sensor sequences respectively, and $\mathcal{E}_i^t$ and $\mathcal{E}_j^t$ are two set of event types involved in $E_1, \ldots, E_k$. The event correlation between $\mathcal{E}_i^t$ and $\mathcal{E}_j^t$ is measured by the possibility that instances (in any order) of $\mathcal{E}_i^t$ will occur in Δt (which is a user-defined time threshold) in $\mathcal{E}$ given that instances (in any order) of $\mathcal{E}_j^t$ have already occurred. We denote the event correlation as, $\left(\mathcal{E}_i^t, \mathcal{E}_j^t, \Delta t, p\right)$ in which $\mathcal{E}_i^t$ is the antecedent, $\mathcal{E}_j^t$ is the consequent, and p is the possibility.

Example (Event Correlation): Fig. 1 implies an event correlation ({H-CF, H-AVD, L-CHL}, {H-OAP, H-DP}, 3 h, 1.0). Herein, H-AVD, H-OAP, and H-DP is over high hot air valve degree, over high inlet air pressure and over high differential pressure event type respectively. L-CHL represents over low coal hopper level event type, and 1.0 is the possibility.

In this paper, our main goal is to discover event correlations from a set of sensor sequences.

4   Event Detection

The primacy of discovering event correlations from a set of sensor sequences is to detect events, i.e., values deviating from normal behaviors in each sensor sequence. It is a mature area and many traditional techniques and algorithms can be borrowed. This paper adopts three of them. Owing to the page limitation, innovations of these methods are beyond the scope of this paper.

A range-based event detection algorithm customizes value bounders for individual sensor on experiences. We learn these experiences from inspectors, sensor/device instructions and maintenance records. The values beyond the customized bounders are regarded as events. Note that such events are usually severe failures.

Outliers are widely known as the values which deviate sufficiently from most ones to consider that they were generated by a different generative process [13]. Herein, we mainly focus on the outliers in a sensor sequence, which is a one-dimension space. As pointed out by literature [13], Gaussian Mixture Model (GMM) with one component can perform well in low dimensional dataset. Thus, we select GMM to generate events from each sensor sequence. We use an open source data mining software written in Java, which is called ELKI[1]. It supports various of outlier detection algorithms, including GMM.

A discord is the subsequence which are most dissimilar with others in a sequence [14]. This paper adopts a novel technique, named Matrix Profile. It can discover top k discords in a given sequence. The authors of this technique provided an open source code written in Matlab[2]. Matlab Engine API for Java allows us to invoke their code from Java.

Any result of the above algorithms will be transformed into an event, no matter a data point or a subsequence. Events from one sensor sequence is sorted by time to form an event sequence.

5   Event Correlation Discovery

5.1   The Framework of Our Algorithm

The core of this paper is how to discover event correlations from a set of event sequences generated by the above three algorithms. Our main idea is to transform event correlation discovery into a constrained frequent co-occurrence pattern mining problem. Essentially, an event correlation requires instances of two event type sets frequently occur closely in time, i.e., within Δt. In other words, an event correlation is a relationship among the objects in a frequent co-occurrence pattern. It inspires us to mine frequent co-occurrence patterns so as to discover event correlations.

The challenge is how to identify the time delay between two related event type sets. It actually reflects how long that a set of events will be affected by its related events. Unfortunately, traditional frequent co-occurrence pattern mining algorithms cannot directly solve such problem. They only focused on the occurrence frequency of a group of unordered objects [15, 16]. Hence, we try to design an algorithm to discover a

---

[1] https://elki-project.github.io/.

[2] http://www.cs.ucr.edu/∼eamonn/MatrixProfile.html.

constrained frequent co-occurrence pattern. Such pattern consists of two object groups, where intra-group objects are unordered and inter-group objects are time-ordered, and all objects span no more than Δt. We call such pattern as time-constrained frequent co-occurrence pattern.

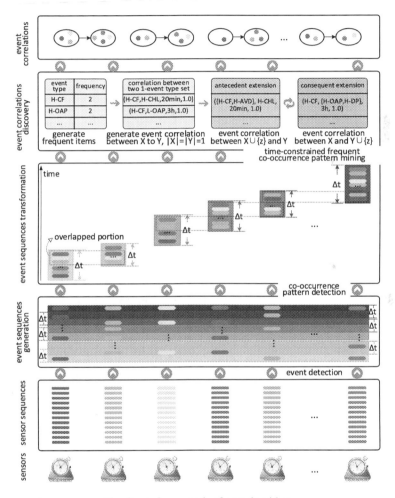

Fig. 2. A framework of our algorithm.

Figure 2 presents the framework of our approach. Historical sensor records coming from individual sensors are saved in time order and form sensor sequences. Event detection algorithms can detect event sequences from sensor sequences. The events from different sequences are clustered by a predefined time threshold Δt. Events in each cluster is sorted by timestamps to generate a new event sequence. We call this step as event sequence transformation. Events in a new sequence forms a co-occurrence pattern. Thus, we discover event correlations by mining time-constrained frequent co-occurrence patterns among the new event sequences.

5.2   Event Correlation Discovery

Time-constrained Frequent Co-occurrence Pattern Mining. In this section, we explain what a time-constrained frequent co-occurrence pattern is, what the differences between the novel pattern mining and traditional frequent co-occurrence pattern mining are, and how to mine the novel patterns. Firstly, we remind the traditional concepts of frequent co-occurrence pattern mining.

Frequent Co-occurrence Pattern Mining. We list some related concepts in this area. A group of objects $\mathcal{O} = \{o_1, o_2, \ldots, o_m\}$ from a sequence s_i is a co-occurrence pattern, if $\max\{T(\mathcal{O})\} - \min\{T(\mathcal{O})\} \leq \Delta t$, where $T(\mathcal{O}) = \{t_{o_1}, t_{o_2}, \ldots, t_{o_m}\}$, t_{o_j} is the occurrence time of o_j $(j = 1, 2, \ldots, m)$ in s_i, and Δt is a predefined time threshold. The co-occurrence pattern $\mathcal{O}$ becomes a frequent co-occurrence pattern, if it occurs in no less than k sequences.

Researchers tried to generate all co-occurrence patterns and count them to discover frequent ones [15, 16]. Obviously, traditional algorithms cannot handle the time constraint we mentioned above.

Time-Constrained Frequent Co-occurrence Pattern. A group of objects $\mathcal{O} = \{o_1, o_2, \ldots, o_m\}$ is **a time-constrained co-occurrence pattern**, short for *TCP*, if it satisfies the following conditions: (1) object o_i is from sequence $s_i (i = 1, 2, \ldots, m)$; (2) $\max\{T(\mathcal{O})\} - \min\{T(\mathcal{O})\} \leq \Delta t$, where $T(\mathcal{O}) = \{t_{o_1}, t_{o_2}, \ldots, t_{o_m}\}$, and t_{o_i} is the occurrence time of o_i; (3) $\mathcal{O}$ can be divided into two parts, which is denoted as a division $d : \mathcal{O} = \mathcal{O}_{ant} \cup \mathcal{O}_{cons}$, and objects in $\mathcal{O}_{cons}$ follow those in $\mathcal{O}_{ant}$. A *TCP* may have many valid divisions. To tell the difference, a *TCP* $\mathcal{O}$ for a valid division d can be denoted as $TCP(\mathcal{O}, d)$. Following this concept, $\mathcal{O}$ becomes **a time-constrained frequent co-occurrence pattern**, short for *TFCP*, if there is at least one division d and $TCP(\mathcal{O}, d)$ occurs for no less than k times in the sequence set $\{s_1, s_2, \ldots, s_m\}$. Similarly, *TFCP* $\mathcal{O}$ for division d can be denoted as $TFCP(\mathcal{O}, d)$. Herein, an object refers to an event type. $TFCP(\mathcal{O}, d)$ implies that there are no less than k instance groups, each of which corresponds to $\mathcal{O}$. In each group, instances of $\mathcal{O}_{cons}$ follow instances of $\mathcal{O}_{ant}$. From now on, the event correlation discovery is actually equal to mining all valid divisions of *TFCPs*.

We list some examples of above concepts based on Fig. 1. Let k be 2. An event type set {H-CF, H-AVD} has no valid divisions so that it is not a *TFCP*. On the other hand, an event type set {H-CF, H-AVD, L-CHL} is a *TFCP* with the division of {H-CF, H-AVD} ∪ {L-CHL}.

Differences Between Time-Constrained Frequent Co-occurrence Pattern Mining and Traditional Frequent Co-occurrence Pattern Mining. Our *TFCP* mining task is significantly different from the traditional one. The first difference is, each object in a *TCP* comes from a unique sequence. But in the traditional co-occurrence pattern, all objects have a unique source. The multi-source objects greatly improve the difficulty of the mining task. In our task, to avoid missing results, two adjacent *TCPs* probably have a large portion of repeated objects, as Fig. 2 shows. The overlapped portion may cause repeated counting of a *TFCP*'s frequency. It urges us to carefully check the frequency so as to guarantee the correctness of results.

Another difference is a *TCP* is supposed to be divided into two groups, where intra-group objects are unordered and inter-group objects are time-ordered. This time constraint raises the complexity of our task. Assume that the frequency of a *TFCP* $\mathcal{O} = \{o_1, o_2, \ldots, o_m\}$ is l. To find out all valid divisions by traditional ideas, we have to count the frequency for any possible division of $\mathcal{O}$, The number of possible divisions is $2 * (C_m^2 + \ldots + C_m^{m/2})$, where $m/2$ will return the closest integer greater than or equal to $m/2$, not to mention the number of object groups. Owing to this difference, our task is unable to be simply solved by the well-known generation and counting strategy.

Time-Constrained Frequent Co-occurrence Pattern Mining. To find all valid divisions of each *TFCP* from a sequence set, we have to generate all *TCPs* firstly. This can be done by clustering objects by the time threshold Δt across different sequences. Objects in a cluster is sorted by their timestamps so that they form a new sequence called a **transformed sequence**. Based on the above analysis, two adjacent clusters may have a large portion of repeated objects. If the later cluster is completely contained by the previous one, the *TCP* in the later cluster will not cause any new *TFCP*. Thus, such cluster should be removed from the further computation.

How to generate *TFCPs* and their valid divisions from all the *TCPs* above is the best question we concentrate on. Before presenting our idea, we present some observations which inspire us during developing our algorithm.

Theorem 1. Given a transformed sequence set $\mathcal{R}$, $\mathcal{O}$ is the complete object set of $\mathcal{R}$. Let an object group $\mathcal{O}' \subsetneq \mathcal{O}$ be a *TFCP* for division $d' : \mathcal{O}' = \mathcal{O}'_{ant} \cup \mathcal{O}'_{cons}$, where $\mathcal{O}' = \{o'_1, o'_2, \ldots, o'_m\}$. If $\hat{\mathcal{O}}$ be a *TFCP* for division $\hat{d} : \hat{\mathcal{O}}_{ant} \cup \mathcal{O}'_{cons}$, where $\hat{\mathcal{O}}_{ant} = \{o\} \cup \mathcal{O}'_{ant}$ and $o \in \mathcal{O} - \mathcal{O}'$, the frequency of $TFCP(\hat{\mathcal{O}}, \hat{d})$ is smaller or equal to that of $TFCP(\mathcal{O}', d')$.

Proof. Denote the frequency of the *TFCP* $\mathcal{O}'$ be $f_{\mathcal{O}'}$. For the ith time $\mathcal{O}'$ occurs in $\mathcal{R}$ (repeated occurrences in adjacent transformed sequences are counted as one time), $T_i(\mathcal{O}') = \{t_{i,o'_1}, t_{i,o'_2}, \ldots, t_{i,o'_m}\}$ is the ith timestamps of objects in $\mathcal{O}'$ ($1 \leq i \leq f_{\mathcal{O}'}$). If $f_{\hat{\mathcal{O}}} > f_{\mathcal{O}'}$, there must be $T_{f_{\mathcal{O}'}+1}(\hat{\mathcal{O}}) = \{t_{f_{\mathcal{O}'}+1,o}, t_{f_{\mathcal{O}'}+1,o'_1}, t_{f_{\mathcal{O}'}+1,o'_2}, \ldots, t_{f_{\mathcal{O}'}+1,o'_m}\}$, which satisfies that $T_i(\mathcal{O}') \neq T_{f_{\mathcal{O}'}+1}(\hat{\mathcal{O}}) - \{t_{f_{\mathcal{O}'}+1,o}\}$, where $1 \leq i \leq f_{\mathcal{O}'}$. $TFCP(\hat{\mathcal{O}}, \hat{d})$ is a *TFCP* for $\hat{d}$ so that $\mathcal{O}'$ is a *TCP* for d' with $T_{f_{\mathcal{O}'}+1}(\hat{\mathcal{O}}) - \{t_{f_{\mathcal{O}'}+1,o}\}$. Consequently, the frequency of the $TFCP(\mathcal{O}', d')$ must be larger than $f_{\mathcal{O}'}$. Hence, the theorem is proved.

For $\hat{d} : \mathcal{O}'_{ant} \cup \hat{\mathcal{O}}_{cons}$, where $\hat{\mathcal{O}}_{cons} = \{o\} \cup \mathcal{O}'_{cons}$, the above theorem can also be proved in the same manner. This theorem illustrates that we can generate valid divisions by extending existing ones. Given an existing division $d : \mathcal{O} = \mathcal{O}_{ant} \cup \mathcal{O}_{cons}$, we can extend it by adding an object to $\mathcal{O}_{ant}$ or $\mathcal{O}_{cons}$. The former one is called as antecedent extension, while the latter one is called as consequent extension.

After we make an antecedent/consequent extension to generate a new division, we check its frequency to judge its validity. For each valid division $d : \mathcal{O} = \mathcal{O}_{ant} \cup \mathcal{O}_{cons}$, we compute the possibility by the following formula: $p(\mathcal{O}_{ant}|\mathcal{O}_{cons}) = \frac{f_{\mathcal{O}}}{f_{\mathcal{O}_{cons}}}$, where $f_{\mathcal{O}_{ant}}$ is the frequency of $\mathcal{O}_{ant}$ and $f_{\mathcal{O}}$ is the frequency of $\mathcal{O}$. The condition possibility is responsible for measuring the relationship between $\mathcal{O}_{ant}$ and $\mathcal{O}_{cons}$ as Definition 5 shows.

Based on the above idea, we propose an algorithm called *EventCorrelator* to discover event correlations on top of *TFCP* mining.

The Implementation of *EventCorrelator* Algorithm. *EventCorrelator* algorithm takes a set of event sequences detected from sensor sequences as inputs. The time threshold Δt and frequency threshold k are another two input parameters of this algorithm. *EventCorrelator* outputs event correlations in format of $\left(\mathcal{E}_i^t, \mathcal{E}_j^t, \Delta t, p\right)$, where $\mathcal{E}_i^t, \mathcal{E}_j^t$ corresponds to a *TFCP* $\mathcal{E}^t$ for division $d: \mathcal{E}^t = \mathcal{E}_i^t \cup \mathcal{E}_j^t$ in a bijective manner. The main steps of *EventCorrelator* is listed as follows.

1. Cluster events across the input event sequence set by Δt.
2. Sort events in each cluster into a transformed sequence so as to create a transformed sequence set $\mathcal{R}$.
3. Scan $\mathcal{R}$ to generate an object set $\mathcal{O}$. Each object in $\mathcal{O}$ has an equal or higher frequency than k.
4. Generate all possible divisions with two objects from $\mathcal{O}$ in form of $d: \{o_i, o_j\} = \{o_i\} \cup \{o_j\}$. Note that $\{o_j\} \cup \{o_i\}$ is also a possible division.
5. Test the validity of each possible division by comparing its frequency with k.
6. For each valid division, calculate the condition possibility, and output this result.
7. Extend each valid division by adding an object and repeat steps 5, 6, and 7. The recursion stops when there is no valid division.

There are some skills during the extensions to avoid generating repeated divisions. We always make an antecedent extension firstly and then make a consequent extension by adding an object. All objects are added in lexicographical order during antecedent/consequent extensions. It indicates that we only add a larger object to $\mathcal{O}_{ant}$ or $\mathcal{O}_{cons}$ for a division $d: \mathcal{O} = \mathcal{O}_{ant} \cup \mathcal{O}_{cons}$. Besides, once we get a new division by extending a valid division d, we only check those transformed sequences contain d instead of the whole set $\mathcal{R}$.

6   The Service-Based Predictive Maintenance Workflow

6.1   Proactive Data Service and Service Hyperlink

Our previous works proposed a proactive data service model to encapsulate sensor records into a service [11, 12]. It can serve as the fundamental unit to form an IoT application. When building a service, a user customizes its functionality by customizing the input sensor records as well as operations. Event handler invokes operations for different inputs. Event definition is responsible for defining output event type and format. In this way, each service processes its inputs and generates high-level events. A created service can be encapsulated into a Restful-like API so that other services or applications can use it conveniently.

Our service has an important component called service hyperlink. It is an abstraction of event correlations and responsible for indicating target services for an output event. Propagation of anomalies among sensors and devices, as Fig. 1 shows, can be depicted as event routing among services in the software layer. A propagation path probably involves many sensors from different devices. And event correlations among sensors are dynamically interwoven. Rather than starting from scratch, service hyperlink can keep valuable event correlations from the historical sensor sequence set, and serve as a reusable knowledge segment. Event correlation reuse will certainly improve the quality and efficiency of event routing among services. Via service hyperlinks, our services can run proactively to correlate and collaborate with events to serve IoT applications.

Definition 6 (Service Hyperlink): Let $\mathcal{E}_i^t$ and $\mathcal{E}_j^t$ be two set of event types, which come from two set of proactive data services $\mathcal{S}_i$ and $\mathcal{S}_j$ respectively, and $\left(\mathcal{E}_i^t, \mathcal{E}_j^t, \Delta t, p\right)$ be their event correlation. Given a possibility threshold p_{min}, if $p \geq p_{min}$, there is a service hyperlink $L(\mathcal{S}_i, \mathcal{S}_j, \Delta t) = (Vertices, Edges)$, where $Vertices = \{Sr | Sr \in \mathcal{S}_i \bigvee Sr \in \mathcal{S}_j\}$, and $Edges = \{(Sr_p, Sr_q) | (Sr_p \in \mathcal{S}_i \wedge Sr_q \in Vertices) \vee (Sr_p, Sr_q \in \mathcal{S}_j)\}$. We call services in $\mathcal{S}_i$ as the source services, services in $\mathcal{S}_j$ as the target services, and Δt as the valid time.

6.2 A Service-Based Predictive Maintenance Approach

Learning Event Routing Paths Among Proactive Data Services. A service hyperlink $L(\mathcal{S}_i, \mathcal{S}_j, \Delta t)$ encapsulates an event correlation between $\mathcal{E}_i^t$ and $\mathcal{E}_j^t$. It indicates that the instances of $\mathcal{E}_i^t$ can be routed from the source services in $\mathcal{S}_i$ to the target services in $\mathcal{S}_j$. Assume that we create five services for coal feed sensor, hot air valve degree sensor, coal hopper level sensor, inlet air pressure sensor, and differential pressure of grinding bowl sensor mentioned in Fig. 1 respectively. For simplification, we denote them as CF-Sr, AVD-Sr, CHL-Sr, OAP-Sr and DP-Sr. We build a service hyperlink between source services {CF-Sr, AVD-Sr, CHL-Sr} and target services {OAP-Sr, DP-Sr}. It encapsulates an event correlation ({H-CF, H-AVD, L-CHL}, {H-OAP, H-DP}, 3 h, 1.0). As the first coal blockage failure shows, an H-AVD event, denoted as e_{H-AVD}, is firstly generated by AVD-Sr and will be routed to other four services. CF-Sr generates an H-CF event (e_{H-CF}) after it receives e_{H-AVD} in the valid time. Thus, it routes e_{H-AVD} and e_{H-CF} to other services. Similarly, CHL-Sr generates an L-CHL event (e_{L-CHL}) after receiving e_{H-AVD} and e_{H-CF}. Then, instances of the antecedent in the event correlation will trigger predictive maintenance of coal mill D device, which the consequent relates to.

As the above example shows, if we can learn how events are routed among proactive data services, we can make predictive maintenance of a specific device. Splicing service hyperlinks is an effective way to learn the routing paths.

A maintenance record is a 4-tuple, denoted as $r = \langle rid, start\_time, end\_time, failure\_desc \rangle$. Each maintenance record may correspond to an event type set $\mathcal{E}_0^t$ included by service set $\mathcal{S}_0$ according to the failure description. Denote $SHLs = \{L(\mathcal{S}_i, \mathcal{S}_0, \Delta t)\}$ be the service hyperlinks related to $\mathcal{E}_0^t$. Each service hyperlink $L(\mathcal{S}_i, \mathcal{S}_0, \Delta t) \in SHLs$ encapsulates an event correlation $(\mathcal{E}_i^t, \mathcal{E}_0^t, \Delta t, p)$. Then discover all service hyperlinks encapsulating an event correlation with consequent $\mathcal{E}_i^t$ in the same manner. The process is stopped if there are no corresponding service hyperlinks. Then, we splice the service hyperlinks into a routing path. For maintenance records with same anomaly/failure type, we get the maximum common sub path as the anomaly propagation path of this type of anomaly/failure.

Service-based Predictive Maintenance. Once a proactive data service Sr_i holds instances occurring in the valid time of $\mathcal{E}_i^t$, it matches $\mathcal{E}_i^t$ with all event type sets in each path learned above. If there is any event type set $\mathcal{E}_j^t$ matches up with $\mathcal{E}_i^t$, the service Sr_i will make a warning of predictive maintenance for the physical device corresponding to consequent of $\mathcal{E}_j^t$ and route these instances along with the corresponding service hyperlink in the path. The target service repeats the process in the same way.

7 Experiments

7.1 Experiment Setup

Datasets: The following experiments use a real sensor sequence set from a coal power plant. The set contains sensor sequences from 2015-07-26 23:58:30 to 2016-08-17 07:55:00. Totally 629 sensors deployed on 21 devices are involved and each sensor generates one record per second. Firstly, we test the effects of *EventCorrelator* algorithm on some one-day sets. We analyze that how parameters affect the service hyperlink number. Secondly, we test the effectiveness of our predictive maintenance approach. We divide the set into two parts. The training set is from 2015-07-26 23:58:30 to 2016-01-31 23:59:59. It is responsible for learning event routing paths. The testing set is from 2016-02-01 00:00:00 to 2016-08-17 07:55:00. It is used for making warnings to perform predictive maintenance. We use real maintenance records of the plant power from 2015-07-26 23:58:30 to 2016-08-17 07:55:00 to verify our warnings. Notably, in this paper, we only consider the records with failures occurring both in training set and testing set.

Environments: The experiments are done on a PC with four Intel Core i5-2400 CPUs 3.10 GHz and 4.00 GB RAM. The operating system is Windows 7 Ultimate. All the algorithms are implemented in Java with JDK 1.8.0.

7.2 Experiment Results

Effects of *EventCorrelator* Algorithm. In this experiment, we try to verify how key parameters affect service hyperlink number generated by encapsulating discovered event correlations. Hence, parameter p_{min} is a key parameter we concentrate on. Time threshold Δt is another significant parameter.

We randomly select 40 days from the whole sequence set, which spans 387 days. The selected sets are more than 10% of the whole set. For each one-day set, we get 629 sensor sequences. Each sensor sequence generates an event sequence. Then we invoke

EventCorrelator algorithm under different values of p_{min} and Δt to compute service hyperlinks from 629 event sequences. We record the number of results under different parameters' value. The average number for the 40 one-day event sequence sets are drawn in Fig. 3.

As we expect, Fig. 3 shows that the average number of service hyperlinks decreases with the rise of p_{min} and increases with the rise of Δt. **Firstly, we study how average service hyperlinks number decreases with the rise of p_{min}.** A descending number between two adjacent p_{min} values, such as 0.5 and 0.6, is gained by subtracting service hyperlink number under $p_{min} = 0.6$ from that under $p_{min} = 0.5$. Generally, under each value of Δt, there is a growing trend to the descending number between adjacent p_{min} values. Exceptionally, the descending number reaches minimal value between $p_{min} = 0.6$ and 0.7. The average descending number over different values of Δt can prove the above conclusion. It is around 9296, 8163, 10268, 14805 between $p_{min} = 0.5$ and 0.6, $p_{min} = 0.6$ and 0.7, $p_{min} = 0.7$ and 0.8, $p_{min} = 0.8$ and 0.9 respectively. Besides, for each two adjacent values of p_{min}, the descending number increases slightly with the growth of Δt. For example, the descending number between $p_{min} = 0.5$ and 0.6 is around 9045, 9072, 9097, 9138, 9180, 9203, 9285, 9360, 9466, 9504, 9569, 9628 under $\Delta t = 5$ min, 10 min, 15 min, ..., 60 min respectively.

Secondly, we look into the figure to study how average service hyperlink number increases with the rise of Δt. An ascending number between adjacent Δt values, such as $\Delta t = 5$ min and 10 min, is gained by subtracting service hyperlink number under $\Delta t = 5$ min from that under $\Delta t = 10$ min. The ascending number between adjacent Δt values becomes smaller on the whole. The peak of ascending number appears between $\Delta t = 10$ min and 15 min under each value of p_{min}. For instance, under $p_{min} = 0.5$, the ascending number is around 419, 431(maximal value), 357, 356, 300, 357, 328, 319, 320, 290, 272(minimal value) between $\Delta t = 5$ min and 10 min, $\Delta t = 10$ min and 15 min, ..., $\Delta t = 55$ min and 60 min respectively. However, the minimal value does not always occur between $\Delta t = 55$ min and 60 min. It can also appear between $\Delta t = 50$ min and 55 min or between $\Delta t = 40$ min and 45 min.

Effects of Our Solution in Predictive Maintenance. In this experiment, we create a service for each sensor in the training set. Sensor sequences in the training set are input into corresponding services respectively. We customize service operations as the three event detection algorithms to generate events. For each maintenance record during the time period of the training set, we compute its routing path as follows.

Each sensor sequence in the testing set is simulated as a stream and input into the corresponding service. By a sliding window, each service detects events and judges whether it should make a warning of predictive maintenance and route the events. After all streams are processed, we count the warning results to analyze the effects. Details of the process can be found in Sect. 5.3.

To measure the effects, we use the following indicators. **Precision** is the number of correct results divided by the number of all results. **Recall** is the number of correct results divided by the number of results that should have been returned.

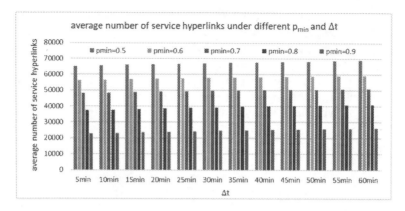

Fig. 3. Average number of service hyperlinks under different values of p_{min} and Δt.

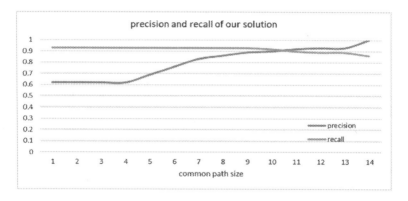

Fig. 4. The precision and recall of our solution.

Common path size between routing paths in testing set and in training set will affect the precision and recall of our solution. Final results are drawn in Fig. 4. As the figure shows, **our precision shows a growing trend towards the common path size**. At first, our solution discovers many event routing paths in testing set, each of which has a common path with some path learned in training set. It leads to a large result set, which causes a low precision at first. Obviously, paths in testing set leading to no anomaly/failure probably have a common path with small size. Thus, when the common path size reaches 4, our solution does not make warnings of predictive maintenance for them anymore, which causes the growth of precision. Looking into the results and we find some interesting conclusions. Event routing paths in the testing set leading to some anomaly/failure, even with same type, probably have different common path sizes with the paths learned in the training set. Consequently, both of the result set and the correct result set becomes smaller with the rise of common path size. But the descending speed of the correct result set size is slower than that of result set size. Therefore, we have an ascending precision curve.

On the other hand, **the recall curve seems smooth and steady at first and falls slightly at the end.** It is also caused by the slow descending of correct result set. Some event routing paths in testing set, related to some anomaly/failure, do not have a large enough common path with paths learned in training set. So they can no longer trigger warnings of predictive maintenance under large common path size, e.g., 11, 12, 13, 14. But actually, they have issued a warning before.

8   Related Works

Event correlation discovery is a hot topic [5–10]. It can be used in various areas like process discovery [5, 6], anomaly detection [7, 8], healthcare monitoring [9, 10] and so on. In the field of business process discovery, event correlation challenge is well known as the difficulty to relate events that belong to the same case. Pourmirza and et al. proposed a technique called correlation miner, to facilitates discovery of business process models when events are not associated with a case identifier [5, 6]. Some studies used event correlation to detect anomalies. Friedberg and et al. proposed a novel anomaly detection approach. It keeps track of system events, their dependencies and occurrences, and thus learns the normal system behavior over time and reports all actions that differ from the created system model [7]. Fu and et al. focused on temporal correlation and spatial correlation among failure events. They developed a model to quantify the temporal correlation and characterize spatial correlation. Failure events are clustered by correlations to predict their future occurrences [8]. Other works applied event correlation in healthcare monitoring. Forkan and et al. concentrated on vital signs, which are used to monitor a patient's physiological functions of health. The authors proposed a probabilistic model to make predictions of future clinical events of an unknown patient in real-time using the learned temporal correlations of multiple vital signs from many similar patients [9, 10].

Recently, some researchers focus on event dependencies. Song and et al. mined activity dependencies (i.e., control dependency and data dependency) to discover process instances when event logs cannot meet the completeness criteria [17]. In this paper, the control dependency indicates the execution order and the data dependency indicates the input/output dependency in service dependency. A dependency graph is utilized to mine process instances. In fact, the authors do not consider the dependency among events. Plantevit and et al. presented a new approach to mine temporal dependencies between streams of interval-based events [18]. Two events have a temporal dependency if the intervals of one are repeatedly followed by the appearance of the intervals of the other, in a certain time delay.

9   Conclusion

In this paper, we propose an event correlation based approach on predictive maintenance. We firstly focus on discovering the correlations between two event type sets, where intra-group instances are unordered and inter-group instances are time-ordered, and all instances occur closely in time. We transform the discovery problem into a time-constrained frequent co-occurrence mining problem and develop the *EventCorrelator* algorithm. Our previous works proposed a proactive data service model with an important component called service hyperlink. We encapsulate discovered correlations into service hyperlink and learn anomaly propagation paths by splicing service hyperlinks. Finally, we perform predictive maintenance based on these paths.

Acknowledgement. Funding: This work was supported by National Natural Science Foundation of China (No. 61672042).

References

1. Qiu, H., Liu, Y., Subrahmanya, N.A., Li, W.: Granger causality for time-series anomaly detection. In: 12th IEEE International Conference on Data Mining, pp. 1074–1079. Institute of Electrical and Electronics Engineers Inc., Brussels (2012)
2. Sipos, R., Fradkin, D., Moerchen, F., Wang, Z.: Log-based predictive maintenance. In: 20th ACM SIGKDD International Conference on Knowledge Discovery and Data Mining, pp. 1867–1876. Association for Computing Machinery, New York (2014)
3. Yan, Y., Luh, P.B., Pattipati, K.R.: Fault diagnosis of HVAC air-handling systems considering fault propagation impacts among components. IEEE Trans. Autom. Sci. Eng. **14** (2), 705–717 (2017)
4. Ye, R., Li, X.: Collective representation for abnormal event detection. J. Comput. Sci. Technol. **32**(3), 470–479 (2017)
5. Pourmirza, S., Dijkman, R., Grefen, P.: Correlation miner: mining business process models and event correlations without case identifiers. Int. J. Coop. Inf. Syst. **26**(2), 1–32 (2017)
6. Pourmirza, S., Dijkman, R., Grefen, P.: Correlation mining: mining process orchestrations without case identifiers. In: Barros, A., Grigori, D., Narendra, N.C., Dam, H.K. (eds.) ICSOC 2015. LNCS, vol. 9435, pp. 237–252. Springer, Heidelberg (2015). https://doi.org/10.1007/978-3-662-48616-0_15
7. Friedberg, I., Skopik, F., Settanni, G., Fiedler, R.: Combating advanced persistent threats: from network event correlation to incident detection. Comput. Secur. **48**, 35–57 (2015)
8. Fu, S., Xu, C.: Quantifying event correlations for proactive failure management in networked computing systems. J. Parallel Distrib. Comput. **70**(11), 1100–1109 (2010)
9. Forkan, A.R.M., Khalil, I.: PEACE-Home: probabilistic estimation of abnormal clinical events using vital sign correlations for reliable home-based monitoring. Pervasive Mob. Comput. **38**, 296–311 (2017)
10. Forkan, A.R.M., Khalil, I.: A probabilistic model for early prediction of abnormal clinical events using vital sign correlations in home-based monitoring. In: 14th IEEE International Conference on Pervasive Computing and Communications, pp. 1–9. Institute of Electrical and Electronics Engineers Inc., Sydney
11. Han, Y., Liu, C., Su, S., Zhu, M., Zhang, Z., Zhang, S.: A proactive service model facilitating stream data fusion and correlation. Int. J. Web Serv. Res. **14**(3), 1–16 (2017)
12. Zhu, M., Liu, C., Wang, J., Su, S., Han, Y.: An approach to modeling and discovering event correlation for service collaboration. In: Maximilien, M., Vallecillo, A., Wang, J., Oriol, M. (eds.) ICSOC 2017. LNCS, vol. 10601, pp. 191–205. Springer, Cham (2017). https://doi.org/10.1007/978-3-319-69035-3_13
13. Domingues, R., Filippone, M., Michiardi, P., Zouaoui, J.: A comparative evaluation of outlier detection algorithms: experiments and analyses. Pattern Recogn. **74**, 406–421 (2018)

14. Yeh, C.M., et al. : Time series joins, motifs, discords and shapelets: a unifying view that exploits the matrix profile. Data Min. Knowl. Discov., 1–41 (2017)
15. Yagci, A.M., Aytekin, T., Gurgen, F.S.: Scalable and adaptive collaborative filtering by mining frequent item co-occurrences in a user feedback stream. Eng. Appl. Artif. Intell. **58**, 171–184 (2017)
16. Yu, Z., Yu, X., Liu, Y., Li, W., Pei, J.: Mining frequent co-occurrence patterns across multiple data streams. In: 18th International Conference on Extending Database Technology, pp. 73–84. OpenProceedings.org, University of Konstanz, University Library, Brussels, Belgium (2015)
17. Song, W., Jacobsen, H.A., Ye, C., Ma, X.: Process discovery from dependence-complete event logs. IEEE Trans. Serv. Comput. **9**(5), 714–727 (2016)
18. Plantevit, M., Robardet, C., Scuturici, V.M.: Graph dependency construction based on interval-event dependencies detection in data streams. Intell. Data Anal. **20**(2), 223–256 (2016)

Using Crowdsourcing for Fine-Grained Entity Type Completion in Knowledge Bases

Zhaoan Dong[1], Ju Fan[1(✉)], Jiaheng Lu[2,1], Xiaoyong Du[1], and Tok Wang Ling[3]

[1] DEKE, MOE and School of Information, Renmin University of China,
Beijing, China
fanj@ruc.edu.cn
[2] Department of Computer Science, University of Helsinki, Helsinki, Finland
[3] School of Computing, National University of Singapore, Singapore, Singapore

Abstract. Recent years have witnessed the proliferation of large-scale Knowledge Bases (KBs). However, many entities in KBs have *incomplete* type information, and some are totally untyped. Even worse, fine-grained types (e.g., BasketballPlayer) containing rich semantic meanings are more likely to be incomplete, as they are more difficult to be obtained. Existing machine-based algorithms use predicates (e.g., birthPlace) of entities to infer their missing types, and they have limitations that the predicates may be insufficient to infer fine-grained types. In this paper, we utilize crowdsourcing to solve the problem, and address the challenge of controlling crowdsourcing cost. To this end, we propose a hybrid machine-crowdsourcing approach for fine-grained entity type completion. It firstly determines the types of some *"representative"* entities via crowdsourcing and then infers the types for remaining entities based on the crowdsourcing results. To support this approach, we first propose an embedding-based influence for type inference which considers not only the distance between entity embeddings but also the distances between entity and type embeddings. Second, we propose a new difficulty model for entity selection which can better capture the uncertainty of the machine algorithm when identifying the entity types. We demonstrate the effectiveness of our approach through experiments on real crowdsourcing platforms. The results show that our method outperforms the state-of-the-art algorithms by improving the effectiveness of fine-grained type completion at affordable crowdsourcing cost.

Keywords: Crowdsourcing · Entity type completion · Knowledge base

1 Introduction

The last decades have witnessed the booming of large-scale and open-accessible Knowledge Bases (KBs) such as DBpedia [12], Freebase [2], and YAGO [21].

This work is partially supported by National Natural Science Foundation of China (No. 61602488, No. 61632016 and No. 61472427) and Academy of Finland (No. 310321).

Y. Cai et al. (Eds.): APWeb-WAIM 2018, LNCS 10988, pp. 248–263, 2018.
https://doi.org/10.1007/978-3-319-96893-3_19

These KBs contain thousands of millions of real-world *entities* that fall into different *types*, e.g., Person, Place, Sport. Due to their large coverage and high quality, the KBs have been successfully used to support many applications, such as web search, question answering and entity linking. However, many entities in the KBs have *incomplete* type information, and some are even totally untyped [18]. In DBPedia, for example, there are over 4 million entities assigned with about 4 million types, which means that only one type per entity in average [10]. Even worse, many entities in DBPedia only have *coarse-grained* types such as Thing, while *fine-grained* types such as GolfPlayer are missing since they are difficult to be obtained when constructing the KBs. This incompleteness of entity types affects the usefulness and usability of the KBs. For example, entity Hedy_Lamarr[1] in DBpedia is assigned with only general types including Thing, Agent and Person but none of the fine-grained ones: neither Actor nor Inventor. Therefore, Hedy_Lamarr is bound to be missed in the answer of following question: *Who is not only a famous **Actor**, but also an **Inventor**?*

To complete the type information in KBs, some *machine-based* approaches have been proposed. For example, *SDType*, which is reported as the state-of-the-art method [15,19], exploits the predicates, i.e., links between entities, to infer missing types. Intuitively, if a predicate occurs in entities of one specific type, it would be assigned with a large weight. In contrast, if it occurs in entities of many different types, it will be assigned with a low weight. Obviously, high weighted predicates are more likely to identify fine-grained types. For example, in DBPedia, predicates teachingStaff and numberOfClassrooms have a high weight to infer type School. *SDType* computes a confidence score for every possible type of an entity based on predicate weights. Then, if the score of a type is larger than some threshold, say 0.5, it completes the type for the entity.

However, since the KBs are often incomplete and noisy, it is difficult for *SDType* to infer the correct *fine-grained* types if an entity misses the correct highly weighted predicates or has some wrong predicates. For example, in DBPedia with version 3.8, the BasketballPlayer entity Ron_Harper has only 4 predicates, college, draftTeam, birthPlace and nationality. Given the entity denoted by x, *SDType* computes a score of "*inference ability*" for each p of these predicates to infer type C, which is denoted as $Prob(C(x)|p)$. For example, as shown in Table 1, $Prob$(BasketballPlayer(x)|draftTeam) = 0.281 means 28.1% entities having predicate draftTeam belong to type BasketballPlayer. We can see that the scores of the four predicates of Ron_Harper are quite small, which is insufficient to infer type BasketballPlayer for the entity. A detailed analysis on computing $Prob(C(x)|p)$ and limitations of *SDType* is referred to Sect. 3.

To overcome the limitation of machine-based approaches, we propose to utilize crowdsourcing that leverages intelligence of the crowd to solve the entity type completion problem. The main motivation is that human is much better than machine to identify entity types, even though predicates of entities may be missing. For example, Fig. 1 shows an example crowdsourcing task for Ron_Harper. We can see that it is not difficult for human to identify the correct type(s) for the

---

[1] http://dbpedia.org/page/Hedy_Lamarr.

Table 1. Statistics of top 10 predicates linked to `BasketballPlayer`

	p	$Prob(C(x)|p)$	w_p
1	draftTeam	0.281	0.078
2	highschool	0.241	0.337
3	college	0.224	0.422
4	nationality	0.035	0.005
5	ceo	0.023	0.002
6	coach	0.023	0.002
7	alumni	0.015	0.004
8	league	0.014	0.246
9	birthPlace	0.009	0.021
10	formerTeam	0.006	0.248

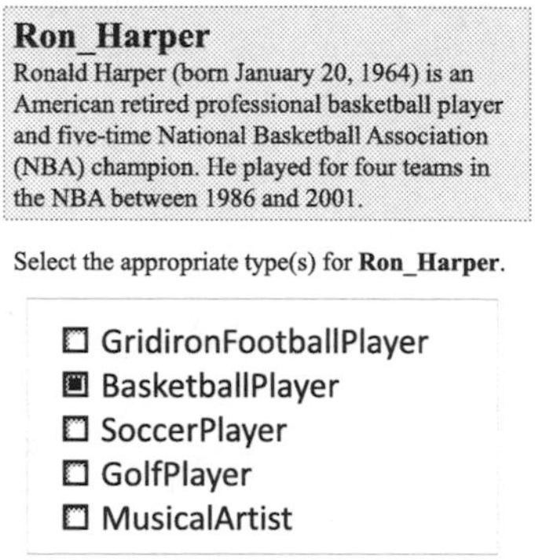

Ron_Harper

Ronald Harper (born January 20, 1964) is an American retired professional basketball player and five-time National Basketball Association (NBA) champion. He played for four teams in the NBA between 1986 and 2001.

Select the appropriate type(s) for **Ron_Harper**.

☐ GridironFootballPlayer
☑ BasketballPlayer
☐ SoccerPlayer
☐ GolfPlayer
☐ MusicalArtist

Fig. 1. An example of micro-task

given entities. This is also verified by our experiments that the crowdsourcing result is much better than machine-based approaches.

Unfortunately, the challenge in utilizing crowdsourcing for entity type completion is the crowdsourcing cost, since we need to pay rewards to the crowd for their answers. Especially, in a large-scale knowledge base, it would be extremely expensive if all the entities with their possible types are crowdsourced. Therefore, we devise a hybrid framework which combines the intelligence of crowdsourcing workers with the algorithmic approaches. We firstly select some *"representative"* entities. Next, we publish the selected entities for crowdsourcing. At last, we infer and determine the entities' types based on the crowdsourcing results.

To support the hybrid framework, we develop the following techniques. First, we propose an embedding-based influence for type inference which considers not only the distance between entities but also the distance between entities and types when inferring entity types. Second, we propose a new difficulty model for entity selection which can better describe the uncertainty of the machine algorithm to determine the correct types of entities. We demonstrate the effectiveness of the method through experiments on real datasets. The results show that our hybrid method outperforms the baseline machine algorithm by recalling more fine-grained entity types with small extra crowdsourcing cost.

The remainder of this paper is organized as follows: Sect. 2 presents an overview of our approach. Sections 3, 4 and 5 introduce the techniques on generating candidate types, type inference and entity selection respectively. The experimental results are reported in Sect. 6 and related works are reviewed in Sect. 7. We finally conclude the paper in Sect. 8.

2   An Overview of Entity Type Completion

This section presents an overview of entity type completion in KBs. We first formally define the problem and then introduce our crowdsourcing-based approach.

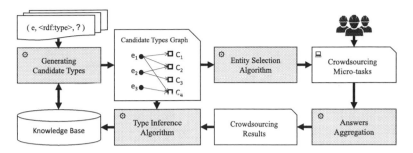

Fig. 2. An overview of the hybrid method

2.1 Problem Formalization

We denote a knowledge base $\mathcal{K} = \{\mathcal{E}, \mathcal{T}, \mathcal{R}, \mathcal{K}_{\mathcal{R}}, \mathcal{K}_{\mathcal{T}}\}$, where $\mathcal{E}$ is the set of entities (e.g., Ron_Harper), $\mathcal{T}$ is the set of entity types (e.g., BasketballPlayer) and $\mathcal{R}$ is the set of predicates (e.g., birthPlace). $\mathcal{K}_{\mathcal{R}} = \{p(e, o) \mid e, o \in \mathcal{E} \land p \in \mathcal{R}\}$ contains the known predicate instances (e.g., birthPlace(Ron_Harper,Dayton)). $\mathcal{K}_{\mathcal{T}} = \{C(e) \mid e \in \mathcal{E} \land C \in \mathcal{T}\}$ is a set of known type assertions indicating that entity e is an instance of type C (e.g., Person(Ron_Harper)). With the notations, we formulate the entity type completion problem as follows:

Definition 1 (Entity Type Completion). *Given a set of entities $\mathcal{A} \subseteq \mathcal{E}$ and a set of entity types $\mathcal{T}$, it determines whether an entity-type pair (e, C) is true or not, where $e \in \mathcal{A}$, $C \in \mathcal{T}$ and $C(e) \notin \mathcal{K}_{\mathcal{T}}$. Then, if (e, C) is true, we can add the new found type assertion $C(e)$ to $\mathcal{K}_{\mathcal{T}}$: $\mathcal{K}_{\mathcal{T}} \leftarrow \mathcal{K}_{\mathcal{T}} \cup \{C(e)\}$.*

2.2 Framework of Our Crowdsourcing-Based Approach

To address the problem, we propose a crowdsourcing-based approach that utilizes the intelligence of crowdsourcing workers. Moreover, as the crowdsourcing budget is limited, we devise a machine-crowdsourcing hybrid framework that is illustrated in Fig. 2.

The approach takes as input a knowledge base $\mathcal{K}$ that contains all the known predicate instances $\mathcal{K}_{\mathcal{R}}$ and type assertions $\mathcal{K}_{\mathcal{T}}$. Given a set of entities missing types, it first employs a machine-based algorithm to generate the candidate types for them. Then, an *Entity Selection Algorithm* selects the most *"representative"* entities under a given crowdsourcing budget. On the one hand, the selected entities should be *"uncertain"* for the algorithm to identify the correct types. On the other hand, the selected entities should be more useful to infer more type assertions for unselected entities. Next, we generate micro-tasks for the selected entities and publish them to crowdsourcing platform, e.g., Amazon Mechanical Turk (AMTurk)[2], where human workers could help to identify the right types. Then, the answers collected from the crowds will be aggregated. Finally, the

---

[2] https://www.mturk.com.

Type Inference Algorithm infers the types for all target entities based on both the crowdsourcing results and the inferred results.

We note that there are some existing studies [5–7,9,11] that devise machine-crowdsourcing hybrid approaches in other applications, e.g., web table matching, information extraction and academic knowledge acquisition(see Sect. 7). However, we are the first to devise the hybrid framework in entity type completion. The variety of knowledge bases requires us to develop new techniques on the following components. We discuss how to generate candidate types in Sect. 3, introduce the inference algorithm in Sect. 4, and present how to select "representative" entities in Sect. 5.

3   Generating Candidate Types

We use the state-of-the-art machine-based type completion method *SDType* [19] to generate candidate types. The basic building blocks of SDType are conditional probabilities, e.g., the probability of an entity being of type C if it has a predicate p. Additionally, each predicate p is assigned a weight w_p which reflects the discriminating power of the predicate. Note that p is treated differently with its reverse predicate p^{-1}, i.e., they are assigned different weights respectively. According to the probability distribution of all predicates associated with each entity, *SDType* computes a confidence score for each entity-type pair.

$$w_p = \sum_{C \in \mathcal{T}} (Prob(C) - Prob(C|p))^2 \tag{1}$$

$$score(e_i, C_j) = \frac{\sum_{p \in Pred(e_i)} Prob(C_j(e_i)|p) \cdot w_p}{\sum_{p \in Pred(e_i)} w_p} \tag{2}$$

where $Prob(C_j(e_i)|p)$ indicates how likely an entity e_i having predicate p is of class C_j, and $Prob(C|p)$ indicates the prior probability of type C, i.e, how many entities that belongs to the type C with predicate p.

Limitations of *SDType*. We utilize the example in Table 1 to analyze limitations of *SDType*. In DBPedia, the `BasketballPlayer` entity `Ron_Harper` has only 4 predicates, `college`, `draftTeam`, `birthPlace` and `nationality`. All of them appear in the top-10 predicates linked with type `BasketballPlayer` respect to the conditional probability. As is shown in Table 1, $Prob(C(x)|p)$ indicates how likely an entity having predicate p is of class C, and w_p is the weight of predicate p. For example, `Prob(BasketballPlayer(x)|draftTeam)` = 0.281 means 28.1% entities having predicate `draftTeam` belong to `BasketballPlayer`. Although the predicate `college` has the maximum weight as 0.422, it is hard to determine the correct type `BasketballPlayer` because the conditional probability is just 0.224. As a result, the confidence value of the entity-type pair <`Ron_Harper, BasketballPlayer`> is just 0.233. In fact, if the confidence threshold is set to 0.5, Table 1 indicates that none of the `BasketballPlayer` entities could obtain the correct type via *SDType* because no predicate has a conditional probability greater than 0.5.

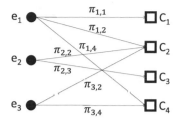

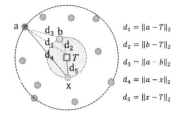

Fig. 3. A toy candidate types graph

Fig. 4. Embedding based influence

Candidate Types Graph. The target entities and their candidate types are represented in a Candidate Types Graph. A *Candidate Types Graph* is a bipartite graph $\mathcal{G} = \{\mathcal{A}, \mathcal{T}^*, \Pi\}$, where $\mathcal{A}$ is the set of target entities, $\mathcal{T}^*$ is the set of candidate types and $\Pi = \{<e_i, C_j, s_{i,j}> | e_i \in \mathcal{A} \wedge C_j \in \mathcal{T}^*\}$ is the set of all possible entity-type pairs. And $\pi_{i,j} = <e_i, C_j, s_{i,j}>$ indicates the probability that the type assertion $C_j(e_i)$ holds is $s_{i,j}$, where $1 \leq i \leq |\mathcal{A}|$ and $1 \leq j \leq |\mathcal{T}^*|$. We also represent all candidate types of entity e as $\mathcal{C}(e)$. For example, Fig. 3 is a toy Candidate Types Graph that consists of 3 entities, 4 candidate types and 7 candidate entity type pairs.

4   Inferring Types Using Crowdsourcing Results

4.1   Type Inference Algorithm

We adopt the concept determination algorithm proposed in [7] as our type inference algorithm. As shown in Algorithm 1, it takes the Candidate Types Graph $\mathcal{G}$ and the crowdsourcing validated entity-type pairs $\mathcal{S}^q$ as input. For each entity-type pair, the algorithm updates its score based on two evidences. One is the initial score from *SDType* whose prior probability is denoted as α. The other is the influences from the approved entity-type pairs by crowdsourcing with prior probabilities $1 - \alpha$. Finally, the algorithm outputs updated Candidate Types Graph $\mathcal{G}'$ with the revised confidence scores for each entity-type pairs. The major difference is that we propose a new method to compute the influences $Inf(\pi|\mathcal{A}^q)$ in line 6 of Algorithm 1.

4.2   Influence Between Entity-Type Pairs

In [7], the authors proposed a concept-based method to compute the *inter-table influence* between two columns as the cosine similarity of their concept vectors. The same idea can be adopted for entity type completion tasks, that is, similar entities are more likely to belong to the same type.

Algorithm 1. *Type Inference Algorithm*

Input
    $\mathcal{G} = \{\mathcal{A}, \mathcal{T}^*, \Pi\}$: *a initial Candidate Types Graph.*
    $\mathcal{S}^q = \{<e^q, C>\}$: *the approved entity-type pairs for selected entities $\mathcal{A}^q$.*
Output
    $\mathcal{G}' = \{\mathcal{A}, \mathcal{T}^*, \Pi'\}$: *the updated Candidate Types Graph.*
1: Initialize $\Pi' = \Phi$
2: **for** *each* $\pi = <e, C, s> \in \Pi$ **do**
3:      **if** $e \in \mathcal{A}^q$ **then**
4:         Update the score of π with crowdsourcing answer, i.e., $s' = 1.0$
5:      **else**
6:         Update the score of π: $s' = \alpha \cdot s + (1 - \alpha) \cdot Inf(\pi | \mathcal{A}^q)$
7:      **end if**
8:      $\Pi' \leftarrow \Pi' \cup \{<e, C, s'>\}$
9: **end for**
10: *return* $\mathcal{G}' = \{\mathcal{A}, \mathcal{T}^*, \Pi'\}$

Type Vector Based Influence: Specifically, we generate a type vector for each entity e based on Candidate Types Graph, which is denoted as $\vec{e}$. Each dimension in the vector represents a candidate type and the value is the confidence score $s_{i,j}$ of the edge $<e_i, C_j>$. Therefore, the influence from an entity-type pair $<e_m, C>$ to $<e_n, C>$ can be computed as the cosine similarity of two entities' type vectors, i.e., $\vec{e_m}$ and $\vec{e_n}$.

$$Inf(<e_n, C>|<e_m, C>) = CosineSimilarity(\vec{e_m}, \vec{e_n}) \tag{3}$$

Embedding-based Influence: Unlike the type vectors which are constructed based on the edges in Candidate Types Graph, embedding is a latent representation for knowledge bases. Embedding-based algorithms, such as TransE [3], embed entities and relations into relatively low dimensional representations (i.e., embeddings) so that semantic related entities can be close to each other. For example, entities of the same type are usually close to each other in the embeddings space. Additionally, entity types can also be embedded into the same space, so that one type embedding can be close to the entities from that type [10]. Therefore, we propose an embedding-based method to compute the influences between entity-type pairs, e.g., from an entity-type pair $<e_m, C>$ to $<e_n, C>$.

$$Inf(<e_n, C>|<e_m, C>) = \begin{cases} \frac{1}{1+\|\vec{e_m}, \vec{e_n}\|_2} & if \|\vec{e_m}, \vec{C}\|_2 \geq \|\vec{e_n}, \vec{C}\|_2 \\ 0 & if \|\vec{e_m}, \vec{C}\|_2 < \|\vec{e_n}, \vec{C}\|_2 \end{cases} \tag{4}$$

where $\vec{e_m}$, $\vec{e_n}$ denote the embeddings of e_m and e_n respectively, and $\vec{C}$ is the type embedding of C in the same embedding space. We use 2-norm to measure the distance between entity and type embedding. Unlike type vector based influence which only considers the similarity between two entities, embedding based influence not only considers the distance between embeddings of two entities but also the distances from type embedding to each entity.

Specifically, there are two meanings in Eq. 4. On the one hand, if the type of an entity is already determined to be one type, we can infer the type of entity closer to the same type. For example, in Fig. 4, if a belongs to type T, $Inf(<b,T>|<a,T>) = 1/(1 + d_3)$, which depends on the distance between a and b. So that, $Inf(<b,T>|<a,T>) > Inf(<x,T>|<a,T>)$ because b is closer to a than x. On the other hand, we cannot infer the type of entity far away from the type. For example, in Fig. 4, we cannot infer the type of a based on b, i.e., $Inf(<a,T>|<b,T>) = 0$.

4.3   Aggregating the Influences

We use the same method used in [7] to aggregate the influences from the selected entities, where $\xi(e_m)$ denotes all the candidate type assertions of entity e_m and $\xi(\mathcal{A}^q)$ represents the approved entity-type pairs of selected entities $\mathcal{A}^q$. Firstly, we assume that the influences from those edges in $\xi(e_m)$ to $\pi_{n,j}$ are independent with each other. Therefore, the influence from an entity to an entity-type pair is aggregated as:

$$Inf(\pi_{n,j}|e_m) = 1 - \prod_{\pi_{m,i} \in \xi(e_m)} (1 - Inf(\pi_{n,j}|\pi_{m,i})) \qquad (5)$$

which could be interpreted as the probability that $\pi_{n,j}$ is influenced by at least one edge of $\xi(e_m)$. Then we compute the influence from selected entities to an entity-type pair as:

$$Inf(\pi_{n,j}|\mathcal{A}^q) = 1 - \prod_{\pi_{m,i} \in \xi(\mathcal{A}^q)} (1 - Inf(\pi_{n,j}|\pi_{m,i})) \qquad (6)$$

Similarly, the influence from an entity e_m to e_n indicates the probability that at least one edge in $\xi(e_n)$ is influenced by entity e_m.

$$Inf(e_n|e_m) = 1 - \prod_{\pi_{n,j} \in \xi(e_n)} (1 - Inf(\pi_{n,j}|e_m)) \qquad (7)$$

Finally, we obtain the influence from selected entities $\mathcal{A}^q$ to an entity e_n as:

$$Inf(e_n|\mathcal{A}^q) = 1 - \prod_{\pi_{n,j} \in \xi(e_n)} (1 - Inf(\pi_{n,j}|\mathcal{A}^q)) \qquad (8)$$

5   Selecting Entities for Crowdsourcing

The fundamental challenge in the hybrid approach is to determine which entities should be selected for crowdsourcing. In [7], the authors proposed an expected utility function which considers both task difficulty and influence. They developed a greedy-based algorithm based on the expected utilities. Similarly, for entity type completion tasks, we define the expected utility of the selected entities $\mathcal{A}^q$ as:

$$E[(\mathcal{A}^q)] = \sum_{e \in \mathcal{A}} \mathcal{D}(e) \cdot Inf(e|\mathcal{A}^q) \tag{9}$$

We modify the greedy algorithm proposed in [7] and apply it to select entities for entity type completion. The major difference is that we use our new difficulty model and embedding based influence model when computing the utilities of the selected entities. Thus, we mainly introduce our new difficulty model below.

5.1 Entity Difficulty Model

The entity difficulty model illustrates how certain the machine algorithm is when identifying the types for one entity. For example, the entropy of types distribution is a common measure of uncertainty which is already used in [7]. The intuition is that, the more uncertain the machine algorithm is, the more likely it will be to make mistakes. Therefore, we need crowdsourcing to complete the types of those entities which are most uncertain for machine algorithm.

For *SDType*, we find it is more uncertain because: (1) the probabilities of an entity's candidate types are almost identical; (2) the weights of related predicates are very low; (3) the maximum of scores is very low; (4) the entity has too many candidate types. Based on the observations, we propose a new difficulty model which takes all of the above factors into consideration.

Entropy. Similar to [7], we firstly consider the entropy which reflects the distribution of the probabilities of an entity's candidate types. On the one hand, if a type has clearly higher score than others, the entropy is low. On the other hand, if it is close to a uniform distribution, the entropy is high.

$$\mathcal{D}_1(e) = - \sum_{C \in \mathcal{C}(e)} \frac{score(e, C)}{S} \cdot \log \frac{score(e, C)}{S} \tag{10}$$

where $S = \sum_{C \in \mathcal{C}(e)} score(e, C)$ is used for normalization.

Average Weight of Predicates. *SDType* uses all predicates linked with the entity as indicators for its types. If each predicate in $Pred(e)$ has a large weight w_p, i.e., each of them has a great discriminating power, then it is easy to identify correct types for entity e. Otherwise, it is difficult.

$$\mathcal{D}_2(e) = \frac{1}{|Pred(e)|} \cdot \sum_{p \in Pred(e)} w_p \tag{11}$$

Max Score of Candidate Types. The intuition is that, if one type has significant higher score than that of other types, it is easy to determine the answer, otherwise, difficult.

$$\mathcal{D}_3(e) = 1 - \max\{score(e, C)|t \in \mathcal{C}(e)\} \tag{12}$$

Number of Candidate Types. Obviously, it is difficult to infer the correct type if there are too many candidate types. Thus,

$$\mathcal{D}_4(e) = |\mathcal{C}(e)| \tag{13}$$

Finally, the difficulty of an entity e is defined as follows:

$$\mathcal{D}(e) = \mathcal{D}_1(e) \times \mathcal{D}_2(e) \times \mathcal{D}_3(e) \times \mathcal{D}_4(e) \tag{14}$$

6   Experiments

We implement the algorithms using `Scala 2.11` including *SDType*, the *Entity Selection Algorithm* and *Type Inference Algorithm*. The codes run in a pseudo-distributed `spark-2.0.0` on a single PC with a 2.6 GHz Intel core i5 processor and 16 GB RAM.

6.1   Datasets

We firstly extract $9,970,687$ predicate instances of 629 predicates about $2,283,173$ entities from DBPedia3.8[3]. For types, we extract $7,727,665$ type assertions and transform the hierarchical structure in DBpedia Ontology[4] into a tree structure[5] where the root type ($Level = 0$) is "`Thing`". It should be mentioned that, in this paper, we mainly focus on the fine-grained entity type completion. In particular, we evaluate completion for types with $Level = 4$.

DBP-904: We randomly extract about 1000 type assertions, i.e., $<e, C>$ pairs from $7,727,665$ types where the type C is from $Level = 4$ and has at least 100 instances. Then we filter out some entities according to the pruning strategy for candidate types (see Sect. 6.2). At last, 904 entities with their fine-grained types are retained.

DBP-4987: We first extract predicates having at least 1000 instances and their subjects and objects must have types of $Level = 4$. For each extracted predicate, we sample 50 instances, then we obtain 5324 entities. We also filter out those entities whose ground truth does not appear in the candidate type list. Finally, we obtain 4987 entities with their type assertions. There are 7054 predicate instances of 107 predicates in total.

---

[3] http://wiki.dbpedia.org/services-resources/datasets/previous-releases/dataset-38.
[4] http://wiki.dbpedia.org/services-resources/ontology.
[5] http://mappings.dbpedia.org/server/ontology/classes/.

Embeddings of Entities and Types: We generate entity embeddings using Fast-TransE[6] which is an efficient implementation of TransE. We run the code with the parameters: $nepoches = 5000$, $nbatches = 100$, $dimension = 250$, $alpha = 0.001$, and $threads = 8$. The remaining parameters have the default values as presented in the source code. Based on the entity embeddings, we use the algorithm proposed in [10] to generate type embeddings in the same semantic vector space.

6.2   Crowdsourcing on Amazon Mechanical Turk

Pruning Candidate Types: In *SDType*, the number of candidate types for different entities is quite different. For example, some may have more than 100 candidate types since one predicate often provides many candidate types. Unfortunately, this will affect the quality of crowdsourcing because human workers are easy to be bored with such a long list.

To tackle this problem, we set a threshold η for $Prob(C(e)|p)$ to prune the entity-type pairs with low probabilities. For example, when $\eta = 0$, we get 92 candidate types for entity Hirooka_Station. When $\eta = 0.1$, only 3 of them are retained. Thus, it is easy for human workers to select the right types. In our experiments, we empirically set $\eta = 0.05$ which can keep the number of candidate types within an acceptable range. After being pruned, the average number of the candidate types is 6 and the maximum is 15.

Micro-Tasks and Answer Aggregation: We generate micro-tasks for selected entities and publish them on Amazon Mechanical Turk (AMTurk). In order to reduce the crowdsourcing cost, a Human Intelligence Task (HIT) is designed to contain 10 micro-tasks. In our experiments, each HIT is assigned to 5 workers and we spend $0.1 for each assignment. As is shown in Fig. 1, each micro-task contains a short description of an entity and a list of candidate types, human workers are asked to select correct types from the list. For crowdsourcing answers aggregration, we employ the codes of *Get-Another-Label* algorithm[7], which is a variation of *Expectation-Maximization* (EM) [4].

6.3   Evaluation on Entity Difficulty Model

We evaluate our entity difficulty model on DBP-904. Firstly, all testing entities are sorted according to their difficulty in ascending order and then equally separated into 10 buckets. Figure 5 shows that pure crowdsourcing method obtains high and stable recall of entity-type pairs, while the *SDType* algorithm performs worse on two datasets when the difficulty increases. This shows that the proposed entity difficulty model can effectively capture the uncertainty of entities.

---

[6] https://github.com/thunlp/Fast-TransX.
[7] https://github.com/ipeirotis/Get-Another-Label/wiki.

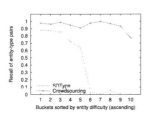

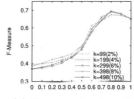

(a) Type vector based

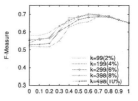

(b) Embedding based

Fig. 5. Evaluation on the entity difficulty model.

Fig. 6. Determining the prior probability α.

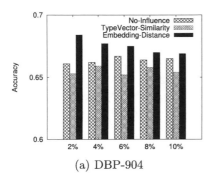

(a) DBP-904

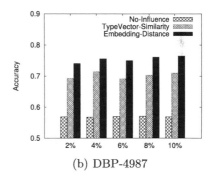

(b) DBP-4987

Fig. 7. Comparison of different influence models.

6.4 The Prior Probability α

To set an appropriate prior probability α in Algorithm 1, we perform an experiment on DBP-4987. We set the α from 0.0 to 1.0 with step 0.1 and examine the F-Measure with different amount of randomly selected entities, because F-Measure considers both the Recall and Precision of the algorithm. On one hand, we hope to recall more positive entity type pairs. On the other hand, we want the false positives to be as few as possible. As shown in Fig. 6, we find that $\alpha = 0.8$ is the best for type vector based inference while $\alpha = 0.7$ for embedding based method.

6.5 Comparison of Influence Models

To evaluate the embedding based influence, we first randomly select $x\%$ of entities and publish them on AMTurk for crowdsourcing. Then, we infer the types of all target entities using the following three methods respectively: (1). *No-Influence* does not consider the influences, i.e., it directly merges the crowdsourcing results for selected entities and that of *SDType* for unselected. (2). *TypeVector-Similarity*: Inferring with the type vector based influence. (3). *Embedding-Distance*: Inferring with the embeddings based influence.

As shown in Fig. 7, *Embedding-Distance* method outperforms the other two methods in Accuracy. The increasement is significant on DBP-4987. Com-

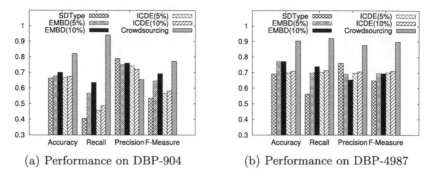

(a) Performance on DBP-904   (b) Performance on DBP-4987

Fig. 8. Comparison of different inference methods.

pared with *No-Influence*, it is increased by 30% in average across various number of selected entities and about 8% compared with *TypeVector-Similarity*. This improvement is mainly attributed to the embedding-based representation learned, which can better capture global structural information in the KBs.

6.6   Comparison with the Existing Methods

In this section, we compare our hybrid method with three existing methods. Firstly, we briefly describe the methods to be compared as follows. (1) *SDType*: the state-of-the-art pure machine algorithm [15,19]. (2) *ICDE*: the hybrid method proposed in [7] for web table matching. We implement this algorithm for entity type completion task. In comparison, we examine the performance separately when 5% and 10% entities are selected. (3) *EMBD*: our hybrid method for entity type completion based on *embedding based influence*. Similarly, we examine the performance with two budgets 5% and 10%, separately. (4) *Crowdsourcing*: the pure crowdsourcing method, i.e., all the target entities are crowdsourced.

As shown in Fig. 8, the pure crowdsourcing method achieves the best performance. For example, the values of Recall are greater than 0.9 on both DBP-904 and DBP-4987. However, it is too expensive to crowdsource all the entities. The only special case is that on DBP-904, its Precision is lower than others. This is mainly because we allow human workers to select multiple types for one entity. As a result, some false positive entity-type pairs occur in the crowdsourcing results.

On DBP-904, EMBD significantly outperforms *SDType* and ICDE. For example, compared with *SDType*, EMBD (5%) increases about 40% (from 0.40 to 0.57) on the Recall and about 21% (from 0.54 to 0.65) on F-Measure. Compared with ICDE (5%), EMBD (5%) increases about 24% (from 0.46 to 0.57) on the Recall and about 14% (from 0.57 to 0.65) on F-Measure.

On DBP-4987, although the advantage of EMBD is not so significant on F-Measure, it still outperforms *SDType* on Accuracy and Recall. For example, EMBD (5%) increases about 12% (from 0.69 to 0.77) on the Accuracy and about

24% (from 0.56 to 0.70) on Recall. In a word, EMBD outperforms *SDType* and ICDE on the whole by recalling more fine-grained entity type assertions with affordable crowdsourcing cost.

7 Related Work

Entity Type Completion is an important subproblem of knowledge base completion. Paulheim [18] classified the related methods into two categories: *internal methods* and *external methods*.

Internal methods only use the data in current knowledge base as input. A straightforward internal method is classical *ontology reasoning*. However, *RDFS reasoning* is prone to propagate errors since most knowledge bases are usually incomplete and noisy [19]. Paulheim and Bizer proposed *SDType* [19,20]. which performs like a weighted voting, where each predicate casts a vote on its object's type, using the statistical distributions to weight the votes.

External methods utilize external sources outside the KBs, e.g., text corpora or links to other KBs. For example, Tipalo system [8] parses the abstracts of entities which often follow similar patterns and map them to the WordNet and DOLCE ontologies to find types. Palmero Aprosio et al. [1] exploit cross-language links between different language versions of DBpedia as features for type completion. Sleeman and Finin [22] use SVM to predict entity types in DBpedia and Freebase. Crowdsourcing can be viewed as an external method since it utilize human knowledge which is a kind of external sources beyond the KBs.

Hybrid Machine-Crwodsourcing Methods have attracted many attentions in recent years. For example, Lofi and Maarry [14] extensively investigated the commonly reoccurring challenges and solutions for hybrid algorithmic-crowdsourcing workflows and propose a set of design patterns. Kondreddi et al. [11] proposed Higgins, a novel system architecture that effectively integrates combines Human Computing (HC) inputs with machine based Information Extraction (IE). Mozafari et al. [16] advocated integrating machine learning into crowdsourced databases. Fan et al. [7] proposed a hybrid framework which assigns the most "beneficial" column-to-concept matching tasks to human workers and then infer the best matches for the remain columns utilizing the crowdsourcing results. In this paper, we extend their algorithms with new features including an embedding-based influence model and a new entity difficulty model and apply them to entity type completion. Dong et al. proposed a platform for academic knowledge discovery and acquisition called PANDA [5,6], which exploits a hybrid algorithmic-crowdsourcing framework to identify and extract Knowledge Cells [9] such as Figures, Tables, Definitions, Algorithms, etc., from academic literature.

Knowledge Graph Embedding, in recent years, has become an active area of research for knowledge base construction and completion. One of the most successful model is TransE [3], which learns the embeddings of entities and relations in a neural-based approach. Various methods such as TransR [13], HolE [17] are also proposed. Since TransE is the most simple and popular method, it is chosen to train the embeddings.

8   Conclusion

In this paper, we have addressed the problem of *fine-grained* entity type completion in knowledge bases. We proposed a hybrid method integrating the intelligence of crowdsourcing and the speed of machine algorithm. To discover more type assertions with affordable crowdsourcing cost, we proposed a new entity difficulty model for crowdsourcing entity selection and an embedding-based influence for type inference, which considers not only the distances between entities but also the distances between entities and types. The experimental results on two real datasets illustrated the potential of our hybrid method. One promising future work is to learn some useful rules to filter out the false positives, which is expected to further improve the precision of type completion.

References

1. Palmero Aprosio, A., Giuliano, C., Lavelli, A.: Automatic expansion of DBpedia exploiting wikipedia cross-language information. In: Cimiano, P., Corcho, O., Presutti, V., Hollink, L., Rudolph, S. (eds.) ESWC 2013. LNCS, vol. 7882, pp. 397–411. Springer, Heidelberg (2013). https://doi.org/10.1007/978-3-642-38288-8_27
2. Bollacker, K., Evans, C., Paritosh, P., Sturge, T., Taylor, J.: Freebase:a collaboratively created graph database for structuring human knowledge. In: SIGMOD Conference, pp. 1247–1250 (2008)
3. Bordes, A., Usunier, N., Garcia-Duran, A., Weston, J., Yakhnenko, O.: Translating embeddings for modeling multi-relational data. In: International Conference on Neural Information Processing Systems, pp. 2787–2795 (2013)
4. Dawid, A.P., Skene, A.M.: Maximum likelihood estimation of observer error-rates using the em algorithm. J. Roy. Stat. Soc. **28**(1), 20–28 (1979)
5. Dong, Z., Lu, J., Ling, T.W.: PANDA: a platform for academic knowledge discovery and acquisition. In: 2016 International Conference on Big Data and Smart Computing (BigComp), pp. 10–17. IEEE (2016)
6. Dong, Z., Lu, J., Ling, T.W., Fan, J., Chen, Y.: Using hybrid algorithmic-crowdsourcing methods for academic knowledge acquisition. Cluster Comput. **20**(4), 3629–3641 (2017). https://doi.org/10.1007/s10586-017-1089-8
7. Fan, J., Lu, M., Ooi, B.C., Tan, W.C., Zhang, M.: A hybrid machine-crowdsourcing system for matching web tables. In: IEEE International Conference on Data Engineering, pp. 976–987 (2014)
8. Gangemi, A., Nuzzolese, A.G., Presutti, V., Draicchio, F., Musetti, A., Ciancarini, P.: Automatic typing of DBpedia entities. In: Cudré-Mauroux, P., et al. (eds.) ISWC 2012. LNCS, vol. 7649, pp. 65–81. Springer, Heidelberg (2012). https://doi.org/10.1007/978-3-642-35176-1_5
9. Huang, F., Li, J., Lu, J., Ling, T.W., Dong, Z.: PandaSearch: a fine-grained academic search engine for research documents. In: ICDE 2015 (2015)
10. Kejriwal, M., Szekely, P.: Supervised typing of big graphs using semantic embeddings, p. 3 (2017)
11. Kondreddi, S.K., Triantafillou, P., Weikum, G.: Combining information extraction and human computing for crowdsourced knowledge acquisition. In: ICDE, pp. 988–999 (2014)
12. Lehmann, J.: DBpedia: a large-scale, multilingual knowledge base extracted from Wikipedia. Seman. Web **6**(2), 167–195 (2015)

13. Lin, Y., Liu, Z., Sun, M., Liu, Y., Zhu, X.: Learning entity and relation embeddings for knowledge graph completion. In: Twenty-Ninth AAAI Conference on Artificial Intelligence, pp. 2181–2187 (2015)
14. Lofi, C., Maarry, K.E.: Design patterns for hybrid algorithmic-crowdsourcing work-flows. In: CBI, pp. 1–8 (2014)
15. Melo, A., Völker, J., Paulheim, H.: Type prediction in noisy RDF knowledge bases using hierarchical multilabel classification with graph and latent features. Int. J. Artif. Intell. Tools 26(2), 1760011 (2017)
16. Mozafari, B., Sarkar, P., Franklin, M.J., Jordan, M.I., Madden, S.: Scaling up crowd-sourcing to very large datasets: a case for active learning. Proc. VLDB Endow. (PVLDB) 8(2), 125–136 (2014)
17. Nickel, M., Rosasco, L., Poggio, T.: Holographic embeddings of knowledge graphs. In: Thirtieth AAAI Conference on Artificial Intelligence, pp. 1955–1961 (2016)
18. Paulheim, H.: Knowledge graph refinement: a survey of approaches and evaluation methods. Seman. Web 8, 1–20 (2016). (Preprint) survey
19. Paulheim, H., Bizer, C.: Type inference on noisy RDF data. In: Alani, H., et al. (eds.) ISWC 2013. LNCS, vol. 8218, pp. 510–525. Springer, Heidelberg (2013). https://doi.org/10.1007/978-3-642-41335-3_32
20. Paulheim, H., Bizer, C.: Improving the quality of linked data using statistical distributions. Int. J. Seman. Web Inf. Syst. 10(2), 63–86 (2014)
21. Rebele, T., Suchanek, F., Hoffart, J., Biega, J., Kuzey, E., Weikum, G.: YAGO: a multilingual knowledge base from Wikipedia, wordnet, and geonames. In: Groth, P., et al. (eds.) ISWC 2016. LNCS, vol. 9982, pp. 177–185. Springer, Cham (2016). https://doi.org/10.1007/978-3-319-46547-0_19
22. Sleeman, J., Finin, T.: Type prediction for efficient coreference resolution in hetero-geneous semantic graphs. In: IEEE Seventh International Conference on Semantic Computing, pp. 78–85 (2014)

Improving Clinical Named Entity Recognition with Global Neural Attention

Guohai Xu, Chengyu Wang, and Xiaofeng He[✉]

School of Computer Science and Software Engineering,
East China Normal University, Shanghai, China
`guohai.explorer@gmail.com, chywang2013@gmail.com, xfhe@sei.ecnu.edu.cn`

Abstract. Clinical named entity recognition (NER) is a foundational technology to acquire the knowledge within the electronic medical records. Conventional clinical NER methods suffer from heavily feature engineering. Besides, these methods treat NER as a sentence-level task and ignore the long-range contextual dependencies. In this paper, we propose an attention-based neural network architecture to leverage document-level global information to alleviate the problem. The global information is obtained from document represented by pre-trained bidirectional language model (Bi-LM) with neural attention. The parameters of pre-trained Bi-LM which makes use of unlabeled data can be transferred to NER model to further improve the performance. We evaluate our model on 2010 i2b2/VA datasets to verify the effectiveness of leveraging global information and transfer strategy. Our model outperforms previous state-of-the-art method with less labeled data and no feature engineering.

Keywords: Clinical named entity recognition · Neural attention
Language model

1 Introduction

The clinical text in electronic medical records has the potential to make a significant impact in many aspects of healthcare research such as drug analysis, disease inference, clinical decision support, and more. To analyze such clinical free text, one sequence labeling application namely NER plays a crucial role to identify medical entities at first step. Table 1 shows a clinical snippet containing such medical entities.

NER is still a challenging task in the clinical domain due to the distinctive characteristics of language. Dictionary-based methods fail to tag abbreviated phrases and acronyms which are common in clinical text. Rule-based systems are laborious to implement and trend to miss a number of misspellings that have their specific meaning. To overcome these limitations, various machine learning algorithms have been proposed to improve the performance. However, traditional machine learning approaches rely heavily on hand-crafted features,

© Springer International Publishing AG, part of Springer Nature 2018
Y. Cai et al. (Eds.): APWeb-WAIM 2018, LNCS 10988, pp. 264–279, 2018.
https://doi.org/10.1007/978-3-319-96893-3_20

it is especially tough to design features in the clinical-specific domain where specialized knowledge is needed. In the past few years, due to the simple but effective pre-trained word embedding [3,20,23], neural network models with as input distributed word representations achieve competitive performance against traditional models. Thus, current sequence labeling models typically include a RNN-based network that encodes each token into context vector and a CRF layer that decodes the representation to make predictions [10,15,21].

Table 1. A snippet of clinical text containing medical concepts, such as disease entities (in red), test entities (in blue) and treatment entities (in green).

The patient is a 58 year old right hand dominant white male with a long history of *hypertension* and *adult onset diabetes mellitus*. On physical examination, patient is in no acute distress, afebrile, *blood pressure* 134/80, *heart rate* 80 and regular, no bruits.

   The patient was managed with *Vasotec*, *Nifedipine* and *Clonidine* with *blood pressure* under good control at the time of discharge, average 125 *systolic*, 70 *diastolic*, *heart rate* of 72. The patient was started on 2.5 of *Micronase* with *resulting sugars* as low as 63, decreased to 1.25 mg q.day.

Above-mentioned methods in practice treat NER as a sentence-level task where sentences in the same document are viewed as independent. However, clinical documents which are generated by physician to record the process of patients' treatments are centered on one or a few diseases. As shown in Table 1, the medical entities are topic-related to describe the condition of patients, for example, "*Vasotec*" (treatment entity) is used to control the "*blood pressure*" (test entity) due to his "*hypertension*" (disease entity). Thus, the long-range contextual dependencies are useful to improve the performance of sentence-level NER methods. Besides, ignoring the long-range contextual dependencies will lead to tagging non-consistency problem that the same mentions separated in different sentences from a document are tagged with different labels.

   In this paper, we propose an attention-based stacked bidirectional long short-term memory with conditional random field (Att-BiLSTM-CRF) for clinical named entity recognition. Our model leverages global information within document and makes use of unlabeled data to achieve better performance. Inspired by the work of Peters et al. [24], we first pre-train a word embedding model and a bidirectional neural language model (Bi-LM) on unlabeled corpus in unsupervised learning (Sect. 3.2). Thus, the pre-trained Bi-LM can represent the sentences from document containing the global information. Then, we adopt stacked BiLSTM to encode the input sentence which consists of word embeddings, and incorporate all the representation of sentences within the document which the input sentence in with neural attention (Sect. 3.3). Finally, we use a CRF layer [14] to decode the representations to make sequence decision. The main contributions of this paper can be summarized as follows:

- We propose an attention-based neural network architecture namely Att-BiLSTM-CRF to incorporate global information to alleviate the problem of ignoring long-range contextual dependencies for clinical NER task.
- We transfer the parameters of pre-trained Bi-LM which makes use of unlabeled data to BiLSTM and show the advantages of transfer strategy than random initialization.
- Combining the global neural attention and pre-trained Bi-LM, our model outperforms previous state-of-the-art method on 2010 i2b2/VA datasets [25] with less labeled data and no feature engineering.

The rest of this paper is organized as follows: Sect. 2 discusses related research. Section 3 formulates the task and describes the architecture. Section 4 describes the datasets, training, experiments and results. Section 5 summarizes the paper.

2   Related Work

Our method is based on two lines of research which are sequence labeling and how to improve it with global information. Therefore, we mainly outline the recent work on NER and previous efforts in clinical domain. Then we will review the related work which aims to capture global information.

2.1   Named Entity Recognition

NER is a widely studied sequence labeling task, and many different approaches have been proposed. Among them, neural network models have been rapidly growing in popularity as they can be trained end-to-end with no feature engineering and task-specific resources. Taking inspiration from research of feed-forward network presented by Collobert et al. [3], Huang et al. [10] use a BiLSTM over a sequence of word embeddings and other hand-crafted spelling features with a CRF layer on top. Chiu and Nichols [4] also propose a similar model, but instead use CNN to learn character-level features. Lample et al. [15] also employ a similar architecture, but utilize LSTM to learn character-level features instead. Similar to Chiu and Nichols [4], Ma and Hovy [21] also use CNN to model character-level information, but without using any data preprocessing and achieving better NER performance. To relieve the limitation of relatively little labeled data, Peters et al. [24] explore a general semi-supervised approach which uses pre-trained neural bidirectional LM to augment context sensitive representation from large unlabeled corpus to improve previous methods.

Our architecture is based on the success of BiLSTM-CRF model [10,15,21], and is further modified to better incorporate global information with neural attention. Our model employs stacked BiLSTM to effectively model the context and excluding character-level information for simplicity. Furthermore, the Bi-LM can make use of unlabeled data and a simple transfer strategy can further improve the performance.

In clinical domain, there are a number of traditional machine learning algorithms based on hand-crafted features and domain-dependent knowledge or resources. Uzuner et al. [25] overview performance of systems on 2010 i2b2/VA challenge in detail. Among the all submitted systems in the evaluations, de Bruijin et al. [6] ranked first, and they trained a hidden semi-Markov model based on unsupervised feature representations obtained by Brown clustering and other text-oriented features. Subsequent work can be roughly divided into two directions. On the one hand, researchers focus attention on better feature representations. Jonnalagadda et al. [11] explore the use of distributed semantics derived empirically from unannotated text to improve the performance of clinical NER. Wu et al. [26] systematically compare two word neural embedding algorithms and show that low-cost distributed feature representations can be better than Brown clustering. On the other hand, researchers concentrate on appropriate data-preprocessing. Fu and Ananiadou [8] show that truecasing and annotation combination can best increase the NER system performance. Boag et al. [1] develop a lightweight tool by cascading CRF and SVM classifiers for clinical NER. Until recently, Chalapathy et al. [5] explore the effectiveness of BiLSTM-CRF based on off-the-shelf word embedding without any hand-crafted features. In contrast, the most advantage of our architecture is requiring no task-specific knowledge or feature engineering, and meanwhile achieving better performance with augmented global information.

2.2   Leveraging Global Information

Several studies have noticed the importance of global information to aid sentence-level NER. Finkel et al. [7] take non-local information into account while preserving tractable inference with Gibbs sampling. Krishnan and Manning [13] propose a two-stage model for exploiting non-local dependency. They use first CRF-based NER model using local features to make predictions and then train second CRF based on the output of the first CRF to maintain label consistency. Recently, Liu et al. [16] propose an extension to CRFs by integrating external memory to capture long-range contextual dependencies. Luo et al. [18] regard the whole document as input into BiLSTM-based NER model with self-attention mechanism. However, the method is only effectively applied to short text because RNN-based (including LSTM) models perform poorly as the length of input sentence increases [2,17].

Inspired by these earlier work, we also leverage global information to improve performance of clinical NER. In contrast, we propose a neural network architecture to combine the local and global information with neural attention. The stacked BiLSTM has its advantage over encoding sequential inputs than plain linear-chain CRF based on hand-crafted features. The performance of our method is not suffer from the variant length of document.

3   Neural Network Architecture

In this Section, we first provide the task definition and flow of our method for the problem of clinical NER. Then we illustrate the approach to pre-train the word embedding model and Bi-LM which is a key component in our architecture. Finally, we describe attention-based neural network architecture from bottom to up in detail.

3.1   Overview

Task Definition. We formally describe the Clinical NER task as follows: Given a sentence, $s = (w_1, w_2, ..., w_n)$ where n is the length of the sentence, find the medical entities $o = (y_1, y_2, ..., y_n)$ where y is the predefined label. The problem is a typical sequence labeling task. We use the BIO format to tag the entities. In particular, there are three medical entity categories: Disease (Dise for short), Test (Test for short), Treatment (Trea for short). If the word is the first word in medical entities, the word is labeled B-X (X is the entity category). The word is labeled I-X if the word is inside but not the first position of the medical entities. Otherwise, the word is labeled O.

For instance, which is shown in Fig. 1, the input sentence is (*a, long, history, of, hypertension*), then the model can output the sequence tag (O, O, O, O, B-Dise).

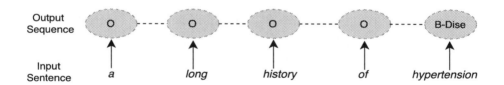

Fig. 1. An example for sequence labeling task.

In contrast to previous work, we additionally leverage the global information from the document $D = (s_1, s_2, ..., s_m)$ where the input sentence s is located to improve the performance. Thus, all of the representation of sentences in document will be utilized to complement the single input sentence. In a nutshell, the input to our model not only contains the single sentence, but also incorporates all of the sentences from the same document.

Flow of the Method. As illustrated in Fig. 2, the main components in our architecture are *Pre-Training, Encoder, Neural Attention, Decoder* respectively. First of all, we use unlabeled corpus to pre-train word embedding model and Bi-LM. Secondly, the first BiLSTM takes the word embeddings of single sentence as input, and then the pre-trained Bi-LM represents all the sentences within the same document which the input sentence in. Next, the second BiLSTM integrates

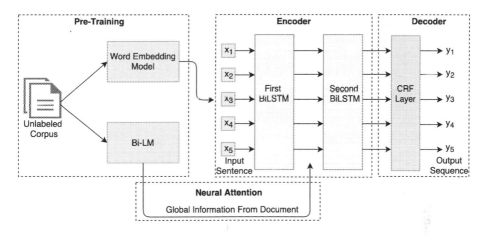

Fig. 2. The flow of our method for clinical NER.

the outputs of first BiLSTM and representations from Bi-LM that includes the global information from document with neural attention. Lastly, the CRF layer plays a decoding role to make sequence decision over the encoding of input.

3.2 Pre-training

Word Embedding Model. Word embedding is ubiquitous in NLP tasks since Mikolov et al. [20] propose an efficient method called Word2Vec for learning distributed representation of words. It is commonly believed that the word embedding captures useful semantic and syntactic information. Therefore, we use skip-gram algorithm [20] to train word embedding as input instead of heavily hand-crafted features.

Bi-LM. The Bi-LM is a vital component in our neural network architecture. On the one hand, pre-trained Bi-LM encodes the representation of sentences to enable the BiLSTM to look beyond the local context of sentence and extent to the global context of document. On the other hand, Bi-LM can make use of unlabeled data and its learned parameters can be transferred to first BiLSTM in NER model to improve performance. Now we describe the Bi-LM in detail.

Language model is proposed to learn a probability distribution over sequences of token pertaining to a language. Instead of count-based N-grams language model, we choose neural language model which has been shown to better retain long term dependencies. We use LSTM to model joint probabilities over word sequences which represented by word embeddings. Give a word sequence $(w_1, w_2, ..., w_n)$, LM computes the probability of the next word given all the previous words at each step. Here it can be called forward LM since we obtain the next word depending on the forward words, and LSTM is called forward LSTM as well. Thus, the overall probability can be written as:

$$p(w_1, w_2, ..., w_n) = \prod_{i=2}^{n} p(w_i | w_1, w_2, ..., w_{i-1}) \tag{1}$$

At each step, forward LSTM encode the history $(w_1, w_2, ..., w_{i-1})$ into a fixed dimensional vector $\overrightarrow{\mathbf{h}}_{i-1}^{LM}$ which is the hidden state of forward LSTM at position $i-1$ actually. Then, a softmax layer predicts the probability of next word w_i in the vocabulary. We train the forward LM model which maximizes the likelihood of given sentences in corpus.

A backward LM can be implemented in an analogous way if we reverse the word sequence. Thus, we obtain the similar overall probability:

$$p(w_n, w_{n-1}, ..., w_1) = \prod_{i=n-1}^{1} p(w_i | w_n, w_{n-1}, ..., w_{i+1}) \tag{2}$$

The backward LM predicts the previous word given the future sequence. Also, we utilize a backward LSTM to build the backward LM.

The forward and backward LSTM share the same input layer (word embedding layer) and output layer (softmax layer). After pre-training, the pre-trained Bi-LM can be used to represent sentences of document in training corpus. We concatenate the last cell state of forward and backward LSTM to represent the input sentence, i.e., $\mathbf{s} = [\overrightarrow{\mathbf{c}}_n^{LM}; \overleftarrow{\mathbf{c}}_1^{LM}]$.

Transfer Strategy. In NLP, pre-trained word embedding like Word2Vec [20] and GloVe [23] has been common initialization for the input layer of neural network models. The word vectors obtained from training on large amounts of unlabeled corpus achieve better performance than random initialization on a variety of NLP tasks. However, the form of transfer learning is not limited to word vectors, but also includes weights from pre-trained recurrent neural networks [22,27].

Inspired by above ideas, we propose a transfer strategy to further improve the performance of NER model. We let Bi-LM and first BiLSTM in Encoder component of NER model have the same architecture. Therefore, the parameters of pre-trained Bi-LM can be shared to the first BiLSTM. The well-trained Bi-LM from large, unlabeled corpus can help the NER model have a better initialization, thus leads to better performance.

3.3   Att-BiLSTM-CRF Model

Encoder. As depicted in Fig. 3, this architecture is similar to the ones presented by Huang et al. [10], Lample et al. [15] and Ma et al [21]. In contrast, we use stacked BiLSTM to encode sequential input for incorporating global information with neural attention.

For a given sentence $s = (w_1, w_2, ..., w_n)$ containing n words in a document $D = (s_1, s_2, ..., s_m)$ including m sentences. At first, the sentence is represented as a sequence of vectors $\mathbf{X} = (\mathbf{x}_1, ..., \mathbf{x}_t, ..., \mathbf{x}_n)$ through the embedding layer.

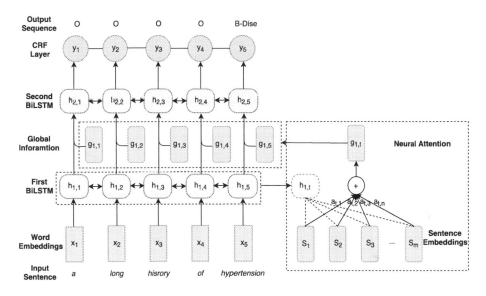

Fig. 3. The architecture of our Att-BiLSTM-CRF model.

Next, a forward LSTM in first BiLSTM computes a representation $\overrightarrow{\mathbf{h}}_{1,t}$ of the left context of the sentence at each word t, and a backward LSTM computes a representation $\overleftarrow{\mathbf{h}}_{1,t}$ of the same sequence in reverse. Then, the representation of each word t is obtained by concatenating its left and right context representations, $\mathbf{h}_{1,t} = [\overrightarrow{\mathbf{h}}_{1,t}; \overleftarrow{\mathbf{h}}_{1,t}]$.

In most previous NER methods, the representation of each word will be followed by a transformation layer and CRF layer to make prediction without considering the long-range contextual dependencies. While we introduce the Neural Attention component to leverage all the sentences in the document D. We use pre-trained Bi-LM to represent all the sentences which can be regard as global context. Then we apply the neural attention to seek the related global context based on the representation of each word which can be regard as local context. The global context in the document can supply extra useful information to each word. As a result, the extended representation of each word consists of the local context in sentence and the global context in document.

Every sentence in document D can be represented by pre-trained Bi-LM, thus we get a another sequence of vectors $\mathbf{D} = (\mathbf{s_1}, ..., \mathbf{s_j}, ..., \mathbf{s_m})$ for sentences. Firstly, we use an attention matrix A to calculate the similarity between the local context in sentence and global context in the document. The attention weight value $a_{t,j}$ in attention matrix A is computed by comparing the local context $\mathbf{h}_{1,t}$ with each sentence embedding $\mathbf{s}_j$:

$$a_{t,j} = \frac{exp(score(\mathbf{h}_{1,t}, \mathbf{s}_j))}{\sum_k exp(score(\mathbf{h}_{1,t}, \mathbf{s}_k))} \tag{3}$$

Above *score* is referred as a bilinear function which is borrowed from Bahdanau et al. [2] and Luong et al. [19]:

$$score(\mathbf{h}_{1,t}, \mathbf{s}_j) = \mathbf{h}_{1,t}^T \mathbf{W_a} \mathbf{s}_j \tag{4}$$

here the weight matrix $\mathbf{W_a}$ is a parameter of the model. Secondly, the global context $\mathbf{g}_{1,t}$ is computed as a weighted sum of each sentence embedding $\mathbf{s}_j$:

$$\mathbf{g}_{1,t} = \sum_{j=1}^{m} a_{t,j} \mathbf{s}_j \tag{5}$$

Thirdly, we concatenate the global context and local context into a vector $[\mathbf{h}_{1,t}; \mathbf{g}_{1,t}]$ to represent each word. Next, the extended representation of each word become a sequential of intermediate representation, which can be sent into second BiLSTM.

Decoder. After process of encoding, it is simple to use a linear layer to predict a score for each possible label independently based on the output of the second BiLSTM. But there are strong dependencies across output labels, for example, I-Dise cannot follow B-Test. Therefore, instead of modeling tagging decisions independently, we add another CRF layer to decode the best label path in all possible label paths. Followed by Lample et al. [15], we only consider the relations between labels in neighborhoods and jointly decode the best chain of labels.

We consider $\mathbf{P} \in \mathbb{R}^{n \times k}$ to be the matrix scores output by the second BiLSTM, where the n is length of input sentence and the k is the number of distinct labels. The element $P_{i,j}$ in the matrix is the score of jth label of the ith word in the sentence. We introduce a label transition matrix $\mathbf{T}$, where element $T_{i,j}$ represents a score of a transition from the label i to label j. After that, the whole input sentence $\mathbf{X}$ gets a sequence of predictions $\mathbf{y} = (y_1, y_2, ..., y_n)$ from model, we can define its score to be

$$s(\mathbf{X}, \mathbf{y}) = \sum_{i=1}^{n} (T_{y_{i-1}, y_i} + P_{i, y_i}) \tag{6}$$

where the transition matrix $\mathbf{T} \in \mathbb{R}^{(k+2) \times (k+2)}$ is the parameter of our model. In above equation, y_0 and y_n are the start and end labels of a given sentence. Therefore, the transition matrix $\mathbf{T}$ is a square matrix of size $k + 2$.

During training, we use the maximum conditional likelihood estimation. First, as shown in Eq. (7), a softmax function is used to normalize the above score over all possible label paths $\tilde{\mathbf{y}}$ to form the conditional probability of the path $\mathbf{y}$. Then, the log-likelihood of the conditional probability of the correct tag sequence is given in Eq. (8). We train the model to maximize the log-likelihood of the probability of all the correct tag sequences in labeled data to obtain the final parameters.

$$p(\mathbf{y}|\mathbf{X}) = \frac{exp(s(\mathbf{X}, \mathbf{y}))}{\sum_{\tilde{\mathbf{y}}} exp(s(\mathbf{X}, \tilde{\mathbf{y}}))} \tag{7}$$

$$\mathcal{L} = log(p(\mathbf{y}|\mathbf{X})) \tag{8}$$

During inference, as given in Eq. (9), the best label path $\mathbf{y}^*$ is predicted through computing the maximum score among all the possible label paths. Because we only consider the interactions between two successive labels, dynamic programming such as Viterbi algorithm can be applied to effectively computes the scores.

$$\mathbf{y}^* = argmax_{\tilde{\mathbf{y}}}\; s(\mathbf{X}, \tilde{\mathbf{y}}) \tag{9}$$

4   Experiments

4.1   Datasets

In this paper, we use datasets from 2010 i2b2/VA Natural Language Processing Challenges for Clinical Records[1] containing a concept extraction task focused on identifying medical concepts from realistic clinical narratives. Because of the restrictions introduced by Institutional Review Board (IRB), only part of original datasets is available. The challenge requires the systems to predict the exact boundary of medical concepts and classify them into specified category including problem, test, treatment and other. Table 2 summarizes the statistics of labeled datasets which we have used in our experiments. In addition, we get a number of unlabeled clinical notes from MIMIC-III corpus[2] [12] for pre-training word embedding model and Bi-LM.

Table 2. A basic statistics of datasets.

	Training data	Test data	Unlabeled data
# Documents	170	256	5000
# Sentences	16315	27626	1042534
# Mentions	16525	31161	-

4.2   Model Training

Preparation and Evaluation. We split the training data into two parts, 130 documents (about 80%) for training set and 40 documents (about 20%) for development set. We tune the hyperparameters of our model on development set and report the results on the test set. Note that, to compare to other existing methods (Sect. 4.5), the final training is done on both the training and development sets. We don't do any feature engineering except using a special token for numbers. For evaluation, we do exact matching of entity mentions to compute micro-precision, micro-recall and micro-F_1.

---

[1] https://i2b2.org/NLP/DataSets/Main.php.
[2] https://mimic.physionet.org.

Model Architecture Details. Dimensions of word embedding are set 300. For language model, the hidden state of LSTM has 300 dimensions. For first BiLSTM in NER model, the hidden state of LSTM also has 300 dimensions. In consideration of the transfer strategy, the first BiLSTM and Bi-LM have identical parameter setting. For second BiLSTM in NER model, as it concatenates the output of first BiLSTM and the representations of global information, the dimensions of the hidden state of LSTM are 600.

Training Details. For word embedding model, we use skip-gram algorithm [20] to obtain word vectors on unlabeled data. For Bi-LM, the input embedding layer is initialized with the weights from word embedding model and other parameters are initialized with Xavier initialization [9]. Once the pre-training is done, we use pre-trained Bi-LM to represent the sentences in document and the parameters of Bi-LM also can be transferred to first BiLSTM in NER model. For NER model, the input embedding layer is also initialized with the weights from word embedding model and other parameter are initialized with Xavier initialization as well. We use SGD with momentum of 0.9 to train the NER model. We train our networks using back-propagation algorithm updating parameter on a batch size of 10. The initial learning rate is 0.01 and decay the learning rate by multiplying it by 0.9 if the F_1 score does not improve on development set for one epoch. We use a gradient clipping of 5.0 to avoid gradient exploding problem. We train the model for 30 epochs and use early stopping to avoid over-fitting.

4.3   Effectiveness of Leveraging Global Information

In this part, we verify the effectiveness of global neural attention augmented BiLSTM (Att-BiLSTM) compared with plain BiLSTM. In previous work, most methods treat NER as a sentence-level task. In Contrast, we incorporate the global information in document to capture the long-range contextual dependencies. As shown in Table 3, in irrespective of the impact of CRF, we perform the contrast experiments based on only BiLSTM to evaluate the ability of presentations for each input word. From the results, we see that the number of layers affects the performance. In both of Att-BiLSTM and BiLSTM, the stacked BiLSTM outperforms the BiLSTM with single layer. Also, the global neural attention gives an improvement over the plain BiLSTM due to the leveraging of global information. We observe that the F_1 of stacked Att-BiLSTM is 81.62%, which is an absolute improvement of 1.01% over the plain stacked BiLSTM with no global neural attention.

To be honest, global neural attention don't show obvious effects when the Att-BiLSTM only has one layer. It is because the final tagging predictions mainly depend on local context for each word, while global context only supplements extra information. Therefore, our model need another layer to encode the sequential intermediate vectors containing global context and local context. In other words, the architecture needs second BiLSTM to learn the differences between the two contexts.

Table 3. Performance of leveraging global information.

Model	Layers	Precision (%)	Recall (%)	$F_1(\%)$
BiLSTM	1	77.53	81.63	79.53
	2	80.50	80.71	80.61
Att-BiLSTM	1	78.25	81.17	79.68
	2	80.62	82.65	**81.62**

4.4 Effectiveness of Transfer Strategy

In this part, we verify the effectiveness of transfer strategy. The baselines are stacked BiLSTM and stacked Att-BiLSTM obtained from above experiments. In baseline methods, we initialize the parameters of their stacked BiLSTM with Xavier initialization [9] which has been regarded as an effective initialization strategy. In comparison to Xavier initialization, we initialize the parameters of first BiLSTM from the parameters of pre-trained Bi-LM. The results is showed in Table 4, the simple transfer strategy gives an additional improvement over baselines. For stacked BiLSTM, the F_1 gets an absolute improvement of 0.53%. Also for stacked Att-BiLSTM, the absolute improvement is 0.71% in F_1 score.

Table 4. Performance of transfer strategy.

Model	Transfer	Precision (%)	Recall (%)	$F_1(\%)$
BiLSTM	No	80.50	80.71	80.61
	Yes	80.16	82.15	81.14
Att-BiLSTM	No	80.62	82.65	81.62
	Yes	81.58	83.08	**82.33**

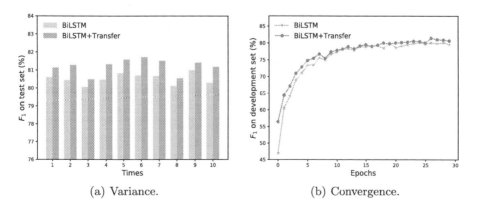

(a) Variance.   (b) Convergence.

Fig. 4. Comparison between plain stacked BiLSTM and stacked BiLSTM with transfer strategy.

To further verify the effectiveness of transfer strategy, we train the model with different random seeds. At first, we respectively train the stacked BiLSTM and stacked BiLSTM with transfer strategy for ten times in different random seeds. From the results in Fig. 4(a), it shows that the transfer strategy always increases the performance of plain stacked BiLSTM more or less. We compute the mean F_1 score of stacked BiLSTM with transfer strategy is 81.22%, and its variance is 0.15%. In contrast, the mean F_1 score of plain BiLSTM is only 80.52%, and its variance is 0.08%. Then we randomly select one example to draw the convergence of the two models. As depicted in Fig. 4(b), transfer strategy accelerates the model training especially at the first several epochs. Also the transfer strategy helps the model achieve the better performance at last. Above comparisons prove the effectiveness of transfer strategy, we believe that it can promote other similar models which contain LSTM.

4.5   Comparison to Other Methods

In this part, we compare the performance of our model with other existing methods on the 2010 i2b2/VA datasets. The results are shown in Table 5, the name of other methods followed by Chalapathy et al. [5]. We have implied the main ideas of other methods in related work (Sect. 2.1). Form the results, our model obtains the state-of-the-art performance than others. Although we only get nearly 0.5% F_1 score higher than the previous state-of-the-art method which is the best submission from the 2010 i2b2/VA challenge, their model is based on original dataset which has more than twice labeled data than ours.

To understand the importance of leveraging global information and transfer strategy, we implement the common BiLSTM-CRF model as baseline. The results confirm that leveraging global information increases F_1 score by 0.53% (from 84.66% to 85.19%) and increases F_1 score by 1.05% (from 84.66% to 85.71%) with additional transfer strategy. We conclude that our model relieves

Table 5. Performance comparison with other existing methods on the 2010 i2b2/VA datasets. * indicates models trained with the use of original larger labeled data.

Model	Precision (%)	Recall (%)	F_1(%)
Distributional semantics CRF * [11]	85.60	82.00	83.70
Hidden semi-markov model * [6]	86.88	83.64	**85.23**
Truecasing CRFSuite [8]	80.83	71.47	75.86
CliNER [1]	79.50	81.20	80.00
Binarized neural embedding CRF [26]	85.10	80.60	82.80
Glove-BiLSTM-CRF [5]	84.36	83.41	**83.88**
BiLSTM-CRF	86.21	83.17	84.66
Att-BiLSTM-CRF	85.51	84.87	85.19
Att-BiLSTM-CRF + Transfer	86.27	85.15	**85.71**

the problem of ignoring the long-range contextual dependencies and the pre-trained Bi-LM makes use of unlabeled data to further improve the performance.

5   Conclusion

In this paper, we propose an attention-based neural network architecture to leverage document-level global information to alleviate the problem of ignoring long-range contextual dependencies for clinical NER task. In addition, we explore a transfer strategy to further make use of unlabeled data using pre-trained Bi-LM. Our results of experiments show that the transfer strategy consistently improve the performance. Owing to the above two advantages, our model achieves the state-of-the-art performance on public 2010 i2b2/VA datasets.

Although we use clinical data to verify the effectiveness of our method, the Att-BiLSTM-CRF model can be adapted to other domain where global context is useful. Moreover, the transfer strategy using Bi-LM has generalization performance.

Acknowledgments. This work was supported by the National Key Research and Development Program of China under Grant No. 2016YFB1000904.

References

1. Boag, W., Wacome, K., Naumann, T., Rumshisky, A.: CliNER: a lightweight tool for clinical named entity recognition. AMIA Joint Summits on Clinical Research Informatics (poster) (2015)
2. Bahdanau, D., Cho, K., Bengio, Y.: Neural machine translation by jointly learning to align and translate. In: Proceedings of the 3rd International Conference on Learning Representations (2015)
3. Collobert, R., Weston, J., Bottou, L., Karlen, M., Kavukcuoglu, K., Kuksa, P.: Natural language processing (almost) from scratch. J. Mach. Learn. Res. **12**, 2493–2537 (2011)
4. Chiu, J.P.C., Nichols, E.: Named entity recognition with bidirectional LSTM-CNNs. In: Proceedings of TACL, pp. 357–370 (2016)
5. Chalapathy, R., Borzeshi, E.Z., Piccardi, M.: Bidirectional LSTM-CRF for clinical concept extraction. In: Proceedings of the Clinical Natural Language Processing Workshop ClinicalNLP, pp. 7–12 (2016)
6. de Bruijn, B., Kiritchenko, C.C., Martin, J.D., Zhu, X.D.: Machine-learned solutions for three stages of clinical information extraction: the state of the art at i2b2 2010. J. Am. Med. Inf. Assoc. **18**(5), 557–562 (2011)
7. Finkel, J.R., Grenager, T., Mannning, C.D.: Incorporating non-local information into information extraction systems by Gibbs sampling. In: Proceedings of the 43rd Annual Meeting of the Association for Computational Linguistics, pp. 363–370 (2005)
8. Fu, X., Ananiadou, S.: Improving the extraction of clinical concepts from clinical records. In: Proceedings of the 4th Workshop on Building and Evaluating Resources for Health and Biomedical Text Processing (2014)

9. Glorot, X., Bengio, Y.: Understanding the difficulty of training deep feedforward neural networks. In: Proceedings of the Thirteenth International Conference on Artificial Intelligence and Statistics, pp. 249–256 (2010)

10. Huang, Z.H., Xu, W., Yu, K.: Bidirectional LSTM-CRF models for sequence tagging. CoRR, abs/1508.01991 (2015)

11. Jonnalagadda, S., Cohen, T., Wu, S.T., Gonzalez, G.: Enhancing clinical concept extraction with distributional semantics. J. Biomed. Inf. **45**(1), 129–140 (2012)

12. Johnson, A.E., Pollard, T.J., Shen, L., Li-wei, H.L., Feng, M., Ghassemi, M., Moody, B., Szolovits, P., Celi, L.A., Mark, R.G.: MIMIC-III, a freely accessible critical care database. Sci. Data **3**, 160035 (2016)

13. Krishnan, V., Manning, C.D.: An effective two-stage model for exploiting non-local dependencies in named entity recognition. In: Proceedings of the 21st International Conference on Computational Linguistics and 44th Annual Meeting of the Association for Computational Linguistics, pp. 1121–1128 (2006)

14. Lafferty, J.D., McCallum, A., Pereira, P.: Conditional random fields: probabilistic models for segmenting and labeling sequence data. In: Proceedings of the 18th International Conference on Machine Learning, pp. 282–289 (2001)

15. Lample, G., Ballesteros, M., Subramanian, S., Kawakami, K., Dyer, C.: Neural architectures for named entity recognition. In: Proceedings of the 2016 Conference of the North American Chapter of the Association for Computational Linguistics: Human Language Technologies, pp. 260–270 (2016)

16. Liu, F., Baldwin, T., Cohn, T.: Capturing long-range contextual dependencies with memory-enhanced conditional random fields. In: Proceedings of the Eighth International Joint Conference on Natural Language Processing, pp. 555–565 (2017)

17. Lai, S.W., Xu, L.H., Liu, K., Zhao, J.: Recurrent convolutional neural networks for text classification. In: Proceedings of the Twenty-Ninth AAAI Conference on Artificial Intelligence, pp. 2267–2273 (2015)

18. Luo, L., Yang, Z.H., Yang, P., Zhang, Y., Wang, L., Lin, H.F., Wang, J.: An attention-based BiLSTM-CRF approach to document-level chemical named entity recognition. Bioinformatics **1**, 8 (2017)

19. Luong, T., Pham, H., Manning, C.D.: Effective approaches to attention-based neural machine translation. In: Proceedings of the 2015 Conference on Empirical Methods in Natural Language Processing, pp. 1412–1421 (2015)

20. Mikolov, T., Sutskever, I., Chen, K., Corrado, G.S., Dean, J.: Distributed representations of words and phrases and their compositionality. In: Proceedings of the 26th Annual Conference on Advances in Neural Information Processing Systems, pp. 3111–3119 (2013)

21. Ma, X.Z., Hovy, E.H.: End-to-end sequence labeling via bi-directional LSTM-CNNs-CRF. In: Proceedings of the 54th Annual Meeting of the Association for Computational Linguistics, pp. 1064–1074 (2016)

22. McCann, B., Bradbury, J., Xiong, C.M., Socher, R.: Learned in translation: contextualized word vectors. In: Proceedings of the 30th Annual Conference on Advances in Neural Information Processing Systems, pp. 6297–6308 (2017)

23. Penningto, J., Socher, R., Manning, C.D.: GloVe: global vectors for word representation. In: Proceedings of the 2014 Conference on Empirical Methods in Natural Language Processing, pp. 1532–1543 (2014)

24. Peters, M.E., Ammar, W., Bhagavatula, C., Power, R.: Semi-supervised sequence tagging with bidirectional language models. In: Proceedings of the 55th Annual Meeting of the Association for Computational Linguistics, pp. 1756–1765 (2017)

25. Uzuner, O., South, B.R., Shen, S.Y., DuVall, S.L.: 2010 i2b2/VA challenge on concepts, assertions, and relations in clinical text. J. Am. Med. Inf. Assoc. **18**(5), 552–556 (2011)
26. Wu, Y.H., Xu, J., Jiang., M., Zhang., Y.Y., Xu, H.: A study of neural word embeddings for named entity recognition in clinical text. In: Proceedings of the 2015 American Medical Informatics Association Annual Symposium, pp. 1326–1333 (2015)
27. Yang, Z.L., Salakhutdinov, R., Cohen, W.W.: Transfer learning for sequence tagging with hierarchical recurrent networks. In: Proceedings of the 5th International Conference on Learning Representations (2017)

Exploiting Implicit Social Relationship for Point-of-Interest Recommendation

Haifeng Zhu[1], Pengpeng Zhao[1(✉)], Zhixu Li[1], Jiajie Xu[1], Lei Zhao[1], and Victor S. Sheng[2]

[1] Soochow University, Suzhou, China
hfzhu@stu.suda.edu.cn, {ppzhao,zhixuli,xujj,zhaol}@suda.edu.cn
[2] University of Central Arkansas, Conway, USA
ssheng@uca.edu

Abstract. The emergence of Location-based Social Network (LBSN) services allows users to share their check-ins, providing an excellent opportunity to build personalized Point-of-Interest (POI) recommender systems. Social network data which contains important context information has been demonstrated to have a significant effect on improving recommendation performances. However, explicit social relationships are usually partially available or even unavailable. The gap between the importance of social relationships and their partial availability or unavailability motivates us to study POI recommendation with implicit social relationships, which can well characterize users' preferences for POIs on both space and content. In this paper, we first extract implicit social relationships and estimate connection strengths by analyzing co-occurrences in both space and time with people's history check-in data. Then, we propose a new model named Implicit Social Relationship Enhanced POI Recommendation (ImSoRec) to incorporate implicit and explicit social relationships for POI recommendation. We conducted extensive experiments on two large-scale real-world location-based social networks datasets, and our experimental results show that our proposed ImSoRec model outperforms the state-of-the-art methods.

Keywords: Recommendation · POI recommendation
Implicit social relationship

1 Introduction

Recommender systems play a crucial role in mitigating information overload problem by suggesting relevant information to users. The pervasiveness of GPS-enabled mobile devices and the popularity of location-based social networks (LBSNs) contribute massive data that present the movements of people in the real world at a high resolution, a.k.a. spatio-temporal data. For example, both Twitter and Foursquare reported that they received millions of spatio-temporal records per day as geo-tagged tweets or check-ins. Users can build connections with each other and share their experience and check-in information associated

© Springer International Publishing AG, part of Springer Nature 2018
Y. Cai et al. (Eds.): APWeb-WAIM 2018, LNCS 10988, pp. 280–297, 2018.
https://doi.org/10.1007/978-3-319-96893-3_21

with a Point-of-Interest (POI) in LBSNs. This newly available spatio-temporal
data is useful for developing personalized POI recommender systems.

Social relationship data contains important context information for recommender systems. Specifically, social links between users can be employed to derive
user similarities as the input of collaborative filtering model. For example, a user
may like books that his or her friend has read and given a high rating. Similarly, friends often go to places like restaurants together, or users may visit
POIs recommended by friends based on the fact that friends are more likely to
share common interests. Therefore, recently, there are POI recommender systems exploiting social relationships as complement data to improve the recommendation performance [8,17]. Ye et al. [17] propose a linear model to predict
users' preferences to a location, which combines users' interests, social friends'
interests, and geographical influence together. Li et al. [8] develope a POI recommendation framework, which takes social relationships, users' preferences for
POI categories, and distance between users into consideration at the same time.
Researchers have shown that social relationship has significant effects on improving the accuracy of POI recommendation. However, explicit social relationship
data are usually partially available or even unavailable. Taking Yelp as an example, users followed each other on Facebook rarely have following relationship on
Yelp. Or, even if they have social relationship on social networks, they may not
have similar choice for POIs because of different regions and long distance.

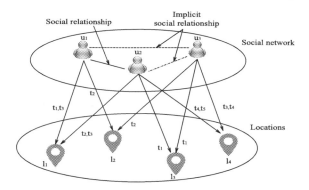

Fig. 1. A typical LBSN contains social network and check-ins from users.

The gap between the importance of social relationships and their partial
availability or unavailability motivates us to study implicit social relationships
of users for POI recommendation. Figure 1 shows a typical example of a location-
based social network. User u_1 has an explicit social connection with u_2. They
checked at POI l_1 two times respectively, once at the same time t_3. Therefore,
to a certain degree, we can infer that u_1 and u_2 have similar interests on POIs
which can be used to improve the recommendation quality. On the other hand,
u_1 and u_3 visited l_2 at the same time t_2, and u_2 and u_3 visited POIs l_3 and l_4
simultaneously twice. Intuitively, these users share similar content and spatial

preferences for POIs and social relationships may exist between them. However, there is no explicit relationship between them. That is, u_1 and u_3, u_2 and u_3 may have implicit social relationships. The implicit social relationships can be considered as a supplement to the explicit social relationships. In addition, in POI recommender systems, the implicit social relationships can be more helpful than the explicit social relationships. Since people with explicit social relationships may have similar content preferences, but they may not have similar spatial preferences because of the different activity centers issue.

In this paper, we investigate the following three challenges: (1) How to capture implicit social relations between users? (2) How to represent the strength of implicit social relations? And (3) how to integrate the implicit social connections into the POI recommendation model. To tackle these three challenges, we propose a new Implicit Social Relationship Enhanced POI Recommendation (ImSoRec) model. Specifically, we first extract implicit social relationships between users. And then location popularity and co-occurrences diversity are utilized to estimate connection strength. Moreover, based on probabilistic matrix factorization (PMF), we incorporate both explicit social trust propagation and implicit social relations that indicate similar visiting behavior patterns into PMF. Our model takes both implicit social relationships and explicit social networks into consideration to improve recommendation performance simultaneously.

Overall, the major contributions of this paper can be summarized as follows.

– We discover that the implicit social relationships are valuable for POI recommendation, which can indeed assist to improve the effectiveness of a recommender system.
– We develop an implicit social relationship extraction method which analyzes the co-occurrences of check-in history data.
– We propose a novel POI recommendation model ImSoRec to integrate the influence of both explicit and implicit social relationships.

2   Problem Statement and System Framework

In this section, we first introduce some basic concepts of the implicit social network. And then we provide the statement of the POI recommendation problem and an overview of our POI recommendation system framework.

2.1   Basic Concepts

Co-occurrence. If two users checked in at the same location within a time-interval, then we say that they have a co-occurrence. That means they are likely to have similar interests. We call that they have an implicit social relationship even though they may not know each other. The time-interval is an application dependent parameter and can be set specifically for each application. Let $r_{i,j}^{l,t} = <i, j, l, t>$ be a co-occurrence of user i and user j in location l at time t. Let

$R^l_{ij} = \bigcup_t r^{l,t}_{i,j}$ be the set of co-occurrences of user i and user j, which occurs in location l. Let R_{ij} be the set of all co-occurrences of user i and user j in all locations. That is, $R_{ij} = \bigcup_l R^l_{i,j} = \bigcup_{l,t} r^{l,t}_{i,j}$.

Co-occurrence Vector. Correspondingly, a co-occurrence vector C_{ij} between user i and user j presents all the co occurrences of users i and j at different locations. For example, $C_{ij} = (c_{ij,1}, c_{ij,2}, \ldots, c_{ij,l}, \ldots, c_{ij,M})$, where $c_{ij,l}$ is the number of co-occurrences between users i and j at location l, which is referred to local frequency, and M is the number of locations.

2.2   Problem Statement

Given a set of users $U = (u_1, u_2, \ldots, u_N)$, a set of locations $L = (l_1, l_2, \ldots, l_M)$ and a set of check-ins in the forms of user-location-time triplets (u, l, t), we infer implicit social relationship and their strength for each pair of users. Then, given all users' check-ins, implicit social relationships and corresponding strengths and explicit social relationships between them, our goal is to predict the preference score $\hat{R}_{u,l}$ of user u regarding a POI l where he/she did not visit before and then return the top-k POIs with the highest scores.

2.3   System Framework Overview

As shown in Fig. 2, the system framework has two major parts: extracting implicit social relationships and POI recommendations. The relation extraction part generates the strength of implicit social relationships from spatio-temporal data for POI recommendation part. This part includes three steps: (a) measuring the diversity of the co-occurrences between two users regarding how many effective locations they present at the same time, (b) measuring the impact weight of each co-occurrence individually (a.k.a. weighted frequency) depending on the popularity of the corresponding location, and (c) computing implicit social strengths based on diversity and weighted frequency of users' co-occurrences. The

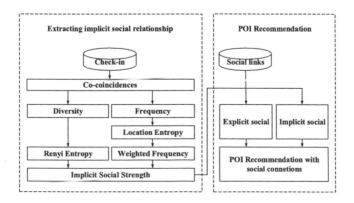

Fig. 2. The overview of the system framework

POI recommendation part incorporates implicit and explicit social strengths by using model-based collaborative filtering techniques, e.g., probabilistic matrix factorization, to estimate scores of users regarding POIs. This POI recommendation part also includes three parts: (a) using PMF model to integrate explicit social relationships to estimate the prediction of users for new POIs, (b) integrating implicit social relationship influence into PMF for POI recommendation, (c) integrating both explicit and implicit social relationship influence into a united model to make more accurate recommendation.

3   Extracting Implicit Social Relationship

What kind of attributes of spatio-temporal data should be measured to infer implicit social relationships and how to quantify their implicit social strengths are challenging. Inspired by Entropy-Based Model (EBM) [12], we infer implicit social relationships and strengths through two independent approaches, i.e., diversity and weighted frequency of co-occurrences. Specifically, diversity measures how diverse the co-occurrences between two users regarding check-in locations, while weighted frequency, the number of co-occurrences of two users, measures the impact of each co-occurrence individually taking the popularity of the location of the co-occurrence into consideration.

3.1   Diversity of Co-occurrences

Let us consider the following co-occurrence vectors for three different pairs of users.

$$C_{12} = (4, 2, 0, 3, 0)$$

$$C_{23} = (2, 1, 1, 3, 2)$$

$$C_{13} = (4, 0, 0, 0, 5)$$

C_{12}, C_{23} and C_{13} are co-occurrence vectors of user u_1 and u_2, u_2 and u_3, and u_1 and u_3, respectively. The element in vectors c_{ijl} is the number of co-occurrences between users i and j at location l. From the three co-occurrences, we can see that user u_1 and u_2 have 9 co-occurrences, and u_2 and u_3 also have 9 co-occurrences. However, in the latter case the co-occurrences are spread over 5 different locations, while in the former case the co-occurrences happened in just 3 different locations. Similarly, u_1 and u_3 co-occurred only in 2 different locations. Hence, C_{23} is more diverse than C_{12}, and C_{12} is more diverse than C_{13}. People tend to visit various places together when they are socially connected [1–3]. This intuition is that the diversity of co-occurrences is useful to infer implicit social relationship of users.

Entropy is common to describe diversity in many research [6]. Following [12], we use Renyi entropy to estimate diversity of the co-occurrences between users, which quantifies the uncertainty of random variables. Therefore, diversity estimation of the co-occurrences of user i and user j based on Renyi entropy is:

$$D_{ij} = exp(H_{ij}^R) = exp\left[\left(-log\sum_l (\frac{c_{ij,l}}{f_{ij}})^q\right)/(q-1)\right]$$

$$= \left[exp\left(-log\sum_l (\frac{c_{ij,l}}{f_{ij}})^q\right)\right]^{\frac{1}{(1-q)}} = \left(\sum_{l,c_{ij,l}\neq 0} (\frac{c_{ij,l}}{f_{ij}})^q\right)^{\frac{1}{(1-q)}}. \qquad (1)$$

where $c_{ij,l}$ is the number of co-occurrences between users i and j at location l, termed local frequency; $f_{ij} = \sum_l c_{ij,l}$ is the total number of co-occurrences of user i and user j, termed frequency. That is, frequency of two users is the sum of all their local frequencies across all locations. And q is the order of diversity controlling the impact of coincidences on diversity.

3.2   Weighted Frequency

While diversity measures the breadth of co-occurrences across locations, weighted frequency, on the other hand, measures the depth of co-occurrences. It weighs each co-occurrence individually depending on the popularity of the location. Co-occurrences in small uncrowded places, such as private houses, often result in more social interaction, as compared to those in crowded places. Therefore, the probability of implicit social relationships strongly depends on the locations of co-occurrences.

Cranshaw et al. [2] first introduces location entropy to describe the popularity of a location. It measures the popularity of a location by taking into account both the number of unique visitors to the location, and the relative proportions of their visits. Specifically, let l be a location, let $V_{u,l} = \{<u,l,t> : \forall t\}$ be the set of visits (a.k.a. check-ins or spatio-temporal records) in location l by user u, let $V_l = \{<u,l,t> : \forall t, \forall u\}$ be the set of all visits in location l by all users. The probability that a randomly picked check-in from V_l belongs to user u is $P_{u,l} = |V_{u,l}|/|V_l|$. Define the probability as a random variable, then its uncertainty is given by the Shannon entropy as follows:

$$H_l = -\sum_{u,P_{u,l}\neq 0} P_{u,l}logP_{u,l} \qquad (2)$$

A high value of the location entropy indicates a popular place with many visitors, but it proves that this place is not special to anyone. Weighed frequency utilizes location entropy to weight co-occurrence $c_{ij,l}$ between user i and j at location l individually depending on the popularity of the location. The formula of weighted frequency is given as follows:

$$F_{ij} = \sum_l c_{ij,l} \times exp(-H_l) \qquad (3)$$

Weighted frequency describes how important the co-occurrences at non-crowded places are to implicit social connections. Crowed locations have high location entropy H_l, resulting in a low value of $exp(-H_l)$, and consequently the impact of $c_{ij,l}$ on F_{ij} is decreased. On the other hand, for non-crowded locations, the value of $exp(-H_l)$ is high and consequently increases the impact of $c_{ij,l}$.

3.3 Implicit Social Strength

Diversity and weighted frequency describe the features of the co-occurrences. Diversity decreases the impact of frequent coincidences while weighted frequency increases the impact of co-occurrences at less crowded places. The less crowded a place, the more impact weighted frequency has. Thus, we measure implicit social strength between users by integrating diversity with weighted frequency of their co-occurrences. Let S_{ij} be the implicit social strength that captures both diversity and weighted frequency, defined as follows.

$$S_{ij} = D_{ij} \times F_{ij} \qquad (4)$$

where D_{ij} and F_{ij} are diversity and weighted frequency of co-occurrences between two users, defined in Eqs. 1 and 3, respectively. As a good practice, social strength S_{ij} is generally normalized to $[0, 1]$.

4 Proposed ImSoRec Model

After having extracted implicit social relationships in Sect. 3, each pair of users with implicit relationships is associated with a social strength. In this section, based on Probabilistic Matrix Factorization (PMF) [11] which can better integrate context information like social relationship, we propose a new POI recommendation model to integrate the implicit and explicit social relationships.

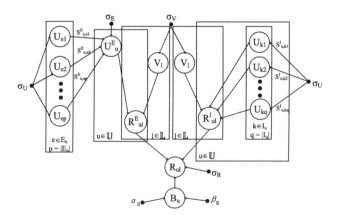

Fig. 3. A graphical model of ImSoRec

4.1 Explicit Social Network Ensemble

As shown in Fig. 3, the left part of the graph model incorporates the explicit social network. Inspiration by the Social Matrix Factorization (SMF) [9], we incorporate social trust propagation into probabilistic matrix factorization, assuming that the user preference of user u will be affected by his or her explicit social relations E_u through social influence. The main idea is that the latent feature vectors of users should be similar to their explicit social relations as shown in Eq. 5, which is in line with the sociological theory.

$$U_u = \frac{\sum_{e \in E_u} S_{ue} U_e}{|E_u|} \tag{5}$$

where U_u is the latent feature vector of user u. S_{ue} is the social influence of U_u and his or her explicit social relations U_e. We use Jaccard's coefficient to measure the strength that users i social connected with user j, which is formally defined as

$$S_{ue} = \frac{\Gamma(i) \cap \Gamma(j)}{\Gamma(i) \cup \Gamma(j)} \tag{6}$$

where $\Gamma(i)$ and $\Gamma(j)$ are the sets of users having explicit social relationships with user i and j, respectively. Therefore, the posterior probability can be obtained as follows, where g($\cdot$) is the sigmoid function, which bounds the range of $U_u^T V_i$ within $[0, 1]$.

$$p(U, V | R, S, \sigma_R^2, \sigma_S^2, \sigma_U^2, \sigma_V^2) \propto p(R|U, V, \sigma_R^2) p(U|S, \sigma_S^2, \sigma_U^2) p(V|\sigma_V^2)$$

$$= \prod_{u=1}^{N} \prod_{i=1}^{M} [\mathcal{N}(R_{ui}|g(U_u^T V_i), \sigma_R^2)]^{I_{ui}^R} \times \prod_{u=1}^{N} \mathcal{N}(U_u | \sum_{e \in E_u} S_{ue} U_e, \sigma_S^2 I)$$

$$\times \prod_{u=1}^{N} \mathcal{N}(U_u|0, \sigma_U^2 I) \times \prod_{i=1}^{M} \mathcal{N}(V_i|0, \sigma_V^2 I) \tag{7}$$

4.2 Implicit Social Relation Ensemble

As shown in Fig. 3, the right part of the graph model incorporates the implicit social relationships. Compared to explicit social relationships in online social networks, in which user preference will be affected by his or her social relationships, implicit social relationships are extracted from check-in data, finding people with similar behavior patterns, may be more helpful to POI recommendations. Therefore, we predict scores of users for POIs by check-ins of his or her implicit social relationships on these POIs. The model is defined as follows.

$$R_{ui} = \frac{\sum_{k \in I_u} R_{ki} S_{uk}}{|I_u|} \tag{8}$$

where I_u is the set of users who have implicit social relationships with user u; the physical meaning of S_{uk} can be interpreted as how much user u has similar content and spatial preferences with user k in the implicit social network.

Similarly, the posterior probability can be calculated as follows.

$$p(U, V|R, S, \sigma_R^2, \sigma_U^2, \sigma_V^2) \propto p(R|U, V, S, \sigma_R^2)p(U|0, \sigma_U^2)p(V|\sigma_V^2)$$
$$= \prod_{u=1}^{N} \prod_{i=1}^{M} [\mathcal{N}(R_{ui}|g(\sum_{k \in I_u} S_{uk}U_k^T V_i), \sigma_R^2)]^{I_{ui}^R} \times \prod_{u=1}^{N} \mathcal{N}(U_u|0, \sigma_U^2 I) \times \prod_{i=1}^{M} \mathcal{N}(V_i|0, \sigma_V^2 I)$$
$$(9)$$

4.3 ImSoRec Model

We consider two kinds of relationships between users, i.e., explicit social relationships and implicit social relationships. For an individual user, explicit social relationships tend to more similar to user preference. On the other hand, implicit social relationship may provide more information for POI recommendations by considering distance and content. Based on this assumption, we propose the ImSoRec model to integrate two different relationships when making POI recommendations. Besides, ImSoRec is capable of learning user-specific preferences between explicit and implicit social relationships by introducing parameters B_u. Referring to Eqs. 7 and 9 for explicit and implicit social relationships, the conditional probability of latent feature vector U can be defined as follows.

$$p(U|S^E, \sigma_U^2, \sigma_E^2) \propto p(U|\sigma_U^2) \times p(U^E|S^E, \sigma_E^2)$$
$$= \prod_{u=1}^{N} \mathcal{N}(U_u|0, \sigma_U^2 I) \times \prod_{u=1}^{N} \mathcal{N}(U_u^E| \sum_{e \in E_u} S_{ue}^E, \sigma_E^2 I)$$
$$(10)$$

where S_{ue}^E and S_{uk}^I denote explicit and implicit social strengths of user u; U_u^E is the user preference matrix of user u influenced by explicit social relationships.

The dot product of U_u and item latent feature vector V_l determines u's explicit-social generated probability on POI l, denoted as R_{ul}^E. An implicit-social generated probability R_{ul}^I is determined by the preferences of all his or her implicit social relations on POI l. Through introducing B_u as the weight of preference for explicit social relationships, $1 - B_u$ is the probability that user u prefers implicit social relationships. Thus, the conditional probability of the observed ratings can be expressed as follows.

$$p(R|U^E, U, V, S^E, S^I, B, \sigma_R^2)$$
$$= \prod_{u=1}^{N} \prod_{i=1}^{M} [\mathcal{N}(R_{ui}|g(B_u U_u^{E^T} V_i + (1 - B_u) \sum_{k \in I_u} S_{uk}^I U_k^T V_i), \sigma_R^2)]^{I_{ui}^R}$$
$$(11)$$

We assume that B follows a Beta distribution, and both U and V follow the same zero mean normal distribution. Through a Bayesian inference, given the observed ratings and two types of social relationships as well as the hyperparameters, the posterior probability of all model parameters can be obtained as follows.

$$p(U^E, U, V, B | R, S^E, S^I, \sigma_R^2, \sigma_E^2, \sigma_U^2, \sigma_V^2)$$

$$\propto p(R | U^E, U, V, S^E, S^I, B, \sigma_R^2) p(U | S^E, \sigma_U^2, \sigma_E^2) p(v | \sigma_V^2) p(B | \alpha_B, \beta_B)$$

$$= \prod_{u=1}^{N} \prod_{i=1}^{M} [\mathcal{N}(R_{ui} | g(B_u {U_u^E}^T V_i + (1 - B_u) \sum_{k \in I_u} S_{uk}^I U_k^T V_i), \sigma_R^2)]^{I_{ui}^R}$$

$$\times \prod_{u=1}^{N} \mathcal{N}(U_u | 0, \sigma_U^2 I) \times \prod_{u=1}^{N} \mathcal{N}(U_u^E | \sum_{e \in E_u} S_{ue}^E, \sigma_E^2 I) \qquad (12)$$

$$\times \prod_{i=1}^{M} \mathcal{N}(V_i | 0, \sigma_V^2 I) \times \prod_{u=1}^{N} Beta(B_u | \alpha_B, \beta_B).$$

We learn the parameters of ImSoRec by maximizing a posterior (MAP) inference. Taking the ln on both sides of Eq. 12, fixing the Gaussian noise variance and beta shape parameters, maximizing the log-posterior over U^E, U, V, B is equivalent to minimizing the following objective function:

$$\mathcal{L}(R, U^E, U, V, B, S^E, S^I)$$

$$= \frac{1}{2} \sum_{u=1}^{N} \sum_{i=1}^{M} I_{ui}^R (R_{ui} - g(R_{ui}^*))^2 + \frac{\lambda_U}{2} \sum_{u=1}^{N} U_u^T U_u + \frac{\lambda_V}{2} \sum_{i=1}^{M} V_i^T V_i$$

$$+ \frac{\lambda_E}{2} \sum_{u=1}^{N} ((U_u^E - \sum_{e \in E_u} S_{ue}^E U_e)^T (U_u^E - \sum_{e \in E_u} S_{ue}^E U_e)) \qquad (13)$$

$$- \lambda_B \sum_{u=1}^{N} ((\alpha_B - 1) \ln B_u + (\beta_B - 1) \ln(1 - B_u))$$

where $\lambda_U = \frac{\sigma_R^2}{\sigma_U^2}, \lambda_V = \frac{\sigma_R^2}{\sigma_V^2}, \lambda_E = \frac{\sigma_R^2}{\sigma_E^2}; \lambda_B = \sigma_R^2; R_{ui}^* = B_u {U_u^E}^T V_i + (1 - B_u) \sum_{k \in I_u} S_{uk}^I U_k^T V_i$ and $B(\cdot, \cdot)$ is a beta function:

$$B(a, b) = \int_0^1 q^{a-1} (1 - q)^{b-1} dq. \qquad (14)$$

A local minimum of the above objective function can be found by taking the derivative and performing gradient descent on parameters U^E, U, V, B separately.

4.4   Incorporation of Geographical Information

For POI recommendation, geographical information is an influential factor to the behaviors of a user. For example, users are usually willing to visit nearby locations, and the willingness to visit a POI decreases as the distance to the POI's location increases. Following [18], we use a power law distribution over the willingness of a user moving from one place to another as the distance function. More specifically, the willingness of a user to visit a POI (dis km away) is defined as $w(dis) = a \cdot dis^\kappa$, where a and κ are parameters.

Given a user u and the set of his/her historical POIs I_u, we can calculate the distance score $p(i\,|I_u)$ for each candidate POI i, and use this value to enhanced POI predictions. That is

$$R_{ui} \propto R_{ui} \times p(i\,|I_u) \propto R_{ui} \times p(i) \prod_{i^* \in I_u} p(i^*\,|i) \tag{15}$$

where $p(i^*\,|i)$ is the user's willingness to check in a POI with distance proportional to the $w(dis)$.

5   Experiment

In this section, we conduct experiments on real-world datasets to evaluate the effectiveness of our proposed recommender model ImSoRec. Specifically, we aim to answer the following three questions:

RQ1 Does ImSoRec incorporating with explicit and implicit social relationships outperform the state-of-the-art POI recommendation methods with social influence?

RQ2 Can implicit social relationships improve the performance of POI recommendation when explicit social networks do not exist?

RQ3 Can implicit social relationships still improve the performance of POI recommendation even when explicit social networks exist?

5.1   Experimental Settings

Datasets. The data used in the experiments was collected by Foursquare [15] and Gowalla [8], two well-known location-based social networks, where users shared their locations through check-ins. The Foursquare and Gowalla datasets were collected from December 2009 to July 2013 and January 2009 to August 2010, respectively. Each dataset has both the spatio-temporal check-in information and an explicit social network. Since the spatial data is heavily concentrated in the United States, we used only the spatio-temporal data within the United States for the experiments.

Since the focus of this article in the first step is on inferring implicit social relationships only from spatio-temporal data, so we remove those users who have less than two co-occurrences with others in two datasets. Then the details of the two datasets are shown in Table 1.

Table 1. Statistics of datasets

Dataset	User	Location	Check-in	Friendship	Sparsity
Foursquare	2443	96874	223110	12782	0.0943%
Gowalla	14650	105905	1612709	119246	0.1040%

We randomly select 70% of check-in data for each individual user for training. We run the experiment five times to get an average result.

Parameter Settings. The best experimental parameters were selected through multiple cross validation. When extracting implicit social relationships from spatio-temporal data, the time-interval is set to one hour, the order of diversity q is set to 0.1. We normalize the explicit and implicit social relationships of each user u respectively so that $\sum_{e \in E_u} S_{ue}^E = 1$ and $\sum_{k \in I_u} S_{uk}^I = 1$. We set the dimension of the latent factors K as 10 and 20 on Foursquare and Gowalla, respectively. We set a small value for regularization coefficient $\lambda_U = \lambda_V = \lambda_E = \lambda_B = 0.01$.

Baselines. To evaluate the effectiveness of the proposed recommendation model, we compare it with the following five models.

- **PMF** [11], the basic model of our proposed ImSoRec, which minimized the square error loss only using the observed check-ins based on matrix factorization.
- **RegPMF** [10], which modeled the influence of social network by placing a social regularization constraint on learning user-specific feature vectors between friends.
- **USG** [17], which combined geographical influence, social network and user interest with collaborative filtering.
- **SoDimRec** [14], which exploited a community detection algorithm to exploit heterogeneity of social relations and weak dependency between users, and combined them with matrix factorization as weight.
- **ASMF-LA** [8], which defined three kinds of relationships to learn potential POIs for users and integrated them to matrix factorization as augmented square error based matrix factorization.

Evaluation Metrics. As a POI recommender system only recommends limited locations to users, we evaluate our models versus other models in terms of ranking performance, i.e., Precision@k and Recall@k metrics (denoted as Pre@k and Rec@k). They are formally defined as follows.

$$Pre@k = \frac{1}{N} \sum_{i=1}^{N} \frac{S_i(k) \cap T_i}{k} \qquad Rec@k = \frac{1}{N} \sum_{i=1}^{N} \frac{S_i(k) \cap T_i}{T_i}$$

where $S_i(k)$ is a set of top-k unvisited locations recommended to user i excluding those locations in the training data, and T_i is a set of locations that are visited by user i in the testing data.

5.2   Experiment Results

Performance Comparison (RQ1). Here we compare the performance of our proposed model ImSoRec with the state-of-the-art POI recommendation methods.

Table 2. Evaluation of Top-K POI recommendation on the two datasets

Datasets	Metrics	Algorithms					
		PMF	RegPMF	USG	SoDimRec	ASMF	ImSoRec
Foursquare	Pre@1	0.1603	0.1608	0.1674	0.1620	0.1706	**0.1715**
	Rec@1	0.0076	0.0097	0.0094	0.0098	0.0094	**0.0102**
	Pre@5	0.0578	0.0907	0.0992	0.0965	0.1036	**0.1103**
	Rec@5	0.0142	0.0203	0.0213	0.0210	0.0227	**0.0239**
	Pre@10	0.0357	0.0688	0.0886	0.0781	0.0916	**0.0930**
	Rec@10	0.0177	0.0255	0.0304	0.0282	0.0316	**0.0322**
Gowalla	Pre@1	0.3352	0.3638	0.4040	0.3796	0.4036	**0.4160**
	Rec@1	0.0152	0.0168	0.0180	0.0168	0.0179	**0.0184**
	Pre@5	0.1564	0.1703	0.1831	0.1841	0.1930	**0.1998**
	Rec@5	0.0347	0.0380	0.0390	0.0390	0.0408	**0.0416**
	Pre@10	0.1061	0.1163	0.1306	0.1287	0.1369	**0.1414**
	Rec@10	0.0463	0.0509	0.0542	0.0533	0.0567	**0.0572**

As shown in Table 2, both RegPMF and SoDimRec are slightly superior to PMF. One possible explanation is that social network assists to make more accurate recommendation. USG outperforms previous methods due to taking geography information into account, which is a significant factor in POI recommendation. SoDimRec [14] and ASMF-LA [8] are closely related to our work because both of them take supplement of social relationships into consideration. ImSoRec performs better than both of them.

SoDimRec exploits community detection algorithms tending to produce many different clusterings, and it is unclear to decide which one to use. Moreover, SoDimRec ignores different individual preference for heterogeneous social network and weak dependency. In addition, SoDimRec is a traditional recommender system without considering spatial constraints. That is why it shows worse precision and recall than ASMF-LA and ImSoRec do.

ASMF-LA changes the ground truth of check-in matrix. It not only narrows the range of target POIs, but also blurs the target. In addition, ASMF-LA also takes the category of POIs into consideration as one of the selection criteria. Therefore, ASMF-LA performs better than SoDimRec and USG. But ASMF-LA does not consider the time influence. Compared to ASMF-LA, ImSoRec also considers the time factor, i.e., have similar activities at a similar time, can more detailed depict users' preference. In addition, ImSoRec uses a user-specific value B_u to learn individual user's preference for explicit and implicit social relationships, which can make better use of consistency and difference in social connections and implicit social relationships.

Here we give some significant findings during our experiments. In general, the precision and the recall of all models on the Gowalla dataset are higher than those on the Foursquare dataset because the density of the former dataset is a little larger than that of the latter dataset, as shown in Table 1. But on the Foursquare dataset, ImSoRec performs a little better than other models compared to those on the Gowalla dataset with less existing social relationships, which proves that implicit social relationships extracted from check-in data finding people with similar behavior patterns are useful to POI recommendation.

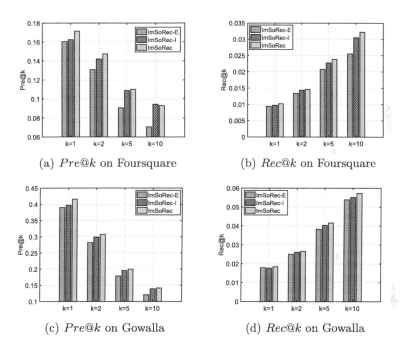

(a) $Pre@k$ on Foursquare   (b) $Rec@k$ on Foursquare

(c) $Pre@k$ on Gowalla   (d) $Rec@k$ on Gowalla

Fig. 4. Effects of explicit and implicit social relationships on ImSoRec

Impact of Explicit and Implicit Social relationships (RQ2 and RQ3). In this subsection, we investigate the effects of explicit and implicit social relationships on the proposed model ImSoRec and its variants to answer question RQ2 and RQ3.

– **ImSoRec-E:** ImSoRec-E is a variant of the proposed method ImSoRec, which only considers the explicit social relationships.
– **ImSoRec-I:** ImSoRec-I is a variant of the proposed method ImSoRec, which only considers the implicit social relationships extracted from check-in data.
– **ImSoRec:** ImSoRec is our proposed model taking advantage of both social relationships and implicit social relationships.

ImSoRec-E and ImSoRec-I are variants of the proposed model ImSoRec, which only considers the explicit or implicit social relationships, respectively. As shown in Fig. 4, ImSoRec-I improves POI recommendation performance greatly comparing to ImSoRec-E both on Foursquare and Gowalla duo to implicit

social relationships. The gap between ImSoRec-I and ImSoRec-E is larger on
Foursquare. Because users on Foursquare have less explicit social relationships,
then implicit social relationships play a better role in POI recommendation, making up for the lack of explicit social relationships. When explicit social networks
exists, ImSoRec can take advantage of both explicit and implicit social relationships. As shown in Fig. 4, ImSoRec improves POI recommendation performance
greatly comparing to ImSoRec-E and slightly outperforms to ImSoRec-I. That
means implicit social relationships can not only be a supplement to explicit
social relationships but also correct explicit social influence on recommendations (Users with explicit social relationships may not have similar preference
or will not visit POIs due to long distance). But in some situation, ImSoRec is
worse than ImSoRec-I. That means explicit social connected users may not have
similar spatial preferences because of the different activity centers issue.

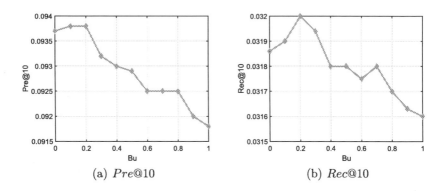

(a) *Pre@10*   (b) *Rec@10*

Fig. 5. The influence of B_u on Foursquare dataset.

Parameter Sensitivity. We chose B_u to analyze the influence of experimental
parameters on the results due to limited space. The ImSoRec model treats B_u
as the user-specific preference for explicit social relationships and implicit social
relationships. Figure 3 suggests that the both relationships affect users' access
to the POI decision-making process. A small value of B_u indicates that implicit
social relationship of user u has a greater impact on the decision-making process
than the explicit social relationship. We vary B_u from 0 to 1 by 0.1 on the
Foursquare dataset to study users' preference for implicit and explicit social
relationships.

Based on the results shown in Fig. 5, we can observe that ImSoRec achieves
the best performance when B_u is set as 0.2. The performance then drops dramatically when B_u increases. Small B_u is equivalent to only consider implicit social
relationships, which means implicit social relationships have a greater impact
on the users compared to explicit social relationships. It occurs possibly due to
that users might be in fact not interested in some locations that users having
explicit social relationships with them have checked-in, or the place is too far

away from her/his home. In addition, the curve is not a smooth drop, indicating that some users may prefer explicit social relationships. Thus a user-specific B_u is necessary.

6 Related Work

This paper makes a forward step for POI recommendation with implicit and explicit social connections. POI recommendation is to recommend a list of top-k most relevant POIs to a user, based on user implicit feedback, such as check-in frequency [19,20]. Collaborative filtering (CF) is widely used in POI recommendation. The state-of-the-art CF is based on matrix factorization and its variants [5,7,11]. Salakhutdinov and Mnih [11] proposed a PMF model in a Bayesian probabilistic framework to include Gaussian noise in observations. Under the Gaussian assumption, maximizing the posterior probability over latent features is equivalent to minimizing the square error.

One main category of POI recommendation sheds light on elaborating social network information. For example, Ye et al. [16,17] proposed user-based collaborative filtering to estimate the unobserved rating by directly using the check-in information of friends. Ma et al. [10] assumed that friends would share similar interests and then placed a social regularization term to constrain the objective functions for learning accurate user feature vectors. Gao et al. [4] proposed to model four types of social correlations (i.e., local friends, distant friends, local non-friends and distant non-friends) by using a geo-social correlation model with users' check-in activities, where the check-in probability was measured as a linear combination of these four geo-social correlations, and the corresponding coefficients were learned by a group of features in a logistic regression like fashion. Tang et al. [13] modeled local and global social relationships for all users. Specifically, in a local context, it models the correlation between users and their friends, while in a global context, it uses the reputation of a user in the whole social network as weight to fit observed ratings.

SoDimRec [14] and ASMF-LA [8] are closely related to our work. Both of them take social relationships into consideration for traditional and POI recommendation. SoDimRec modeled social dimensions for heterogeneity and weak dependency relationships by community detection. It is a supplement to online social relationships by social trust propagation, no essential difference to online social relationships. ASMF-LA defined three kinds of friends to model social relationships, similar history check-ins and neighboring users. It limited the number of potential POIs, and therefore efficiency increased. But it also limited the scope of the target POIs, and did not consider the time influence. However, our work is different from these existing works. Specifically, we learn users' implicit social relationships and strength only from spatio-temporal check-ins for POI recommendation, indicating similar visiting behavior patterns.

7   Conclusion

In this paper, we study how to extract and exploit implicit social relationships and propose an implicit social relationship enhanced matrix factorization model for POI recommendation. Specifically, we first use co-occurrences between users' check-in data with diversity and weighted frequency to describe implicit social relationships and its strengths, a supplement of explicit social relationship, which can capture users' behaviour patterns and preference more accurate. And then we incorporate implicit social strengths and explicit social relationships into our proposed model ImSoRec, which is based on PMF, to learn user-specific preference for explicit or implicit social relationship on POI recommendation. Finally, the experimental results on two real-world datasets show the effectiveness of our model over baseline methods in terms of top-k recommendation accuracy.

Acknowledgements. This research is partially supported by National Natural Science Foundation of China (Grant No. 61572335) and Natural Science Foundation of Jiangsu Province of China (No. BK20151223).

References

1. Crandall, D.J., Backstrom, L., Cosley, D., Suri, S., Huttenlocher, D., Kleinberg, J.: Inferring social ties from geographic coincidences. PNAS **107**(52), 22436–22441 (2010)
2. Cranshaw, J., Toch, E., Hong, J., Kittur, A., Sadeh, N.: Bridging the gap between physical location and online social networks. In: UbiComp, pp. 119–128. ACM (2010)
3. Eagle, N., Pentland, A.S., Lazer, D.: Inferring friendship network structure by using mobile phone data. PNAS **106**(36), 15274–15278 (2009)
4. Gao, H., Tang, J., Liu, H.: gSCorr: modeling geo-social correlations for new check-ins on location-based social networks. In: CIKM, pp. 1582–1586. ACM (2012)
5. Hu, Y., Koren, Y., Volinsky, C.: Collaborative filtering for implicit feedback datasets. In: ICDM 2008, pp. 263–272. IEEE (2008)
6. Jost, L.: Entropy and diversity. Oikos **113**(2), 363–375 (2006)
7. Koren, Y.: Factorization meets the neighborhood: a multifaceted collaborative filtering model. In: SIGKDD, pp. 426–434. ACM (2008)
8. Li, H., Ge, Y., Hong, R., Zhu, H.: Point-of-interest recommendations: learning potential check-ins from friends. In: KDD, pp. 975–984 (2016)
9. Ma, H., King, I., Lyu, M.R.: Learning to recommend with social trust ensemble. In: SIGIR, pp. 203–210. ACM (2009)
10. Ma, H., Zhou, D., Liu, C., Lyu, M.R., King, I.: Recommender systems with social regularization. In: WSDM, pp. 287–296. ACM (2011)
11. Mnih, A., Salakhutdinov, R.R.: Probabilistic matrix factorization. In: NIPS, pp. 1257–1264 (2008)
12. Pham, H., Shahabi, C., Liu, Y.: EBM: an entropy-based model to infer social strength from spatiotemporal data. In: SIGMOD, pp. 265–276. ACM (2013)
13. Tang, J., Hu, X., Gao, H., Liu, H.: Exploiting local and global social context for recommendation. In: IJCAI, pp. 264–269 (2013)

14. Tang, J., Wang, S., Hu, X., Yin, D., Bi, Y., Chang, Y., Liu, H.: Recommendation with social dimensions. In: AAAI, pp. 251–257 (2016)
15. Wang, W., Yin, H., Chen, L., Sun, Y., Sadiq, S., Zhou, X.: Geo-SAGE: a geographical sparse additive generative model for spatial item recommendation. In: SIGKDD, pp. 1255–1264. ACM (2015)
16. Ye, M., Yin, P., Lee, W.-C.: Location recommendation for location-based social networks. In: SIGSPATIAL, pp. 458–461. ACM (2010)
17. Ye, M., Yin, P., Lee, W.-C., Lee, D.-L.: Exploiting geographical influence for collaborative point-of-interest recommendation. In: SIGIR, pp. 325–334. ACM (2011)
18. Yuan, Q., Cong, G., Ma, Z., Sun, A., Thalmann, N.M.: Time-aware point-of-interest recommendation. In: SIGIR, pp. 363–372. ACM (2013)
19. Zhao, P., Xu, X., Liu, Y., Sheng, V.S., Zheng, K., Xiong, H.: Photo2Trip: exploiting visual contents in geo-tagged photos for personalized tour recommendation. In: MM, pp. 916–924. ACM (2017)
20. Zhao, P., Xu, X., Liu, Y., Zhou, Z., Zheng, K., Sheng, V.S., Xiong, H.: Exploiting hierarchical structures for POI recommendation. In: ICDM, pp. 655–664. IEEE (2017)

Spatial Co-location Pattern Mining Based on Density Peaks Clustering and Fuzzy Theory

Yuan Fang, Lizhen Wang$^{(\boxtimes)}$, and Teng Hu

School of Information Science and Engineering, Yunnan University,
Kunming 650091, China
{fangyuan,lzhwang}@ynu.edu.cn, hutengann@sina.com

Abstract. Spatial co-location patterns are the subsets of spatial features whose instances are frequently located together in geographic space. Traditional co-location pattern mining framework usually determines the proximity relationship of spatial instances by a user-specific distance threshold. However, in real life, the proximity relationship is a fuzzy concept and difficult to measure only by an absolute distance threshold. Furthermore, the spatial clique generating process consumes huge computational and spatial costs. In this paper, we propose a new framework for mining co-location patterns based on density peaks clustering and fuzzy theory. The experiments show that our method performs more efficient than the traditional Join-less method and the mining results on two real-world data sets indicate our method is significant and practical.

Keywords: Spatial co-location pattern · Proximity relationship
Fuzzy theory · Density peaks clustering

1 Introduction

Spatial co-location pattern mining aims to find the subsets of spatial features located together frequently in spatial proximity. For example, a co-location pattern {Matsutakes, Abies geogei Orrs} means Matsutakes usually grow under the Abies geogei Orrs. As an important branch of spatial data mining, co-location mining is used in extensive domains [1, 7, 8]. Co-location was originally proposed by Shekhar and Huang [1], this approach designs prevalence metric, namely participation index, to measure the interestingness of co-locations and proposed an Apriori-like method, namely Join-Based algorithm to mine co-location patterns. After that, Partial-Join approach [2] and Join-less approach [3] were successively proposed for improving the co-location mining efficiency.

Most of existed works require a distance threshold to determine proximity relationship between spatial instances, however, this traditional distance measure still exists shortcomings. On the one hand, Inappropriate thresholds will cause such problems: (1) Higher distance threshold may generate proximity relationships not adjacent in the real world, which may increase the computational cost and even leads incorrect results, (2) Lower distance threshold may lost many real proximity relationships. The discussions above show that "proximity" is a relative and fuzzy concept. Even in the same spatial data set, different spatial area may require different proximity

© Springer International Publishing AG, part of Springer Nature 2018
Y. Cai et al. (Eds.): APWeb-WAIM 2018, LNCS 10988, pp. 298–305, 2018.
https://doi.org/10.1007/978-3-319-96893-3_22

distance thresholds. Therefore, it is unreasonable to use an absolute distance threshold to determine the proximity relationships. Thus, how to measure "proximity" is a tough challenge. On the other hand, the proximity relationship identification approach based on the distance threshold will consume huge computational and spatial costs on checking clique relationship of spatial instances. How to improve the efficiency of the co-location pattern mining by avoiding generating spatial clique instances is another challenge in this article.

Motivated by above issues, we propose a new co-location pattern mining frame-work based on density peaks clustering [4] and fuzzy theory. Our contributions are described as follows: (1) we replace the absolute distance threshold by the density peaks clustering to divide the spatial data. (2) Based on fuzzy theory, we define the FPR (fuzzy participation ratio) and the FPI (fuzzy participation index) concepts to measure the prevalence of co-location patterns. (3) A new co-location mining frame work which avoids generating and testing cliques is developed to mine the spatial co-location patterns efficiently. The experimental results show that our method can discover interesting co-location patterns and is more efficient than the traditional co-location mining algorithms (such as Join-less).

This article is organized as follows: the preliminary concepts and definitions are introduced in Sect. 2. The results and contrast experiments are presented in Sect. 3. Section 4 makes a summary and the prospect of the article.

2   Definitions and Mining Framework

2.1   Definitions

Considering the instances in a clique supporting the prevalence of a co-location pattern are overlapped, we firstly introduce the density peak clustering technique to cluster spatial instances, and then calculate the membership of each instance belonging to clusters, so as to achieve a fuzzy partition for each instance in the data set. Firstly, we introduce a fuzzy threshold λ to obtain a λ-cluster (i.e. λ cut set).

Definition 1 (membership). Given a data set $S = \{s_1, s_2, \ldots, s_n\}$ as a sample set, $C = \{c_1, c_2, \ldots, c_k\}$ is a set of k clusters, assume $O = \{o_1, o_2, \ldots, o_k\}$ is a set of k cluster centers, then the membership of instance $s_j(s_j \in S)$ to the cluster $c_i(c_i \in C)$ is defined as:

$$c_i(s_j) = \frac{|s_j - o_i|^{-\frac{2}{b-1}}}{\sum\limits_{t=1}^{k} |s_j - o_t|^{-\frac{2}{b-1}}} \tag{1}$$

where $|s_j - o_i|$ represents the distance of instance s_j to the cluster center o_i; b is a weighting coefficient, also called smoothing factor, we use $b = 2$ in this paper [5].

Definition 2 (λ-cluster). If an instance $s_j \in S$ belongs to a cluster of c_i, its membership must be greater than a given threshold $\lambda \in [0, 1]$, which is the concept of cut set. The set of these instances is called **λ-cluster** c_i, expressed as c_i^λ.

Given a set of spatial instances $S = \{s_1, s_2, \ldots, s_n\}$ and $\lambda \in [0, 1]$, the λ cut set of cluster c_i is:

$$c_i^\lambda = \{s_j \in S | c_i(s_j) \geq \lambda\} \tag{2}$$

As we mentioned in Sect. 1, it is difficult to measure the proximity relationship in real life with a single absolute threshold especially in the datasets with relatively large density differences (such as in urban data mining). For solving this problem, we propose the definition of λ- proximity to replace the absolute threshold to obtain the fuzzy proximity relationship, after that we can use λ-proximity relationship to mine the interesting patterns.

Definition 3 (λ-proximity). Let the set of spatial instance be $S = \{s_1, s_2, \ldots, s_n\}$, λ-cluster c_i be $c_i^\lambda = \{s_1, s_2, \ldots, s_h\}$. For $s_i \in S$, $s_j \in S$, if $s_i \in c_i^\lambda$ and $s_j \in c_i^\lambda$, we called the s_i and s_j satisfy the λ proximity expressed as $\lambda(s_i, s_j)$.

Note that all instances in same λ-cluster satisfy the λ-proximity relationship with each other.

Definition 4 (Fuzzy Clique). Given a spatial instances set $I = \{i_1, i_2, \ldots, i_m\}$, if there are $\{ \lambda(i_j, i_k) | 1 \leq j \leq m, 1 \leq k \leq m\}$, then I is a **Fuzzy clique**.

Note that all instances in the λ-cluster satisfy the λ-proximity with each other, thus these instances form a clique naturally. If features of instances of a clique I' contains all the features of a co-location pattern cp, and no subset of I' where features of instances can contain all the features in cp, I' is called a **row instance** of cp, the set of all row instances of cp is called **table instance** of cp. In the following formula, we use $T(cp)$ to represent the table instance of cp.

Compared with the traditional co-location mining algorithm (such as Join-less), in the generation of higher-size candidate patterns, the traditional methods must regenerate the row-instances and store them which cost huge calculation consumption and storage space. In our method, the row-instance and table-instance of the candidate patterns can be obtained from each λ-cluster directly, so we do not need to recalculate the clique relationship and store row-instances and table-instances.

Definition 5 (fuzzy participation ratio). Given a set of all instance $S = \{s_1, s_2, \ldots s_n\}$, $C = \{c_1, c_2, \ldots, c_k\}$ is a set of k clusters, for an h-size co-location $cp = \{f_1, f_2, \ldots, f_h\}$ and a feature $f_l(f_l \in cp)$ in cp, the fuzzy participation ratio of f_l in cp is defined as:

$$FPR(cp, f_l) = \frac{\sum\limits_{i=1}^{k} \sum\limits_{s_j \in (T(cp) \cap c_i^\lambda)} c_i(s_j)}{|T(\{f_l\})|} \tag{3}$$

where f_l represents the l_{th} feature in cp, c_i represents the i_{th} cluster, $c_i(s_j)$ represents the membership of s_j belonging to c_i, $T(\{f_l\})$ is the table instance of 1-size $\{f_l\}$, which means all instances of f_l in space; $s_j \in (T(cp) \cap c_i^\lambda)$ means s_j is in the table instance of cp, and s_j is also in the λ-cluster c_i.

Definition 6 (fuzzy participation index). Given a co-location pattern $cp = \{f_1, f_2, \ldots, f_h\}$, the FPI (fuzzy participation index) of cp is the minimum FPRs of features in cp expressed as:

$$FPI(cp) = \min_{f_l \in cp}(FPR(cp, f_l)) \tag{4}$$

Given a minimum prevalent threshold min_fprev, if $FPI(cp) \geq min\_fprev$, cp is called a prevalent co-location pattern.

Although the paper [6] had proposed the definitions of FPR and FPI, the definitions are aimed to mine prevalent co-location patterns in a fuzzy data set. Our definitions in this paper are different from that in paper [6], it is to mine the prevalent co-location patterns in traditional data set based on fuzzy method.

Lemma 1 (anti-monotone)
FPR (fuzzy participation ratio) and FPI (fuzzy participation index) decrease monotonously with the size of the co-location patterns.

Proof. If an instance of a feature in cp occurs in the row-instances of cp, and $cp' \subseteq cp$, this instance must be in the row-instances of cp'. So the FPR of a feature is monotonous decreasing.

Also assume $cp = \{f_1, \ldots, f_k\}$,

$$FPI(cp \cup f_{k+1}) = \min_{i=1}^{k+1}\{FPR(cp \cup f_{k+1}, f_i)\} \leq \min_{i=1}^{k}\{FPR(cp \cup f_{k+1}, f_i)\}$$
$$\leq \min_{i=1}^{k}\{FPR(cp, f_i)\} = FPI(cp)$$

So the FPI of a co-location pattern is also monotonous decreasing.

2.2 Mining Framework

Figure 1 is the framework of mining all prevalent co-location patterns based on the defined FPR and FPI (we call DPC-MCP), firstly we use DPC algorithm to cluster the instances, then calculate the membership of each instance to all clusters and obtain the λ-clusters under the threshold of λ, and then because of the anti-monotone of FPR and FPI (Lemma 1), we can generate the $k + 1$-size candidate co-location patterns from k-size prevalent co-location patterns and calculate the FPRs of all features in candidate co-location patterns, finally we can obtain the FPI of each candidate co-location pattern and compare it with threshold min_fprev, if the FPI value is no less than min_fprev, the candidate pattern is a prevalent co-location pattern, repeat these steps we can obtain all prevalent co-location patterns.

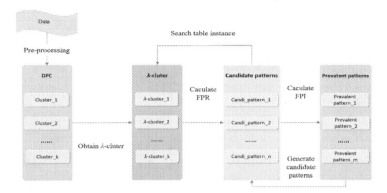

Fig. 1. The framework of DPC-MCP

3   Experiments on Real-Life Data Sets

This section evaluates the performance of the proposed algorithm on two real-life data sets. The data set real-1 concerns the rare plant data of the three Parallel Rivers of Reserved Areas in Yunnan Province, China. There are 31 features and only 355 instances in a 130000 m × 80000 m area, and real-2 is a POI data set, where there are fewer features than real-1 set, but 23025 instances in a 22000 m × 14000 m area. In the Table 1, we give the parameters and results of both Join-less and DPC-MCP on real-1 set and real-2 data set. We have optimized implementation of Join-less and, compared with other algorithms for mining traditional co-location patterns, it seems Join-less can deal with more data for the same run-time.

Table 1. The parameters and results of Join-less and DPC-MCP on two data sets

Parameters and results	Real-1 data set		Real-2 data set	
	Join-less	DPC-MCP	Join-less	DPC-MCP
Distance threshold d(m)	8,000	✕	50	✕
Cut off distance d_c(m)	✕	4,000	✕	50
Proximity threshold λ	✕	0.27	✕	0.15
Threshold min_prev	0.3	✕	0.28	✕
Threshold min_fprev	✕	0.7	✕	0.674
Mining results	{B,C,W}	{B,C,W}	{A,C,D}	{A,C,D}
	✕	{H,W,Y}	{E,I,M}	{E,I,M}
	{a, c, e}	✕	✕	{A,D,K}

Table 1 shows the different parameters and mining results of Join-less and DPC-MCP on two real data sets. The sign "✕" means null value.

Figure 2 shows two data set clustering results on different cut off distance, Fig. 2(a) and (b) show clustering results on d_c equals 4,000 m and 3,000 m respectively in real-1 data set; Fig. 2(c) and (d) give clustering results on d_c equals 50 m and 60 m

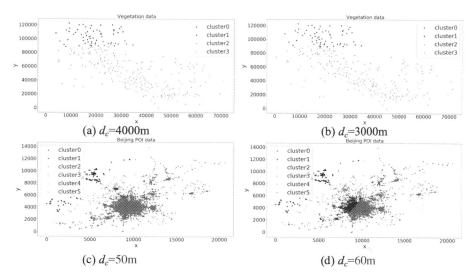

Fig. 2. (a) and (b) are the clustering results of the real-1 data set at different cut off distance, and (c) and (d) are the results of the real-2 by DPC

respectively in real-2 data set. Figure 2 shows that the clustering results vary little with different cut off distance d_c, it can be seen that the DPC-MCP is insensitive to cut off distance. That's one of the reasons why we choose DPC as our clustering algorithm.

We give some examples of the mining results of the two algorithms at different thresholds on the real-1 data set in Table 1. The pattern {B, C, W} can be discovered by both algorithms. The pattern {H, W, Y} cannot be mined by Join-less. However, in real world, the plant H, W and Y need the similar growth environments, and they most probably grow in the proximity area. The reason why Join-less cannot obtain the pattern such as {H, W, Y} is that the distance threshold d is too small to mine these patterns. Unfortunately, we cannot easily increase the distance threshold, if do that, some unwanted co-location patterns would be discovered.

Also, in our DPC-MCP framework, if the number of instances of a feature is too few, it is thought not to be representative and not taken into account in mining process. Our algorithm cannot obtain the pattern {a, c, e} which discovered by Join-less, since the instances' amount of these features (such as a, c, e and R) are too few, the instances of feature "e" are only 3 in the real-1 data set. More and then, these instances of the features are far away from the centers of the clusters, so its membership to every cluster is too small, and the λ cut set prune these instances of the features.

We also give some examples of mining results on the real-2 data set in Table 1. The feature A, D, and K represent Chinese restaurant, hotel, and train station respectively in POI real-2 data set. In real life, the prevalent appearance of pattern {A, D, K} is acceptable. The reason why the feature K is absence in the pattern mined by Join-less is that the proximity distance threshold d is too small to mine the feature K.

Figure 3 shows the mining results and time consuming by DPC-MCP and Join-less respectively on the real-1 data set. Figure 3(a) and (b) show the numbers of patterns mined by DPC-MCP and Join-less respectively, the x-axis represents different *min_prev* (or *min_fprev*), and the *y*-axis represents the number of patterns; the different colors of bar represents different λ threshold in Fig. 3(a) and (b), different colors of the bar represent different proximity threshold *d*. Figure 3(c) and (d) are running time of the DPC-MCP and Join-less on different thresholds respectively.

On the real-1 data set in Fig. 3, at different thresholds, DPC-MCP can obtain more prevalent patterns and the time consuming is much less than Join-less (the highest size of the pattern by DPC-MCP is higher than that of Join-less).

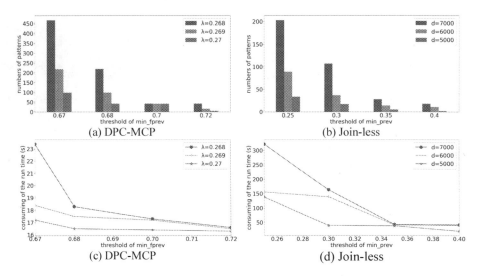

Fig. 3. The compared experiments of DPC-MCP and Join-less on the real-1 data set. (Color figure online)

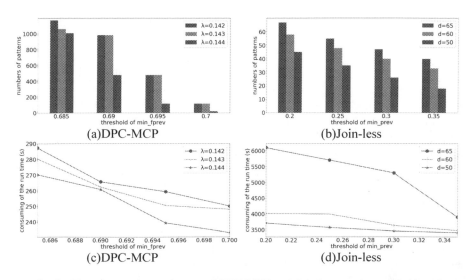

Fig. 4. The compared experiments of DPC-MCP and Join-less on the real-2 data set

The experiments in Fig. 4 on the real-2 data set also show that DPC-MCP can obtain more prevalent patterns which cannot be mined by Join-less (because of the distance threshold is too small to obtain the proximity relationship) and the time consuming is much less than Join-less (the highest size of the pattern by DPC-MCP is higher than that of Join-less) in most of cases.

4 Conclusion and Future Work

In this paper, we propose a new framework for efficiently mining prevalent co-location pattern. In this framework, we redefine proximity relationship, row-instance, table-instance, participation ratio and participation index, and our proposed DPC-MCP algorithm can efficiently mine interesting co-location patterns. Our method not only produces interesting co-location patterns without a strict distance threshold, but also need much less calculating consumption than traditional mining framework (such as Join-less). The experiments on real data sets show the proposed DPC-MCP method is more efficient than Join-less and the mining results of DPC-MCP are significant and practical. We also find DPC-MCP is much sensitive to the thresholds (such as λ and min_fprev). In the future work we plan to explore how to select the appropriate λ and min_fprev values in different data sets.

Acknowledgements. This work is supported by the National Natural Science Foundation of China (61472346, 61662086, 61762090), the Natural Science Foundation of Yunnan Province (2015FB114, 2016FA026), and the Project of Innovative Research Team of Yunnan Province.

References

1. Huang, Y., Shekhar, S., Xiong, H.: Discovering colocation patterns from spatial data sets: a general approach. IEEE Trans. Knowl. Data Eng. **16**(12), 1472–1485 (2004)
2. Yoo, J.S., Shekhar, S.: A partial join approach for mining co-location patterns. In: The 12th Annual ACM International Workshop on Geographic Information Systems, pp. 241–249 (2004)
3. Yoo, J.S., Shekhar, S.: A joinless approach for mining spatial colocation patterns. IEEE Trans. Knowl. Data Eng. **18**(10), 1323–1337 (2006)
4. Rodriguez, A., Laio, A.: Clustering by fast search and find of density peaks. Science **344** (6191), 1492 (2014)
5. Bezdek, J.C., Ehrlich, R., Full, W.: FCM: the fuzzy c-means clustering algorithm. Comput. Geosci. **10**(2), 191–203 (1984)
6. Ouyang, Z., Wang, L., Wu, P.: Spatial co-location pattern discovery from fuzzy objects. Int. J. Artif. Intell. Tools **26**, 1750003 (2017). 20 p.
7. Wang, L., Bao, X., Zhou, L.: Redundancy reduction for prevalent co-location patterns. IEEE Trans. Knowl. Data Eng. **30**(1), 142–155 (2018)
8. Wang, L., Bao, X., Chen, H., Cao, L.: Effective lossless condensed representation and discovery of spatial co-location patterns. Inf. Sci. **436**, 197–213 (2018)

A Tensor-Based Method for Geosensor Data Forecasting

Lihua Zhou, Guowang Du, Qing Xiao$^{(\boxtimes)}$, and Lizhen Wang

School of Information, Yunnan University, Kunming 650500, China
{lhzhou, Xiaoqing, lzhwang}@ynu.edu.cn,
bingwei2642@qq.com

Abstract. In recent years, geosensor data forecasting has received considerable attention. However, the presence of correlation (i.e. spatial correlation across several sites and time correlation within each site) poses difficulties to accurate forecasting. In this paper, a tensor-based method for geosensor data forecasting is proposed. Specifically, a tensor pattern is first introduced into modelling the geosensor data, which can take advantage of geosensor spatial-temporal information and preserve the multi-way nature of geosensor data, and then a tensor decomposition based algorithm is developed to forecast future values of time series. The proposed approach not only combines and utilizes the multi-mode correlations, but also well extracts the underlying factors in each mode of tensor and mines the multi-dimensional structures of geosensor data. Experimental evaluations on real world geosensor data validate the effectiveness of the proposed methods.

Keywords: Geosensor data forecasting · Tensor decomposition
CP-WOPT model

1 Introduction

With the rapid growth of digital sources of information, enormous amounts of geophysical time series are being continually generated and collected. Accurate forecasts of geosensor data can be useful for decision-makers, and thus it is receiving increasing attention from researchers in recent years [1, 2, 3].

Some forecasting approaches, such as sARIMA, cARIMA and cVAR model proposed by Pravilovic et al. [4, 5, 6], have considered temporal and spatial correlations simultaneously in the process of forecasting geosensor data and achieved promising results. However, the spatial information and temporal information contained in geosensor data have not been fully utilized, for example, spatio-temporal dissimilarity including both time series dissimilarity and spatial distance as separate contributions in the cVAR does not incorporate the inner-correlations of geosensor data. These inner-correlations of geosensor data may have an impact on the predicted performance.

To combine and utilize the multi-mode correlations, a series of studies have been carried out in multi-dimension data [7, 8, 9]. Tensor-based methods have been proved to be a good analytical tool for dealing with the multidimensional data, because tensor decomposition can capture the global structure of the data. Therefore, we propose a

© Springer International Publishing AG, part of Springer Nature 2018
Y. Cai et al. (Eds.): APWeb-WAIM 2018, LNCS 10988, pp. 306–313, 2018.
https://doi.org/10.1007/978-3-319-96893-3_23

tensor-based method for geosensor data forecasting. Specifically, we first formulate geosensor data as multi-way array, i.e., tensor pattern, keeping the multi-dimension characteristics and covering enough spatial and temporal information, and then we use tensor decomposition to forecast future values of time series. The proposed approach not only combines and utilizes the multi-mode correlations through preserving the multi-way nature of the geosensor data, but also well extracts the underlying factors in each mode of tensor and mines the multi-dimensional structures of geosensor data, by CP decomposition with weighted optimization (CP-WOPT) that has been testified to provide a good imputation performance for dealing with missing value [7].

In summary, the specific contributions of this paper are highlighted as follows:

(1) A tensor pattern is introduced to model geosensor data that can preserve the multi-way nature of geosensor data, and thus the inner-correlations of geosensor data can be incorporated into the process of forecasting.
(2) A CP-WOPT decomposition based forecasting method (GDF-TD) is proposed to forecast the future values of time series. The proposed method well extracts the underlying factors in each mode of tensor and mines the multi-dimensional structures of geosensor data.
(3) Extensive numerical study on twelve real geosensor data sets has been conducted to validate the performance of our proposed approach.

The remaining sections of this paper are organized as follows: a brief overview of related work about geosensor data forecasting and tensor decomposition is given in Sect. 2. Section 3 presents the notation used in this paper and introduces our proposed method. Experiments on real geosensor data and results are presented in Sect. 4. Finally, we conclude the paper in Sect. 5.

2   Related Works

Egrioglu et al. [10] determined a time series model by accounting for temporal information, estimated the model parameters and provided accurate point estimates of future values of time series, but they disregarded the spatial dimension of data. Pokrajac and Obradovic [11] used a generalization of the standard spatial auto-regression and included a disturbance term modelled as a temporal auto-regression to produce spatio-temporal forecasts. Kamarianakis and Prastacos [12] modeled a geosensor time series as a linear combination of past observations and disturbances at neighboring sites and considered space-time auto-regressive integrated moving average models. Ohashi and Torgo [13] computed technical indicators for each time stamp as summaries of certain properties of the time series in a neighbourhood, and then used these computed technical indicators and the past values of the series up to a certain time window of fixed length to determine forecasts of each series. Pravilovic et al. [4, 5] integrated cluster analysis to the ARIMA model [14]. The proposed sARIMA and cARIMA model determine the number of ARIMA coefficients for a specific time series automatically, without any human intervention. Saengseedam and Kantananth [15] proposed linear mixed models (LMMs) with spatial random effects in a Bayesian framework, where the spatial correlation is taken into account with a conditional auto regressive (CAR) prior distribution for spatial effects. Pravilovic et al. [6] proposed a spatio-temporal cluster-based vector auto regressive model (cVAR) including a clustering phase and a forecasting phase. The clustering phase divides time series into clusters based on the spatial location of the time series, as well as by the time-stamped values of the time series. The forecasting phase constructs spatially-coupled variables from the clustered time series.

Tensor decomposition is a method for decomposing a Pth-order tensor into another Pth-order tensor of smaller size, termed as the "core tensor," and P factor matrices. Tucker decomposition [16] and CP decomposition [17] are two particular tensor decompositions that can be considered to be higher-order generalization of the matrix singular value decomposition (SVD) and principal component analysis (PCA) [8]. The CP decomposition factorizes a tensor into a sum of component rank-one tensors,

$$\mathcal{X} = \sum_{r=1}^{R} \mathbf{a}_r^{(1)} \circ \mathbf{a}_r^{(2)} \circ \cdots \circ \mathbf{a}_r^{(P)}, \text{ where } R \text{ is a positive integer and } \mathcal{X} \in \mathfrak{R}^{I_1 \times \cdots \times I_P}, \mathbf{a}_r^{(i)} \in \mathfrak{R}^{I_i}$$

for $i = 1, \ldots, P$, $r = 1, \ldots, R$, the symbol "$\circ$" represents the *vector outer product*. It captures multi-linear structure. Acar et al. [7] formulated CP as a weighted least squares problem modelling only the known data entries, and then developed the CP-WOPT (CP Weighted OPTimization) algorithm using gradient-based optimization to solve the weighted least squares formulation of the CP problem. The Tucker decomposition factorizes a tensor into a core tensor multiplied (or transformed) by a matrix along each mode, $\mathcal{X} \approx \mathcal{S} \times_1 \mathbf{U}^{(1)} \times_2 \mathbf{U}^{(2)} \times \ldots \times_P \mathbf{U}^{(P)}$, where $\mathcal{X} \in \mathfrak{R}^{I_1 \times \cdots \times I_P}$, $\mathcal{S} \in \mathfrak{R}^{J_1 \times \cdots \times J_P}$, $\mathbf{U}^{(i)} \in \mathfrak{R}^{I_i \times J_i}$ for $i = 1, \ldots, P$, the symbol "$\times_n$" represents the *n-mode product* of a tensor. The tensor $\mathcal{S}$ is the core tensor and its entries show the level of interaction between the different components, and $\mathbf{U}^{(i)}, i = 1, \ldots, P$ are the factor matrices that can be thought of as the principal components in each mode.

3   Proposed Method

Let K be a set of geo-locations over a given spatial domain, Y be a numeric geophysical variable, and the time axis be discretized in equally-spaced time points denoted as $t = 1, 2, \ldots, T$; $y(k, t)$ denote the sequence of geo-referenced measures of Y collected at a certain projected geolocation $k \in K$ for each time point $t = 1, 2, \ldots, T$ and $\mathbb{D}(K, Y, T)$ be a geophysical time series dataset. Given $\mathbb{D}(K, Y, T)$, the problem of geosensor data forecasting is to learn a model from $\mathbb{D}(K, Y, T)$ and then use the model to forecast new data points $\hat{y}(k, T + 1), \ldots, \hat{y}(k, T + N)$, $\forall k \in K$ for a suitable forecasting horizon N.

In this section, we present a tensor decomposition based geosensor data forecasting methodology, called GDF-TD, which inputs dataset $\mathbb{D}(K, Y, T)$ and consists of a pipeline of four algorithmic steps:

(1) Setting a time window and selecting time series data within the time window;
(2) Using the selected time series data to construct a tensor;
(3) Factorizing the constructed tensor;
(4) Using the factorized tensor to forecast future data points.

3.1   Setting a Time Window

A time window $\mathbf{W}(t, W)$ means a time interval ending at time t with size W, then the observed values of $y(k, t)$ within $\mathbf{W}(t, W)$ form a sub-sequence $y_{\mathbf{W}}(k, t)$, which is defined as:

$$\forall k \in K, \ y_{\mathbf{W}}(k, t) = \begin{cases} \{y(k, t - W + 1), \ldots, y(k, T)\}, & t \leq T \\ \{y(k, t - W + 1), \ldots, y(k, T), \underbrace{0, \ldots, 0}_{W - t + T}\}, & t > T \ \& \ W > t - T \end{cases} \tag{1}$$

$W - t + T$ zeros are added into $y_{\mathbf{W}}(k, t)$ when $t > T$ because there are not observed values for variable Y at time point $T + 1, \ldots, t$. The geophysical time series dataset within the time window $\mathbf{W}(t, W)$ is denoted as $\mathbb{D}_{\mathbf{W}}(K, Y)$.

3.2   Constructing a Tensor

A *tensor* is a multidimensional array, and the number of dimensions is the order of the tensor, also known as ways or modes (Kolda and Bader [8]). Tensors of order $P \geq 3$ are denoted by Euler script letter $(\mathcal{X}, \mathcal{Y})$, matrices are denoted by boldface capital letters $(\mathbf{A}, \mathbf{B})$, vectors are denoted by boldface lowercase letters $(\mathbf{a}, \mathbf{b})$, and scalars are denoted by capital or lowercase letters (A, B, a, b). Columns of a matrix are denoted by boldface lower letters with a subscript ($\mathbf{a}_1, \mathbf{a}_2$ are first two columns of $\mathbf{A}$). Entries of a matrix or a tensor are denoted by lowercase letters with subscripts, i.e., the $(i_1, i_2, \ldots, i_P)$ entry of an P-way tensor $\mathcal{X} \in \Re^{I_1 \times I_2 \times \cdots \times I_P}$ is denoted by $x_{i_1, i_2, \ldots, i_P}$, where $1 \leq i_k \leq I_k, 1 \leq k \leq P$.

To forecast new data points $\hat{y}(k, T + 1), \ldots, \hat{y}(k, T + N), \forall k \in K$ for a suitable forecasting horizon N, the time window $\mathbf{W}(T + N, W)$ can be selected, and the data in $\mathbb{D}_{\mathbf{W}}(K, Y)$ are formulated as a four-way tensor $\mathcal{Z} \in \Re^{I_1 \times I_2 \times I_3 \times I_4}$, where $I_4 = |K|$, $I_2 = N, I_1 \times I_3 = W/I_2$, each three-way tensor $\mathcal{Z}_i \in \Re^{I_2 \times I_3 \times I_4}$, $i = 1, \ldots, I_1$ is called as a sub-tensor of $\mathcal{Z} \in \Re^{I_1 \times I_2 \times I_3 \times I_4}$. All entries in $\mathcal{Z}_i$ are observation, thus $\mathcal{Z}_i \in \Re^{I_2 \times I_3 \times I_4}$, $i = 1, \ldots, I_1 - 1$ is complete. But the sub-tensor $\mathcal{Z}_{I_1} \in \Re^{I_2 \times I_3 \times I_4}$ is incomplete, because it contains entries needed to be forecasted (the last column of each frontal slice in $\mathcal{Z}_{I_1}$).

3.3   Factorizing a Tensor

To decompose tensor $\mathcal{Z} \in \Re^{I_1 \times I_2 \times I_3 \times I_4}$, the nonnegative weight tensor $\mathcal{W}$ of the same size as $\mathcal{Z}$ is defined as:

$$w_{i_1, i_2, i_3, i_4} = \begin{cases} 1 & i_1 = 1, \ldots, I_1 - 1, i_2 = 1, \ldots, I_2, i_3 = 1, \ldots, I_3, i_4 = 1, \ldots, I_4 \\ 1 & i_1 = I_1, i_2 = 1, \ldots, I_2, i_3 = 1, \ldots, I_3 - 1, i_4 = 1, \ldots, I_4 \\ 0 & i_1 = I_1, i_2 = 1, \ldots, I_2, i_3 = I_3, i_4 = 1, \ldots, I_4 \end{cases} \tag{2}$$

The four-way objective function is defined by:

$$f_{W_{CP}}(\mathbf{A}^{(1)}, \dots, \mathbf{A}^{(4)}) = \frac{1}{2} \left\| \mathcal{Z} - [\![\mathbf{A}^{(1)}, \mathbf{A}^{(2)}, \dots, \mathbf{A}^{(4)}]\!] \right\|_W^2 \tag{3}$$

In Eq. (3), the notation $\left[\![\mathbf{A}^{(1)}, \mathbf{A}^{(2)}, \mathbf{A}^{(3)}, \mathbf{A}^{(4)}]\!\right]$ defines a *four*-way tensor of size $I_1 \times I_2 \times I_3 \times I_4$ whose elements are given by $\left([\![\mathbf{A}^{(1)}, \mathbf{A}^{(2)}, \mathbf{A}^{(3)}, \mathbf{A}^{(4)}]\!] \right)_{i_1 i_2 i_3 i_4} = \sum_{r=1}^{R} \prod_{n=1}^{4} a_{i_n r}^{(n)}$ for all $i_n \in \{1, \dots, I_n\}$ and $n \in \{1, \dots, 4\}$ (Kolda and Bader [8]).

The goal of CP-WOPT decomposition is to find matrices $\mathbf{A}^{(n)} \in \Re^{I_n \times R}$ for $n \in \{1, 2, 3, 4\}$ that minimize the weighted objective function in (3). This optimization problem can be solved by using gradient-based optimization method. The updating rules for $\mathbf{A}^{(n)} (n \in \{1, 2, 3, 4\})$ are shown in Theorem 1.

Theorem 1. Let $\mathcal{X}$ be a Pth-order tensor with size $I_1 \times I_2 \times \dots \times I_P$ and rank R (components). If fixed $\mathbf{A}^{(k)}, k = \{1, \dots, P\}, k \neq n$, the objective function $f_W(\mathbf{A}^{(1)}, \dots, \mathbf{A}^{(P)})$ is nonincreasing under the updating rule:

$$a_{i_n r}^{(n)} = a_{i_n r}^{(n)} \frac{\sum\limits_{\substack{k=1 \\ k \neq n}}^{P} \sum\limits_{i_k=1}^{I_k} w_{i_1,\dots,i_P}^2 x_{i_1,\dots,i_P} \prod\limits_{\substack{m=1 \\ m \neq n}}^{P} a_{i_m r}^{(m)}}{\sum\limits_{\substack{k=1 \\ k \neq n}}^{P} \sum\limits_{i_k=1}^{I_k} w_{i_1,\dots,i_P}^2 \left(\sum\limits_{l=1}^{R} \prod\limits_{m=1}^{P} a_{i_m l}^{(m)} \right) \prod\limits_{\substack{m=1 \\ m \neq n}}^{P} a_{i_m r}^{(m)}} \tag{4}$$

The Proof is omitted due to the limitation of spaces. Fixed $\mathbf{A}^{(1)}, \dots, \mathbf{A}^{(k-1)}, \mathbf{A}^{(k+1)}, \dots, \mathbf{A}^{(P)}, \mathbf{A}^{(k)}$ can be updated with Eq. (4).

3.4   Forecasting

After tensor $\mathcal{Z} \in \Re^{I_1 \times I_2 \times I_3 \times I_4}$ is factorized as $\mathbf{A}^{(n)} \in \Re^{I_n \times R}$ for $n \in \{1, 2, 3, 4\}$, a approximate tensor $\hat{\mathcal{Z}} \in \Re^{I_1 \times I_2 \times I_3 \times I_4}$ of $\mathcal{Z}$ can be computed. Let $\mathbf{a}_i^{(n)}$, $i = 1, \dots, r$ be the ith column of $\mathbf{A}^{(n)}$, then $\hat{\mathcal{Z}} = \sum\limits_{r=1}^{R} \mathbf{a}_r^{(1)} \circ \mathbf{a}_r^{(2)} \circ \mathbf{a}_r^{(3)} \circ \mathbf{a}_r^{(4)}$. The entries in the last column of each frontal slice in the sub-tensor $\hat{\mathcal{Z}}_{I_1}$ of $\hat{\mathcal{Z}}$ are regarded as forecasted values, i.e. $\hat{y}(k, T+j) = z_{I_1 j I_3 k}, j = 1, \dots, N, k = 1, \dots, K$.

4   Experimental Evaluations

In this paper, we conduct experiments on twelve geosensor data sets used by Pravilovic et al. [6]. For each data set, the time series are split into training and testing data sets. The training data set is used to construct and factorize tensor, while the testing data set is used to evaluate the performance of algorithms. In this paper, we use *root mean square error* (RMSE) between the forecasted values $\hat{y}(k, t)$ and the real values $y(k, t)$ as the performance metric. We evaluate the performances of the GDF-TD algorithm by comparing to other techniques. Pravilovic et al. [6] reported forecasting results of their cVAR model and auto.ARIMA [18], sARIMA [4] and cARIMA [5] model under different parameter α on twelve geosensor data sets. auto.ARIMA model neglects spatial autocorrelation, sARIMA and cARIMA model account for spatial correlation in a univariate time series setting, cVAR model summarize the dynamic structure of spatial correlation over time. We denote the lowest errors of these models as LowE for simplicity in this paper. In the experiments, we initialize factor matrices randomly with fixed seeds. We compare our results obtained by tensor decomposition with LowE. The results are shown in Table 1, where the lowest errors are in bold.

Table 1. The RMSE averaged per geosensor on twelve data sets (LowE denotes the lowest errors of cVAR, auto.ARIMA, sARIMA and cARIMA). The lowest errors in LowE and Tensor are in bold, the numbers in brackets represent the dimension size of the tensor and the tensor rank specified in the process of decomposition.

Data title	Phenomenon	Average RMSE	
		LowE	Tensor
TCEQ	Wind speed	0.31	**0.14** (3*24*5*26, $R = 2$)
	Air temperature	0.21	**0.16** (3*24*5*26, $R = 1$)
	Ozone concentration	0.53	**0.19** (3*24*5*26, $R = 6$)
MESA	NOx concentration	0.18	**0.11** (4*12*5*20, $R = 5$)
NREL	Wind speed	0.39	**0.34** (3*12*4*1326, $R = 1$)
SAC	Air temperature	0.15	**0.08** (3*12*4*900, $R = 1$)
NREL/NSRDB	Global solar radiation	0.17	**0.15** (2*7*4*1071, $R = 2$)
	Direct solar radiation	0.45	**0.32** (2*7*4*1071, $R = 1$)
	Diffuse solar radiation	0.30	**0.27** (2*7*4*1071, $R = 1$)
NCDC	Air temperature	0.12	**0.07** (2*12*4*72, $R = 1$)
	Precipation	0.26	**0.20** (2*12*4*72, $R = 1$)
	Solar energy	0.13	**0.08** (2*12*4*72, $R = 1$)

In brackets of the column "Tensor" of Table 1, $I_1 * I_2 * I_3 * I_4$ represents the dimension sizes with respect to orders of a tensor, and $R = x$ means the rank of the tensor is x. From Table 1, we can observe that our tensor decomposition-based approach yields the lower RMSEs than LessE over all twelve geosensor data sets. The average improvement in accuracy is 33%. In particular, the greatest accuracy gain is achieved on the series of TCEQ_Ozone (64%). It indicates that our proposed method is effective for geosensor data forecasting.

5 Conclusion

In this paper, tensor concept is introduced into forecasting geosensor data. A CP-WOPT decomposition based forecasting method has been proposed. It formulates the geosensor data into a multi-way data set and forecasts future values of time series by factorizing tensors. The multi-way data set covers geosensor spatial-temporal information and gives full play on the multi-mode correlations. The decomposition of tensors extracts the underlying factors in each mode of tensor and mines the multi-dimensional structures of geosensor data.

Acknowledgement. This research was supported by the National Natural Science Foundation of China (61762090, 61262069, 61472346, and 61662086), The Natural Science Foundation of Yunnan Province (2016FA026, 2015FB114), the Project of Innovative Research Team of Yunnan Province, and Program for Innovation Research Team (in Science and Technology) in University of Yunnan Province (IRTSTYN).

References

1. Yang, B., Guo, C., Jensen, C.S.: Travel cost inference from sparse, spatio temporally correlated time series using Markov models. PVLDB **6**(9), 769–780 (2013)
2. Yu, R., Cheng, D., Liu, Y.: Accelerated online low rank tensor learning for multivariate spatiotemporal streams. ICML **2015**, 238–247 (2015)
3. Sun, Y., Yuan, N.J., Wang, Y., et al.: Collaborative intent prediction with real-time contextual data. ACM Trans. Inf. Syst. **35**(4), 30 (2017)
4. Pravilovic, S., Appice, A., Malerba, D.: An intelligent technique for forecasting spatially correlated time series. In: Baldoni, M., Baroglio, C., Boella, G., Micalizio, R. (eds.) AI*IA 2013. LNCS (LNAI), vol. 8249, pp. 457–468. Springer, Cham (2013). https://doi.org/10.1007/978-3-319-03524-6_39
5. Pravilovic, S., Appice, A., Malerba, D.: Integrating cluster analysis to the ARIMA model for forecasting geosensor data. In: Andreasen, T., Christiansen, H., Cubero, J.-C., Raś, Zbigniew W. (eds.) ISMIS 2014. LNCS (LNAI), vol. 8502, pp. 234–243. Springer, Cham (2014). https://doi.org/10.1007/978-3-319-08326-1_24
6. Pravilovic, S., Bilancia, M., Appice, A., Malerba, D.: Using multiple time series analysis for geosensor data forecasting. Inf. Sci. **380**(2017), 31–52 (2017)
7. Acar, E., Dunlavy, D.M., Kolda, T.G.: Mϕrup, M.: Scalable tensor factorizations for incomplete data. Chemom. Intell. Lab. Syst. **106**(1), 41–56 (2011)
8. Kolda, T.G., Bader, B.W.: Tensor decompositions and applications. SIAM Rev. **51**(3), 455–500 (2009)
9. Tan, H.C., Feng, G.D., Feng, J.S., Wang, W.H., Zhang, Y.J.: A tensor-based method for missing traffic data completion. Transp. Res. Part C **28**, 15–27 (2013)
10. Egrioglu, E., Yolcu, U., Aladag, C., Bas, E.: Recurrent multiplicative neuron model artificial neural network for non-linear time series forecasting. Procedia – Soc. Behav. Sci. **109**(8), 1094–1100 (2014)
11. Pokrajac, D., Obradovic, Z.: Improved spatial-temporal forecasting through modelling of spatial residuals in recent history. In: SDM, Chicago, IL, USA, 5–7 April 2001, pp. 1–17 (2001)

12. Kamarianakis, Y., Prastacos, P.: Space–time modeling of traffic flow. Comput. Geosci. **31** (2), 119–133 (2005)
13. Ohashi, O., Torgo, L.: Wind speed forecasting using spatio-temporal indicators. In: ECAI, France, 27–31 August 2012, pp. 975–980 (2012)
14. Asteriou D., Hall S.: ARIMA models and the box-jenkins methodology, In: Applied Econometrics, 2nd edn., pp. 265–286. Palgrave MacMillan (2011)
15. Saengseedam, P., Kantanantha, N.: Spatio-temporal model for crop yield forecasting. J. Appl. Stat. **44**(3), 427–440 (2017)
16. Tucker, L.R.: Implications of factor analysis of three-way matrices for measurement of change. In: Harris, C.W. (ed.) Problems in Measuring Change, pp. 122–137. University of Wisconsin Press (1963)
17. Kiers, H.A.: Towards a standardized notation and terminology in multiway analysis. J. Chemom. **14**(3), 105–122 (2000)
18. Hyndman, R.J., Khandakar, Y.: Automatic time series forecasting: the forecast package for R. J. Stat. Softw. **27**(3), 1–22 (2008)

Query Processing

Aggregate k Nearest Neighbor Queries in Metric Spaces

Xin Ding[1], Yuanliang Zhang[1], Lu Chen[2], Keyu Yang[1], and Yunjun Gao[1(✉)]

[1] College of Computer Science, Zhejiang University, Hangzhou, China
{dingxin,yuanlz,kyyang,gaoyj}@zju.edu.cn
[2] Department of Computer Science, Aalborg University, Aalborg, Denmark
luchen@cs.aau.dk

Abstract. Aggregate k nearest neighbor (AkNN) queries are useful in many areas, such as multimedia retrieval and resource allocation, to name but a few. Most of existing works on AkNN query only focus on Euclidean space or specific metric space, which employ properties of particular data to accelerate the query. However, due to the complex data types involved and the needs for flexible similarity criteria seen in real applications, properties of particular data cannot be used for general case. Hence, in this paper, we investigate AkNN search in metric spaces, termed as metric AkNN (MAkNN) search, as metric spaces can support any type of data and flexible similarity criteria as long as satisfying triangle inequality. To efficiently answer MAkNN queries, we develop several pruning techniques and corresponding algorithms based on SPB-tree. Extensive experiments using three real data sets verify the efficiency of our MAkNN algorithms.

Keywords: Metric space · Aggregate k nearest neighbor query
Algorithm

1 Introduction

Aggregate k nearest neighbor (AkNN) retrieval is an interesting type of spatial queries, which finds k objects similar to all the specified query objects using an aggregate similarity criterion. It is useful in a variety of applications, such as resource allocation, recommender systems, etc. Here, we give two examples below.

Resource Allocation. Consider the carpooling, i.e., carpoolers want to take the same taxi to save money. An AkNN query can be utilized to help find candidate taxis for the carpoolers with smallest aggregate distances. Here, with the objective to save time, the aggregate distance summarizes all the distances from the taxi to each carpooler.

Table 1. Symbols and description

Notation	Description
q	A query object
Q or O	The set of objects in metric spaces
P	The set/table of pivots
o or p	An object in O, a pivot in P
$|Q|, |O|, |P|$	The cardinality of Q, O or P
$d()$	The distance function for the generic metric space
$D()$	The L_∞-norm metric for the mapped vector space
$d_{agg}(Q, o)$	The aggregate distance between Q and o in generic metric space
$\phi(o)$	The data point for o in the mapped vector space
$SFC(o)$	The space-filling curve value of an object o
$MAkNN(Q, O, k)$	The result set of an MAkNN query w.r.t. the query set Q and the object set O
$curAND_k$	The current k-th nearest neighbor distance

Recommender Systems. An image recommender system can generate personalized recommendations (i.e., the images that the user may be interested in) based on the images the user already reviewed. Here, the aggregate distance could be the minimum distance between the image to be recommended and the images reviewed.

Considering the wide range of data types in the above application scenarios, e.g., taxis and images, a generic model is desirable that is capable of accommodating not just a single type, but a wide spectrum. In addition, the distance metrics for comparing the similarity of objects, such as road network distance used for taxis and L_p-norm used for images, are not restricted to the Euclidean distance (i.e., L_2-norm). To accommodate a wide range of similarity notions, we investigate AkNN retrieval in metric spaces, termed as metric AkNN (MAkNN) search, where *no* detailed representations of objects are required and where any similarity notion that satisfies the *triangle inequality* can be accommodated.

Most of existing works on AkNN search focus on Euclidean space or particular metric space (e.g., road network, graph), where properties of particular data (e.g., geometric property for Euclidean space) are used to improve the query efficiency. However, these properties cannot be used for the general case, i.e., these approaches cannot answer MAkNN search efficiently. Motivated by this, we develop several pruning lemmas based on the triangle inequality property of metric spaces, and present corresponding algorithms. To sum up, the key contributions of this paper are as follows:

- We develop several pruning lemmas based on SPB-tree for sum, min, and max aggregate functions to accelerate the search.
- We present an efficient algorithm designed for MAkNN search by integrating the designed pruning lemmas.
- We conduct extensive experiments using three real data sets to verify the efficiency of our proposed algorithms, compared with a baseline algorithm extended from the state-of-the art MAkNN framework.

The rest of this paper is organized as follows. Section 2 reviews related work. Section 3 describes the SPB-tree. Section 4 defines MAkNN search and presents corresponding algorithms. Considerable experimental results and findings are reported in Sect. 5. Finally, Sect. 6 concludes the paper with some directions for future work.

2   Related Work

In this section, we survey existing work on metric access methods, and AkNN search algorithms. Table 1 summarizes the notations frequently used throughout this paper.

2.1   Metric Access Methods

Two broad categories of metric access methods (MAMs) exist, namely, compact partitioning methods and pivot-based approaches, to accelerate query processing in metric spaces. Compact partitioning methods partition the space as compact as possible, and try to prune unqualified regions during search. Many indexes, e.g., BST [1], GHT [2], GANT [3], SAT [4], M-tree [5] family, D-Index [6], LC [7], BP [8] exist. Pivot-based methods store pre-computed distances from every object in the database to a set of pivots, and then utilize these distances and the triangle inequality to prune objects during search. Many indexes, e.g., LAESA [9], EP [10], BKT [11], FQT [12], MVPT [13], the Omni-family [14] exist.

Although pivot-based methods clearly outperform compact partitioning approaches in terms of the number of distance computations (i.e., CPU cost) [14–17], they generally have high I/O cost because objects are not well clustered on disk. Recently, hybrid methods that combine compact partitioning with the use of pivots have appeared in the literature. PM-tree [18] uses cut-regions defined by pivots to accelerate query processing on the M-tree. M-Index [19] generalizes the iDistance technique for metric spaces, which compacts the objects by using pre-computed distances to their closest pivots. SPB-tree [20] utilizes the two mapping phase to further improve the efficiency. Hence, in this paper, we use SPB-tree as the underlying index.

2.2   AkNN Search Algorithm

Aggregate k nearest neighbor (AkNN) retrieval generalizes kNN search, which considers multiple query objects. Consequently, the distances from each query object to an object must be aggregated (min, max or sum) according to an optimization goal, in order to offer the similarity measure employed to rank answered objects. Many works [21,22] only focus on AkNN in Euclidean space, where geometric properties are used to accelerate the search. In addition, AkNN in particular metric space (e.g., road network [23], graphs [24], trajectories [25]) are also investigated. However, all these approaches cannot solve our MAkNN search problem, due to the general case we focus on.

Razente et al. [26] study circumscription-constrained aggregate similarity (CCAS) queries in metric spaces, where the region circumscribed by the query objects limits the search space. However, algorithms developed for CCAS queries can not be efficiently extended to solve MAkNN search. This is because, they utilize the circumscription-constrained region to significantly prune search space. Without the circumscription constraint, they have to scan the whole object set to obtain the final query result, which is costly. In addition, Ranzente et al. [27] also develop a framework for MAkNN search that can be adaptive to all kinds of MAMs. Flowing the framework of [27], we develop a baseline algorithm (BL) based on the-state-of-the art MAM SPB-tree.

3   The SPB-tree

In this section, we describe the SPB-tree used as the underlying index.

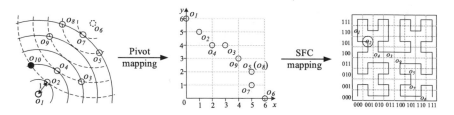

Fig. 1. Pivot mapping and space-filling curve mapping

3.1   Construction Framework

The construction framework of a SPB-tree is based on a two-stage mapping. The first stage maps the objects in a metric space to data points in a vector space using well-chosen pivots. The vector space offers more freedom than the metric space when designing search approaches, since it is possible to utilize the geometric information that is unavailable in the metric space. The second stage uses the space-filling curve (SFC) to map the data points in the vector space into integers in an one-dimensional space. Finally, a B$^+$-tree with MBB information is employed to index the resulting integers.

Pivot Mapping. Given a pivot set $P = \{p_1, p_2, \ldots, p_n\}$, a metric space (M, d) can be mapped to a vector space (R^n, L_∞). Specifically, an object o in the metric space is represented as a point $\phi(o) = \langle d(o, p_1), d(o, p_2), \ldots, d(o, p_n) \rangle$ in the vector space. For instance, consider the example in Fig. 1, where $O = \{o_1, o_2, \ldots, o_9\}$ and L_2-*norm* is used. If $P = \{o_1, o_6\}$, O can be mapped to a two-dimensional vector space, in which the x-axis denotes $d(o_i, o_1)$ and the y-axis represents $d(o_i, o_6)$, $1 \leq i \leq 9$.

Given objects o_i, o_j, and p in a metric space, $d(o_i, o_j) \geq |d(o_i, p) - d(o_j, p)|$ according to the triangle inequality. Hence, for a pivot set P, $d(o_i, o_j) \geq \max\{|d(o_i, p_i) - d(o_j, p_i)| \mid p_i \in P\} = D(\phi(o_i), \phi(o_j))$, in which $D(\ )$ is the L_∞-norm. Consequently, we can conclude that the distance in the mapped vector space is a *lower bound* on that in the metric space. For example, in Fig. 1, $d(o_2, o_3) > D(\phi(o_2), \phi(o_3)) = 2$.

Space-Filling Curve Mapping. Given a vector $\phi(o)$ after pivot mapping and assume that the range of $d(\ )$ in the metric space is *discrete* integers (e.g., edit distance), SFC can directly map $\phi(o)$ to an integer $SFC(\phi(o))$. Consider the SFC mapping examples in Fig. 1, where SFC value $SFC(\phi(o_2)) = 18$ for the Hilbert curve. As a default, we use the Hilbert curve for SPB-tree. If the range of $d(\ )$ in the metric space is continuous real numbers, we can partition the range of $d(\ )$ into discrete integers.

3.2 Indexing Structure

An SPB-tree used to index an object set in a generic metric space contains three parts, i.e., the pivot table, the B$^+$-tree, and the random access file (RAF). Figure 2 shows an SPB-tree example to index the object set $O = \{o_1, \ldots, o_9\}$ in Fig. 1. A pivot table stores selected objects (e.g., o_1 and o_6) to map a metric space into a vector space.

A B$^+$-tree is employed to index the SFC values of objects after a pivot mapping. Each leaf entry in the leaf node (e.g., N_3, N_4, N_5, and N_6) of the B$^+$-tree records (1) the SFC value *key*, and (2) the pointer *ptr* to a real object, which is the address of the actual object kept in the RAF. For example, in Fig. 2, the leaf entry E_7 associated with the object o_2 records the Hilbert value 18 and the storage address 0 of o_2. Each non-leaf entry in the root or intermediate node (e.g., N_0, N_1, and N_2) of the B$^+$-tree records (1) the minimum SFC value *key* in its subtree, (2) the pointer *ptr* to the root node of its subtree, and (3) the SFC values *min* and *max* for $\langle L_1, L_2, \ldots, L_{|P|}\rangle$ and $\langle U_1, U_2, \ldots, U_{|P|}\rangle$, to represent the MBB $M(= \{[L_i, U_i] | i \in [1, |P|]\})$ of the root node N of its subtree. Specifically, an MBB M denotes the axis aligned *minimum bounding box* to contain all $\phi(o)$ with $SFC(\phi(o)) \in N$, and thus, L_i and U_i represent the minimum and maximum

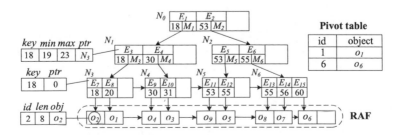

Fig. 2. Example of an SPB-tree

values of $\phi(o)$ on dimension i. For instance, the non-leaf entry E_3 uses min ($=$ $SFC(\langle 0, 5 \rangle) = 19$) and max ($= SFC(\langle 1, 6 \rangle) = 23$) to represent the M_3 ($= \{[0, 1]$, $[5, 6]\}$) of N_3.

RAF is sorted to store the objects in ascending order of SFC values as they appear in the B$^+$-tree. Each RAF entry records (1) an object identifier id, (2) the length len of the object, and (3) the real object obj. In Fig. 2, the RAF entry associated with an object o_2 records the object identifier 2, the object length 8, and the real object o_2.

4   Metric Aggregate k Nearest Neighbor Search

In this section, we first formalize AkNN retrieval in metric spaces, and then propose an efficient algorithm for processing metric AkNN queries based on the SPB-tree.

4.1   Problem Definition

A metric space is a tuple (M, d), in which M is the domain of objects and d is a distance function which defines the similarity between the objects in M. In particular, the distance function d has four properties: (1) *symmetry*: $d(q, o) = d(o, q)$, (2) *non-negativity*: $d(q, o) \geq 0$, (3) *identity*: $d(q, o) = 0$ iff $q = o$, and (4) *triangle inequality*: $d(q, o) \leq d(q, p) + d(p, o)$. Based on the properties of the metric space, AkNN queries in metric spaces have been investigated.

Definition 1. *(MAkNN Query). Given a query object set Q, an object set O, and an integer k , an MAkNN query finds k objects in O with the smallest aggregate distances $d_{agg}(Q, o)$, i.e., MAkNN$(Q$, O, $k) = \{o_i | o_i \in O \land 1 \leq i \leq k \land \forall o_j (\neq o_i) \in O, d_{agg}(Q, o_j) \geq d_{agg}(Q, o_i)\}$. In particular, $d_{agg}(Q, o)$ can be computed as $f(d(q_1, o)$, $d(q_2, o)$, $\ldots, d(q_{|Q|}, o))$, in which the aggregate function f might be sum, min, or max.*

Consider two English word sets $Q = \{$ "defoliate", "defoliates"$\}$ and $O = \{$ "citrate", "defoliation", "defoliating", "defoliated"$\}$, for which the edit distance is the similarity measurement. Suppose $k = 2$, an MAkNN query $MAkNN$ $(Q, O, 2)$ finds the two words in O having the smallest aggregate distances from Q. If f is sum function, the query result is $\{$ "defoliated", "defoliation"$\}$; if f is min function, the query result is $\{$ "defoliated", "defoliation"$\}$; and if f is max function, the query result is $\{$ "defoliated", "defoliating"$\}$. It is worth noting that $MAkNN(Q, O, k)$ may be not unique due to the distance tie. Nonetheless, the target of our presented algorithms is to find one possible instance.

4.2   MAkNN Query Processing

MAkNN search generalizes the form of MkNN queries, in which there are multiple (instead of one) query objects. Consider a running example of MAkNN

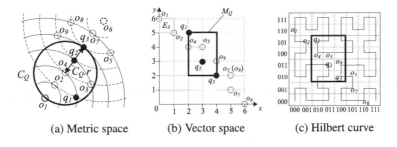

(a) Metric space   (b) Vector space   (c) Hilbert curve

Fig. 3. Illustration of $MAkNN(Q, O, k)$

retrieval depicted in Fig. 3, where $Q = \{q_i | 1 \leq i \leq 3\}$ and $O = \{o_j | 1 \leq j \leq 9\}$. Assume that $k = 2$ and L_2-norm is utilized, the result of $MAkNN(Q, O, 2)$ is $\{o_4, o_3\}$ if sum function is used to compute the aggregate distance; the query result is $\{o_4, o_7\}$ if min function is used; and the query result is $\{o_4, o_3\}$ if max function is used. To solve MAkNN search, a simple method BL is to use SPB-tree and follow the framework [27] developed for MAkNN retrieval. In particular, BL traverses the B$^+$-tree entries in ascending order of their minimum aggregate distances to Q in the mapped vector space. As discussed in Sect. 3, the distance in the mapped vector space is the lower bound distance of the original metric space, we develop Lemma 1 below for MAkNN search, to avoid unnecessary verifications of B$^+$-tree entries.

Lemma 1. *Given a query set Q and a B^+-tree entry E, E can be safely pruned if $MIND_{agg}(Q, E) \geq curAND_k$, where $MIND_{agg}(Q, E)$ denotes the minimum aggregate distance between E and Q in the mapped vector space, and $curAND_k$ represents the current k-th aggregate NN distance from Q.*

Proof. Since the aggregate function is monotonically increasing, the aggregate distance in the mapped vector space is still the lower bound distance of that in the original metric space. Then, we can get that $mind_{agg}(E, Q) \geq MIND_{agg}(E, Q)$, with $mind_{agg}(E, Q)$ denoting the minimum aggregate distance between E and Q in the original metric space. If $MIND_{agg}(E, Q) \geq curAND_k$, then for each $o \ (\in E)$, $d_{agg}(o, Q) \geq mind_{agg}(E, Q) \geq curAND_k$. Consequently, E can be discarded safely. $\square$

Note that, $curAND_k$ used in Lemma 1 is obtained and updated during MAkNN search. In particular, after computing the aggregate distance of an object, we can update immediately the result set and $curAND_k$ if necessary. Consider the example depicted in Fig. 3 with the corresponding SPB-tree in Fig. 2. Assume that $curAND_k = 1$ and min function is used, E_3 and E_6 can be safely pruned as $MIND_{agg}(E_3, M_Q) = MIND_{agg}(E_6, M_Q) = curAND_k$.

Since $MIND_{agg}(E, Q)$ is computed as $f(MIND(E, q_1), MIND(E, q_2), \ldots, MIND(E, q_{|Q|}))$, it is costly (because it needs $|Q|$ computations of $MIND$).

Motivated by this, we build MBB M_Q $(= \{[L_{Qi}, U_{Qi}] \mid 1 \leq i \leq |P|\})$ for Q in the mapped vector space, to reduce $MIND_{agg}(E, Q)$ computation cost. Back to the running example illustrated in Fig. 3, the thick black rectangle in Fig. 3(b) represents MBB M_Q $(= \{[2, 4], [2, 5]\})$ for Q in the mapped vector space using $P = \{o_1, o_6\}$. Let $MIND_i(E, q_t)$ be the minimum distance between E and q_t $(\in Q)$ on dimension i $(1 \leq i \leq |P|)$, and $MIND_i(E, Q)$ be the minimum aggregate distance between E and Q on dimension i.

$$
\begin{aligned}
MIND_{agg}&(E, Q) \\
&= f(MIND(E, q_1), \ldots, MIND(E, q_{|Q|})) \\
&= f(max\{MIND_i(E, q_1)|1 \leq i \leq |P|\}, \\
&\quad \ldots, max\{MIND_i(E, q_{|Q|})|1 \leq i \leq |P|\}) \\
&\geq max\{f(MIND_i(E, q_1), \ldots, \\
&\quad MIND_i(E, q_{|Q|}))|1 \leq i \leq |P|\} \\
&= max\{MIND_i(E, Q)|1 \leq i \leq |P|\}
\end{aligned}
\tag{1}
$$

According to Eq. (1), the lower bound distance of $MIND_{agg}(E, Q)$, termed as $EMIND_{agg}(E, Q)$, can be computed as $max\{MIND_i(E,Q)|1 \leq i \leq |P|\}$. To obtain $EMIND_{agg}(E, Q)$, we only need to compute $MIND_i(E, Q)$ on each dimension i, with the detailed computations stated below for sum, min, or max function, respectively.

Sum function. If $L_{Ei} \geq U_{Qi}$ (as shown in Fig. 4(a)), $MIND_i(E, Q) = \sum_{1 \leq t \leq |Q|} MIND_i(E, q_t) = \sum_{1 \leq t \leq |Q|} (L_{Ei} - d(q_t, p_i)) = |Q| \times L_{Ei} - \sum_{1 \leq t \leq |Q|} d(q_t, p_i)$. If $U_{Ei} \leq L_{Qi}$ (as depicted in Fig. 4(b)), then $MIND_i(E, Q) = \sum_{1 \leq t \leq |Q|} MIND_i(E, q_t) = \sum_{1 \leq t \leq |Q|} (d(q_t, p_i) - U_{Ei}) = \sum_{1 \leq t \leq |Q|} d(q_t, p_i) - |Q| \times U_{Ei}$. Otherwise, i.e., M_Q and M_E are intersected on dimension i, $MIND_i(E, Q)$ is estimated as 0.

Note that, $\sum_{1 \leq t \leq |Q|} d(q_t, p_i)$ used in $MIND_i(E, Q)$ computation for sum function can be obtained and stored for reuse when building M_Q. Hence, for sum function, the computational cost of $EMIND_{agg}(E, Q)$ is O(1), which is much smaller than O($|Q|$) of $MIND_{agg}(E, Q)$ computation. For example, in Fig. 3, and assume that sum function is used on dimension x, as $U_{E_3x} < L_{Qx}$, $MIND_x$ $(E_3, Q) = d(q_1, o_1) + d(q_2, o_1) + d(q_3, o_1) - 3 \times U_{E_3x} = 6$. Thus, we can get that $EMIND_{agg}(E_3, Q) = max\{MIND_x$ $(E_3, Q), MIND_y(E_3, Q)\} = 6$, which is a tight lower bound of $MIND(E_3, Q)$ $(= 6)$.

Min function. If $L_{Ei} \geq U_{Qi}$ (as shown in Fig. 4(a)), then $MIND_i(E, Q) = min_{1 \leq t \leq |Q|} MIND_i(E, q_t) = L_{Ei} - U_{Qi}$. If $U_{Ei} \leq L_{Qi}$ (as depicted in Fig. 4(b)), then $MIND_i(E, Q) = min_{1 \leq t \leq |Q|} MIND_i(E, q_t) = L_{Qi} - U_{Ei}$. Otherwise, i.e., M_Q and M_E are crossed on dimension i, $MIND_i(E, Q)$ is estimated as 0.

Similarity, the $EMIND_{agg}(E, Q)$ computational cost is also reduced to O(1) for min function. Back to the example shown in Fig. 3 and suppose that min function is used, since $U_{E_3x} < L_{Qx}$ on dimension x, $MIND_x(E_3, Q) = L_{Qx} - $

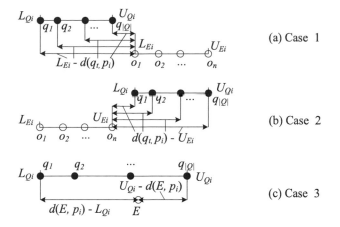

Fig. 4. $MIND_i(Q, E)$ computation

$U_{E3x} = 1$. Hence, we can get that $EMIND_{agg}(E_3, Q) = \max\{MIND_x(E_3, Q), MIND_y(E_3, Q)\} = 1$, which is a tight lower bound of $MIND(E_3, Q) (= 1)$.

Max function. If $L_{Ei} \geq U_{Qi}$ (as shown in Fig. 4(a)), then $MIND_i(E, Q) = \max_{1 \leq t \leq |Q|} MIND_i(E, q_t) = L_{Ei} - L_{Qi}$. If $U_{Ei} \leq L_{Qi}$ (as depicted in Fig. 4(b)), then $MIND_i(E, Q) = \max_{1 \leq t \leq |Q|} MIND_i(E, q_t) = U_{Qi} - U_{Ei}$. Otherwise, i.e., M_Q and M_E are intersected on dimension i, $MIND_i(E, Q)$ is estimated as 0. Note that, for the case when M_E is intersected with M_Q, if E is a leaf entry (as illustrated in Fig. 4(c)), then $MIND_i(E, Q) = \max\{d(E, p_i) - L_{Qi}, U_{Qi} - d(E, p_i)\}$.

For max function, the $EMIND_{agg}(E, Q)$ computational cost is also reduced to O(1). Back to the example depicted in Fig. 3 and assume that max function is used, on dimension x, as $U_{E3x} < L_{Qx}$, $MIND_x(E_3, Q) = U_{Qx} - U_{E3x} = 3$; on dimension y, $MIND_y(E_3, Q) = d(E_{10}, o_6) - L_{Qy} = 2$. Thus, we can get that $EMIND_{agg}(E_3, Q) = \max\{MIND_x(E_3, Q), MIND_y(E_3, Q)\} = 3$, which is a tight lower bound of $MIND(E_3, Q) (= 3)$. For object o_3, on dimension y, since $L_{Qy} < d(o_3, o_6) < U_{Qy}$, $MIND_y(o_3, Q) = \max\{d(o_3, o_6) - L_{Qi}, U_{Qi} - d(o_3, o_6)\} = 2$. therefore, we can get that $EMIND_{agg}(o_3, Q) = \max\{MIND_x(o_3, Q)), MIND_y(o_3, Q)\} = 2$, which is also a tight lower bound of $MIND(o_3, Q) (= 2)$.

Based on $EMIND_{agg}(E, Q)$ derived, we develop a lemma to avoid unnecessary $MIND_{agg}(Q, E)$ computations.

Lemma 2. *Given a query set Q and a B^+-tree entry E, E can be safely pruned if $EMIND_{agg}(Q, E) \geq curAND_k$.*

Proof. Since $EMIND_{agg}(Q, E) \leq MIND_{agg}(Q, E)$, if $EMIND_{agg}(Q, E) \geq curAND_k$, then $MIND_{agg}(Q, E) \geq curAND_k$. Hence, E can be safely pruned due to Lemma 1, which completes the proof. □

Consider the example depicted in Fig. 3 with the corresponding SPB-tree in Fig. 2. Assume that *sum* function is used and $curAND_k = 5$, E_3 can be safely discarded due to $EMIND_{agg}(E_3, M_Q) > curAND_k$.

Lemma 2 utilizes MBB to reduce the computational cost in the mapped vector space. In order to further reduce the computational cost of the aggregate distance $d_{agg}(Q, o)$ between the object o and the query set Q, we can also build a minimum bounding circle (MBC) for Q in original metric space. The MBC C_Q is centered at $C_Q.o$ with the radius $C_Q.r$ equaling to the maximum distance $d(q, C_Q.o)$ $(q \in Q)$. Consider the example illustrated in Fig. 3(a), the thick black circle, centered at object o_4 with the radius $C_Q.r = d(o_4, q_3)$, denotes the MBC for Q. With the assistant of MBC, we can get the lower bound $ed_{agg}(Q, o)$ of $d_{agg}(Q, o)$, with the detailed derivation stated as follows for *sum*, *min*, and *max* function, respectively.

***Sum* function.** According to the triangle inequality,

$$
\begin{aligned}
d_{agg}(Q, o) &= \sum_{1 \leq t \leq |Q|} d(o, q_t) \\
&\geq \sum_{1 \leq t \leq |Q|} |d(o, C_Q.o) - d(C_Q.o, q_t)| \\
&\geq \left| \sum_{1 \leq t \leq |Q|} (d(o, C_Q.o) - d(C_Q.o, q_t)) \right| \\
&= \left| d(o, C_Q.o) \times |Q| - \sum_{1 \leq t \leq |Q|} d(C_Q.o, q_t) \right|
\end{aligned}
\tag{2}
$$

Hence, $ed_{agg}(Q, o)$ can be computed as $|d(o, C_Q.o) \times |Q| - \sum_{1 \leq t \leq |Q|} d(C_Q.o, q_t)|$ for *sum* function. Note that, $\sum_{1 \leq t \leq |Q|} d(C_Q.o, q_t)$ can be computed and stored for reuse when building MBC C_Q. For example, in Fig. 3(a), suppose that *sum* function is used, $ed_{agg}(Q, o_6) = 3 \times d(o_6, o_4) - \sum_{1 \leq t \leq 3} d(o_4, q_t) = 9 - d(o_4, q_1) = 7.2$, which is a lower bound value of $d_{agg}(Q, o_6) (= 5 + d(q_1, o_6) = 10)$.

***Min* function.** Based on the triangle inequality,

$$
\begin{aligned}
d_{agg}(Q, o) &= \min_{1 \leq t \leq |Q|} d(o, q_t) \\
&\geq \min_{1 \leq t \leq |Q|} |d(o, C_Q.o) - d(C_Q.o, q_t)| \\
&\geq \min_{1 \leq t \leq |Q|} (d(o, C_Q.o) - d(C_Q.o, q_t)) \\
&= d(o, C_Q.o) - \max_{1 \leq t \leq |Q|} d(C_Q.o, q_t) \\
&= d(o, C_Q.o) - C_Q.r
\end{aligned}
\tag{3}
$$

Thus, $ed_{agg}(Q, o)$ can be computed as $d(o, C_Q.o) - C_Q.r$ for max function. Back to the example shown in Fig. 3(a), and assume that min function is used, $ed_{agg}(Q, o_6) = d(o_6, o_4) - C_Q.r = 2$, which is a tight lower bound of $d_{agg}(Q, o_6)$ $(= 2)$.

Max function. According to the triangle inequality,

$$
\begin{aligned}
d_{agg}(Q, o) &= \max_{1 \leq t \leq |Q|} d(o, q_t) \\
&\geq \max_{1 \leq t \leq |Q|} |d(o, C_Q.o) - d(C_Q.o, q_t)| \\
&= max\{ \max_{1 \leq t \leq |Q|} d(C_Q.o, q_t) - d(o, C_Q.o), \\
&\quad d(o, C_Q.o) - \min_{1 \leq t \leq |Q|} d(C_Q.o, q_t)\} \\
&= max\{C_Q.r - d(o, C_Q.o), d(o, C_Q.o) - \\
&\quad min\{d(C_Q.o, q_t)|1 \leq t \leq |Q|\}\}
\end{aligned} \tag{4}
$$

Therefore, $ed_{agg}(Q, o)$ can be computed as $max\{d(o, C_Q.o) - C_Q.r, d(o, C_Q.o) - min\{d(C_Q.o, q_t)|1 \leq t \leq |Q|\}\}$ for min function. Note that, $min\{d(C_Q.o, q_t)|1 \leq t \leq |Q|\}$ can be computed and stored for reuse when building C_Q. Back to the example depicted in Fig. 3(a), and suppose that max function is used, $ed_{agg}(Q, o_6) = d(o_6, o_4) - d(o_4, q_2) = 3$, which is a lower bound value of $d_{agg}(Q, o_6)(= d(q_1, o_6) = 5)$.

According to Eqs. 2–(4), it only needs one distance computation for $ed_{agg}(Q, o)$ calculation, instead of $|Q|$ distance computations for $d_{agg}(Q, o)$ calculation, which reduces significantly the computational cost. Thus, we develop a new lemma based on $ed_{agg}(Q, o)$ derived, to avoid unnecessary computations of $d_{agg}(Q, o)$.

Lemma 3. *Given a query set Q and an object o, o can be safely pruned if $ed_{agg}(Q, o) \geq curAND_k$.*

Proof. As $ed_{agg}(Q, o) \leq d_{agg}(Q, o)$, $d_{agg}(Q, o) \geq curAND_k$ if $ed_{agg}(Q, o) \geq curAND_k$. Hence, o can be safely pruned due to the definition of the aggregate kNN query, which completes the proof. □

Consider the example shown in Fig. 3 with the corresponding SPB-tree in Fig. 2. Assume that max function is used and $curAND_k = 5$, object o_6 can be safely discarded due to $EMIND_{agg}(o_6, M_Q) > curAND_k$, without any further verification.

Algorithm 1 Aggregate kNN Algorithm (AkNNA)

Input: a query set Q, an integer k, an object set O indexed by a SPB-tree
Output: the result set $MAkNN(Q, O, k)$ of an aggregate kNN query
1: $curAND_k = \infty$, $C_Q = H = \varnothing$ // H stores the intermediate and leaf entries of
$\qquad\qquad\qquad\qquad\qquad$ B$^+$-tree in ascending order of $MIND_{agg}(Q, E)$
2: compute $\phi(q)$ for each object q in Q using P and obtain M_Q
3: push the root entries of B$^+$-tree onto H
4: **while** $H \neq \varnothing$ **do**
5: $\quad$ de-heap the top entry E from H
6: $\quad$ **if** $MIND_{agg}(M_Q, E) \geq curAND_k$ **then** break // Lemma 1
7: $\quad$ **if** E is a non-leaf entry **then**
8: $\quad\quad$ **for** each sub entry $e \in E$ **do**
9: $\quad\quad\quad$ **if** $EMIND_{agg}(M_Q, e) < curAND_k$ **then** // Lemma 2
10: $\quad\quad\quad\quad$ **if** $MIND_{agg}(M_Q, e) < curAND_k$ **then** // Lemma 1
11: $\quad\quad\quad\quad\quad$ push e onto H
12: $\quad$ **else** // E is a leaf entry
13: $\quad\quad$ **if** $C_Q \neq \varnothing$ **then**
14: $\quad\quad\quad$ **if** $ed_{agg}(Q, e.ptr) \geq curAND_k$ **then** continue // Lemma 3
15: $\quad\quad$ **if** $d_{agg}(Q, e.ptr) < curAND_k$ **then**
16: $\quad\quad\quad$ insert $e.ptr$ into $MAkNN(Q, O, k)$
17: $\quad\quad\quad$ update $curAND_k$ and C_Q if necessary
18: **return** $MAkNN(Q, O, k)$

To achieve the strongest pruning power of Lemma 3, i.e., the lower bound $ed_{agg}(Q, o)$ must approach to $d_{agg}(Q, o)$ as much as possible, we need to tight the MBC. In other words, we need to choose an MBC center to obtain the minimal MBC radius. A simple way to obtain the optimal center is to perform an $MAkNN(Q, O, 1)$ query using max function. However, it is costly to perform an additional aggregate NN query. Therefore, we can update the center of MBC using the object o ($\in O$) during MAkNN search when verifying whether o is contained in the final result.

Based on Lemmas 1 to 3, we present an efficient *Aggregate kNN Algorithm* (AkNNA), with the pseudo-code depicted in Algorithm 1. To begin with, AkNNA sets $curAND_k$ to infinity, and initializes the MBC C_Q and min-heap H to empty. Then, it computes $\phi(q)$ for each $q \in Q$ using P, and obtains the MBB M_Q in the mapped vector space. Next, the algorithm pushes the root entries of a B$^+$-tree into H. In the sequel, a while-loop is performed until H is empty (lines 4–17). In every while-loop, AkNNA de-heaps the top entry E from H, and stops searching if $MIND_{agg}(Q, E)$ is no smaller than $curAND_k$ by Lemma 1. If E is a non-leaf entry, the algorithm pushes all the qualified sub entries of E into H according to Lemmas 1 and 2 (lines 8–11). Otherwise (i.e., E is a leaf entry), if C_Q exists, AkNNA computes $ed_{agg}(Q, e.ptr)$ and prunes object $e.ptr$ without any further verification using Lemma 3 (lines 13–14). Thereafter, if $d_{agg}(Q, e.ptr)$ is smaller than $curAND_k$, the algorithm inserts $e.ptr$ into the result set $MAkNN(Q, O, k)$ (line 16), and updates $curAND_k$ and C_Q if necessary (line 17). In the end, the final query result set $MAkNN(Q, O, k)$ is returned.

Example 1. We illustrate AkNNA using the example depicted in Fig. 3 with the corresponding SPB-trees shown in Fig. 2. Assume that $k = 2$ and *sum* function is utilized. First of all, $curAND_k$ is initialized to infinity, and C_Q and the min-heap H are set to empty. Then, AkNNA computes $\phi(q_1) = \langle 2, 5 \rangle$, $\phi(q_2) = \langle 3, 3 \rangle$, and $\phi(q_3) = \langle 4, 2 \rangle$ using P, obtains MBB $M_Q = \{[2, 4], [2, 5]\}$, and pushes the root entries into H $(= \{E_1, E_2\})$. Next, it performs a while-loop. In the first loop, AkNNA pops the top entry E_1 from H. Since E_1 is a non-leaf entry, the algorithm pushes its qualified sub entries E_3 and E_4 into H $(= \{E_4, E_2, E_3\})$, due to $EMIND_{agg}$ and $MIND_{agg}$ of E_3 and E_4 from Q are smaller than $curAND_k$. Similarly, in the second loop, AkNNA pops E_4 and pushes the qualified sub leaf entries into $H(=\{E_9, E_{10}, E_2, E_3\})$. Then, A$k$NNA pops the leaf entry E_9 and inserts o_4 into $MAkNN(Q, O, 2)$ as $d_{agg}(o_4, Q) < curAND_k$. After that, $C_Q.o$ and $C_Q.r$ are set as o_4 and 2, respectively. In the sequel, it pops and evaluates entries in H similarly until $MIND_{agg}(E_3, Q) > curAND_k$, after which $MAkNN(Q, O, 2) = \{o_4, o_3\}$. Finally, A$k$NNA stops and returns $MAkNN(Q, O, 2)$ as the final result set. $\qquad\square$

5   Performance Study

In this section, we experimentally evaluate the performance of MAkNN retrieval algorithms based on the SPB-tree. We implemented the algorithms in C++. All experiments were conducted on an Intel Core 2 Duo 2.93 GHz PC with 3 GB RAM.

5.1   Experimental Setup

We employ three real datasets, namely, *Words*, *Color*, and *DNA*, as depicted in Table 2. *Words*[1] contains proper nouns, acronyms, and compound words taken from the Moby Project, and the edit distance is used to compute the distance between two words. *Color*[2] denotes the color histograms extracted from an image database, and L_5-norm is utilized to compare the color image features. *DNA*[3] consists of 1 million DNA data, and the cosine similarity is used to measure its similarity under the tri-gram counting space.

We investigate the efficiency of MAkNN retrieval algorithms under various parameters, which are listed in Table 3. Note that, in every experiment, only one factor varies, whereas the others are fixed to their default values. The main

Table 2. Statistics of the datasets used

Dataset	Cardinality	Dim.	Ins. Dim.	Measurement
Words	611,756	1–34	4.9	*Edit distance*
Color	112,682	16	2.9	L_5-*norm*
DNA	1,000,000	108	6.9	*Cosine similarity under tri-gram counting space*

---

[1] *Words* is available at http://icon.shef.ac.uk/Moby/.
[2] *Color* is available at http://www.sisap.org/Metric_Space_Library.html.
[3] *DNA* is available at http://www.ncbi.nlm.nih.gov/genome.

Table 3. Parameter ranges and default values

Parameter	Setting	Default
k	1, 2, 4, 8, 16, 32	8
query set cardinality $|Q|$	4, 16, 64, 256, 1024	64
query set area A_Q of the whole space	2%, 4%, 8%, 16%, 32%	8%

performance metrics include the number of page accesses (PA), the number of distance computations (*compdists*), and the CPU time. Each measurement we report is the average of 500 queries.

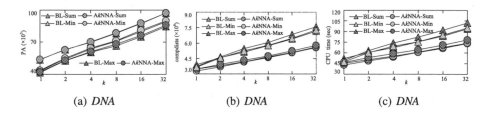

(a) *DNA*   (b) *DNA*   (c) *DNA*

Fig. 5. AkNN query performance vs. k

5.2   Results on AkNN Queries

We verify the performance of our proposed algorithms (i.e., BL and AkNNA) in answering MAkNN queries in metric spaces. BL is a baseline method directly extended from MkNN framework [27] using SPB-tree. We inspect the influence of various parameters, containing (1) the area of query set A_Q, (2) the cardinality of query set $|Q|$, and (3) the value of k, i.e., the number of aggregate NNs required.

Figures 5, 6, and 7 show the experimental results w.r.t. k, A_Q, and $|Q|$, respectively. The first observation is that, AkNNA achieves better performance in terms of the number of distance computations and the CPU time, but has similar number of page accesses as BL. This is because, AkNNA employs Lemmas 2 and 3 to save the distance computational cost and avoid unnecessary distance computations, while BL only uses Lemma 1. However, the I/O cost of MAkNN search is related with the search region. In other words, the I/O cost is mostly related with the distribution of the query set and the dataset, which can hardly be reduced by Lemmas 2 and 3. Thus, BL and AkNNA have similar I/O cost. The second observation is that, the query cost increases with A_Q and k, due to the growth of search space. Note that, the query cost of AkNNA, including the number of distance computations and the CPU time, approaches to that of BL as A_Q grows. The reason is that, with the growth of A_Q, the minimum bounding box and minimum bounding circle for the query set becomes larger, and thus, the pruning power of Lemmas 2 and 3 decreases. In addition, the number of distance computations and the CPU time increase with $|Q|$. This is because, the

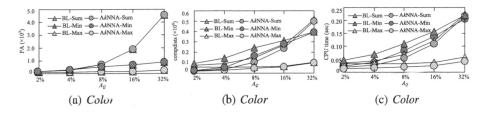

Fig. 6. AkNN query performance vs. query set area A_Q

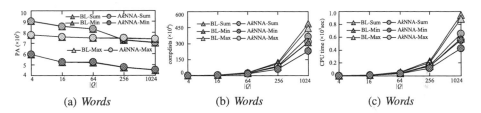

Fig. 7. AkNN query performance vs. query set cardinality $|Q|$

aggregate distance computation needs more distance computations and becomes more costly as the number of query objects $|Q|$ ascends. Nevertheless, the I/O cost drops as $|Q|$ grows, since the search region decreases due to the dropping k-the aggregate NN distance (AND_k) value for min and max functions, and $AND_k/|Q|$ value for sum function.

6   Conclusions

Metric aggregation k nearest neighbor (MAkNN) search is useful in many areas of computer science, such as multimedia retrieval, resource allocation, and so forth, because it can support various data types and flexible similarity measurements as long as the measurements satisfy the triangle inequality. To answer MAkNN efficiently, we develop several pruning lemmas that utilizes the triangle inequality and present efficient algorithms based on SPB-tree. Extensive experiments show that, our MAkNN search algorithm is more efficient than the baseline algorithm extended from the state-of-the art MAkNN search framework. In the future, we plan to extend the MAkNN search algorithms to various distributed environments.

Acknowledgments. This work was supported in part by the 973 Program of China under Grant No. 2015CB352502, the NSFC under Grant No. 61522208, the NSFC-Zhejiang Joint Fund under Grant No. U1609217, and the ZJU-Hikvision Joint Project.

References

1. Kalantari, I., McDonald, G.: A data structure and an algorithm for the nearest point problem. IEEE Trans. Softw. Eng. **9**(5), 631–634 (1983)
2. Uhlmann, J.K.: Satisfying general proximity/similarity queries with metric trees. Inf. Process. Lett. **40**(4), 175–179 (1991)
3. Brin, S.: Near neighbor search in large metric spaces. In: VLDB, pp. 574–584 (1995)
4. Navarro, G.: Searching in metric spaces by spatial approximation. VLDB J. **11**(1), 28–46 (2002)
5. Ciaccia, P., Patella, M., Zezula, P.: M-tree: an efficient access method for similarity search in metric spaces. In: VLDB, pp. 426–435 (1997)
6. Dohnal, V., Gennaro, C., Savino, P., Zezula, P.: D-index: distance searching index for metric data sets. Multimed. Tools Appl. **21**(1), 9–33 (2003)
7. Chavez, E., Navarro, G.: A compact space decomposition for effective metric indexing. Pattern Recogn. Lett. **26**(9), 1363–1376 (2005)
8. Almeida, J., Torres, R.D.S., Leite, N.J.: BP-tree: an efficient index for similarity search in high-dimensional metric spaces. In: CIKM, pp. 1365–1368 (2010)
9. Mico, L., Oncina, J., Carrasco, R.C.: A fast branch & bound nearest neighbour classifier in metric spaces. Pattern Recogn. Lett. **17**(7), 731–739 (1996)
10. Ruiz, G., Santoyo, F., Chavez, E., Figueroa, K., Tellez, E.S.: Extreme pivots for faster metric indexes. In: SISAP, pp. 115–126 (2013)
11. Burkhard, W., Keller, R.: Some approaches to best-match file searching. Commun. ACM **16**(4), 230–236 (1973)
12. Baeza-Yates, R.A., Cunto, W., Manber, U., Wu, S.: Proximity matching using fixed-queries trees. In: CPM, pp. 198–212 (1994)
13. Bozkaya, T., Ozsoyoglu, M.: Distance-based indexing for high-dimensional metric spaces. In: SIGMOD, pp. 357–368 (1997)
14. Traina Jr., C., Filho, R.F.S., Traina, A.J.M., Vieira, M.R., Faloutsos, C.: The Omni-family of all-purpose access methods: asimple and effective way to make similarity search more efficient. VLDB J. **16**(4), 483–505 (2007)
15. Ares, L.G., Brisaboa, N.R., Esteller, M.F., Pedreira, O., Places, A.S.: Optimal pivots to minimize the index size for metric access methods. In: SISAP, pp. 74–80 (2009)
16. Chavez, E., Navarro, G., Baeza-Yates, R.A., Marroquin, J.L.: Searching in metric spaces. ACM Comput. Surv. **33**, 273–321 (2001)
17. Mosko, J., Lokoc, J., Skopal, T.: Clustered pivot tables for I/O-optimized similarity search. In: SISAP, pp. 17–24 (2011)
18. Skopal, T., Pokorny, J., Snasel, V.: PM-tree: pivoting metric tree for similarity search in multimedia databases. In: ADBIS, pp. 803–815 (2004)
19. Novak, D., Batko, M., Zezula, P.: Metric index: an efficient and scalable solution for precise and approximate similarity search. Inf. Syst. **36**(4), 721–733 (2011)
20. Chen, L., Gao, Y., Li, X., Jensen, C.S., Chen, G.: Efficient metric indexing for similarity search. In: ICDE (2015, to appear)
21. Papadias, D., Tao, Y., Mouratidis, K., Hui, C.K.: Aggregate nearest neighbor queries in spatial databases. ACM Trans. Database Syst. (TODS) **30**(2), 529–576 (2005)
22. Li, F., Yi, K., Tao, Y., Yao, B., Li, Y., Xie, D., Wang, M.: Exact and approximate flexible aggregate similarity search. VLDB J. **25**(3), 317–338 (2016)

23. Wang, H., Zheng, K., Su, H., Wang, J., Sadiq, S., Zhou, X.: Efficient aggregate farthest neighbour query processing on road networks. In: Wang, H., Sharaf, M.A. (eds.) ADC 2014. LNCS, vol. 8506, pp. 13–25. Springer, Cham (2014). https://doi.org/10.1007/978-3-319-08608-8_2

24. Liu, Z., Wang, C., Wang, J.: Aggregate nearest neighbor queries in uncertain graphs. World Wide Web **17**(1), 161–188 (2014)

25. Abbasifard, M.R., Naderi, H., Fallahnejad, Z., Alamdari, O.I.: Approximate aggregate nearest neighbor search on moving objects trajectories. J. Central South Univ. **22**(11), 4246–4253 (2015)

26. Razente, H.L., Barioni, M.C.N., Traina, A.J.M., Traina Jr., C.: Constrained aggregate similarity queries in metric spaces. In: SBBD, pp. 145–159 (2007)

27. Razente, H.L., Barioni, M.C.N., Traina, A.J.M., Faloutsos, C., Traina Jr., C.: A novel optimization approach to efficiently process aggregate similarity queries in metric access methods. In: CIKM, pp. 193–202. ACM (2008)

Finding the K Nearest Objects over Time Dependent Road Networks

Muxi Leng[1], Yajun Yang[1(✉)], Junhu Wang[2], Qinghua Hu[1], and Xin Wang[1]

[1] School of Computer Science and Technology, Tianjin University, Tianjin, China
{mxleng,yjyang,huqinghua,wangx}@tju.edu.cn
[2] School of Information and Communication Technology, Griffith University,
Brisbane, Australia
j.wang@griffith.edu.cn

Abstract. K nearest neighbor (kNN) search is an important problem and has been well studied on static road networks. However, in real world, road networks are often time-dependent, i.e., the time for traveling through a road always changes over time. Most existing methods for kNN query build various indexes maintaining the shortest distances for some pairs of vertices on static road networks. Unfortunately, these methods cannot be used for the time-dependent road networks because the shortest distances always change over time. To address the problem of kNN query on time-dependent road networks, we propose a novel voronoi-based index in this paper. Moreover, we propose an algorithm for pre-processing time-dependent road networks such that the waiting time is not necessary to be considered. We confirm the efficiency of our method through experiments on real-life datasets.

1 Introduction

With the rapid development of mobile devices, k nearest neighbor (kNN) search on road networks has become more and more important in location-based services. Given a query location and a set of objects (e.g., restaurants) on a road network, it is to find k nearest objects to the query location. kNN search problem has been well studied on static road networks. However, road networks are essentially time-dependent but not static in real world. For example, the Vehicle Information and Communication System (VICS) and the European Traffic Message Channel (TMC) are two transportation systems, which provide real-time traffic information to users. Such road networks are time-dependent, i.e., travel time for a road varies with taking "rush hour" into account.

The existing works propose various index techniques for answering k nearest object query on road networks. The main idea behind these indexes is to partition the vertices into several clusters, and then the clusters are organized as a voronoi diagram or a tree (e.g., R-tree, G-tree, etc.). All these methods pre-compute and maintain the shortest distances for some pairs of vertices to facilitate kNN query. Unfortunately, these indexes cannot be used for time-dependent road networks.

Y. Cai et al. (Eds.): APWeb-WAIM 2018, LNCS 10988, pp. 334–349, 2018.
https://doi.org/10.1007/978-3-319-96893-3_25

The reason is that the minimum travel time between two vertices often varies with time. For example, u and v are in the same cluster for one time period but they may be in two distinct clusters for another time period because of the minimum travel time varying with time. Therefore, the existing index techniques based on the static shortest distance cannot handle the case that the minimum travel time is time-dependent. Moreover, the waiting time is allowed on time-dependent road networks, i.e., someone can wait a time period to find another faster path. When the waiting time is considered, it is more difficult to build an index for kNN query by existing methods because it is difficult to estimate an appropriate waiting time for pre-computing the minimum travel time between two vertices.

Recently, there are some works about kNN query on time-dependent graphs [4–6,13]. Most of these works utilize A* algorithm to expand the road networks by estimating an upper or lower bound of travel time. There are two main drawbacks of these methods. First, in these works, the FIFO (first in first out) property is required for the networks and the waiting time is not allowed. Second, the indexes proposed by these works are based on the estimated value of travel time. However, these indexes cannot facilitate query effectively for large networks because the deviation are always too large between the estimated and actual travel time.

In this paper, we study k nearest object query on time dependent networks. A time dependent road network is modeled as a graph with time information. The weight of every edge is a time function $w_{i,j}(t)$ which specifies how much time it takes to travel through the edge (v_i, v_j) if departing at time point t. The main idea of our method is to pre-compute minimum travel time functions (or mtt-function for short) instead of concrete values for some pairs of vertices and then design a "*dynamic*" voronoi-based index based on such functions. Here "dynamic" means that in a time-dependent network it can be easily decided which cluster a vertex should be in for any given time point t. Different to previous works, our index can facilitate query effectively for large networks. Moreover, our method does not require the FIFO property for networks and we allow waiting time on every vertex.

The main contributions of this paper are summarized as below. First, we propose an algorithm to process $w_{i,j}(t)$ for every edge such that the waiting time is not necessary to be considered. Let G_T and G_T^* be the original graph and the graph after processing $w_{i,j}(t)$. We can prove that a shortest path with consideration of waiting time on G_T is one-one mapped to a shortest path without waiting time on G_T^*. Furthermore, we show how to compute the mtt-function for two vertices. Second, we propose a novel voronoi-based index for time-dependent road networks and an algorithm to answer kNN query using our index. Finally, we confirm the efficiency of our method through extensive experiments on real-life datasets.

The rest of this paper is organized as follows. Section 2 gives the problem statement. Section 3 describes how to process $w_{i,j}(t)$ and compute the mtt-function. Section 4 explains how to build the voronoi-based index for time-dependent networks and Sect. 5 proposes the kNN query algorithm. The experimental results are presented in Sect. 6. The related work is in Sect. 7. Finally, we conclude this paper in Sect. 8.

2   Problem Statement

Definition 1 (Time-Dependent Road Network): *A time-dependent road network is a simple directed graph, denoted as $G_T(V, E, W)$ (or G_T for short), where V is the set of vertices; $E \subseteq V \times V$ is the set of edges; and W is a set of non-negative value functions. For every edge $(v_i, v_j) \in E$, there is a time-function $w_{i,j}(t) \in W$, where t is a time variable. A time function $w_{i,j}(t)$ specifies how much time it takes to travel from v_i to v_j, if one departs from v_i at time point t.*

In this paper, we assume that $w_{i,j}(t) \geq 0$. The assumption is reasonable, because the travel time cannot be less than zero in real applications. Our work can be easily extended to handle undirected graphs. An undirected edge (v_i, v_j) is equivalent to two directed edges (v_i, v_j) and (v_j, v_i), where $w_{i,j}(t) = w_{j,i}(t)$.

The are several works that study how to construct time function $w_{i,j}(t)$, which is always modeled as a piecewise linear function [7,8,11] and it can be formalized as follows:

$$
w_{i,j}(t) = \begin{cases}
a_1 t + b_1, & t_0 \leq t < t_1 \\
a_2 t + b_2, & t_1 \leq t < t_2 \\
\cdots \\
a_p t + b_p, & t_{p-1} \leq t \leq t_p
\end{cases}
$$

Given a path p, the travel time of p is time-dependent. In order to minimize the travel time, some waiting time ω_i is allowed at every vertex v_i in p. That is, when arriving at v_i, one can wait a time period ω_i if the travel time of p can be minimized. We use $\mathsf{arrive}(v_i)$ and $\mathsf{depart}(v_i)$ to denote the arrival time at v_i and departure time from v_i, respectively. For each v_i in p, we have

$$\mathsf{depart}(v_i) = \mathsf{arrive}(v_i) + \omega_i$$

Let $p = v_1 \rightarrow v_2 \rightarrow \cdots \rightarrow v_h$ be a given path with the departure time t and the waiting time ω_i for each vertex v_i, then we have

$$\mathsf{arrive}(v_1) = t$$
$$\mathsf{arrive}(v_2) = \mathsf{depart}(v_1) + w_{1,2}(\mathsf{depart}(v_1))$$
$$\cdots$$
$$\mathsf{arrive}(v_h) = \mathsf{depart}(v_{h-1}) + w_{h-1,h}(\mathsf{depart}(v_{h-1}))$$

Thus the travel time of path p is $w(p) = \mathsf{arrive}(v_h) - t$. Given two vertices v_i and v_j in G_T, the minimum travel time from v_i to v_j with departure time t is defined as $m_{i,j}(t) = \min\{w(p) | p \in P_{i,j}\}$, where $P_{i,j}$ is the set of all the paths from v_i to v_j in G_T. Obviously, $m_{i,j}(t)$ is also a function related to the departure time t. We call $m_{i,j}(t)$ the **minimum travel time function** (or mtt-function shortly) from v_i to v_j. Let $|V|$ be n, in the following, we use $m_{i,n+j}(t)$ to represent mtt-function from a vertex v_i to an object o_j, in order to distinguish from $m_{i,j}(t)$ from v_i to a vertex v_j. Note that an object o_i is also a vertex regarded as v_{n+i} in the network.

Next, we give the definition of kNN query over time-dependent road networks.

Definition 2 (k Nearest Objects on Time-Dependent Road Networks):
Given a time-dependent road network $G_T(V, E, W)$, a set of the objects $O = \{o_1, o_2, \cdots\}$, a query point $v_q \in V$ and a departure time t_d, k nearest objects query of v_q is to find a k-size subset $O(v_q) \subseteq O$, such that $m_{q,n+j}(t_d) \geq \max\{m_{q,n+i}(t_d)|o_i \in O(v_q)\}$ for every object $o_j \in O \setminus O(v_q)$.

3   Minimum Travel Time Function

We pre-compute mtt-functions for some pairs of vertices and then build the index to facilitate kNN query over time-dependent road networks. In this section, we first describe how to process the time function $w_{i,j}(t)$ for every edge in G_T such that the waiting time is not necessary to be considered when computing mtt-function and then explain how to compute mtt-function without waiting time.

3.1   Pre-processing Time Function for Every Edge

Given a path p, the waiting time ω_i is allowed for any vertex $v_i \in p$. However, it is not easy to find an appropriate value of ω_i for every $v_i \in p$ to minimize the travel time of p. In this section, we propose an algorithm to convert time function $w_{i,j}(t)$ to a new function $w_{i,j}^*(t)$ for every edge $(v_i, v_j) \in E$. We call $w_{i,j}^*(t)$ the "no waiting time function" of edge (v_i, v_j) (or nwt-function for short). The waiting time can be considered as zero when nwt-function is used to compute the minimum travel time of path p. The nwt-function $w_{i,j}^*(t)$ is defined by the following equation.

$$w_{i,j}^*(t) = \min_{\omega_i}(\omega_i + w_{i,j}(t + \omega_i)) \tag{1}$$

The following theorem guarantees the nwt-function $w_{i,j}^*(t)$ can be used to compute the minimum travel time for any path p in G_T without waiting time.

Theorem 1. *Given two time-dependent graphs $G_T(V, E, W)$ and $G_T^*(V, E, W^*)$, where W^* is the set of nwt-functions of all edges in E, for any path p in G_T, the minimum travel time of p in G_T **with consideration of waiting time** equals to the minimum travel time of p in G_T^* **without waiting time**.*

PROOF: Let $p = v_1 \rightarrow v_2 \rightarrow \cdots \rightarrow v_h$ be a given path with the departure time t. ω_i^* is the waiting time on v_i ($1 \leq i \leq h$) minimizing the travel time of p in G_T. We have $\mathsf{depart}(v_i) = \mathsf{arrive}(v_i) + \omega_i^*$ and $\mathsf{arrive}(v_{i+1}) = \mathsf{depart}(v_i) + w_{i,i+1}(\mathsf{depart}(v_i))$. Similarly, we have $\mathsf{depart}^*(v_i) = \mathsf{arrive}^*(v_i)$ and

---

Algorithm 1. Nwt-Function $(G_T(V, E, W))$

Input: $G_T(V, E, W)$.
Output: W^*.

1: $W^* \leftarrow \emptyset$;
2: **for** every $w_{i,j}(t) \in W$ **do**
3:     $\phi \leftarrow w_{i,j}(t_p)$, $w_{i,j}^*(t_p) \leftarrow w_{i,j}(t_p)$;
4:     **for** $k = p$ to 1 **do**
5:         $a^* \leftarrow -1$, $b^* \leftarrow t_k + \phi$;
6:         $w_{i,j}^*(t) \leftarrow a^*t + b^*$ for $t \in [t_{k-1}, t_k)$;
7:         $w_{i,j}^*(t) \leftarrow \min\{w_{i,j}^*(t), w_{i,j}(t) | t \in [t_{k-1}, t_k)\}$;
8:         $\phi \leftarrow \min\{w_{i,j}^*(t_{k-1}), w_{i,j}^-(t_{k-1})\}$;
9:     $W^* \leftarrow W^* \cup \{w_{i,j}^*(t)\}$;
10: **return** W^*

$\mathsf{arrive}^*(v_{i+1}) = \mathsf{depart}^*(v_i) + w_{i,i+1}^*(\mathsf{depart}^*(v_i))$ for G_T^*. We only need to prove $\mathsf{arrive}(v_h) = \mathsf{arrive}^*(v_h)$. It can be easily proved by induction on v_i. We omit it due to the space limitation. $\square$

The algorithm to compute nwt-function is shown in Algorithm 1. For every $w_{i,j}(t) \in W$, Algorithm 1 computes $w_{i,j}^*(t)$ backward from $[t_{p-1}, t_p]$ to $[t_0, t_1]$ iteratively. In each iteration, $w_{i,j}^*(t)$ for $t \in [t_{k-1}, t_k)$ is computed. Algorithm 1 first sets $w_{i,j}^*(t)$ as $a^*t + b^*$, where $a^* = -1$ and $b^* = t_k + \phi$. ϕ is the minimum value between $w_{i,j}^*(t_k)$ and $w_{i,j}^-(t_k)$. $w_{i,j}^-(t_k)$ is the left limit value of $w_{i,j}(t)$ on t_k. Note that $w_{i,j}^*(t_k)$ and ϕ have been computed in the last iteration, i.e., the iteration for computing $w_{i,j}^*(t)$ on $[t_k, t_{k+1})$. ϕ is initialized as $w_{i,j}(t_p)$. Next, Algorithm 1 updates $w_{i,j}^*(t)$ as $\min\{w_{i,j}^*(t), w_{i,j}(t)\}$ for $t \in [t_{k-1}, t_k)$ and then ϕ is updated as $\min\{w_{i,j}^*(t_{k-1}), w_{i,j}^-(t_{k-1})\}$. The algorithm terminates when $w_{i,j}^*(t)$ has been computed for $t \in [t_0, t_1)$.

The time and space complexities analysis for Algorithm 1 are given below. Let n and m be the number of the vertices and edges in G_T respectively. For every edge (v_i, v_j), Algorithm 1 needs to compute $w_{i,j}^*(t)$ on $[t_{k-1}, t_k)$ iteratively from $k = p$ to 1. For every time interval $[t_{k-1}, t_k)$, $w_{i,j}^*(t)$ can be computed in constant time. Therefore, the time complexity of Algorithm 1 is $O(mp)$. Moreover, Algorithm 1 needs to maintain $w_{i,j}^*(t)$ and then the space complexity is also $O(mp)$.

Example 1. We illustrate how to compute $w_{i,j}^*(t)$ by an example in Fig. 1. As the solid black line in Fig. 1(a), $w_{i,j}(t)$ is a piecewise linear function:

$$w_{i,j}(t) = \begin{cases} t + 5, & 0 \leq t < 10 \\ 15, & 10 \leq t < 20 \\ -2t + 55, & 20 \leq t \leq 25 \end{cases}$$

In the first iteration, ϕ is initialized as $w_{i,j}(25) = 5$ and then $b^* = 25 + \phi = 30$. As the dashed red line in the right-side of Fig. 1(a), we find $a^*t + b^* = -t + 30$ is

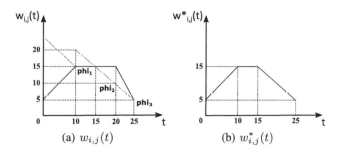

Fig. 1. Computing $w^*_{i,j}(t)$ (Color figure online)

always less than $w_{i,j}(t)$ on $[20, 25]$, then $w^*_{i,j}(t) = -t + 30$ for $t \in [20, 25]$ and ϕ is updated as 10. Similarly, in the second iteration, $w^*_{i,j}(t)$ on $[10, 20)$ is computed as $\min\{15, -t + 30\}$, i.e., $w^*_{i,j}(t) = 15$ for $t \in [10, 15)$ and $w^*_{i,j}(t) = -t + 30$ for $t \in [15, 20)$. Then ϕ is updated as $\min\{w^*_{i,j}(10), w^-_{i,j}(10)\} = 15$. In the final iteration, as the dashed red line in the left-side of Fig. 1(a), $a^*t + b^* = -t + 25$ is always larger than $t + 5$ on $[0, 10)$, we have $w^*_{i,j}(t) = t + 5$ for $t \in [0, 10)$. Then $w^*_{i,j}(t)$ is given below and depicted in Fig. 1(b)).

$$w^*_{i,j}(t) = \begin{cases} t + 5, & 0 \le t < 10 \\ 15, & 10 \le t < 15 \\ -t + 30, & 15 \le t \le 25 \end{cases}$$

The following theorem guarantees the correctness of Algorithm 1.

Theorem 2. *The $w^*_{i,j}(t)$ computed by Algorithm 1 is exactly the* nwt-*function $w^*_{i,j}(t)$ given by Eq. (1).*

PROOF: We proved it by induction on p.

Basis. We need to prove that $w^*_{i,j}(t)$ on time interval $[t_{p-1}, t_p]$ can be correctly computed by Algorithm 1. First, ω_i can only be zero when $t = t_p$, then we have $w_{i,j}(t_p) = w^*_{i,j}(t_p)$ and $\phi_p = w_{i,j}(t_p)$. Next, we consider the case of $t \in [t_{p-1}, t_p)$. By the definition of $w^*_{i,j}(t)$, we have

$$\begin{aligned} w^*_{i,j}(t) &= \min_{\omega_i}(\omega_i + w_{i,j}(t + \omega_i)) \\ &= \min_{\omega_i}(\omega_i + a_p(t + \omega_i) + b_p) \\ &= \min_{\omega_i}((a_p + 1)\omega_i + a_p t + b_p) \\ &= \min_{\omega_i}((a_p + 1)\omega_i + w_{i,j}(t)) \end{aligned}$$

For $t \in [t_{p-1}, t_p)$, if $a_p \ge -1$, $w^*_{i,j}(t)$ cannot decrease with ω_i increasing. It means $(a_p + 1)\omega_i + w_{i,j}(t)$ is minimum when $\omega_i = 0$ and then $w^*_{i,j}(t) = w_{i,j}(t)$. If $a_p < -1$, $w^*_{i,j}(t)$ will decrease with ω_i increasing and thus $(a_p + 1)\omega_i + w_{i,j}(t)$

is minimum when $\omega_i = t_p - t$, which is the longest waiting time on v_i for $t \in [t_{p-1}, t_p)$. Then we have

$$w_{i,j}^*(t) = (a_p + 1)(t_p - t) + a_p t + b_p = -t + t_p + \phi_p$$

Obviously, $w_{i,j}(t) \leq -t + t_p + \phi_p$ when $a_p \geq -1$ and $w_{i,j}(t) \geq -t + t_p + \phi_p$ when $a_p < -1$. Then we have $w_{i,j}^*(t) = \min\{w_{i,j}(t), -t + t_p + \phi_p\}$ for $t \in [t_{p-1}, t_p]$.

Induction. Assume the correct $w_{i,j}^*(t)$ can be computed by Algorithm 1 for $t \in [t_k, t_p]$, then we need to prove it also can be correctly computed for $t \in [t_{k-1}, t_k)$. We consider the following two cases: (1) $\omega_i \geq t_k - t$; and (2) $\omega_i < t_k - t$.

For case (1), the departure time $t + \omega_i \in [t_k, t_p]$ because $\omega_i \geq t_k - t$. By the assumption, nwt-function $w_{i,j}^*(t)$ has been correctly computed for $t \in [t_k, t_p]$, then $w_{i,j}^*(t_k)$ is the minimum travel time for edge (v_i, v_j) with departure time t_k. Therefore, $w_{i,j}^*(t)$ for $t \in [t_{k-1}, t_k)$ can be computed by the following equation:

$$w_{i,j}^*(t) = t_k - t + w_{i,j}^*(t_k)$$

For case (2), because $\omega_i < t_k - t$, then $t + \omega_i \in [t_{k-1}, t_k)$. Similar to the proof of basis, we have

$$w_{i,j}^*(t) = \min\{w_{i,j}(t), -t + t_k + w_{i,j}^-(t_k)\}$$

Note that, when $a_k < -1$, $w_{i,j}^*(t) = -t + t_k + w_{i,j}^-(t_k)$ because $w_{i,j}(t)$ may be noncontinuous at t_k. Therefore, we have

$$w_{i,j}^*(t) = \min\{w_{i,j}(t), -t + t_k + w_{i,j}^-(t_k), -t + t_k + w_{i,j}^*(t_k)\}$$

The proof is completed.   □

3.2   Computing Minimum Travel Time Function

We adopt a Dijkstra-based algorithm proposed in [7] to compute mtt-function for two vertices v_i and v_j in G_T. This algorithm is only used for the case that the waiting time is not allowed. After converting $w_{i,j}(t)$ to nwt-function $w_{i,j}^*(t)$ for every edge in G_T by Algorithm 1, this algorithm can be used for time-dependent graphs with waiting time.

The main idea of this Dijkstra-based algorithm is to refine a function $g_{i,j}(t)$ iteratively for every $v_j \in V$, where $g_{i,j}(t)$ represents the earliest arrival time on v_j if departing from v_i at time point t. In every iteration, algorithm selects a vertex $v_x \in V$ and then refine $g_{i,x}(t)$ by extending a time domain I_x to a larger I_x', where $I_x = [t_0, \tau_x]$ is a subinterval of the whole time domain T. $g_{i,x}(t)$ is regarded as well-refined in I_x if it specifies the earliest arrival time at v_x from v_i for any departure time $t \in I_x$. The algorithm repeats time-refinement process till $g_{i,j}(t)$ of destination v_j has been well-refined in the whole time domain T and then mtt-function $m_{i,j}(t)$ can be computed as $m_{i,j}(t) = g_{i,j}(t) - t$. The more details about this Dijkstra-based algorithm is given in [7]. As shown in [7], the time and space complexities are $O((n \log n + m)\alpha(T))$ and $O((n + m)\alpha(T))$ respectively, where $\alpha(T)$ is the cost required for each function (defined in interval T) operation.

4   The Novel Voronoi-Based Index

We propose a novel voronoi-based index for kNN query over time-dependent road networks. In static road networks, the voronoi diagram divides the network (or space) into a group of disjoint subgraphs (or sub-spaces) where the nearest object of any vertex inside a subgraph is the object generating this subgraph. However, in time-dependent road networks, the nearest object of a vertex may be dynamic. The nearest object of a vertex v may be o_i for departure time $t \in [t_1, t_2]$ but it may be o_j for $t \in [t_3, t_4]$. The main idea of our novel voronoi-based index is also to divide the vertex set V into some vertex subsets V_i and every subset V_i is associated with one object $o_i \in O$. Different to static road networks, our voronoi-based index are time-dependent, that is, every vertex v inside a subset is with a time interval indicating when the object o_i is nearest to v. Next, we describe what is the novel voronoi-based index and how to construct it.

4.1   What Is the Voronoi-Based Index?

Given a vertex v and an object o_i, $I_i(v)$ is called v's **maximum time interval** about o_i if it satisfies the following two conditions: (1) o_i is the nearest object of v for any departure time $t \in I_i(v)$; and (2) there does not exist another $I'_i(v) \supset I_i(v)$ satisfying the condition (1). Note that $I_i(v)$ may not be a continuous time interval, that is, if o_i is nearest to v for two disjoint departure time intervals $[t_1, t_2]$ and $[t_3, t_4]$, then $[t_1, t_2] \cup [t_3, t_4] \subseteq I_i(v)$. The voronoi-based index maintains a set C_i for every object $o_i \in O$, where C_i is a set of the tuples $(v, I_i(v))$ for all the vertices v with non-empty $I_i(v)$, i.e.,

$$C_i = \{(v, I_i(v)) | v \in V \wedge I_i(v) \neq \emptyset\}$$

We call C_i the **closest vertex-time pair set** of o_i. For simplicity, we say v is a vertex in C_i if $(v, I_i(v)) \in C_i$. Next, we give the definition of the border vertex.

Definition 3 (Border Vertex): *A vertex v_x in C_i is called a **border vertex** of C_i if there exist $v_y \in N^+(v_x)$ such that $(v_y, I_y) \notin C_i$ for any $I_y \supseteq f_{x,y}(I_i(v_x))$, where $N^+(v_x)$ is the outgoing neighbor set of v_x and $f_{x,y}(I_i(v_x))$ is the time interval mapped from $I_i(v_x)$ by the function $f_{x,y}(t) = t + w^*_{x,y}(t)$.*

The border vertex v_x of C_i indicates there exist a time point $t \in f_{x,y}(I_i(v_x))$ such that o_i is not the nearest object of v_y if one departs at time point t.

We use B_i to denote the set of all the border vertices of C_i. For every C_i, D_i is the set of mtt-functions $m_{x,n+i}(t)$ for all vertices v_x in C_i, that is,

$$D_i = \{m_{x,n+i}(t) | v_x \text{ is a vertex in } C_i\}$$

and M_i is a matrix of size $|C_i| \times |B_i|$ to maintain mtt-function $m_{x,y}(t)$ for all pairs of vertex v_x and border vertex v_y in C_i, i.e.,

$$M_i = \{m_{x,y}(t) | v_x \in C_i \wedge v_y \in B_i\}$$

The voronoi-based index is $\{C, B, D, M\}$, where C, B, D and M are the collections of all C_i, B_i, D_i and M_i respectively.

4.2   How to Construct the Voronoi-Based Index?

We have explained how to compute mtt-function in Sect. 3. Next, we describe how to compute C_i and B_i for every $o_i \in O$.

For every vertex $v_x \in V$, $I_i(v_x)$ is initialized as the whole time domain T. We refine $I_i(v_x)$ iteratively by removing the sub-intervals on which $m_{x,n+i}(t)$ is larger than $m_{x,n+j}(t)$ for another object o_j. It means o_i is not the nearest object of v_x when departure time is in these sub-intervals. For every $o_j \in O$ ($o_j \neq o_i$), let $T_j(v_x)$ denote the maximum time interval on which $m_{x,n+j}(t) < m_{x,n+i}(t)$, $I_i(v_x)$ is updated as $I_i(v_x) - T_j(v_x)$. After removing $T_j(v_x)$ for every other object o_j, if $I_i(v_x)$ is not empty, then the pair $(v_x, I_i(v_x))$ is inserted into C_i.

For every vertex v_x in C_i, if there exists an outgoing neighbor v_y of v_x, such that v_y is not in C_i or $f_{x,y}(I_i(v_x)) \nsubseteq I_i(v_y)$, then v_x must be a border vertex of C_i and it is inserted into B_i.

Algorithm 2. kNN-QUERY (G_T^*, v_q, t_d, k)

Input: time-dependent graph G_T^*, query vertex v_q, departure time t_d and k
Output: the k nearest neighbor set $O(v_q)$

1: $O(v_q) \leftarrow \emptyset$, $Q \leftarrow \{C_q\}$; $E_q \leftarrow \{v_q\}$
2: **while** $|O(v_q)| < k$ **do**
3:     $C_i \leftarrow$ DEQUEUE (Q), $O(v_q) \leftarrow O(v_q) \cup \{o_i\}$;
4:     **for each** $v_y \in B_i$ **do**
5:        **for each** $v_x \in E_i$ **do**
6:           $m_{q,y} \leftarrow \min\{m_{q,y}, m_{q,x} + m_{x,y}(t_d + m_{q,x})\}$;
7:        **for each** $v_z \in N^+(v_y)$ **do**
8:           **if** $m_{q,z} > m_{q,y} + w_{y,z}^*(t_d + m_{q,y})$ **then**
9:              $m_{q,z} \leftarrow m_{q,y} + w_{y,z}^*(t_d + m_{q,y})$;
10:             Let C_j be the set including v_z when $t = t_d + m_{q,z}$;
11:             **if** $C_j \notin O(v_q)$ **then**
12:                $E_j \leftarrow E_j \cup \{v_z\}$;
13:                **if** $m_{q,n+j} > m_{q,z} + m_{z,n+j}(t_d + m_{q,z})$ **then**
14:                   $m_{q,n+j} \leftarrow m_{q,z} + m_{z,n+j}(t_d + m_{q,z})$;
15:                **if** $C_j \notin Q$ **then**
16:                   ENQUEUE(Q, C_j);
17:                **else**
18:                   UPDATE(Q, C_j);
19: **return** $O(v_q)$

---

5   Query Processing

Algorithm 2 describes how to find the k nearest objects for a query vertex v_q with departure time t_d. In Algorithm 2, $O(v_q)$ is a set to maintain the objects that have been found so far and Q is a priority queue to maintain a candidate set of C_i whose o_i is possible to be an object in kNN set. All $C_i \in Q$ are sorted in an ascending order by the minimum travel time $m_{q,n+i}$ from v_q to o_i. The top C_i in Q is with the minimum $m_{q,n+i}$ and it can be easily done using Fibonacci Heap. $O(v_q)$ and Q are initialized as $\emptyset$ and $\{C_q\}$ respectively, where C_q contains v_q for the departure time t_d, i.e., $(v_q, I_q(v_q)) \in C_q$ and $t_d \in I_q(v_q)$. $O(v_q)$ is expanded iteratively by inserting objects one by one from Q until $|O(v_q)| = k$. In each iteration, if $|O(v_q)| < k$, Algorithm 2 first dequeues the top C_i from Q with the minimum $m_{q,n+i}$. The object o_i of C_i must be one of k nearest objects of v_q. It can be guaranteed by Theorem 3. Then o_i will be inserted into $O(v_q)$. For every border vertex v_y in C_i, Algorithm 2 computes $m_{q,y}$ as $\min\{m_{q,x} + m_{x,y}(t_d + m_{q,x})|v_x \in E_i\}$, where E_i is the entry set of C_i. The "entry" means any path entering into C_i must go through a vertex in E_i. E_i will be updated when Algorithm 2 runs. For every $v_z \in N^+(v_y)$, if $m_{q,z} > m_{q,y} + w_{y,z}^*(t_d + m_{q,y})$, then $m_{q,z}$ will be updated as $m_{q,y} + w_{y,z}^*(t_d + m_{q,y})$. Next, if v_z is in C_j ($C_j \neq C_i$ and $C_j \notin O(v_q)$) at the time point $t_d + m_{q,z}$, then v_z will be inserted into E_j as an entry of C_j. For the object o_j of C_j, $m_{q,n+j}$ will be updated as $m_{q,z} + m_{z,n+j}(t_d + m_{q,z})$ when $m_{q,n+j} > m_{q,z} + m_{z,n+j}(t_d + m_{q,z})$. If C_j is not in Q, then C_j will be enqueued into Q. Otherwise, C_j has been in Q and Q will be updated by C_j with new $m_{q,n+j}$. Algorithm 2 terminates when the size of $O(v_q)$ is k.

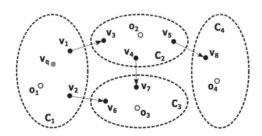

Fig. 2. Query processing

Example 2. We use the example in Fig. 2 to illustrate the kNN querying process for $k = 3$. In this example, v_q is the query vertex and it is in C_1 for the departure time t_d. Q and $O(v_q)$ are initialized as $\{C_1\}$ and $\emptyset$ respectively. In the first iteration, C_1 is dequeued from Q and then o_1 is inserted into $O(v_q)$. Because v_1 is a border vertex of C_1 and v_3 is an outgoing neighbor of v_1, Algorithm 2 computes $m_{q,1}(t_d)$ and $m_{q,3}(t_d) = m_{q,1}(t_d) + w_{1,3}^*(t_d + m_{q,1}(t_d))$. Note that v_3 is in C_2 when $t = t_d + m_{q,3}(t_d)$ and then it is an entry of C_2. Therefore, C_2 is enqueued into Q. Similarly, C_3 is also enqueued into Q and $Q = \{C_2, C_3\}$. Assume that o_2 nearer to v_q than o_3, in the second iteration, C_2 is dequeued and $O(v_q)$ is updated as $\{o_1, o_2\}$. In the same way, C_4 will be enqueued into Q in this iteration. In the final iteration, C_3 will be dequeued due to o_3 is nearer to v_q and then $O(v_q) = \{o_1, o_2, o_3\}$. Because $|O(v_q)| = 3$, Algorithm 2 terminates and returns $O(v_q)$.

The next theorem guarantees the correctness of Algorithm 2.

Theorem 3. *In Algorithm 2, the object o_i of C_i dequeued from Q in the k-th iteration must be the k-th nearest object of query vertex v_q for the departure time t_d.*

PROOF: We prove it by induction on k.

Basis. Obviously, C_q is dequeued from Q in the first iteration. By the definition of C_q, o_q is the nearest object of v_q when the departure time is t_d.

Induction. Assume that the i-th nearest neighbor of v_q is dequeued from Q in the i-th iteration for $i < k$. We need to prove it also hold for $i = k$. We prove it by contradiction. Let C_k be the closest vertex-time pair set dequeued from Q in the k-th iteration and o_k is the object of C_k. Suppose that the k-th nearest object of v_q is $o_{k'}$ and $o_{k'} \neq o_k$. Let p be the shortest path from v_q to $o_{k'}$ with the departure time t_d. Because $k > 1$, then $C_{k'}$ is not C_q and there must exist an entry v_e of $C_{k'}$ in p. Let v_b be the predecessor of v_e in p, then v_b must be a border vertex of C_b at time point $t_d + m_{q,b}$ and $C_b \neq C_{k'}$. There are two cases for the object o_b of C_b: (1) o_b is not in the k nearest object set of v_q; and (2) o_b is in the k nearest object set of v_q.

For case (1), by the definition of C_b, o_b is the nearest neighbor of v_b at time point $t_d + m_{q,b}$, then we have

$$m_{q,b} + m_{b,n+b}(t_d + m_{q,b}) < m_{q,b} + m_{b,n+k'}(t_d + m_{q,b})$$

Thus o_b is nearer to v_q than $o_{k'}$ when the departure time is t_d. It means o_b must be in the k nearest object set of v_q, which is a contradiction.

For case (2), Let o_b be the i-th $(i < k)$ nearest object of v_q, by the inductive assumption, C_b is dequeued from Q in i-th iteration. According to the Algorithm 2, $C_{k'}$ is enqueued into Q in this iteration. Therefore, $C_{k'}$ will be dequeued from Q in k-th iteration instead of C_k, which is a contradiction. The proof is completed   □

The time and space complexities of Algorithm 2 are given below. Let b and e be the average size of B_i and E_i respectively. In every iteration, Algorithm 2 upadates $m_{q,y}$ as $\min\{m_{q,y}, m_{q,x} + m_{x,y}(t_d + m_{q,x})\}$ for every border vertex v_y in C_i. It will cost $O(be)$ time. For every outgoing neighbor v_z of border vertex v_y, Algorithm 2 needs to compute $m_{q,z}$ and then it will cost $O(bd)$ time, where d is the average out-degree of the vertices in G_T. Therefore, the time complexity of Algorithm 2 is $O(kb(d + e))$. On the other hand, because Algorithm 2 needs to maintain $m_{q,y}$ and $m_{q,z}$, then the space complexity is $O(k(b + e))$.

6   Experiements

We compare our voronoi-based index method (marked as VI) with FTTI (Fast-Travel-Time Index) method [13] and TLNI (Tight-and-Loose-Network Index) method [6] on the real-life datasets. FTTI and TLNI are the state of the art index-based methods for kNN query over time-dependent road networks. Note that FTTI and TLNI are used on G_T^* in which every edge is an nwt-function $w_{i,j}^*(t)$ because FTTI and TLNI do not allow the waiting time. Although some algorithms are proposed in recent works [1,4], they are only to find the nearest object (i.e., $k = 1$) and they cannot be used for general kNN query on time-dependent graphs. All the experiments are conducted on a 2.6 GHz Intel Core i7 CPU PC with the 16 GB main memory, running on Windows 7.

6.1 DataSets and Experiment Setup

We tested the voronoi-based index method on California road network (CARN) with 196,5206 vertices and 553,3214 edges. We extracted five time-dependent graphs with different size using the CARN dataset. The number of vertices ranges from 100k to 500k. The time domain is set as $T = [0, 2000]$, i.e., the departure time t can be selected from $[0, 2000]$ for any vertex. Here, 2000 means 2000 time units. For every $w_{i,j}(t)$, we split the time domain T to p subintervals and assign a linear function randomly for every sub-interval and then $w_{i,j}(t)$ is a piecewise linear function.

6.2 Experimental Results

Exp-1. Impact of Network Size: In this group of experiments, we study the impact of time-dependent network size. The number of the vertices increases from 100k to 500k and the number of objects is fixed at 10k. We investigate the querying time for $k = 7$. The number of piecewise intervals of $w_{i,j}(t)$ is set as 4. As shown in Fig. 3(a) and (b), the querying time of our method is always less than FTTI and TLNI. Specifically, the querying time of TLNI is always much more than our method even though TLNI has the smallest index size. The reason is TLNI index only maintain the vertices for an object o_i that the upper bound of travel time to o_i are less than the lower bound to the other objects. It cannot facilitate query effectively in large networks.

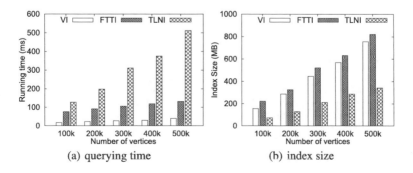

Fig. 3. Impact of the network size

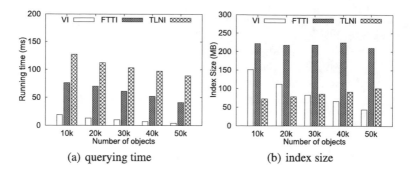

Fig. 4. Impact of the object set size

Exp-2. Impact of Object Set Size: In this group of experiments, the number of the vertices is fixed at 100k and the number of objects ranges from 10k to 50k. As shown in Fig. 4(a) and (b), the querying time of our method are always less than FTTI and TLNI. Moreover, the querying time and index size decrease with the increasing of the object set size. There are two reasons as follows: (1) the average size of C_i and B_i decrease if the object set size increases; (2) the increasing of object size results in that the objects become nearer to v_q and then querying time decreases.

Exp-3. Impact of the Time Domain: In Fig. 5, we study the impact of time domain. In this group of experiments, the number of vertices and objects are fixed at 100k and 10k respectively. The time domain ranges from $[0, 1000]$ to $[0, 3000]$. We investigate the querying time for $k = 7$. As shown in Fig. 5(a) and (b), the querying time and index size of our method are not affected by the expanding of time domain. However, for FTTI and TLNI, the querying time increases with the the expanding of time domain. It is because they need to maintain the estimated value about travel time in index to facilitate kNN query. If the time domain becomes larger, the deviation between the estimation and actual travel time will become larger too. It cannot facilitate query effectively.

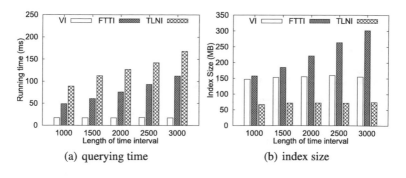

Fig. 5. Impact of the length of time interval

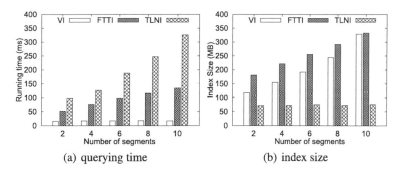

Fig. 6. Impact of the number of piecewise interval of time function

Exp-4. Impact of the Number of Piecewise Intervals: In Fig. 6, we investigate the impact of the number of piecewise intervals of $w_{i,j}(t)$. In this group of experiments, the number of piecewise intervals of $w_{i,j}(t)$ increases from 2 to 10. The number of the vertices and objects are fixed at 100k and 10k, respectively. As shown in Fig. 6(a) and (b), the querying time and index size always increase with the increasing of the number of piecewise intervals. The reason is that the more piecewise intervals of $w_{i,j}(t)$ results in more piecewise intervals of mtt-function and then the more border vertices will be maintained in the index.

Exp-5. Impact of k: In Fig. 7, we study the querying time by varying k from 1 to 10 on two different networks with 10k vertices and 50k vertices respectively. In this group of experiments, the number of objects are fixed at 10k and 50k for two different networks respectively. As shown in Fig. 7(a) and (b), the querying time always increases marginally with the increasing of k for our index method.

7   Related Work

kNN query has been well-studied on static road networks. Most of the existing works propose various index techniques. The main ideas of these methods are to partition the vertices into several clusters, and then the clusters are organized as a voronoi diagram or a tree (e.g., R-tree) [9,10,12,14–16,18–20]. These methods pre-compute and maintain the shortest distances for some pairs of vertices to facilitate kNN query. Unfortunately, these index techniques cannot be used for the time-dependent road networks because the minimum travel time between two vertices always varies with time.

kNN query has also been studied on time-dependent road networks [1,3–6,13]. Most of these works are based on A* algorithm. The authors in [1,4] study the problem to find nearest (i.e., $k = 1$) object on time-dependent networks. In [4], A virtual node v is inserted into the graph G with the zero-cost edges connecting to all the objects. The nearest object can be found on the shortest path from the query vertex to v. The authors in [2] study problem of finding k POIs that minimize the aggregated travel time from a set of query points.

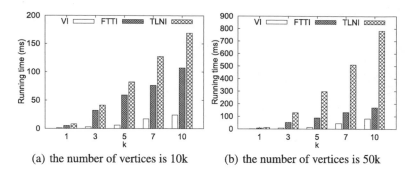

(a) the number of vertices is 10k   (b) the number of vertices is 50k

Fig. 7. Impact of k

The index-based methods are proposed in [6,13]. In [6], A* algorithm is utilized to expand the road networks by estimating an upper or lower bound of travel time. An index is built to facilitate kNN query using these estimated bounds. In [13], time domain is divided to several sub-intervals. For every sub-interval, C nearest objects of every vertex are found by an estimation of minimum travel time. There are two main drawbacks of these methods. First, in these works, the FIFO (first in first out) property is required for networks and waiting time is not allowed. Second, the indexes proposed by these works are based on the estimated value of travel time. However, these indexes cannot facilitate query effectively for the large networks because the deviations are always too large between the estimated and actual travel time.

Recently, there are some works about the shortest path query between two given vertices over time-dependent graphs [7,17]. However, these works does not study any index that can be used in kNN query over time-dependent road networks. The method in [7] is used to compute mtt-function between two vertices in our paper.

8 Conclusion

In this paper, we study the problem of k nearest objects query on time-dependent road networks. We first give an algorithm for processing time-dependent road networks such that the waiting time is not necessary to be considered and then propose a novel voronoi-based index to facilitate kNN query. We explain how to construct the index and complete the querying process using our index. We confirm the efficiency of our method through extensive experiments on real-life datasets.

Acknowledgments. This work is supported by the grant of the National Natural Science Foundation of China No. 61402323, 61572353 and the Australian Research Council Discovery Grant DP130103051.

References

1. Chucre, M.R.R.B., do Nascimento, S.M., de Macêdo, J.A.F., Monteiro, J.M., Casanova, M.A.: Taxi, please! A nearest neighbor query in time-dependent road networks. In: MDM, pp. 180–185 (2016)
2. Costa, C.F., Machado, J.C., Nascimento, M.A., de Macêdo, J.A.F.: Aggregate k-nearest neighbors queries in time-dependent road networks. In: SIGSPATIAL, pp. 3–12 (2015)
3. Costa, C.F., Nascimento, M.A., de Macêdo, J.A.F., Machado, J.C.: A*-based solutions for KNN queries with operating time constraints in time-dependent road networks. In: MDM, pp. 23–32 (2014)
4. Cruz, L.A., Lettich, F., Júnior, L.S., Magalhães, R.P., de Macêdo, J.A.F.: Finding the nearest service provider on time-dependent road networks. In: ECML-PKDD, pp. 21–31 (2017)
5. Cruz, L.A., Nascimento, M.A., de Macêdo, J.A.F.: K-nearest neighbors queries in time-dependent road networks. JIDM **3**(3), 211–226 (2012)
6. Demiryurek, U., Kashani, F.B., Shahabi, C.: Efficient k-nearest neighbor search in time-dependent spatial networks. In: DEXA, pp. 432–449 (2010)
7. Ding, B., Yu, J.X., Qin, L.: Finding time-dependent shortest paths over large graphs. In: EDBT, pp. 205–216 (2008)
8. George, B., Shekhar, S.: Time-aggregated graphs for modeling spatio-temporal networks. J. Data Semant. **11**, 191–212 (2006)
9. Hu, H., Lee, D.L., Xu, J.: Fast nearest neighbor search on road networks. In: EDBT, pp. 186–203 (2006)
10. Huang, X., Jensen, C.S., Saltenis, S.: The islands approach to nearest neighbor querying in spatial networks. In: SSTD, pp. 73–90 (2005)
11. Kanoulas, E., Du, Y., Xia, T., Zhang, D.: Finding fastest paths on a road network with speed patterns. In: ICDE, p. 10 (2006)
12. Kolahdouzan, M.R., Shahabi, C.: Voronoi-based K nearest neighbor search for spatial network databases. In: VLDB, pp. 840–851 (2004)
13. Komai, Y., Nguyen, D.H., Hara, T., Nishio, S.: kNN search utilizing index of the minimum road travel time in time-dependent road networks. In: SRDS, pp. 131–137 (2014)
14. Lee, K.C.K., Lee, W., Zheng, B.: Fast object search on road networks. In: EDBT, pp. 1018–1029 (2009)
15. Wei-Kleiner, F.: Finding nearest neighbors in road networks: a tree decomposition method. In: EDBT, pp. 233–240 (2013)
16. Yang, S., Cheema, M.A., Lin, X., Zhang, Y., Zhang, W.: Reverse k nearest neighbors queries and spatial reverse top-k queries. VLDB J. **26**(2), 151–176 (2017)
17. Yang, Y., Gao, H., Yu, J.X., Li, J.: Finding the cost-optimal path with time constraint over time-dependent graphs. Proc. VLDB Endow. **7**, 673–684 (2014)
18. Zheng, Y., Guo, Q., Tung, A.K.H., Wu, S.: Lazylsh: approximate nearest neighbor search for multiple distance functions with a single index. In: SIGMOD, pp. 2023–2037 (2016)
19. Zhong, R., Li, G., Tan, K., Zhou, L.: G-tree: an efficient index for KNN search on road networks. In: CIKM, pp. 39–48 (2013)
20. Zhu, H., Yang, X., Wang, B., Lee, W.: Range-based obstructed nearest neighbor queries. In: SIGMOD, pp. 2053–2068 (2016)

Reverse Top-k Query on Uncertain Preference

Guohui Li[1], Qi Chen[2], Bolong Zheng[2,3], and Xiaosong Zhao[2(✉)]

[1] School of Software Engineering, Huazhong University of Science and Technology,
Wuhan, China
guohuili@hust.edu.cn
[2] School of Computer Science and Technology,
Huazhong University of Science and Technology, Wuhan, China
{chenqijason,zxs}@hust.edu.cn, zblchris@gmail.com
[3] Department of Computer Science, Aalborg University, Aalborg, Denmark

Abstract. As a reverse rank-aware query, reverse top-k query returns
the user preferences which make the given object belong to the top-k result set. This paper studies the reverse top-k query on uncertain
preferences for the first time. A user's uncertain preference consists of
several probable preference instances, which reflects the user's potential
consumption tendency. In this paper, we design an optimization algorithm *BBUPR* based on the proposed RUI-tree index. Our experiment
results show that *BBUPR* outperforms the other algorithms.

Keywords: Uncertain preference · Reverse top-k query · RUI-tree

1 Introduction

Given a user set, a data set and a query object, reverse top-k query returns a
set of user preferences for which the query object is in top-k query result set.
Reverse top-k query is put forward by Vlachou [3] for market analysis and product placement. Recently, uncertainty as an inherent attribute in some fields, such
as market analysis, drug trial, has attracted a lot of research attention. Therefore, how to execute a query on uncertain data set has been becoming research
hotspot [1,2,6]. However, these existing studies just consider the uncertainty of
the queried data, but ignore the uncertainty of users' preferences.

Different from traditional single weight representation, we use a weight list to
denote a user's uncertain preference. Each user's preference consists of several
preference instances and each preference instance represents a user's possible
preference weight. Table 1 shows two users' uncertain preferences on the restaurants and each preference instance reflects the importance of the features in the
user's view. For example, Jason has two independent preference instances. Compared to the first instance, the rating is more important in the second instance.
Each preference instance is assigned a value to describe the probability that the
corresponding instance takes effect.

© Springer International Publishing AG, part of Springer Nature 2018
Y. Cai et al. (Eds.): APWeb-WAIM 2018, LNCS 10988, pp. 350–358, 2018.
https://doi.org/10.1007/978-3-319-96893-3_26

Table 1. Example of user uncertain preference

User	Price	Ratings	Probability
Jason	0.7	0.3	0.8
Jason	0.5	0.5	0.2
Tom	0.6	0.4	0.3
Tom	0.5	0.5	0.4
Tom	0.3	0.5	0.3

In this paper, we first discuss the revere top-k query on user uncertain preferences and propose a novel query named uncertain preference reverse top-k query($UPRTop$-k). The main contributions are summerised as follows:

- We define the problem of $UPRTop$-k and design a $UPBBR$ algorithm to efficiently handle the $UPRTop$-k query.
- We propose a novel RUI-tree to index users' uncertain preferences, and introduce its advantage over the existing index structures in detail.
- A series of experiments are conducted to evaluate our algorithms, then we give the experimental results analysis and future work direction.

2   Related Work

Uncertain preference reverse top-k query is an extension of reverse top-k query. Vlachou *et al.* [3] formally define reverse top-k query and introduce two versions of query types, respectively monochromatic and bichromatic reverse top-k queries. Considering the enormous time cost, branch and bound algorithm (BBR) is raised [4], which handles the reverse top-k queries efficiently based on the R-tree [5]. To further solve queries on uncertain data some probabilistic algorithms are proposed [5,6]. Wang and Yan [1] first discusses the reverse top-k query on uncertain data and proposes GM algorithm. Then, a novel approach ALS [2] is proposed to handle the same question. However, all the previous studies on probabilistic reverse top-k query neglect the uncertainty of user's preference.

3   Problem Definitions

Let S denote a D-dimensional data set with cardinality $|S|$ and U denote a user set with cardinality $|U|$. W is a preference weight set and we use $W_u = \{w_i\}_{|u|}$ to denote the D-dimensional uncertain preference of user u, where $w_i \in W$ and $|u|$ is the instance number. The aggregated score $f_{w_i}(q)$ for data point q under w_i is defined as a weighted sum of the individual scores: $f_{w_i}(q) = \sum_{d-1}^{n} w_i[d] \times q[d]$, where $w_i[d]$ and $q[d]$ are values on d-th dimension($1 \leq d \leq D$). Next, we first give the necessary definitions to formalize the problem statement. $RTop$-$k(q)$ is the reverse top-k query on point q which is defined in Vlachou's work [3].

Definition 1 *Uncertain Preference Reverse Top-k Query.* *Given a query point q, an integer k, and a threshold τ. UPRT-k(q) returns the users, each of which meets the condition that PRT-k(u, q) ≥ τ where PRT-k(u, q) is defined below:*

$$PRT\text{-}k(u,q) = \sum_{w_i \in \mathcal{W}_u} Pr(w_i \in RTop\text{-}k(q))$$

Definition 2 *Inclusion.* *Given two users u and v, u includes v, denoted by $u \overset{Inc}{=} v$, if the rectangle bound of u covers that of v, i.e., $\forall d, u.l[d] < v.l[d]$ and $u.u[d] > v.u[d]$ (u.l[d] and u.u[d] denote the lower boundary and upper boundary of u's preference on d-th dimension respectively, $1 \le d \le D$).*

4   Algorithms for UPRTop-*k*

4.1   NA and NA* Algorithms

Algorithm 1 illustrates the steps of naive scan algorithm (*NA*). For a user u, a variable $Pr(u)$ records the cumulative probability of the preference instances in the query result during the calculation. The *NA* algorithm executes reverse top-k query for all the preference instances of u. L is a list to store the weights in the result of $RTop\text{-}k(q)$ which are used to avoid the repeated computation. In lines 11–12, the algorithm breaks the inner loop and adds the current user u into the result set R when $Pr(u) \ge \tau$.

Algorithm 1. NA

    Input: q, k, τ
    Output: R
1 $R = \emptyset$ $L = \emptyset$
2 **for** *each* $u \in \mathcal{U}$ **do**
3   **for** *each* $w_i \in \mathcal{W}_u$ **do**
4   **if** $w_i \notin L$ **then**
5   **if** $q \in Top\text{-}k(w_i)$ **then**
6   add w_i to the L; $Pr(u)+ = Pr(w_i)$;
7   **else**
8   continue;
9   **else**
10            $Pr(u)+ = Pr(w_i)$;
11   **if** $Pr(u) \ge \tau$ **then**
12   add u to the R; break;

13 **return** R;

As depicted in Algorithm 1, a user will be added to the result set R iff $Pr(u)$ goes beyond τ. Apparently, an improvement strategy is to ascertain the relationship between $Pr(u)$ and τ as soon as possible. Note that the value of $Pr(u)$ increases continuously during the calculation, but there exists a maximum value $PF(u)$, *i.e.*, the final value after calculating all the weights in $\mathcal{W}_u$. So we try to define an upper limit of $Pr(u)$, denoted by $PU(u)$ as follows,

Algorithm 2. Creating RUI

```
    Input: data units
    Output: RUI-tree
1   B ← Root    C = ∅;
2   while B in not empty() do
3   |   for all bᵢ ∈ B do
4   |   |   if bᵢ is not a duplicate then
5   |   |   |   C.add(extend(bᵢ));
6   |   for all cᵢ ∈ C do
7   |   |   E ← findIncNode(B, cᵢ);
8   |   |   if E.size == 1 and b₁ ∈ E then
9   |   |   |   insert cᵢ as a child of b₁;
10  |   |   else
11  |   |   |   j = findInsertPosition(E);
12  |   |   |   if bⱼ.ChNum > N then
13  |   |   |   |   partition the children of bⱼ and creatIndexUnit for each group;
14  |   |   |   else
15  |   |   |   |   insert cᵢ as a child of bⱼ;
16  |   |   |   |   for all bₖ ∈ E and k ≠ j do
17  |   |   |   |   |   insert *cᵢ as a child of bⱼ;
18  |   B ← C and C = ∅;
```

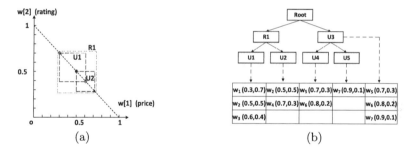

Fig. 1. Example of an RUI-tree

$$Pr(u) \leq PF(u) \leq PU(u)$$
$$PU(u) = 1 - \sum Pr(w_j) \quad \forall w_i, q \notin TopK(w_i)$$

Compared with Algorithm 1, we add an extra conditional statement $PU(u) \leq \tau$ to reduce the loop times. The NA^* algorithm breaks the loop and tests the next user when the inequality becomes true. Moreover, a list L_{dis} is used to store those weights which are discard. When the current weight $w_i \in L_{dis}$, the algorithm skips the subsequent processing and tests the next weight. To further improve the efficiency, we introduce a novel index structure to organize the data.

4.2   RUI-tree

To efficiently solve the queries on uncertain preferences, we first propose RUI-tree (Rectangle Unit Inclusion-tree). As well as R-tree [5], RUI-tree also uses rectangle to index the data and there are three characteristics for an N-branches RUI-tree:

(1) The branch node is a data node or a normal node, where a data node has an extra pointer to the data list except the pointers to its children.
(2) A node may have several duplicates and all the duplicate nodes must be leaf nodes which can't be extended any more.
(3) Each node of RUI-tree has N children nodes at most.

In Fig. 1a, U1 and U2 represent the preferences of Jason and Tom in Table 1 respectively. U1 and U2 have a common preference instance w_2 and they are both covered by the rectangle R1. Figure 1b shows a binary RUI-tree which indexes four users. U3 is a data node which has two children U4 and U5. Different from the normal node R1, it has an extra pointer to the preference weights of u_3.

The Construction of RUI-tree. The Algorithm 2 chooses the unit which has the smallest $ChNum$ as the best insert position. In line 12, if the $ChNum$ of a node goes beyond N, all its children will be divided into two groups. For each group, a normal node is created to index the nodes in the group. Figure 2 gives an example of building a binary RUI-tree.

As depicted in Algorithm 2, the creating algorithm first adds $Root$ into list B. The extension for a data unit is to find the units which are included by it and not included by other extended units. If there is only one unit b_1 which includes the c_i, then insert the c_i as a child of e_1. Otherwise, the algorithm chooses a appropriate position to insert c_i, and inserts its duplicates into other positions.

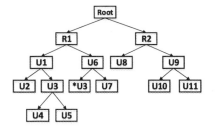

Fig. 2. Example of an RUI-tree

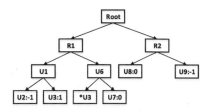

Fig. 3. An RUI-tree after pruning

The Pruning of RUI-tree. To implement our model, we build an RUI-tree over user's uncertain preferences and an R-tree over queried data. Then, we directly use the $INTOPk$ in [4] to prune the RUI-tree. There are three parameters as input for $INTOPk$, an MBR m, a integer k and a query point q. The $INTOPk$ returns $1/-1$ for the weights in m are all accepted/discarded. If the $INTOPk$ can't give a definite conclusion, it will return 0. Each node of RUI-tree is assigned with a sign which is initialized as 0. If the given node s is accepted/discarded, all its descendants are also accepted/discarded. When the $INTOPk$ can't give a definite judgment, the pruning algorithm extends the node and continues to test the extended nodes. Figure 3 gives an example of RUI-tree after pruning.

Algorithm 3. *BBUPR*

Input: q, k, τ
Output: R

1 $t = Root$ $Q.enqueue(t)$;
2 **while** Q *is not empty* **do**
3 $\quad$ $Q.enqueue(t.extend())$;
4 $\quad$ $S.push(t.extend())$;
5 $\quad$ $t = Q.dequeue()$;

6 **while** S *is not empty* **do**
7 $\quad$ $s = S.pop()$;
8 $\quad$ **if** s *is not a leaf node* **then**
9 $\quad\quad$ **for** *each* $ch \in s.chileren$ **do**
10 $\quad\quad\quad$ **if** $ch.sign == 0$ **then**
11 $\quad\quad\quad\quad$ $s.AccW \leftarrow ch.AccW$; $s.DisW \leftarrow ch.DisW$;
12 $\quad\quad\quad$ **else if** $ch.sign == 1$ **then**
13 $\quad\quad\quad\quad$ $s.AccR \leftarrow ch.AccR$;
14 $\quad\quad\quad$ **else**
15 $\quad\quad\quad\quad$ $s.DisR \leftarrow ch.DisR$;

16 $\quad$ **if** s *is a unit node* **then**
17 $\quad\quad$ $Pr(s) = 0$ $PU(s) = 1$;
18 $\quad\quad$ **for** *each* $w_i \in \mathcal{W}_s$ **do**
19 $\quad\quad\quad$ **if** $w_i \in s.AccR$ *or* $w_i \in s.AccW$ **then**
20 $\quad\quad\quad\quad$ $Pr(s)+ = Pr(w_i)$;
21 $\quad\quad\quad$ **else if** $w_i \in s.DisR$ *or* $w_i \in s.DisW$ **then**
22 $\quad\quad\quad\quad$ $PU(s)- = Pr(w_i)$;
23 $\quad\quad\quad$ **else**
24 $\quad\quad\quad\quad$ **if** $q \in Top\text{-}k(w_i)$ **then**
25 $\quad\quad\quad\quad\quad$ $Pr(s)+ = Pr(w_i)$;   add w_i to the $AccW$;
26 $\quad\quad\quad\quad$ **else**
27 $\quad\quad\quad\quad\quad$ $PU(s)- = Pr(w_i)$;   add w_i to the $s.DisW$;
28 $\quad\quad$ **if** $PU(s) < \tau$ **then**
29 $\quad\quad\quad$ break;
30 $\quad\quad$ **if** $Pr(s) \geq \tau$ **then**
31 $\quad\quad\quad$ add s to the R; break;

4.3   UPBBR

BBUPR runs from down to up as depicted in Algorithm 3. In lines 2–5, the algorithm pushes the nodes of RUI-tree into the stack S from the leaves to the Root. s returns the top element of stack and $\mathcal{W}_s$ denotes the preference weights in node s. For each branch node, four lists are arranged to save these useful information. *AccW* and *DisW* are two lists to store the discrete weights accepted and discarded respectively. *AccR* and *DisR* record the rectangle regions in which all the weights are accepted and discarded respectively. These valuable information in the lists are delivered from the children nodes to the parent s and the algorithm gives the final judgment for s by leveraging these information. For those weights which are not in the lists, it executes *RTop-k* queries.

5   Experiments

5.1   Experimental Setup

Our platform is a computer with Intel core i5-4590 CPU@3.30 GHz and 8 GB RAM. All the simulation programs are developed in C++. Two kinds of data sources are examined, synthetic data and real data.

Synthetic Data: We first generate the weight data with two different distributions, namely uniform (UN) and clustered (CL). The user preference instances are sampled uniformly from the weight data and the instance number is a random integer randomly between 1 and 15. Three kinds of queried data are generated respectively uniform (UN), correlated (CO), anti-correlated (ACO).

Real Data: The real data comes from Yelp[1], which consists of 22,870 users' check-in data on 22,014 locations in Las Vegas. A user's preference instance represents his preference on the locations in one category such as restaurant, bar *etc.* and the corresponding visit frequency is regarded as the probability. By means of sentiment analysis methods in [7], we extract the user preferences and location scores on 4 features, environment, traffic, food and service.

5.2   Experimental Result

We conduct experiments by varying the different parameters. The default setup is: $D = 3$, $|S| = 10K$, $|W| = 10K$, $k = 50$, $\tau = 0.6$, $\mathcal{S}$ and $\mathcal{U}$ follow UN distribution.

Cardinality Test. Figurre 4a shows that *NA* and *NA** perform much worse than *UPBBR* and *BBR* with increasing $|\mathcal{S}|$. Notice that *UPBBR* maintains its advantage over the other algorithms as $|\mathcal{S}|$ increases and shows its superiority obviously when $|\mathcal{S}| \geq 10000$. Moreover, the performance of *UPBBR* is influenced by the cardinality of $\mathcal{S}$ only slightly, which shows its good scalablity. Compared to $|\mathcal{S}|$, all the algorithms are more sensitive to the increase of $|\mathcal{U}|$. Because it is more difficult to build an index on the user data than the queried data.

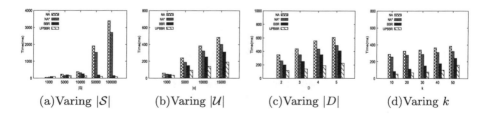

(a) Varing $|\mathcal{S}|$ (b) Varing $|\mathcal{U}|$ (c) Varing $|D|$ (d) Varing k

Fig. 4. Comparative performance for parameter varing

Dimension Test. Figure 4c presents the comparative performance of all the algorithms for varying dimension D. The time cost of the algorithms have different proportional increases with the increasing dimension D. Except for the calculation for the ranking scores, *BBR* and *UPBBR* have extra time cost on the index construction. The growth rate of time cost for *UPBBR* is slightly below *NA* and *NA**, and higher than the *BBR*. But the total time cost of *UPBBR* is far below the other algorithms.

Parameter k Test. In Fig. 4d, we test our algorithms by varying the parameter k. We can see that there exists a positive correlation between the cost time and the value of k for all the algorithms. And the cost time increase much slower than the value of k.

Data Distribution Test. Figure 5 depicts the performance of the algorithms on UN, CO and AC data, when the user set $\mathcal{U}$ follows the UN and CL distribution respectively. *BBR* and *UPBBR* get the best results on CO queried data and perform worst on AC data. The reason is that there are more domination relationships between the data in CO dataset.

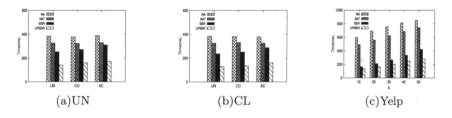

(a)UN   (b)CL   (c)Yelp

Fig. 5. Comparative performance for different datasets

Real Data Test. Figure 5c shows that *BBR* and *UPBBR* are more efficient than *NA* and *NA** on Yelp dataset clearly. Specifically, the performance advantage of *UPBBR* enhances with the increase of k. The results on Yelp keep in accordance with the experiments on the synthetic data. Therefore, we conclude that the *UPBBR* consistently improves the efficiency of *UPRTop-k* query.

6   Conclusion

In this paper, we first study the reverse top-k query on uncertain preferences. To address the problem, we propose an novel index structure named RUI-tree to support the efficient query. Experimental results show that the improved *UPBBR* algorithm outperform the other methods when the data set tends to be large scale. In the future, we will do some work on algorithm optimization and user's uncertain preference mining.

References

1. Wang, X., Yan, J.: Reverse top-k query on uncertain data. Comput. Sci. (Chin.) **39**, 191–194 (2012)
2. Jin, C., Zhang, R., Kang, Q., Zhang, Z., Zhou, A.: Probabilistic reverse top-k queries. In: Bhowmick, S.S., Dyreson, C.E., Jensen, C.S., Lee, M.L., Muliantara, A., Thalheim, B. (eds.) DASFAA 2014, pp. 406–419. Springer, Cham (2014). https://doi.org/10.1007/978-3-319-05810-8_27
3. Vlachou, A., Doulkeridis, C., Kotidis, Y., Norvag, K.: Monochromatic and bichromatic reverse top-k queries. In: Proceedings TKDE (2011)
4. Vlachou, A., Doulkeridis, C., Kjetil, N., Kotidis, Y.: Branch-and-bound algorithm for reverse top-k queries. In: Proceedings SIGMOD (2013)
5. Guttman, A.: R-trees: a dynamic index structure for spatial searching. In: Proceedings SIGMOD, pp. 47–57 (1984)
6. Soliman, M.A., Ilya, I.F., K.C., Chang, C.: Top-k query processing in uncertain databases. In: Proceedings ICDE (2007)
7. Zhang, Y., Lai, G., Ma, S.: Explicit factor models for explainable recommendation based on phrase-level sentiment analysis. In: Proceedings SIGIR 2014, pp. 83–92 (2014)

Keyphrase Extraction Based on Optimized Random Walks on Multiple Word Relations

Wenyan Chen[1,2], Zheng Liu[1,2(✉)], Wei Shi[3], and Jeffrey Xu Yu[3]

[1] Jiangsu Key Laboratory of Big Data Security and Intelligent Processing,
Nanjing, China
[2] School of Computer Science, Nanjing University of Posts and Telecommunications,
Nanjing, China
zliu@njupt.edu.cn
[3] The Chinese University of Hong Kong, Sha Tin, Hong Kong

Abstract. Extracting keyphrases from documents helps to reduce the document information and further assist in information retrieval. In this paper, we construct a multi-relational graph by considering heterogeneous latent word relations (the co-occurrence and the semantic) in a document. Then we optimize the random walks on the multi-relational graph to determine the importance of each node to further generate keyphrases. Experimental results show that our method outperforms the previous methods.

Keywords: Keyphrase extraction · Multi-relational graph
Optimized random walks

1 Introduction

Keyphrases are the thematic and representative words and phrases in documents. Extracting keyphrases from lengthy documents will help readers understand and grasp their main topics quickly. Meanwhile, such keyphrase extraction could assist tasks in natural language processing and information retrieval such as text categorization [13], text clustering [2], etc.

In the literature, keyphrase extraction could be categorized into either supervised methods or unsupervised methods. The former one models keyphrase extraction as a binary classification task to divide the candidate phrases into keyphrases or non-keyphrases [3,11]. These methods construct a classifier by using the training data in which phrases are manually labeled as keyphrase or non-keyphrase by domain experts. However, in these methods the training data and the test data must lie in the same domain. Once the domain changes, the classifier needs to be retrained, which is time-consuming.

On the other hand, unsupervised methods can perform without prior knowledge. These methods mostly consider keyphrase extraction as a ranking problem,

© Springer International Publishing AG, part of Springer Nature 2018
Y. Cai et al. (Eds.): APWeb-WAIM 2018, LNCS 10988, pp. 359–367, 2018.
https://doi.org/10.1007/978-3-319-96893-3_27

most of which are based on graphs [4,10]. In these methods, documents are modeled as graphs where nodes represent words and edges indicate the relationship between words, e.g., the lexical or semantic relations. While most existing graph-based methods focus on a single type of relation between words, which is not sufficient to cover all the document information.

In this paper, we propose a graph-based approach for keyphrase extraction. We construct multi-relational graphs based on heterogeneous word relations and rank the words using the random walking model on tensors. We design an optimized random walk model to bias random surfers to visit essential words and relations more often. Finally, the keyphrases are extracted based on the scores of the candidate phrase by adding the importance score of words. Experiments demonstrate the effectiveness of our proposed method.

The remainder of this paper is organized as follows. Section 2 describes the multi-relational graph construction. And Sect. 3 explains our proposed optimized random walks. Section 4 explains the strategy of keyphrase selection. And we present our experimental evaluation in Sect. 5. Section 6 introduces the related work and finally Sect. 7 concludes the paper.

2   The Construction of Multi-relational Graphs

In the multi-relational graph construction for documents we consider two different word relations, i.e., the co-occurrence relationship and the semantic relationship for simplicity. It is worth noting that our proposed method could be extended to more than two word relations straightforwardly.

We apply the method used in [8] to construct the co-occurrence graph and semantic graph, respectively. Due to space constraints, we only present the adjacency matrix A^c for co-occurrence graph and A^s for semantic graph respectively.

$$A^c_{uv} = w_{uv} = count(u, v), \quad A^s_{uv} = w_{uv} = cos(u, v). \tag{1}$$

We then unite these two graphs into one multi-relational graph $G = (V, E)$. The united graph G could be represented by a tensor $A = (a_{ijk})$, where $i = 1, \ldots, m$, $j = 1, \ldots, m$ and $k = 1, \ldots, n$. m is the number of nodes and n is number of possible relations, in this paper, $n = 2$. A is non-negative due to $a_{ijk} \geq 0$.

$$a_{ijk} = \begin{cases} A^c_{ij}, & k = 1; \\ A^s_{ij}, & k = 2. \end{cases} \tag{2}$$

3   The Computation of the Word Importance by Optimized Random Walks

3.1   A Random Walk Model on Multi-relational Graphs

Let us first explain the random walk model on the multi-relational graph G. Suppose a random surfer is visiting node v_j at time $t-1$, since there are multiple relations in the graph, the random surfer will randomly visit any neighbors of v_j using r_k, where r_k is one of the relations. So the transition probability of the random surfer is $p(v_i|v_j, r_k)$, which we call it intra-relation transition probability, then the stationary probability of visiting a node v_i is

$$p(v_i) = \sum_{j=1}^{m} \sum_{k=1}^{n} p(v_i|v_j, r_k) \times p(v_j, r_k). \tag{3}$$

Same as the transition probability in PageRank, we could define the intra-relation transition probability on tensor A as

$$p(v_i|v_j, r_k) = \frac{a_{ijk}}{\sum_{i=1}^{m} a_{ijk}}. \tag{4}$$

It should be noted that if a_{ijk} is equal to 0 for all $1 \leq i \leq m$, this is called a dangling node [7], and the value of $p(v_i|v_j, r_k)$ is set to $1/m$.

In general, it may be difficult to obtain the joint probability $p(v_j, r_k)$. Assume that the word distribution is independent from the relations, we have $p(v_j, r_k) = p(v_j) \times p(r_k)$. So the stationary probability $p(v_i)$ can be expressed as follows.

$$p(v_i) = \sum_{j=1}^{m} \sum_{k=1}^{n} p(v_i|v_j, r_k) \times p(v_j) \times p(r_k). \tag{5}$$

And we could obtain $p(r_k)$ as the same way,

$$p(r_k) = \sum_{i=1}^{m} \sum_{j=1}^{m} p(r_k|v_j, v_i) \times p(v_i, v_j) = \sum_{i=1}^{m} \sum_{j=1}^{m} p(r_k|v_j, v_i) \times p(v_i) \times p(v_j), \tag{6}$$

Similar to the aforementioned intra-relation transition probability, we call $p(r_k|v_j, v_i)$ the inter-relation transition probability. In MultiRank [5], the inter-relation transition probability tensor is defined as

$$p(r_k|v_j, v_i) = \frac{a_{ijk}}{\sum_{k=1}^{n} a_{ijk}}. \tag{7}$$

Now let O_{ijk} denote $p(v_i|v_j, r_k)$ and R_{ijk} denote $p(r_k|v_j, v_i)$, based on Eqs. 5 and 6, the tensor form of the stationary landing probability is

$$\mathbf{p} - \mathbf{Opr}, \quad \mathbf{r} = \mathbf{Rp}^2, \tag{8}$$

where $\mathbf{p} = (p_1, \ldots, p_m)$ and $\mathbf{r} = (r_1, \ldots, r_n)$ are vectors of the corresponding stationary probabilities, which we consider as the importance scores of words and relations, respectively.

3.2   Optimized Random Walks on Multi-relational Graphs

The intra-relation transition probability defined above is the same as one in Pagerank while using the same strategy for inter-relation transition probability is not appropriate since edge weights in multi-relational graphs have various meanings in different domains with different distributions.

In this paper, we propose to model both intra-relation transition probability and inter-relation transition probability as an optimization problem. We aim to bias random surfers to visit essential words and relations more often by optimizing the random walk based on trained transition probabilities.

Assume each node v_i in the multi-relation graph has a feature vector φ_{ik}, where k indicates that the features in φ_{ik} is related to relation r_k. In computing the intra-relation transition probability, we define the edge strength for intra-relation as $f_{w_1}(\varphi_{ik}, \varphi_{jk})$ parameterized by w_1. Let $N_k(i)$ denote the neighbors of node v_i in relation r_k, the intra-relation probability $p(v_i|v_j, r_k)$ is

$$O_{ijk} = p(v_i|v_j, r_k) = \frac{f_{w_1}(\varphi_{ik}, \varphi_{jk})}{\sum_{v_l \in N_k(i)} f_{w_1}(\varphi_{ik}, \varphi_{lk})}. \tag{9}$$

Similarly, we denote the edge strength for inter-relation as $f_{w_2}(\varphi_{ik}, \varphi_{jk})$ parameterized by w_2. Let K denote the total number of relations, then the inter-relation transition probability $p(r_k|v_j, v_i)$ is defined as follows.

$$R_{ijk} = p(r_k|v_j, v_i) = \frac{f_{w_2}(\varphi_{ik}, \varphi_{jk})}{\sum_{l=1}^{K} f_{w_2}(\varphi_{il}, \varphi_{jl})}. \tag{10}$$

It is obvious that words in keyphrases should be more important than words not in them. So let I denote the set of important words and O denote the other words. Now our task is to optimize the parameters w_1 and w_2 so that the importance of words in I is larger than one in O. Formally, the optimization problem is

$$\min_{w_1, w_2} \|w_1\|^2 + \|w_2\|^2$$
$$\text{s.t.} \forall i \in I, \forall j \in O : p_i > p_j, \tag{11}$$

where p_i and p_j are the corresponding stationary probabilities of v_i and v_j in $\mathbf{p}$, respectively. It is possible that there is no solution that satisfies all the constrains in Eq. 11, so we introduce a loss function $h(\cdot)$ to penalize violated constraints. Then the new optimization problem is as follows.

$$\min_{w_1, w_2} \|w_1\|^2 + \|w_2\|^2 + \lambda \sum_{i \in I, j \in O} h(p_j - p_i), \tag{12}$$

where λ is the regularization parameter which controls the degree of constraint violation. $h(\cdot)$ is the loss function that assigns a non-negative penalty. If the constrain is not violated, then $p_j - p_i \leqslant 0$, $h(p_j - p_i) = 0$. Otherwise $h(p_j - p_i) > 0$.

Let F denote the objective function in Eq. 12. To minimize the value of F, we solve the optimization problem using the gradient-based method. We need to compute the value of the derivative $\frac{\partial h(\delta)}{\partial w_1}$, $\frac{\partial \mathbf{p}}{\partial w_1}$, $\frac{\partial h(\delta)}{\partial w_2}$ and $\frac{\partial \mathbf{p}}{\partial w_2}$, where $\frac{\partial h(\delta)}{\partial w_1}$ and $\frac{\partial h(\delta)}{\partial w_2}$ could be computed based on the derivative of the loss function $h(\cdot)$. And $\frac{\partial \mathbf{p}}{\partial w_1}$ and $\frac{\partial \mathbf{p}}{\partial w_2}$ could be computed by solving the derivatives of Eq. 8, which would derive a series of chain derivatives formulas.

By recursively applying the chain rule, we can use a power-method to compute $\frac{\partial \mathbf{p}}{\partial w_1}$ and $\frac{\partial \mathbf{p}}{\partial w_2}$ since $\mathbf{p}$, $\frac{\partial \mathbf{p}}{\partial w_1}$ and $\frac{\partial \mathbf{p}}{\partial w_2}$ are recursively entangled. The computation of $\frac{\partial \mathbf{r}}{\partial w_1}$ and $\frac{\partial \mathbf{r}}{\partial w_2}$ is similar. Then we could apply a gradient-based method to minimize F to find the optimal parameters w_1 and w_2, with which we could calculate the final intra-relation transition probability and inter-relation transition probability. Then the final stationary probability distribution $\mathbf{p}$ and $\mathbf{r}$ could be obtained to represent the word importance scores.

4   Candidate Phrase Generation and Keyphrase Selection

We employ the same strategy like most previous research [8], which uses the syntactic pattern based on the part-of-speech of words to generate candidate phrases.

Now we explain how to calculate the ranking scores for candidate phrases, and the candidate phrases with the top-k largest phrase scores are selected as the keyphrases. The initial score of the candidate phrase C is computed as follow.

$$InitialScore(C) = \sum_{v_i \in C} \mathbf{p}(v_i). \tag{13}$$

Combine with the phrase frequency $freq(C)$ and the first occurrence position $pos(C)$ in the document, the final phrase score $Score(C)$ for ranking is as follow.

$$Score(C) = \sqrt{\frac{freq(C)}{pos(C)}} \times InitialScore(C). \tag{14}$$

5   Experiments

We demonstrate our experimental results on the DUC2001 dataset [6]. The DUC2001 dataset contains a total number of 308 news articles collected from TREC-9, which are categorized into 30 topics, and each article has been manually assigned to around 10 keyphrases. We split the dataset into two parts for training and testing separately and conduct the cross validation.

5.1   Parameters Selection

- **The feature vectors** φ_{ik}. In the experiment, we use the following features to generate a feature vector φ_{ik}: the number of neighbors of v_i in relation r_k, The max, min and average weights of edges connected to v_i in relation r_k.
- **The Strength function** $f_{w_i}(\cdot)$. We apply a logistic function in the process of modelling both the intra-relation transition probability and inter-relation transition probability, which is defined as follows.

$$f_{w_i}(\varphi_{ik}, \varphi_{jk}) = \frac{1}{(1 + exp(-(\varphi_{ik}, \varphi_{jk}) \cdot w_i))}. \tag{15}$$

We first concatenate these two vectors φ_{ik} and φ_{jk} into a united vector which could reflect the property of corresponding edge in relation r_k. $(\varphi_{ik}, \varphi_{jk}) \cdot w_i$ is the inner product of vector w_i and vector $(\varphi_{ik}, \varphi_{jk})$.

- **The loss function** $h(\cdot)$. We adopt the Wilcoxon-Mann-Whitney (WMW) loss function with width b [12] where b is set as 1 in our experiment.
- **The regularization parameter** λ. Since there is no over-fitting issue found in our experiments, we set $\lambda = 1$ for simplicity.

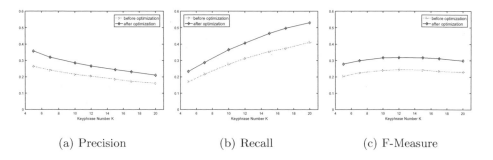

(a) Precision   (b) Recall   (c) F-Measure

Fig. 1. Comparison with no optimized method.

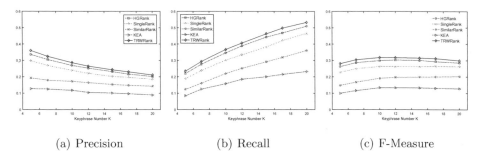

(a) Precision   (b) Recall   (c) F-Measure

Fig. 2. Comparison with other algorithms.

5.2   Experimental Results

Three metrics precision (P), recall (R) and F-measure (F) are used to evaluate the performance of different methods.

Firstly, we compare our approach with the method using MultiRank for word ranking. The results are presented in Fig. 1(a), (b) and (c). These two approach settings are the same except that we calculate the transitional probabilities by optimization. It shows that the optimized random walk outperforms the settings in MultiRank for keyphrase extraction, which proves that the optimized method we designed is reasonable and effective.

We then compare our method, denote as **TRWRank**, with three other unsupervised keyphrase extraction methods, which are HGRank [8], SimilarRank [8]

and SingleRank [10], and one supervised method KEA [11]. Figure 2(a), (b) and (c) shows the precision, recall and F-measure. For HGRank, we set window size as $W = 20$ and similarity threshold $\theta = 0.4$. For SingleRank, we set $W = 20$, and $\theta = 0.4$ in SimilarRank. For TRWRank, we set $W = 20$ and $\theta = 0.4$. As the curves show, our proposed method performs the best among these methods.

Further experiments on different parameter values W and θ are conducted to show the influence of the parameters and to help selecting the best parameters. Figure 3(a) and (b) shows the influence of performance when θ ranges from 0.1 to 0.6. And Fig. 3(c) and (d) shows the influence of W varies between 2 to 20.

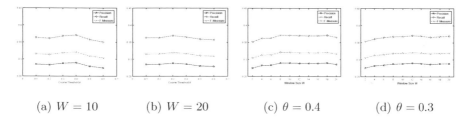

(a) $W = 10$ (b) $W = 20$ (c) $\theta = 0.4$ (d) $\theta = 0.3$

Fig. 3. Effect of cosine threshold θ and window size W.

6   Related Work

Existing graph-based keyphrase extraction methods construct graphs based on the explicit word relations. The word importance score is then calculated to further determine the importance of each phrase. In such methods, the results heavily depend on the word relation used and the ranking method.

The classic work on graph-based method is TextRank [4], which constructs the co-occurrence graph by assigning the initial node weight as $1/n$ where n is the total number of nodes. The initial edge weight is set from 0 to 10 according to the uniform distribution. By ranking the words in the graph using PageRank [7], keyphrases are extracted if the words of the phrase are in the top-k word list. SingleRank [10] also creates the co-occurrence graph for a document and assigns the initial node weight just like TextRank. However, SingleRank employs the number of times that two corresponding nodes appear together in a given window size W as the initial edge weight. Different from the above methods, BetweennessRank [1] employs betweenness to rank words in the constructed co-occurrence graph. In addition, SemantiRank [9] extracts keyphrases based on the semantic graphs constructed using word relationships in WordNet and Wikipedia's page links.

Rather than considering only one single relation, HGRank [8] takes the co-occurrence graph and the semantic graph into account simultaneously to construct a united graph. This method designs a random surfer model on this united graph like PageRank but estimates the intra-relation transition probability and inter-transition probability, and further ranks the nodes in this graph.

7   Conclusion

In this paper, we construct a multi-relational graph based on co-occurrence relations and semantic relations between words to extract keyphrases. We design an optimized random walk to rank the nodes, making the random surfer more often to visit the important nodes through crucial edges. The combination of phrase features further improve the performance. It should be noted that the method we proposed to bias the random walk process may depend on the features and different parameters. Moreover, the initial parameter may also have impact on the result.

Acknowledgements. This work is supported in part by Jiangsu Provincial Natural Science Foundation of China under Grant BK20171447, Jiangsu Provincial University Natural Science Research of China under Grant 17KJB520024, and Nanjing University of Posts and Telecommunications under Grant No. NY215045.

References

1. Boudin, F.: A comparison of centrality measures for graph-based keyphrase extraction. In: Sixth International Joint Conference on Natural Language Processing, IJCNLP 2013, Nagoya, Japan, 14–18 October 2013, pp. 834–838 (2013)
2. Hammouda, K.M., Matute, D.N., Kamel, M.S.: CorePhrase: keyphrase extraction for document clustering. In: Perner, P., Imiya, A. (eds.) MLDM 2005. LNCS (LNAI), vol. 3587, pp. 265–274. Springer, Heidelberg (2005). https://doi.org/10.1007/11510888_26
3. Hulth, A.: Improved automatic keyword extraction given more linguistic knowledge. In: Proceedings of the Conference on Empirical Methods in Natural Language Processing, EMNLP 2003, Sapporo, Japan, 11–12 July 2003 (2003)
4. Mihalcea, R., Tarau, P.: TextRank: Bringing Order into Texts, pp. 404–411. UNT Scholarly Works (2004)
5. Ng, M.K., Li, X., Ye, Y.: Multirank: co-ranking for objects and relations in multi-relational data. In: Proceedings of the 17th ACM SIGKDD International Conference on Knowledge Discovery and Data Mining, San Diego, CA, USA, 21–24 August 2011, pp. 1217–1225 (2011). https://doi.org/10.1145/2020408.2020594
6. Over, P.: Introduction to DUC-2001: an intrinsic evaluation of generic news text summarization systems. In: DUC 2001 Workshop on Text Summarization (2001)
7. Page, L.: The PageRank citation ranking: bringing order to the web. Stanf. Digit. Libr. Work. Pap. **9**(1), 1–14 (1998)
8. Shi, W., Liu, Z., Zheng, W., Yu, J.X.: Extracting keyphrases using heterogeneous word relations. In: Huang, Z., Xiao, X., Cao, X. (eds.) ADC 2017. LNCS, vol. 10538, pp. 165–177. Springer, Cham (2017). https://doi.org/10.1007/978-3-319-68155-9_13
9. Tsatsaronis, G., Varlamis, I., Nørvåg, K.: SemanticRank: ranking keywords and sentences using semantic graphs. In: 23rd International Conference on Computational Linguistics, Proceedings of the Conference, COLING 2010, 23–27 August 2010, Beijing, China, pp. 1074–1082 (2010)
10. Wan, X., Xiao, J.: Exploiting neighborhood knowledge for single document summarization and keyphrase extraction. ACM Trans. Inf. Syst. **28**(2), 8:1–8:34 (2010). https://doi.org/10.1145/1740592.1740596

11. Witten, I.H., Paynter, G.W., Frank, E., Gutwin, C., Nevill-Manning, C.G.: KEA: practical automatic keyphrase extraction. In: Proceedings of the Fourth ACM Conference on Digital Libraries, Berkeley, CA, USA, 11–14 August 1999, pp. 254–255 (1999). https://doi.org/10.1145/313238.313437

12. Yan, L., Dodier, R., Mozer, M.C., Wolniewicz, R.: Optimizing classifier performance via an approximation to the Wilcoxon-Mann-Whitney statistic. In: Machine Learning, Proceedings of the Twentieth International Conference, pp. 848–855 (2003)

13. Youn, E., Jeong, M.K.: Class dependent feature scaling method using naive bayes classifier for text datamining. Pattern Recogn. Lett. **30**(5), 477–485 (2009). https://doi.org/10.1016/j.patrec.2008.11.013

Answering Range-Based Reverse kNN Queries

Zhefan Zhong[1,2], Xin Lin[1,2(✉)], Liang He[1,2], and Yan Yang[1,2]

[1] Shanghai Key Laboratory of Multidimensional Information Processing,
Shanghai, China
zhfzhong@ica.stc.sh.cn
[2] Department of Computer Science and Technology, East China Normal University,
Shanghai 200241, China
{xlin,lhe,yanyang}@cs.ecnu.edu.cn

Abstract. Given a point q, a reverse k nearest neighbor (RkNN) query retrieves all the data points that have q as one of their k nearest neighbors. Despite significant progress on this problem, there is a research gap in finding RkNNs not just for an object, but for a given range, which is a natural extension of the problem. Motivated by this, we develop algorithms for *exact* processing of range-based RkNN with *arbitrary* values of k on *dynamic* datasets, which retrieve all the data points that have any position in the given query range R as one of their k nearest neighbors. The experimental results demonstrate the efficiency and the accuracy of our proposed optimizations and algorithms.

Keywords: Range-based RkNN queries · Location-based services

1 Introduction

Given a dataset D and a point q, a reverse nearest neighbor (RNN) query retrieves all the points $p \in D$ that have q as their nearest neighbor. Although the RNN problem was first proposed in [3], it still has received considerable attention due to its importance in several applications involving decision support, resource allocation, profile-based marketing, etc.

Despite significant progress on this problem, there is a research gap in finding RNNs not just for an object, but for a given range, which is a natural extension of the problem. In this paper, we proposed a range-based reverse nearest neighbor (RRNN) query, it retrieves all the points $p \in D$ that have any position in the query range R as their nearest neighbor. We assume that the shape of range R is rectangle.

Figure 1 shows a range R and nine 2D points, where each point p is associated with a circle covering its nearest neighbor. For example, the NN of p_4 (eg. p_5) is in the circle centered at p_4. Some of these circles (such as circle of p_1) intersect with the range R. Accordingly, $p_1 \in \text{RNN}(R)$ (see Definition 2 in Sect. 3). In this case, we can easily get the $\text{RNN}(R)=(p_1, p_2, p_3, p_4, p_5)$.

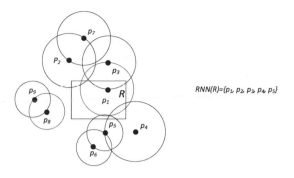

Fig. 1. Range-based RNN examples

As discussed in Sect. 2, all the previous methods for RNN search can not handle a range-based RNN query. Motivated by this, focusing on monochromatic reverse nearest neighbor problems, we develop algorithms to handle range-based reverse k nearest neighbor (RRkNN) queries, which retrieve all the points $p \in D$ that have any position in the query range R as one of their k nearest neighbors. Specifically, we follow the filter-refinement framework, in which the filter step retrieves a set of candidate results that is guaranteed to include all the actual reverse nearest neighbors and the refinement step eliminates the false hits. None of the existing techniques can effectively answer range-based RkNN query accurately. In fact, answering range-based RkNN query is very useful since RkNN query is very common in our daily life.

The rest of the paper is organized as follows. Section 2 surveys related works on RNN search. In Sect. 3, we give preliminaries of range-based RkNN query and illustrate that it is computationally expensive for existing algorithms. In Sect. 4, we present some interesting problem characteristics, and propose a new algorithm with demonstrations to solve the range-based RkNN problem efficiently. In Sect. 5, we report experimental results and we conclude the paper in Sect. 6.

2   Related Work

There exist various versions of RNN problem include (1) continuous RNN [5], in which the database contains linearly moving objects with fixed velocities, and the goal is to retrieve all RNNs of q for a future interval; (2) bichromatic RNN [6], given a set Q of queries, the goal is to find the objects $p \in D$ that are closer to some $q \in Q$ than any other point of Q; (3) stream RNN [4], where data arrives in the form of streams, and the goal is to report aggregated results over the RNNs of a set of query points.

Algorithms for RNN processing can be classified into two categories depending on whether they require preprocessing, or not. The original RNN method [3] pre-computes for each data point p its nearest neighbor NN(p). Then using the RNN-tree, the reverse nearest neighbors of q can be efficiently retrieved by a point location query, which returns all circles that contain q. Similarly, Yang and Lin [7] combine the R-tree and RNN-tree in the RdNN-tree. Another solution based on pre-computation is proposed in [13]. All techniques that rely on pre-processing cannot deal efficiently with updates because each insertion or deletion may affect the vicinity circles of several points. Stanoi et al. [10] solve the problem by utilizing some interesting properties of RNN retrieval, adopts a two-step processing method(filter-refinement framework). Singh et al. [11] finds the kNNs of the query q, eliminates the candidates that are closer to some other candidate than q and then applies boolean range queries on the remaining candidates to determine the actual RNNs. [2] develop algorithms for exact processing of RkNN with arbitrary values of k on dynamic multi-dimensional datasets. In addition, some other techniques like [8,9,12] are helpful for our query, their methods utilize a conventional data-partitioning index on the dataset and do not require any pre-computation. However, all these previous methods for RNN search can not handle a range-based RkNN query.

3   Problem Statement

In this section, we first give preliminaries of range-based RkNN query. Afterwards, we introduce some concepts used for solving this problem.

Given a spatial dataset D of n objects, each object p with 2 attribute values can be represented as a point $p = (p[1], p[2])$ in a 2-dimensional data space. For simplicity, we assume that all attribute values are numeric. A range-based RkNN query is composed of a query range R and an arguments k, retrieves all the points $p \in D$ that have any position in the query range R as one of their kNNs. In this paper, we consider the Euclidean distance as the metric. The query results would then be a set of objects whose kNN vicinity circle [3] overlaps query range R.

Definition 1 (Reverse k Nearest Neighbor Query (RkNNQ)). *Given a dataset D and a query point $q(k)$, the query returns all the points $p \in D$ that have q as one of their k nearest neighbors, denote as $RkNN(q)$.*

Definition 2 (Range-Based Reverse k Nearest Neighbor Query (RRkNNQ)). *Given a dataset D and a query range R, the query returns an answer set denote as $RkNN(R)$, $\forall\, o \in D$, $\forall\, p \in R$, if $o \in RkNN(p)$, then $o \in RkNN(R)$. That is, $RkNN(p) \subseteq RkNN(R)$.*

Compare to existing point-based RkNN query, a range-based RkNN query needs great amount of calculation. One of the straightforward methods is to precompute the kNNs for all the data points, however, it's unreasonable since it is costly to extends to arbitrary values of k.

4   Range-Based RkNN Query

In this section, we present our method to solve range-based RkNN queries. Section 4.1 illustrates some problem characteristics that permit the development of efficient algorithms presented in Sect. 4.2. Section 4.3 presents properties that permit pruning of the search space for arbitrary values of k, then extends our methods for range-based RkNN queries. We assume that dataset D is indexed by R-tree.

4.1   Problem Characteristics

Consider single range-based RNN processing first. As shown in Fig. 2a, divide the space into nine subspaces according to the edge of the query range R, we can obtain the candidate results of RNNs in every subspace separately.

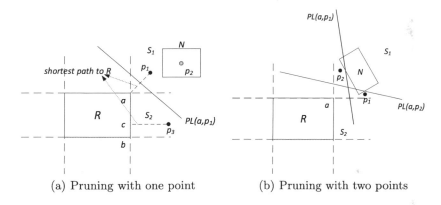

(a) Pruning with one point   (b) Pruning with two points

Fig. 2. Half-plane pruning strategy for subspaces like S_1

Consider the perpendicular bisector $\perp (a, p_1)$ between the vertex of range a and a data point p_1 as shown in Fig. 2a. The bisector divides S_1 into two half-planes: $PL_a(a, p_1)$ that contains a, and $PL_{p_1}(a, p_1)$ that contains p_1. Any point (e.g., p_2) in the $PL_{p_1}(a, p_1)$ cannot be a RNN of R because it is closer to p_1 than a. Similarly, a node MBR (e.g., N) that falls completely in $PL_{p_1}(a, p_1)$ cannot contain any candidate. In some cases, the pruning of an MBR requires multiple half-planes. For example, in Fig. 2b, although N does not fall completely in $PL_{p_1}(a, p_1)$ or $PL_{p_2}(a, p_2)$, it can still be pruned since it lies entirely in the *union* of the two half-planes. In general, if $p_1, p_2, ...p_n$ are n data points, then any node whose MBR falls inside $\bigcup_{i=1 \sim n} PL_{p_i}(a, p_i)$ cannot contain any RNN result.

Similarly, as shown in Fig. 3, in the subspaces like S_2, consider the parabola $Par(L, p_1)$ having L as its directrix and p_1 as its focus. The parabola divides S_2 into two half-planes: $Par_L(L, p_1)$ that contains L, and $Par_{p_1}(L, p_1)$ that contains p_1. Any point (e.g., p_2) in $Par_{p_1}(L, p_1)$ cannot be a RNN of R because it is closer to p_1 than L. And a node MBR (e.g., N) that completely falls in $Par_{p_1}(L, p_1)$ cannot contain any candidate.

While for the data points in query range R, they are born to be RNNs.

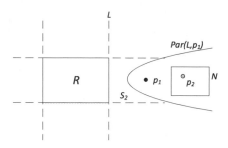

Fig. 3. Half-plane pruning strategy for subspaces like S_2 (pruning with one point)

4.2   The TPR Algorithm

Based on the above discussions, we adopt a two-step framework that retrieves a set of candidate RNNs (filtering step) and then removes the false hits (refinement step). Different from [10,11], our algorithm (hereafter, called TPR) traverses the R-tree in a best-first manner, retrieving potential candidates in ascending order of their distance to the query range R in each subspace respectively, because RNNs are likely to be near R. The concept of half-plane is used to prune node MBRs (data points) that cannot contain (be) candidates. In the refinement step, we applies boolean range queries on the remaining candidates to determine the actual RNNs. Algorithm 1 shows the details of TPR algorithm.

4.3   Range-Based R*k*NN Processing

This section presents properties that help pruning of the search space for arbitrary values of k and extends our TPR algorithm for range-based R*k*NN query.

Figure 4 shows an example with $k = 2$. In Fig. 4a, p_3 is not a R2NN of R, since p_3 is in the *intersection* of $PL_{p_1}(a, p_1)$ and $PL_{p_2}(a, p_2)$. In Fig. 4b, p_3 is not a R2NN of R, since p_3 is in the *intersection* of $Par_{p_1}(L, p_1)$ and $Par_{p_2}(L, p_2)$.

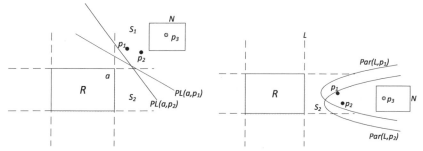

(a) Range 2-pruning algorithm in sub-   (b) Range 2-pruning algorithm in subspaces like S_1 spaces like S_2

Fig. 4. Examples of R2NN queries

Algorithm 1. TPR algorithm

INPUT: dataset D indexed in R-Tree $\mathcal{T}$, query range R
OUTPUT: S_{rnn}

1: Initialize sets S_{cnd} //RNN candidates
2: Divide the space into nine parts around query range R
3: In the subspaces that adjacent to a vertex (edge) of query range like $S_1(S_2)$ **do**
4:     Initialize a min-heap H accepting entries of the form (e, key)
5:     Insert $(\mathcal{T}, 0)$ to H
6:     **while** H is not empty
7:     (e, key)=de-heap H
8:         **if** e can be pruned **then** goto 6
9:         **else** //entry may be or contain a candidate
10:             **if** e is data point p **then** $S_{cnd}=S_{cnd} \cup p$
11:             **else if** e points to a leaf node N
12:                 **for** each point p in N (sorted on $dist(p, R)$)
13:                     **if** p cannot be pruned **then** insert $(p, dist(p, R))$ in H
14:             **else** //e points to an intermediate node N
15:                 **for** each entry N_i in N
16:                     **if** N_i cannot be pruned **then** insert $(N_i, mindist(N_i, R))$ in H
17: **for** each point p in S_{cnd} apply a boolean range query to determine the S_{rnn}
18: insert all the points p in query range R into S_{rnn}
19: return S_{rnn}

Both p_1 and p_2 are closer to p_3 than R. Similarly, a node MBR N can not contain any candidates (i.e., N can be pruned at the filter step). In some cases, several half-planes' *intersections* are needed to prune a node.

5   Performance Evaluation

5.1   Experimental Setup

Implemented Algorithm. For comparison, we implement the TPL algorithm proposed in [2], using average sampling approach to approximate handle the range-based RkNN queries. The more sampling numbers there are, the more accurate the result, and comes with more cost at the same time. For fairness, we implement all the algorithms in each experiment to demonstrate the effects of our proposed optimizations.

Datasets. The experiments are conducted on four datasets: CaliforniaDB (CD), a spatial data in California (www.usgs.gov); NA dataset contains spatial data corresponding to geometric locations in the North America; two synthesized datasets, a normal distribution dataset and a uniform distribution dataset. The CD dataset and the two synthesized datasets have about 200K objects in total and the NA dataset has 569k objects. Each dataset is indexed by an R-tree [1].

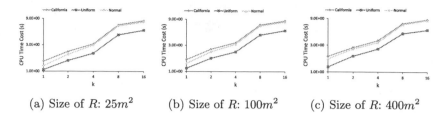

(a) Size of R: $25m^2$ (b) Size of R: $100m^2$ (c) Size of R: $400m^2$

Fig. 5. Varying data distribution, size of R and k (for RkNN)

System Setup and Metrics. We execute our experiments on a PC (Inter(R) Core(TM)2 E7500, 2.93 GHz CPU) running Windows 7 operating system. The simulation codes are written in Java (JDK 1.6).

The query locations are randomly selected in the space. Each experimental result is the average result over 200 queries. For performance metrics, we measure the CPU time and the accuracy (actual RkNN numbers) in the experiment.

5.2   Experiments for Range-Based RkNN Queries

Effect of Data Distribution and Size of R. In the first experiment, we evaluate the effect of varying the data distribution and the size of query range R. As shown in Fig. 5, the difference of CPU time in all datasets with various size of query range R is not that big. That's because our TPR algorithm returns the RkNNs of R within one-time traverse of the R-tree. This observation is confirmed by all experiments (including the real data) despite the different settings.

Effect of k (for RkNN). As expected, the overhead of TPR algorithm grows with k, see Fig. 5, due to the significant increase in CPU time. This is because a larger k degrade the pruning effect of the points or R-tree nodes. Note that the average number of candidates retrieved increases almost linearly with k in the filter step and thus need more computations in the refinement step.

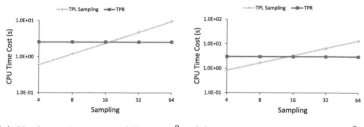

(a) Uniform, $k=4$, size of R: $400m^2$ (b) NA, $k=4$, size of R: $400m^2$

Fig. 6. TPR vs. TPL sampling: CPU time(s)

TPR vs. TPL Sampling. TPL algorithm is intended to solve the point-based RkNN query, we implement the TPL algorithm using average sampling approach for comparison. Figure 6 shows the total cost of TPR and TPL Sampling as a function of the sampling number. In this case, the CPU time of TPL Sampling algorithm increase linearly to the sampling number and the TPR algorithm doesn't need any sampling. When the sampling number come up to about 16, the CPU time of TPR and TPL Sampling are almost the same, but the TPL Sampling can not ensure a correct result. Figure 7 illustrates the accuracy of TPR and TPL Sampling. Our TPR returns the exact RkNNs for the query range and the TPL Sampling needs as many of sampling as possible. As expected, in this experiment the sampling number come up to about 64 can TPL Sampling returns the correct result. Therefore, the TPR algorithm performs better with respect to the accuracy.

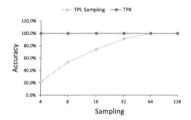

Fig. 7. TPR vs. TPL sampling: accuracy (CD, $k = 4$, size of R: $400 \, \text{m}^2$)

6 Conclusions and Future Work

In this paper we have discussed the problem of range-based RkNN queries. We have proposed algorithms for *exact* processing of range-based RkNN with *arbitrary* values of k on *dynamic* datasets. In particular, we extensively conduct experiments and our experimental results demonstrate that our proposed methods outperform the straightforward method in all aspects, and are superior to existing methods in terms of the efficiency and the accuracy. In the future, we intend to extend this work to the range-based RkNN query with *irregular* range, which retrieves all the points that have any position in the given irregular range as one of their k nearest neighbors.

References

1. Beckmann, N., Kriegel, H.P., Schneider, R., Seeger, B.: The R*-tree: an efficient and robust access method for points and rectangles. In: SIGMOD, pp. 322–331 (1990)
2. Tao, Y., Papadias, D., Lian, X.: Reverse kNN search in arbitrary dimensionality. In: VLDB, pp. 744–755 (2004)

3. Korn, F., Muthukrishnan, S.: Influence sets based on reverse nearest neighbor queries. In: SIGMOD, vol. 29, no. 2, pp. 201–212 (2000)
4. Korn, F., Muthukrishnan, S., Srivastava, D.: Reverse nearest neighbor aggregates over data streams. In: VLDB, pp. 814–825 (2002)
5. Benetis, R., Jensen, C., Karciauskas, G., Saltenis, S.: Nearest neighbor and reverse nearest neighbor queries for moving objects. In: IDEAS, vol. 15, no. 3, pp. 44–53 (2002)
6. Stanoi, I., Riedewald, M., Agrawal, D., Abbadi, A.: Discovery of influence sets in frequently updated databases. In: VLDB, pp. 99–108 (2001)
7. Yang, C., Lin, K.: An index structure for efficient reverse nearest neighbor queries. In: ICDE, pp. 485–492 (2001)
8. Chen, L., Lin, X., Hu, H., Jensen, C.S., Xu, J.: Answering why-not questions on spatial keyword top-k queries. In: ICDE, pp. 279–290 (2015)
9. Chen, L., Xu, J., Lin, X., Jensen, C.S., Hu, H.: Answering why-not spatial keyword top-k queries via keyword adaption. In: ICDE, pp. 697–708 (2016)
10. Stanoi, I., Agrawal, D., Abbadi, A.: Reverse nearest neighbor queries for dynamic databases. In: SIGMOD Workshop, vol. 29, no. 5, pp. 44–53 (2000)
11. Singh, A., Ferhatosmanoglu, H., Tosun, A.: High dimensional reverse nearest neighbor queries. In: CIKM, pp. 91–98 (2003)
12. Li, H., Hu, H., Xu, J.: Nearby friend alert: location anonymity in mobile geosocial networks. IEEE Pervasive Comput. **12**(4), 62–70 (2013)
13. Maheshwari, A., Vahrenhold, J., Zeh, N.: On reverse nearest neighbor queries. In: CCCG, vol. 17, no. 1, pp. 63–95 (2002)

Big Data and Blockchain

EarnCache: Self-adaptive Incremental Caching for Big Data Applications

Yifeng Luo[1,2], Junshi Guo[1], and Shuigeng Zhou[1(✉)]

[1] School of Computer Science, and Shanghai Key Lab of Intelligent Information Processing, Fudan University, Shanghai 200433, China
sgzhou@fudan.edu.cn
[2] School of Data Science and Engineering, East China Normal University, Shanghai 200062, China

Abstract. Memory caching plays a crucial role in satisfying the requirements for (quasi-)real-time processing of exploding data on big-data clusters. As big data clusters are usually shared by multiple computing frameworks, applications or end users, there exists intense competition for memory cache resources, especially on small clusters that are supposed to process comparably big datasets as large clusters do, yet with tightly limited resource budgets. Applying existing on-demand caching strategies on such shared clusters inevitably results in frequent cache thrashing when the conflicts of simultaneous cache resource demands are not mediated, which will deteriorate the overall cluster efficiency.

In this paper, we propose a novel self-adaptive incremental big data caching mechanism, called EarnCache, to improve the cache efficiency for shared big data clusters, especially for small clusters where cache thrashing may occur frequently. EarnCache self-adaptively adjusts resource allocation strategy according to the condition of cache resource competition: turning to incremental caching to depress competition when resource is in deficit, and returning to traditional on-demand caching to expedite data caching-in when resource is in surplus. Extensive experimental evaluation shows that the elasticity of EarnCache enhances the cache efficiency on shared big data clusters, and thus improves resource utilization.

Keywords: Big data · Cache management
Self-adaptive and Incremental caching

1 Introduction

As big data techniques and infrastructures are being applied to facilitate and accelerate the processing of big data with formidable size, people are putting forward eager requests on (quasi-)real-time processing of big datasets yet with exploding volumes, while meeting the (quasi-)real-time processing requests of big datasets is usually held back by the disk-based storage subsystem, because

© Springer International Publishing AG, part of Springer Nature 2018
Y. Cai et al. (Eds.): APWeb-WAIM 2018, LNCS 10988, pp. 379–393, 2018.
https://doi.org/10.1007/978-3-319-96893-3_29

of the expanding tremendous performance gap lying between magnetic disks and processing units. Thus memory caching plays a crucial role in bridging the performance gap between storage subsystems and computing frameworks, and gradually becomes the determinant factor of whether the processing units of big data platforms could work at their wire speed to satisfy the vast and fast data processing requirements. As more and more time-critical applications commence employing memory to cache their big datasets, big data clusters are usually concurrently shared by multiple computing frameworks, applications or end users, just as Fig. 1 shows.

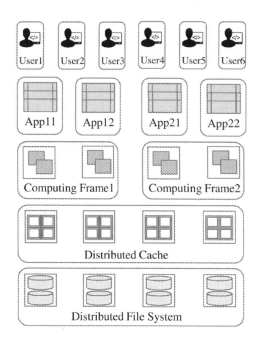

Fig. 1. Big data application hierarchies.

In Web or traditional OLTP database applications, ranges or blocks of the same datasets (or files) usually show vast variance in "hotness" regarding access recency and frequency. While big data applications usually scan their input files as a whole for data processing, and thus all blocks of the same file reveal almost equal hotness. On the other hand, traditional system-level or database-level data caching is executed on small data units (i.e. 8 KB-sized pages), while big data caching is executed on much larger units (i.e. 256 MB-sized blocks). So the cost of caching in/out a data unit in big data scenarios far exceeds that of traditional data caching. Accordingly, traditional caching may have millions of caching slots, which makes hotter data pages less likely to be cached out by colder data pages; while big data caching may only have thousands of slots, which makes comparatively hotter data blocks vulnerable to be cached out by colder data blocks.

Thus there exists intense competition for memory cache resources, especially on small clusters which are supposed to process comparably big datasets as large clusters do, yet with much tightly limited resource budgets.

Intense competition for computational resources would not do much harm to the clusters' running efficiency, while intense competitions for memory cache resources would engender tremendous harm, where frequent cache thrashings would be incurred if cache resource demands are not coordinated, and CPUs may constantly run idle. Applying existing on-demand caching strategies, which cache in data blocks once they are accessed, on small shared clusters inevitably results in frequent cache thrashings, and thus leads to deteriorated overall cluster efficiency. The principal reason behind is that aggressively caching massive numbers of data blocks of big datasets on demand causes constant block replacement in cache, and consequently exacerbates the competition of cache resources when these resources are in strong need. Consequently implementing effective and adaptive management on memory cache resources becomes increasingly important for the efficiency of big data clusters, especially for small/medium enterprises who could only afford non-big (or small) clusters.

Targeting the caching problem existing on non-big clusters, we propose an adaptive cache mechanism, which is named as EarnCache (from sElf-adaptive incremental **Cache**), to coordinate concurrent cache resource demands to prevent exacerbation of cache efficiency, when intense competitions for memory cache resources occur. As big data applications usually access their input data in the Write-Once-Read-Many (WORM) fashion, we only consider read caching in this paper. Major contributions of this paper include: (1) proposing an incremental caching mechanism which could self-adaptively adjust cache allocation strategies according to the competition condition of cache resources; (2) formulating and solving the cache resource allocation and replacement problem as an optimization problem; (3) implementing a prototype of the proposed mechanism, and performing extensive experiments to evaluate the effectiveness of the proposed mechanism.

With EarnCache, applications or end users do not get their datasets cached once they are accessed, but have to incrementally earn cache resources from other applications or end users by accessing their datasets. A dataset is cached gradually as the upper-level application or end user accesses the dataset, and more blocks of the dataset get cached each time it is accessed. In the rest of this paper, we illustrate the system design and the implementation details of EarnCache in Sect. 2. We provide empirical evaluation results in Sect. 3, and present related work in Sect. 4. We finally conclude the paper in Sect. 5.

2   Framework and Techniques

We illustrate how EarnCache works in this section. Firstly we present the overview about the caching mechanism of EarnCache, and then discuss its architecture design, and finally explain the incremental cache-earning policy and its implementation.

2.1   Overview

On a shared non-big cluster with relatively limited cache capacity, cache resource conflicts would be normal. If the cluster is concurrently used by a moderate number of users and the competition for cache resources is mild, hot blocks is less likely to be cached out by cold blocks, and then applying on-demand caching could expedite hot blocks taking over cache resources from cold blocks. If more and more users need to use the cluster concurrently and competition for cache resources gets wild, applying on-demand caching would leave concurrent cache resource demands unmediated, and making hot blocks more vulnerable to being cached out by cold blocks. Then files which are frequently accessed recently could be totally cached out by files which would rarely be accessed for a second time in the near future, and the flushed-out hot files would require to be cached in soon as their next access should occur in the upcoming future. We consequently need to revisit existing on-demand caching mechanisms for big data caching on small clusters, and propose more effective measures to improve the efficiency of data caching on such clusters.

We believe that a good caching strategy for non-big clusters should be self-adaptive to resource competition conditions, depressing competitions and preventing cache thrashings when cache resources are in desperate deficit. Obviously caching big data files on demand as a whole could not provide such self-adaptivity. Not caching-in files entirely on-demand could provide the elasticity of tuning the amount of cache resources allocated for different files, based on their access recency and frequency.

Ideally, more recently frequently accessed files should be assigned with more cache resources, and less recently frequently accessed ones should be assigned with less cache resources. However, it's not possible to know in advance what files would be frequently accessed in the upcoming future, and we could only make predictions based on historical file access patterns, especially the most recent information. Based on files' historical access information, EarnCache implements an incremental caching strategy, where a user should earn cache resources for its files from other concurrent users via accessing these files. Cache resources are incrementally allocated to a file that becomes more frequently accessed, which gradually takes over cache resources, until all blocks of the file have been cached in. The more a file is accessed, the more cache resources it takes over. The incremental caching strategy ensures that files occupying cache resources are recently frequently accessed, and will not be flushed out by files that are only accessed occasionally or randomly.

2.2   Architecture

Files originally reside in the under distributed file system (e.g. Hadoop File System), and EarnCache coordinately caches files across the whole cluster. Earn-Cache consists of a central *master* and a set of *workers* residing on storage nodes as shown in Fig. 2. The *master* is responsible for:

1. determining the cache resource allocation plan for a file, concerning how many cache resources should be allocated to the file based on its recent access information;
2. informing workers of cache resource allocation plans via heartbeats;
3. keeping track of metadata of which storage node a cached block resides on;
4. answering clients' queries on cache metadata.

And a *worker* is responsible for:

1. receiving the resource allocation plan from the master;
2. calculating resource composition plans to determine how many cache resources a cached file should contribute to compose the allocated resources depicted in the cache plan;
3. caching in/out blocks according to the calculated resource composition plans;
4. informing the master of cached blocks via heartbeats;
5. serving clients with cached blocks;
6. transferring in-memory blocks to other workers for remote caching.

As illustrated in Fig. 2, a client accesses a block in the following procedures:

1. the client queries the master where the block is cached;
2. the master tells the client which worker the requested block resides on;
3. the client contacts the worker to access the cached block;
4. the worker serves the client with the block data from cache.

One thing worth noting here is that: the client will not contact any worker to access a block if the block is not cached in any worker node, as the master only keeps track of cached blocks. In this situation, the client has to fetch data directly from the under file system.

2.3   Incremental Caching

As we prefer recently frequently accessed files incrementally taking over resources from less recently frequently accessed files, "recently" should be defined quantitatively before we could design the incremental caching strategy, and other related elements should also be clarified. Table 1 presents the definitions of all notations involved in our incremental caching strategy.

We define a function $h_i(x_i)$ to denote the cache profit gain of the ith file to instruct how cache resources should be allocated across all files falling within the observation window. Then we attempt to maximize the total profit gain of all files falling in the observation window with the profit gain function, just as Eq. 1 shows.

$$\sum_{i=1}^{N} f_i \cdot h_i(x_i) \tag{1}$$

According to definitions in Table 1, we can assume that the time it takes to scan the ith file is:

$$time(x_i) = [a \cdot x_i + b \cdot (1 - x_i)] \cdot d_i \tag{2}$$

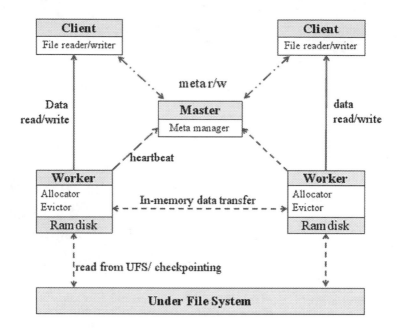

Fig. 2. EarnCache's architecture.

Table 1. Notation definitions

Notation	Definition
W	Predefined window size of the most recently accessed data for observing files falling within
a, b	Scan time per unit data from memory(a) and hdd(b)
N	Total number of files falling in the observation window
d_i	Data size of the ith file
D	Total data size of N files
M	Cache capacity of the whole cluster
f_i	Access frequency of the ith file
F	Total access frequency of N files
x_i	Percentage of data cached for the ith file
$h_i(x_i)$	The ith file's profit gain with x_i data cached

As mentioned above, we use h_i to indicate the ith file's cache profit gain with x_i data cached. For simplicity, we take the file's saved scan time as its cache profit gain, then we can define h_i's deviation at x_i as its gain change over Δx_i, which could be further defined as the percentage of increased saving of the file's scan time with increased cache share at x_i over the total saved scan time at x_i, compared to zero cache share, just formulized as:

$$\frac{\delta h_i}{\delta x_i} = \frac{time(x_i) - time(x_i + \delta x_i)}{time(0) - time(x_i)} = \frac{\delta x_i}{x_i} \tag{3}$$

Thus we can derive that $h_i(x_i) = \ln x_i$, and now our optimization goal becomes:

$$\sum_{i=1}^{N} f_i \cdot \ln x_i \tag{4}$$

subjected to:

$$\sum_{i=1}^{N} x_i \cdot d_i \leq M \tag{5}$$

Note that at any given time, x_i is the only variant contained in the optimization goal, and $f_i \cdot \ln x_i$ is a convex function. After applying Lagrange multiplier method, our optimization goal turns to:

$$L = \sum_{i=1}^{N} f_i \cdot \ln x_i - \lambda(\sum_{i=1}^{N} x_i \cdot d_i - M) \tag{6}$$

Let $\frac{\delta L}{\delta x_i}$ be 0, then we get

$$x_i \cdot d_i = \frac{f_i}{F} \cdot M \tag{7}$$

The above result shows that the amount of memory resources allocated to a file is linear to f_i at a given moment, as all files' access frequencies are determined at that moment, which exactly corresponds to our original intention of incremental caching. One more thing worth noting is that: if the overall size of files falling within the whole observation window is smaller than the cache capacity, and there are cache resources being occupied by files that fall out of the observation window, EarnCache will collect resources from those obsolete files by LRU when there is a caching request, and the requesting file could cache in its blocks once and for all, rather than gradually taking over resources from files falling within the observation window. EarnCache thereby could adaptively devolve to traditional on-demand caching so as to expedite the process of collecting cache resources for actively accessed files when contention for cache resources is light, and evolve to incremental caching to depress competition when resources are in deficit.

2.4   Implementation Details

We implemented EarnCache by implanting our incremental caching mechanism into the modified Tachyon [4]. In EarnCache, we first evenly re-distribute a file's cached data blocks across the whole cluster, so that almost the same amount of blocks are hosted in cache on each cluster node, and all workers can manage their cache resources independently yet still in concert. As uneven data distribution will drag down completion of the whole job, evenly distributing cached data blocks guarantee that tasks running on each node could ideally finish almost simultaneously.

When the ith file needs caching, EarnCache pre-allocates f_i/F fraction of cache resources on each node to the file based on Eq. 7. If resources pre-allocated to the file are more than its aggregated demands, EarnCache has other files in need of cache resources fairly share the spare cache. Each worker checks its available cache resources, and allocates as many as possible to them directly, which could make full use of cache resources. When there are not enough resources available, the worker calls *BlocksToEvict()*, which implements the eviction algorithm with incremental caching, to determine which blocks should be cached out. As all blocks are cached in from the under file system, cached-out blocks need no more backup and thus workers could discard them directly from cache. When block eviction process is done, the worker will inform the master to update the metadata.

Algorithm 1 describes the process of evicting blocks. EarnCache first checks whether the file requesting cache resources has used up its pre-allocated share in Lines 1–3. In the while loop, files who have overcommitted the most cache resources are selected in Lines 7–14. If no such file exists, EarnCache will reject the cache request (Lines 15–17). Otherwise, blocks of these selected files are added to the candidate block set until enough cache resources have been collected (Lines 18–24). As recency and frequency of all blocks within the same file are identical, workers do not differentiate between blocks of the same file when selecting blocks to cache out.

3   Empirical Evaluation

We deploy an HDFS cluster on Amazon EC2 as the under distributed file system to evaluate EarnCache's performance, on which Spark and EarnCache are deployed as the upper-level application tier and the middle-level caching tier respectively. The cluster consists of five Amazon EC2 m4.2xlarge nodes, one of which serves as the master and the other four serve as slaves. Each cluster node has 32 GB of memory, 12 GB memory is reserved as working memory and the remaining 20 GB of memory is employed as cache resources, summing up to 80 GB of overall cache in total.

We mainly evaluate EarnCache's performance by issuing jobs from Spark to scan files in parallel without any further processing, and compare the performance of EarnCache incremental caching, with LRU and LFU on-demand caching, and MAX-MIN fair caching. We set the size of FILE-1, FILE-2 and FILE-3 equally to 40 GB and unequally to 70 GB, 40 GB and 10 GB respectively, and then evaluate EarnCache with different caching strategies and frequency patterns. We set the observation window size of EarnCache to 1000 GB by default. For each experiment, we issue file scanning jobs on three input files, denoted as FILE-1, FILE-2 and FILE-3, with the following three various frequency patterns, denoted as ROUND, ONE and TWO respectively.

Algorithm 1. Eviction Algorithm: BlocksToEvict()

Input: s, requested cache resources; r, the requesting file id; $A=\{a_1, a_2...a_N\}$, a list of files' pre-allocated memory bytes; $C=\{c_1, c_2...c_N\}$, a list of current consumed memory bytes in local node; M, memory capacity of local node

Output: a list of candidate blocks to evict

```
 1: if c_r ≥ a_r then
 2:      algorithm ends as file r has already consumed all its allocated memory
 3: end if
 4: candidate ← {}                          ▷ candidate cached out blocks
 5: mem ← 0                    ▷ free resources obtained from evicting candidates
 6: while mem < s do
 7:      j ← −1
 8:      over_j ← 0
 9:      for a_i in A and i ≠ r do
10:           if c_i − a_i > over_j then
11:                j ← i
12:                over_j ← c_i − a_i
13:           end if
14:      end for
15:      if j = −1 then
16:           return as request failure
17:      end if
18:      find b_j as a block of file j and not in candidate
19:      candidate ← candidate + b_j
20:      mem ← mem + sizeof(b_j)
21:      c_j ← c_j − sizeof(b_j)
22:      if mem ≥ s then
23:           return candidate
24:      end if
25: end while
```

- ROUND Three files are accessed in pattern: FILE-1, FILE-2, FILE-3, ..., where three files are accessed with equal frequency.
- ONE Three files are accessed in pattern: FILE-1, FILE-2, FILE-1, FILE-3, ..., where one file is accessed more frequently than other two files.
- TWO Three files are accessed in pattern: FILE-1, FILE-2, FILE-1, FILE-2, FILE-3, ..., where two files are accessed more frequently than the other file.

Figure 3(a) and (b) show the averaged overall running time of file scanning jobs. Each group of columns involves the scanning of files contained within the whole period of a frequency pattern, namely 3, 4, and 5 files respectively. We can see that EarnCache yields the best performance, which exceeds that of the LRU and LFU on-demand caching by a large margin, and leads the MAX-MIN fair caching by a smaller margin. The reason of EarnCache achieving the best performance is straight-forward, as it prevents cache thrashings and thus more blocks are accessed from memory. We can see that the performance of EarnCache is only slightly better than the MAX-MIN caching strategy, and sometimes they achieve similar performance. This is because files receive similar amounts of cache resources from these two caching strategies, as far as our experimental settings are concerned.

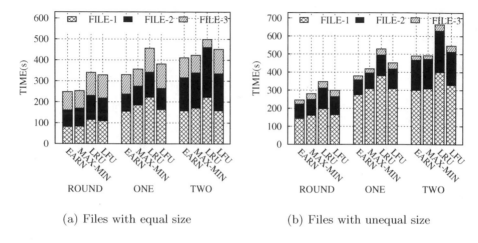

(a) Files with equal size   (b) Files with unequal size

Fig. 3. Running time of file scanning jobs.

In the meanwhile, we also observe that the performance of EarnCache is not as mighty as we have expected, especially compared with the LRU and LFU on-demand caching. The reasons are twofold: (1) EarnCache could not hold all blocks in cache, and thus file scanning jobs are sped up partially; (2) cache-locality is not guaranteed, and a prohibitive number of blocks are accessed from remote cache, rather than local cache. We analyze the distribution of blocks accessed from local cache, remote cache, and the under file system respectively in detail, and the results are shown in Fig. 4. We can see that EarnCache has the largest number of blocks accessed from cache, whether locally or remotely, which means that it yields the highest memory efficiency than other caching strategies. However, we observe that EarnCache has the largest number of blocks accessed from remote cache among the four evaluated strategies. This means EarnCache has the largest potential of performance improvement. If cache-aware task scheduling can be integrated into the upper-level task scheduler, more blocks will be accessed from local cache and EarnCache could obtain much better overall performance.

We showcase the change of cache shares of different files during the process of executing file scanning jobs iteratively, and the results are shown in Fig. 5. We can see that the cache shares of different files with EarnCache remain stable across the whole experimental process, while the LRU and LFU on-demand caching strategies witness cache thrashings with huge variance of cache shares. The MAX-MIN caching statically allocates cache resources based on present files, rather than caching blocks on demand, and thus also witnesses no variance of cache shares and avoid cache thrashings. However, we can also see that the

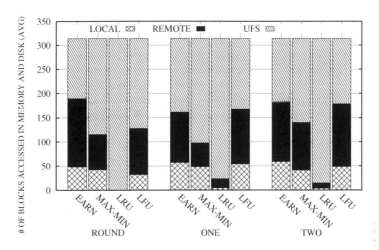

Fig. 4. Distribution of blocks accessed in local cache, remote cache and under file system.

MAX-MIN caching is unable to dynamically re-allocate resources properly when there exist files not receiving any further accesses. To illustrate this, we present the process of resource re-allocation of EarnCache and the MAX-MIN caching in Fig. 6(a) and (b), where two out of the three equal-sized files stop receiving further accesses. We can see that the file remaining accessed gradually takes over cache resources from those obsolete files as time passes, while with MAX-MIN fair caching, the amount of cache resources held by each file does not change. Correspondingly, the running time of each job gradually decreases with EarnCache, yet remains stable with MAX-MIN.

Finally, we experimentally analyze the impact of the predefined observation window size, and the results are presented in Fig. 7. When observation window is set with small sizes, the competition for cache resources could not be coordinated properly, and thus the overall cache efficiency and performance degrades greatly. When the observation window size exceeds 200 GB, which is larger enough compared with the file sizes, EarnCache effectively coordinates cache resources and the performance improves correspondingly.

4   Related Work

There has been extensive work on memory storage and caching, as more and more time-critical applications [19,22] require to store or cache data in memory to gain improved data access performance, such as Ousterhout et al. proposed RAMCloud [2] to keep data entirely stored in memory for large-scale Web applications, and Spark [9,21] enables in-memory MapReduce [3]-style parallel computing by leveraging memory to store and cache distributed (intermediate) datasets. While caching on distributed parallel systems is tremendously different from traditional centralized page-based file system or database caching, and

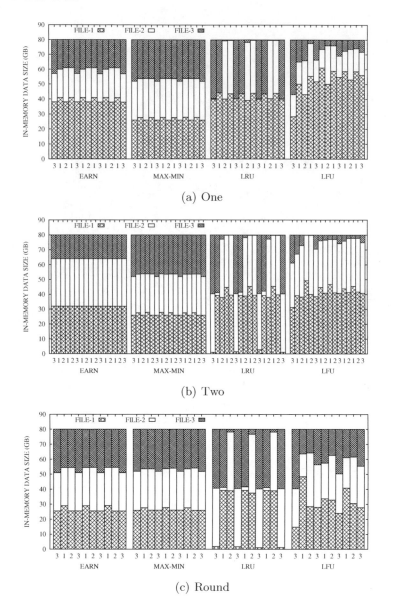

Fig. 5. Change of cache shares as files being accessed.

directly applying centralized caching usually does not help much to improve and sometimes even hurts cache efficiency and performance.

Some previous work focuses on implementing an additional layer on existing distributed file system, which enables applications to cache distributed datasets from the underlying distributed file system. Zhang et al. [1] and Luo et al. [11] respectively proposed the HDCache and RCSS distributed cache system based

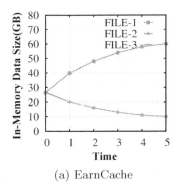 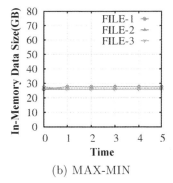

(a) EarnCache   (b) MAX-MIN

Fig. 6. Dynamic resource re-allocation with two files receiving no further accesses.

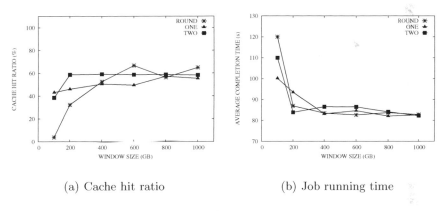

(a) Cache hit ratio   (b) Job running time

Fig. 7. Impact of the observation window size.

on HDFS [6,20], which manages cached data just as HDFS manages disk data. Li et al. [4] further implemented a distributed memory file system for data caching by checkpointing data to the underlying file system. Luo et al. [16] proposed a just-in-time data prefetching mechanism for Spark applications so as to depress the resource demands for caching memory. EARNCache imbeds the incremental caching into Tachyon [4] to coordinate resource competitions and avoid cache thrashings, and improves cache efficiency and resource fairness to a certain degree.

Some work focuses on optimizing data caching for specific frameworks or goals. Zhang et al. [10] proposed to cache MapReduce intermediate data to speed up MapReduce applications. Luo et al. [14,15] optimized cache resource allocations in cloud environments to improve database workload processing efficiency. Ananthanarayanan et al. [7] found the important All-or-Nothing property, which implies that all or none input data blocks of tasks within the same wave should be cached, and then proposed PACMan to coordinate memory caching for parallel jobs. Li et al. [5], Tang et al. [17] and Ghodsi et al. [18] respectively proposed dynamic resource partition strategies to improve fairness, and maximize the overall performance in the meanwhile. Pu et al. [8] extended the MAX-MIN fairness [12,13] with probabilistic blocking, and proposed FairRide to avoid cheating and improve fairness for shared cache resources.

5   Conclusion

In this paper, we propose the EarnCache incremental big data caching mechanism, which adaptively adjusts resource allocation strategy according to resource competition condition. Concretely, when the resources are in deficit, it adopts incremental caching to depress competition, and turns to traditional on-demand-caching to expedite data caching-in when resources are in surplus. On-demand big data cache usually leads to cache thrashings. With EarnCache, files are not cached on demand. Instead, applications or end users incrementally take over cache resources from others by accessing their datasets. EarnCache manages to achieve improved resource utilization and performance with such an incremental caching strategy. Experimental results show that EarnCache can elastically manage cache resources and yields better performance against the LRU, LFU and MAX-MIN cache replacement policies.

Acknowledgements. This work was supported by National Natural Science Foundation of China (NSFC) (No. U1636205), and the Science and Technology Innovation Action Program of Science and Technology Commission of Shanghai Municipality (STCSM) (No. 17511105204).

References

1. Zhang, J., Wu, G., Hu, X., et al.: A distributed cache for Hadoop distributed file system in real-time cloud services. In: Proceedings of GRID, pp. 12–21 (2012)
2. Ousterhout, J., Agrawal, P., Erickson, D., et al.: The case for RAMCloud. Commun. ACM **54**(7), 121–130 (2011)
3. Dean, J., Ghemawat, S., et al.: MapReduce: simplified data processing on large cluster. In: Proceedings of OSDI, pp. 137–150 (2004)
4. Li, H., Ghodsi, A., Zaharia, M., et al.: Tachyon: reliable, memory speed storage for cluster computing frameworks. In: Proceedings of SOCC, pp. 6:1–6:15 (2014)
5. Li, Y., Feng, D., Shi, Z.: Enhancing both fairness and performance using rate-aware dynamic storage cache partitioning. In: Proceedings of DISCS, pp. 31–36 (2013)
6. Shvachko, K., Kuang, H., Radia, S., et al.: The Hadoop distributed file system. In: Proceedings of MSST, pp. 121–134 (2010)
7. Ananthanarayanan, G., Ghodsi, A., Warfield, A., et al.: PACMan: coordinated memory caching for parallel jobs. In: Proceedings of NSDI, pp. 267–280 (2012)
8. Pu, Q., Li, H., Zaharia, M., et al.: FairRide: near-optimal, fair cache sharing. In: Proceedings of NSDI, pp. 393–406 (2016)
9. Zaharia, M., Chowdhury, M., Das, T., et al.: Resilient distributed datasets: a fault-tolerant abstraction for in-memory cluster computing. In: Proceedings of NSDI, pp. 15–28 (2012)
10. Zhang, S., Han, J., Liu, Z., et al.: Accelerating MapReduce with distributed memory cache. In: Proceedings of ICPADS, pp. 472–478 (2009)

11. Luo, Y., Luo, S., Guan, J., et al.: A RAMCloud storage system based on HDFS: architecture, implementation and evaluation. J. Syst. Softw. **86**(3), 744–750 (2013)
12. Ma, Q., Steenkiste, P., Zhang, H.: Routing high-bandwidth traffic in max-min fair share networks. In: Proceedings of SIGCOMM, pp. 206–217 (1996)
13. Cao, Z., Zegura, W.: Utility max-min: an application-oriented bandwidth allocation scheme. In: Proceedings of INFOCOM, pp. 793–801 (1999)
14. Luo, Y., Guo, J., Zhu, J., Guan, J., Zhou, S.: Towards efficiently supporting database as a service with QoS guarantees. J. Syst. Softw. **139**, 51–63 (2018)
15. Luo, Y., Guo, J., Zhu, J., Guan, J., Zhou, S.: Supporting cost-efficient multi-tenant database services with service level objectives (SLOs). In: Candan, S., Chen, L., Pedersen, T.B., Chang, L., Hua, W. (eds.) DASFAA 2017. LNCS, vol. 10177, pp. 592–606. Springer, Cham (2017). https://doi.org/10.1007/978-3-319-55753-3_37
16. Luo, Y., Shi, J., Zhou, S.: JeCache: just-enough data caching with just-in-time prefetching for big data applications. In: Proceedings of ICDCS, pp. 2405–2410 (2017)
17. Tang, S., Lee, B., He, B., et al.: Long-term resource fairness: towards economic fairness on pay-as-you-use computing systems. In: Proceedings of ICS, pp. 251–260 (2014)
18. Ghodsi, A., Zaharia, M., Hindman, B., et al.: Dominant resource fairness: fair allocation of multiple resource types. In: Proceedings of NSDI, pp. 323–336 (2011)
19. Redis. http://redis.io
20. HDFS. http://hadoop.apache.org/hdfs
21. Spark. http://spark.apache.org
22. Memcached. http://danga.com/memcached

Storage and Recreation Trade-Off
for Multi-version Data Management

Yin Zhang[1], Huiping Liu[1], Cheqing Jin[1(✉)], and Ye Guo[2]

[1] School of Data Science and Engineering, East China Normal University,
Shanghai, China
{51164500131,hpliu}@stu.ecnu.edu.cn, cqjin@sei.ecnu.edu.cn
[2] Tongji University, Shanghai, China
guoye@ouyeel.com

Abstract. With the tremendous development of data acquisition technology, massive observation data have been accumulated in scientific disciplines. As the difference between the successive observations only changes slightly, it is critical to utilize multi-version data management technology to compress data to minimize both storage and recreation. However, the existing work on this field only optimizes the total storage and recreation costs, but ignores the recreation cost of some special versions. Consequently, in this paper, we investigate the trade-off among all of three metrics, including total storage cost, total recreation cost, and the maximum recreation cost for each version. We formulate two problems, including (1) discover a storage plan to lower the total recreation and the individual recreation if the total storage is limited; (2) find a storage plan to minimize the total storage with restricted total recreation and individual recreation. To solve above problems, we model all versions with a directed graph and then devise two efficient algorithms based on spanning tree. A series of experiments indicate that our proposals are effective and efficient in dealing with the problems.

Keywords: Multi-version data management
Storage and recreation trade-off · Scientific data management

1 Introduction

With the tremendous development of data acquisition devices and computing ability, massive scientific data are continuously generated and accumulated at higher frequency. For instance, Large Synoptic Survey Telescope (LSST) [2], one of next generation of telescopic sky surveys, can produce the observation of the whole sky every three days, whereas the difference between successive observations at the same area only changes slightly. In order to save the space consumption, multi-version technology can be used to manage such large-scale data.

Supported by the National Key Research and Development Program of China (2016YFB1000905), NSFC (61532021, U1501252, U1401256 and 61402180).

© Springer International Publishing AG, part of Springer Nature 2018
Y. Cai et al. (Eds.): APWeb-WAIM 2018, LNCS 10988, pp. 394–409, 2018.
https://doi.org/10.1007/978-3-319-96893-3_30

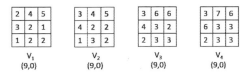

(a) Storage plan 1: all versions are materialized

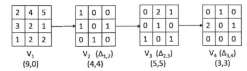

(b) Storage plan 2: only version 1 is materialized.

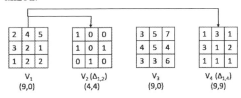

(c) Storage plan 3: only version 1 and version 3 are materialized.

Fig. 1. An instance of three storage plans

Figure 1 shows an example of multi-version data management. The successive observations of the same area are represented as V_1 to V_4. Each version is associated with a pair (a, b), in which a and b denote storage and recreation cost of this version respectively. The total storage cost is the sum of storage cost of all versions. The total recreation cost is the sum of recreation cost of all versions. The maximum recreation cost is the biggest recreation cost of individual version among all versions. Three storage plans for them are depicted in Fig. 1. In plan 1, all versions are materialized, namely all stored completely. The storage cost of plan 1 is $9+9+9+9 = 36$. The whole recreation cost is 0 because all versions can be accessed directly with no recreation cost. In plan 2, only V_1 is materialized and the others are delta against their previous version respectively. For example, V_2 is stored as delta against V_1. V_2 can be recreated by $V_1 + \Delta_{1,2} = 0 + 4 = 4$. The storage cost of plan 2 is $9 + 4 + 5 + 3 = 21$. The whole recreation cost is $0 + (0 + 4) + (0 + 4 + 5) + (0 + 4 + 5 + 3) = 25$. Compared with plan 1, total storage is significantly reduced whereas the total recreation cost is increased. V_4 is the version with the maximum recreation cost $0 + 4 + 5 + 3 = 12$. As for plan 3, only V_1 and V_3 are materialized. V_2 and V_4 are delta against V_1 respectively. The storage cost of plan 3 is $9 + 4 + 9 + 9 = 31$. The whole recreation cost is $0 + (0 + 4) + 0 + (0 + 9) = 13$. V_4 is the version with the maximum recreation cost $0 + 9 = 9$. The storage and recreation costs are between plan 1 and 2. The maximum recreation cost is still high.

Storage and recreation are two oppositions in multi-version data management. More storage may lead to less recreation to cost rebuild versions and vice versa. Hence, the trade-off between storage and recreation is the fundamental. Two problems were studied in [4]: (i) given restricted storage cost, how to minimize the recreation and (ii) given limited recreation cost, how to minimize the storage cost. However, neither of them takes the maximum recreation cost into consideration at the same time, namely the recreation cost of reconstructing some version may be very high. In some cases, only total storage and total recreation are not enough and the other metrics, the maximum recreation cost, should also be considered into the storage-recreation trade-off. For example, In Fig. 1, compared with plan 1, in plan 2, the total storage is reduced whereas the recreation cost of V_4 is increased from 0 to 12, becoming the maximum recreation cost. In plan 3, although the total storage cost and the whole recreation cost are between those of plan 1 and plan 2 respectively, the maximum recreation cost is still high. The recreation or retrieve efficiency of such version is relatively low. Hence, it is relatively necessary to investigate the trade-off that takes storage, recreation and the maximum recreation costs into consideration at the same time. Consequently, we study the trade-off aforementioned and designed two algorithms.

We first model all versions and their derivation relationship by a directed graph and then design two algorithms on the basis of the minimum spanning tree and the shortest path tree. The former is for the minimum storage and the latter is for the minimum recreation.

In particular, we make the following main contributions:

- To the best of our knowledge, we are the first to investigate the trade-off among storage, recreation and the maximum recreation at the same time and formulate two problems.
- We design two efficient algorithms based on the minimum spanning tree and the shortest path tree to deal with aforementioned issues.
- We implement our solutions and conduct comprehensive experiments on both synthetic datasets and real datasets to demonstrate the efficiency and effectiveness of our proposal.

The remainder of this paper is structured as follows. Section 2 reviews the related work of multi-version data management. The problem and relevant conceptions are formulated in Sect. 3. Subsequently, we outline our algorithms and describe them in detail in Sect. 4. In Sect. 5, we evaluate the performance through conducting a series of experiments. We conclude our paper briefly in the last section.

2   Background and Related Work

Array Data: Array data are widely used in scientific data management. More and more scientific observation data are modeled as array data. The operations of array data are built on array algebra and there are specific query languages designed for array data, such as AQL and AFL. The storage of array data can be classified into chunking and arbitrarily tiling. We refer the reader to a survey [12] to a more comprehensive work of array data. Based on prior work, [16] designed ArrayStore, a new storage manager to support parallel iterative processing for array data. Currently, mainstream array databases include SciDB [8], RasDaMan [3] and so on. Iterative processing [15] and various queries such as [5,10] are also supported in SciDB.

After modeling scientific data as array, multi-version data management technology can be also used in scientific data management.

Multi-version Data Management: Data are be organized as a linear chain of versions or an arbitrarily graph in multi-version data management. The version chain [6] has been studied extensively in temporal databases [9,17], such as [13]. Snapshot queries are embedded into array database in [14] by two kinds of time travel operations, retrieving some version given a time point or version ID and exploring a sequence of history array versions. Our work can apply to both linear version chain and graph of version. There are some data version management systems available [18] etc.

Storage and recreation are two critical aspects for this topic. Some previous studies have elaborated on minimizing the total storage cost. With regard to array data, given a long sequence of versions, Seering et al. [13] put forward an algorithm to determine the delta versions and the materialized ones, reducing the total storage cost and maintaining high access efficiency. Relatively few prior work exist on reducing the total recreation cost. [13] optimized the access time of a sequence of multi-versions with regard to array data.

To the best of our knowledge, only [4] conducts research on the trade-off of storage and recreation. However, they do not consider the trade-off that take total storage, total recreation and the maximum recreation into consideration at the same time. In some special cases, the recreation for some single version is relatively high which leads to low efficiency to rebuild or retrieve that version and also decreasing the overall performance. In order to deal with this problem, we formulate two problems to consider them at the same time with the goal of (i) lower the total recreation and the individual recreation given threshold for total storage and (ii) minimize the total storage with restricted total recreation and individual recreation.

3   Preliminary and Problem Definition

In this section, we introduce preliminary concepts and define the problems formally.

We have introduced in Sect. 1 that either making all versions materialized or only making one version materialized is infeasible due to the existence of multiple metrics. Hence, it is necessary to find new solutions. At first, we define two matrices for storage and recreation costs.

$$S = \begin{bmatrix} 600 & 200 & - & 450 \\ - & 580 & 400 & - \\ 750 & - & 350 & - \\ - & 820 & - & 450 \end{bmatrix}, \quad R = \begin{bmatrix} 600 & 200 & - & 450 \\ - & 580 & 2000 & - \\ 2250 & - & 1400 & - \\ - & 1640 & - & 450 \end{bmatrix}$$

Fig. 2. Cost matrix

Definition 1 (Storage Matrix). *Given a version sequence* $V = V_1, V_2, \ldots,$ V_n, *the storage matrix* S *represents storage cost of all versions, in which* $S[i, i]$ *denotes the storage cost for version* V_i *and* $S[i, j]$ *is delta storage cost from version* V_i *to* V_j *where* $i \neq j$.

Definition 2 (Recreation Matrix). *Given a version sequence* $V = V_1, V_2, \ldots,$ V_n, *the recreation matrix* R *represents recreation cost of all versions in which* $R[i, i]$ *denotes the recreation cost for materialized version* V_i *and* $R[i, j]$ *is delta recreation cost from version* V_i *to* V_j *where* $i \neq j$.

Definition 3 (Version Graph). *The version graph* $G(V, E, W)$ *models the versions and their derivation relationship where each vertex denotes one single version and each edge* (V_i, V_j) *indicates derivation relationship from version* V_i *to* V_j *with a weight pair* $(S[i, j], R[i, j]) \in W$. *In order to indicate the storage and recreation cost of materialized versions, a dummy node* V_0 *is set in* G *and edge* (V_0, V_i) *means the version* V_i *is materialized with cost pair* $(S[0, i], R[0, i])$.

An instance of cost matrix is represented in Fig. 2. Only available cost values are represented in the cost matrix. Version graphs are depicted in Fig. 3. For example, G_1 represents the version set V_1 to V_4 and the derivation relationship between them. V_1 is materialized with cost pair (1000, 1000) and V_2 is derived from V_1 with the delta cost (20, 20). The numbers of G_1 in Fig. 3 are fictitious, not built on matrices in Fig. 2. The G_2 is used in 4.

Definition 4 (Storage Plan). *Given a version sequence* $V = V_1, V_2, \ldots, V_n$, *a storage plan* $P = \{p_{(v_1, v'_1)}, p_{(v_2, v'_2)}, \ldots, p_{(v_n, v'_n)}\}$ *describes how to store all versions where* V_i *(i) is materialized if* $v_i = v'_i$ *or (ii) stored as a delta* v'_i *against* v_i.

Total storage cost measures the overall storage cost of P, denoted as $TSC = \sum_{i=1}^{n} S[v_i, v'_i]$. For version V_i, $RStr = \{V_{i_1}, V_{i_2}, \ldots, V_{i_q}\}$ represents the recreation path for V_i with the minimum recreation cost. The recreation cost for V_i is the sum of recreation cost in recreation path of V_i, which is denoted as $RC_i = R[i_1, i_1] + \sum_{m=1}^{q-1} R[i_m, i_{m+1}]$. Total recreation cost measures the overall recreation cost of P, denoted as $TRC = \sum_{i=1}^{n} RC_i$. The maximum recreation cost is denoted as MRC.

However, since it is infeasible to desire a plan that achieves the best result for the above metrics, we propose some new definitions below.

Definition 5 (Edmond-Dijkstra Tree (EDT) Query). *Given a sequence of* $V = V_1, V_2, \ldots, V_n$, *and a threshold* β, *discover a storage plan with the aim of* $\alpha \times \min\{\sum_{i=1}^{n} RC_i\} + (1 - \alpha) \times \min\{\max\{RC_i | 1 \leq i \leq n\}\}$ *under the condition* $TSC \leq \beta$. *The importance of* TRC *and* MRC *can be adjusted by the parameter* α.

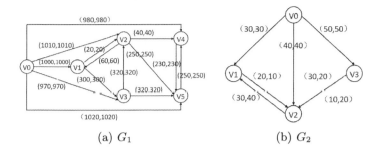

(a) G_1 (b) G_2

Fig. 3. Examples of version graphs

Table 1. Results for EDT and RPT query.

Version graph	Query	V_1	V_2	V_3	V_4	V_5	TSC	TRC	MRC
G_1	EDT	V_1	$\Delta_{1,2}$	V_3	$\Delta_{2,4}$	V_5	3050	2080	2080
G_2	RPT	V_1	$\Delta_{3,2}$	V_3			90	200	70

Definition 6 (Restricted Prim Tree (RPT) Query). *Given a sequence of* $V = V_1, V_2, \ldots, V_n$, *threshold* θ *and* δ, *discover a storage plan with the goal of* $\min\{TSC\}$ *under the condition* $\max\{RC_i | 1 \le i \le n\} \le \theta$ *and* $\sum_{i=1}^{n} RC_i \le \delta$.

Table 1 illustrates a small example of these two queries.

4   Algorithms

In this section, we describe our algorithms for the two problems.

4.1   EDT

We describe the EDT algorithm at high level. Before performing EDT, the threshold of total storage cost is given the version graph G is constructed according to the storage matrix and the recreation one. EDT works on the basis of the minimum spanning tree T_1 and the shortest path tree T_2 of G and the root of T_1 and T_2 is V_0. T_1 is built from *storage matrix* and T_2 is constructed according to *recreation matrix*. EDT consists of three phases. Firstly, obtain the edge set D which contains the edges in T_2 but not in T_1. During the second phase, calculate ρ (we explain the meaning of ρ below) for each single edge in D. During the latter phase, pick the edge from D in a greedy manner with maximum ρ and replace corresponding edge in T_1. Note that the two edges have the same end point. Repeat the iteration until the total storage cost reaches the predefined threshold of total storage cost. The resulting spanning tree T demonstrates the storage plan of the sequence of multi-versions. We then explain ρ below.

$$\rho = \alpha \rho_1 + (1 - \alpha)\rho_2 \tag{1}$$

Algorithm 1. EDT

Input: minimum spanning tree T_1, shortest path tree T_2, storage upper
        bound β, root V_0
Output: a new spanning tree T

1 $T \leftarrow T_1$, $D \leftarrow \emptyset$, $C \leftarrow 0$; // initialization (C is the total storage cost of
   current spanning tree.initialization)
2 $D \leftarrow Getset(T_1, T_2)$; // get the edge set
3 $C \leftarrow Stocost(T_1)$; // obtain the storage cost of current T
4 **while** D *is not empty* **and** $C < \beta$ **do**
5 $\quad$ $\rho_{max} \leftarrow 0$;
6 $\quad$ **foreach** *edge* (a, b) *in* D **do**
7 $\quad\quad$ //find the edge (a,b) from D with the maximum ρ
8 $\quad\quad$ compute ρ by Eq.1;
9 $\quad\quad$ **if** $\rho > \rho_{max}$ **then**
10 $\quad\quad\quad$ $\rho_{max} \leftarrow \rho$;

11 $\quad$ find the $edge(a', b)$ in T_1 ;//find the edge in T_2 with the same end point b
12 $\quad$ $T.remove(a', b)$, $T.add(a, b)$; //update T
13 $\quad$ $C \leftarrow C + S[a, b] - S[a', b]$;//update sotrage cost C
14 $\quad$ $D.remove(a, b)$;
15 **return** T;

Algorithm 2. Getset

Input: T_1, T_2
Output: the edge set (in T_2 not in T_1)
1 **return** $E(T_2) \backslash E(T_1)$;

$$\rho_1 = \frac{reduction\ in\ the\ recreation\ cost}{increase\ in\ the\ storage\ cost} \qquad (2)$$

$$\rho_2 = \frac{reduction\ in\ the\ maximum\ recreation\ cost}{increase\ in\ the\ storage\ cost} \qquad (3)$$

The ρ is explained as Eq. 1. ρ_1 and ρ_2 are in Eqs. 2 and 3 respectively. Suppose there are two edges e_a and e_b. ρ_1 of e_a is greater than ρ_1 of e_b indicates we can reduce more recreation cost with the same increase of storage cost. Similarly, the meaning of the situation where ρ_2 of e_a is bigger than ρ_2 of e_b can be inferred. The importance of ρ_1 and ρ_2 can be adjusted by the parameter α.

The EDT algorithm proceeds as follows. It takes the spanning tree T_1 and the shortest tree T_2 of version graph G as input.

After initialization (line 1), EDT invokes $Getset$ in Algorithm 2 to obtain edge set D (line 2). The result set D includes the edges contained in T_2 but not in T_1. Each edge in D represents one possible choice of recreation path but not contained in storage plan. $stocost()$ in Algorithm 3 are utilized to get the current total storage cost of T_1 via storage matrix S (line 3). Subsequently, we calculate ρ for each edge from D. Notice that the direct or indirect successors

---

Algorithm 3. Stocost

Input: T_1, root V_0
Output: storage cost of T_1
1 **return** $\sum_{e(a,b)\in T_1} S[a,b]$;

---

of the end point also experience the same increment in reduction of recreation cost, corresponding to the numerator of ρ_1. Select the edge (a, b) in D with the maximum ρ in a greedy manner (line 4–10). Next, replace the edge whose end point is b and start point is not a in T_1 with edge (a, b) in D (line 12) and update C (line 13). Repeat the iteration until C exceeds the predefined threshold. Edge (a, b) was deleted from D (line 14). From the overall view, the time complexity is $O(|E|^2)$.

In order to facilitate understanding, we demonstrate an instance on EDT below.

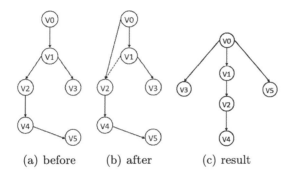

(a) before   (b) after   (c) result

Fig. 4. An example of EDT

Example 1. We illustrate EDT on version graph G_1 in Fig. 4. Let the storage threshold β be 3000 and parameter α be 0.5. Invoke the Edmonds' algorithm and Dijkstra's algorithm to obtain the minimum spanning tree T_1 (as depicted in Fig. 4(a)) and the shortest path tree T_2 respectively. Then we can get the edge set $D = \{(0,2),(0,3),(0,4),(0,5)\}$. Next, we calculate ρ of $(0,2)$, replacing $(1, 2)$ in T_1 with $(0, 2)$ of D shown in Fig. 4(b). $\rho_{(0,2)} = \frac{3\times(1000+20-1010)}{1010-20} + 0.5 \times \frac{1300-1300}{1010-20} = \frac{1}{33}$. In denominator of ρ_1, "3×" because for V_2, the direct successor V_4, indirect successor V_5 and V_2 itself will experience the same change in recreation cost. We can get ρ of other edges in similar way. After we replace $(4,5)$ in T_1 with $(0,5)$ of D, the storage cost of T is 3050, exceeding the β and we get the final T in Fig. 4(c).

Algorithm 4. RPT

Input: version graph G, upper bound for recreation cost δ, threshold for the maximum recreation θ

Output: node set of spanning tree T

1 $REC \leftarrow 0$; $Vset \leftarrow \emptyset$; PQ.enqueue(V_0); $//REC$ is the total recreation cost of current spanning tree T. The current nodes are contained in $Vset$. PQ is a priority queue.

2 **while** PQ *is not empty* **do**

3      $V_i \leftarrow PQ$.top(); //out-queue process $Vset$.add (V_i)

4      $REC \leftarrow REC + V_i.rec$; PQ.dequeue();

5      $Nset \leftarrow Neighbors(V_i)$;

6   **foreach** V_j *in* $Nset$ **do**

7   //cost-updating process

8   **if** V_j *in* $Vset$ **then**

9   // V_j is already in current spanning tree

10   **if** $r_j \geq R[i,j] + r_i$ *and* $S[i,j] \leq c_j$ **then**

11                  $Vset$.Remove(V_j);

12                  $c_j \leftarrow$ S[i,j]; $r_j \leftarrow$ R[i,j]$+r_i$;

13                  $Vset$.add(V_j);

14   **else**

15   //V_j is not in current spanning tree

16   **if** $\theta \geq R[i,j]+r_i$ *and* $REC \leq \delta$ *and* $S[i,j] \leq c_j$ *and* $R[i,j] + r_i \leq r_j$ **then**

17                  $c_j \leftarrow$ S[i,j]; $r_j \leftarrow$ R[i,j]$+r_i$;

18   PQ.enqueue(V_j);

19 **return** $Vset$;

4.2   RPT

This algorithm works in a greedy manner. At each iteration, pick the node with the minimum storage cost from priority queue PQ and update the recreation cost, storage cost and parent if the two cost can become smaller than before. Then the spanning tree will be built through iterations. RPT is similar to Prim [11] but significant difference exists, namely that the already constructed spanning tree may be modified in later iterations in RPT but Prim not. The RPT algorithm is suitable for directed cases. We define the node in this algorithm with the structure $(id, parent, sto, rec)$. To be specific, (1) id is the version id; (2) $parent$ refers to some version from which the current version derives from and each single version only has one parent version in our proposal; (3) sto means the storage cost from parent to the current version; (4) rec refers to the recreation overhead between root version to the current one.

The RPT algorithm proceeds as follows. First, version V_0 is pushed into the priority queue PQ (line 1). In PQ, versions are sorted according to sto

---

Algorithm 5. Neighbors

    Input: node V in $Vset$, $Vset$
    Output: the neighbors of V in $Vset$
1 $Nset \leftarrow \emptyset$;
2 **foreach** *edge V' in $Vset$* **do**
3 | **if** $S[V,V']$ *exists* **and** $R[V,V']$ *exists* **then**
4 | | $Nset.\text{add}(V')$;

5 **return** $Nset$;

---

in ascending order. Then, the algorithm is composed of two phases: out-queue processing phase and cost-updating phase.

In out-queue process (line 2–6), pick the top element, denoted as V_i, from PQ who has the minimum *sto* and insert V_i into the $Vset$ which contains nodes of current spanning tree. Update the total recreation cost. Then identify the neighbor nodes of V_i via the *neighbors* function in Algorithm 5.

Back to RPT algorithm, then we elaborate on the cost-updating phase (line 7–18). The purpose of this phase is to update all of the neighbor nodes of node V_i. Suppose V_j is one of the neighbor nodes of V_i, if V_j is not contained in current spanning tree, we should judge whether the recreation cost of V_j and total recreation cost of the current spanning tree are within the predefined constraint or not first. If the thresholds are met, then we can insert V_j into current spanning tree. In contrast to the circumstances mentioned above, if the neighbor node V_j is already in current spanning tree, we need merely to determine whether the storage and recreation cost of V_j can become smaller than before. If they can, we need to update V_j. Obviously, this is the significant difference mentioned above between RPT and original Prim algorithm. Iterations will not be terminated until all the nodes addressed.

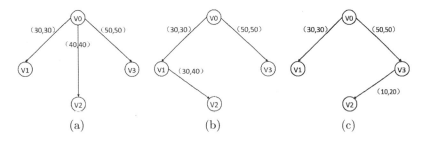

Fig. 5. An example of RPT

We take RPT algorithm on Fig. 5 for instance to facilitate comprehension.

Example 2. Figure 5 illustrates an example of RPT on version graph G_2.

Here, the recreation cost δ is set 200 and the threshold of the maximum recreation cost θ is 70. First, add V_0 to priority queue PQ and dequeue V_0.

Update the neighbors of V_0, namely V_1, V_2 and V_3 and enqueue them (resulting in Fig. 5(a)). Since PQ is sorted according to storage cost in ascending order, we dequeue V_1 and update V_2, the only neighbor of V_1. For V_2, $rec : 30 + 40 = 70$ and $sto : 30 < 40$. Thus, we can also update the parent of V_2 as V_1, as illustrated in Fig. 5(b). Next, dequeue V_2 and update its only neighbor V_3. For V_3, $rec : 70 + 20 > 70$, exceeding the θ. Therefore, we can not update V_3. Now, only V_3 is still in PQ. Dequeue PQ and update its neighbor V_2. For V_2, $rec : 50 + 20 = 70$ and $sto : 10 < 30$. Consequently, update the parent of V_2 as V_3, as shown in Fig. 5(c). Until this step, the total recreation cost is 150, still within the δ.

(Complexity Analysis). Only the root node, not all nodes is put into the priority queue PQ at first. Then the neighbors of the dequeued node are updates and updated nodes are enqueued. The complexity of each "push" operation is $O(\log |V|)$ and scan every single edge once. Thus, the overall time complexity is $O(|E| \log |V|)$.

5   Experiments

5.1   Experimental Setup

In this section, we conduct an extensive series of experiments on real-life and synthetic data sets to evaluate the efficiency and effectiveness of our proposal. All codes written in C++, were conducted on a PC with 16 GB RAM, Intel Core CPU 3.2 GHz i7 processor and the operating system is Windows 10.

Datasets:

- NOAA dataset: (D1) This is a dense collection composed with 1 MB weather satellite images, comprising approximately 14G and from the NOAA of US. We regard observation data of one single time interval (10 min) as one version and about 12,000 versions in total.
- OMCSC dataset: (D2) The Open Mind Common Sence ConceptNet network dataset, a sparse one and is filled with degrees of relationships among many kinds of "concepts". We choose snapshots data from 2015, each version composing of approximately 430,000 data points and 12,000 versions totally.
- Synthetic dataset: (D3) We generate a data set by using R-MAT graph generator in GTgraph [1] with the number of nodes (versions) $n = 10,000$ and the number of edges $m = 800$. Then, we randomly select one node as the start point to run the Breadth-First-Search algorithm until n versions arrive.

Baseline Approaches: We evaluate the performance of EDT by varying the parameter α. For each dataset, we set $\alpha = 0$, 0.5 and 1 respectively. When $\alpha = 0$, EDT optimizes the maximum recreation cost. When $\alpha = 1$, EDT is aimed at optimizing the total recreation cost given restriction on the storage cost. This correspond to the LMG in [4]. When α is set to 0.5, EDT optimizes the

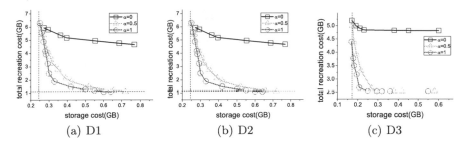

Fig. 6. Total storage and recreation cost of EDT

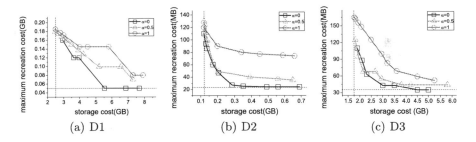

Fig. 7. Total storage and the maximum recreation cost of EDT

total recreation cost and the maximum recreation cost at the same importance. Besides, we compare the trade-off of RPT by varying the number of versions with MP in [4], which does not take the maximum recreation cost into storage-recreation trade-off at the same time. The restriction for the maximum recreation cost is set to 0.005, 0.006, 0.007 and 0.008 respectively. The limitation of the whole recreation cost for RPT is 3.25, 3.48, 4.3 and 4.5 respectively.

Criterion: The total storage and the whole recreation cost are the two criterion in the storage-recreation trade-off. We take another criterion, the maximum recreation cost into trade-off at the same time.

5.2   Experimental Results

Effectiveness of EDT. Figures 6 and 7 illustrate the trade-off of EDT among the total storage, total recreation and the maximum recreation cost upon D1, D2 and D3.

From Fig. 6, when the total storage exceeds the minimum storage (the minimum storage is the storage cost of minimum spanning tree depicted with horizontal pink line parallel to the horizontal axis), the whole recreation cost decreases quickly and afterwards, declines more and more slowly. Meanwhile, for these datasets, EDT with $\alpha = 1$ obtains the smallest recreation cost given the same storage threshold whereas the solutions with $\alpha = 0$ have the largest recreation

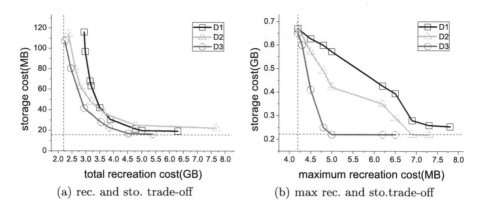

(a) rec. and sto. trade-off   (b) max rec. and sto.trade-off

Fig. 8. Effectiveness evaluation of RPT

cost. This is due to the optimization goal for solution with $\alpha = 1$ is to get the minimum recreation under restricted storage cost. The solutions with $\alpha = 0.5$ have the middle recreation cost between the other two, more approaching to the solutions with $\alpha = 1$.

Next, Fig. 7 illustrates the total storage and the maximum recreation cost of EDT. The solutions with $\alpha = 0$ get the smallest maximum recreation cost given the same threshold of storage cost, because in each iteration, the edge that can reduce the most maximum recreation cost with the same increase in storage cost be picked to replace the corresponding edge in the minimum spanning tree. Besides, some plateaus exist in the plot, because after replacing corresponding edge in the minimum spanning tree, the recreation path for the version with the maximum recreation cost is still the same, not influenced or the influenced versions get the new recreation cost that is still less than the maximum recreation cost. Hence, by combining Figs. 6 and 7, the EDT can achieve the good balance in the storage and recreation trade-off that considers the maximum recreation cost at the same time.

Effectiveness of RPT. Figure 8 investigates the trade-off by RPT among three costs upon D1, D2 and D3.

Figure 8(a) reports the total recreation and storage cost of RPT. From Fig. 8(b), for D1, D2 and D3, we observe that when the recreation cost is a little more than the minimum recreation cost (the minimum total recreation cost is the recreation cost of the shortest path tree by horizontal pink line parallel to the horizontal axis), total storage cost decrease largely and then incline at a slower rate with the increment of the restriction on the total recreation cost.

Figure 8(b) illustrates the maximum recreation and the total storage cost of RPT. In Fig. 8(b), we discover that as the increase of the maximum recreation cost, there exist plateaus for total storage cost. This is due to that the total recreation cost exceeds the predefined thresholds. Hence, the RPT can achieve balance among the three costs mentioned above.

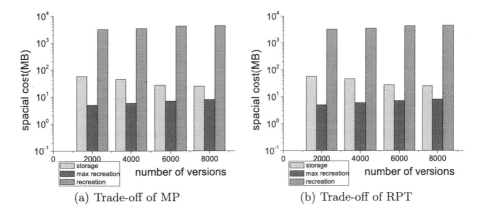

(a) Trade-off of MP

(b) Trade-off of RPT

Fig. 9. MP vs RPT under varying number of versions

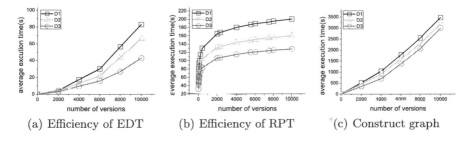

(a) Efficiency of EDT   (b) Efficiency of RPT   (c) Construct graph

Fig. 10. Efficiency evaluation

Figure 9 compares the trade-off result between MP and RPT by varying the number of versions. MP does not require restriction on the total recreation cost since it aims at optimizing the storage cost given a bound on the maximum recreation cost. We can observe that when the number of versions is 2000, restriction on the total recreation cost leads to more storage cost of RPT than that of MP. For 4000 versions, the situation is similar. In the occasion of 6000 versions, MP and RPT have the same total recreation cost and the same maximum recreation overhead, but the recreation cost threshold of RPT slightly rises, because RPT reaches the final balance before the recreation cost reaches the threshold and the balance of MP is the same as the one of RPT. In the last case, for RPT, balance happens when the total recreation cost reaches the predefined the threshold. Both MP and RPT can achieve good trade-off between storage and recreation. Whereas, RPT takes the maximum recreation cost into consideration at the same. Besides minimizing the storage with restricted maximum recreation cost for single version, the total recreation cost for all versions is also limited in some range.

Efficiency. Figure 10 reports the average execution time of EDT and RPT upon D1, D2 and D3 respectively. EDT takes the minimum spanning tree through

Edmonds' Algorithm [7] and the shortest path tree by Dijkstra's Algorithm as input and the construction time of the graph. The α is set to 0.5. For both algorithms, under the same number of versions, the dense dataset D1 requires more time than the sparse one D2 because the dense dataset has more delta than the sparse one, which means more iterations. In addition, we see that the time consumption increases linearly for EDT and logarithmically for RPT in Fig. 10(a) and (b) by varying number of versions.

6 Conclusion

In this paper, we investigate the trade-off in multi-version data management which takes the total storage, recreation and the maximum recreation cost into account simultaneously. We firstly formulate two new problems with different objectives and conditions. Then we devise two efficient algorithms to achieve the goals. Finally, we evaluate our proposals with extensive experiments to illustrate our methods are both effective and efficient. For future work, we plan to further study multi-version data management with regard to the versions which are derived from multiple parent versions.

References

1. GTgraph. http://www.cse.psu.edu/~kxm85/software/GTgraph/
2. Large Synoptic Survey Telescope. http://www.lsst.org/
3. Baumann, P.: Standardizing big earth datacubes. In: 2017 IEEE International Conference on Big Data (Big Data), pp. 67–73. IEEE (2017)
4. Bhattacherjee, S., Chavan, A., Huang, S., Deshpande, A., Parameswaran, A.: Principles of dataset versioning: exploring the recreation/storage tradeoff. VLDB Endow. 8(12), 1346–1357 (2015)
5. Chan, T.N., Yiu, M.L., Hua, K.A.: Efficient sub-window nearest neighbor search on matrix. IEEE Trans. Knowl. Data Eng. 29(4), 784–797 (2017)
6. Chavan, A., Deshpande, A.: DEX: query execution in a delta-based storage system. In: Proceedings of the 2017 ACM International Conference on Management of Data, pp. 171–186. ACM (2017)
7. Cormen, T.H.: Introduction to Algorithms. MIT press, Cambridge (2009)
8. Cudré-Mauroux, P., Kimura, H., Lim, K.T., Rogers, J., Simakov, R., Soroush, E., Velikhov, P., Wang, D.L., Balazinska, M., Becla, J., et al.: A demonstration of SciDB: a science-oriented DBMS. VLDB Endow. 2(2), 1534–1537 (2009)
9. Gosain, A., Saroha, K.: Storage structure for handling schema versions in temporal data warehouses. In: Sa, P.K., Sahoo, M.N., Murugappan, M., Wu, Y., Majhi, B. (eds.) Progress in Intelligent Computing Techniques: Theory, Practice, and Applications. AISC, vol. 518, pp. 501–511. Springer, Singapore (2018). https://doi.org/10.1007/978-981-10-3373-5_50
10. Li, J., Kawashima, H., Tatebe, O.: Efficient window aggregate method on array database system. J. Inf. Process. 24(6), 867–877 (2016)
11. Prim, R.C.: Shortest connection networks and some generalizations. Bell Labs Tech. J. 36(6), 1389–1401 (1957)

12. Rusu, F., Cheng, Y.: A survey on array storage, query languages, and systems. arXiv preprint arXiv:1302.0103 (2013)
13. Seering, A., Cudre-Mauroux, P., Madden, S., Stonebraker, M.: Efficient versioning for scientific array databases. In: 2012 IEEE 28th International Conference on Data Engineering (ICDE), pp. 1013–1024. IEEE (2012)
14. Soroush, E., Balazinska, M.: Time travel in a scientific array database. In: 29th Data Engineering (ICDE), pp. 98–109. IEEE (2013)
15. Soroush, E., Balazinska, M., Krughoff, S., Connolly, A.: Efficient iterative processing in the SciDB parallel array engine. In: 27th International Conference on Scientific and Statistical Database Management, p. 39. ACM (2015)
16. Soroush, E., Balazinska, M., Wang, D.: ArrayStore: a storage manager for complex parallel array processing. In: 2011 ACM SIGMOD International Conference on Management of data, pp. 253–264. ACM (2011)
17. Tansel, A.U., Clifford, J., Gadia, S.K., Jajodia, S., Segev, A., Snodgrass, R.T. (eds.): Temporal Databases: Theory, Design, and Implementation. Benjamin/Cummings, San Francisco
18. Zhang, Y., Xu, F., Frise, E., Wu, S., Yu, B., Xu, W.: DataLab: a version data management and analytics system. In: 2nd International Workshop on BIG Data Software Engineering, pp. 12–18. ACM (2016)

Decentralized Data Integrity Verification Model in Untrusted Environment

Kun Hao[1], Junchang Xin[1,2(✉)], Zhiqiong Wang[3], Zhuochen Jiang[4], and Guoren Wang[5]

[1] School of Computer Science and Engineering, Northeastern University, Shenyang, China
xinjunchang@mail.neu.edu.cn
[2] Key Laboratory of Big Data Management and Analytics, Shenyang, Liaoning, China
[3] School of Sino-Dutch Biomedical and Information Engineering, Northeastern University, Shenyang, China
[4] College of Computer Science and Technology, Jilin University, Changchun, China
[5] School of Computer Science and Technology, Beijing Institute of Technology, Beijing, China

Abstract. Outsourced data, as an significant component of cloud service, has been widely used due to its convince, low overhead and high flexibility. To guarantee the integrity of outsourced data and reduce the computational overhead, data owner (DO) usually adopts a third party auditor (TPA) to execute verification scheme. However, handing over the verification of data to TPA may lead to security vulnerabilities since the TPA is not fully trusted. In this paper, we propose a novel solution for data integrity verification in untrusted outsourced environment. Firstly, we design a decentralized model based on blockchain, consisting by some collaborative verification peers (VPs). Based on our purposed model, we present an advanced data integrity verification algorithm, allowing DO stores and checks verification results by writing and retrieving the blockchain. Moreover, each VP maintains a replication of the entire blockchain to avoid maliciously tampering with. We evaluate our proposed approach on real outsourced data service scenario. Experimental results demonstrate that our proposed approach is efficient and effective.

Keywords: Decentralized · Collaborative · Blockchain
Data integrity · Untrusted environment

1 Introduction

Outsourced data, as an significant component of cloud service, has been widely used due to its convince, low overhead and high flexibility. Nowadays, more and more cloud service providers (CSPs) present outsourced data applications, such as Google App Engine [4], Microsoft Azure Platform [5], Amazon S3 [1] and

Baidu Yun [3]. The success of outsourced data services are driven by that it can operate lots of data centers and offer distributed storage for data owner (DO) to relief their burthen of massive data management.

As a matter of fact, DO will lose physical control over their data and how data is processed or stored. To remedy that, DO usually adopts some strategies to guarantee the integrity and availability of outsourced data. The existing approaches including Proofs of Retrievability (POR) [18,29] which provides DO with the confidence that their data is still retrieved and downloaded if neccessary, and Proofs of Data Possession (PDP) [9] which enables DO to verify that their outsourced data has not undergone any malicious modifications. Both two protocols enable CSP to prove that DO can verify files without downloading entire data. To conduct verification process, as shown is Fig. 1(a), DO can tolerate the computational overhead. A native way to improve the efficiency is to outsource the verification procedure to a third party auditor (TPA) [8]. As shown in Fig. 1(b), when finishing the outsourced data process, the TPA executes verification protocols with CSP on behalf of DO. Additionally, at any point in time, DO can check TPA's work. In such model, DO is not necessary to establish communications with CSP so that reducing the computation overload. However, the TPA is not fully trusted, and handing over verification tasks of data integrity to a untrusted TPA inevitably raises new threat to data security. For example, the untrusted TPA may cheat DO by returning fake verification results. More seriously, TPA and CSP can collude to tamper with the verification records. The reason for these cases is the overdependence of DO for the centralized untrusted auditor party.

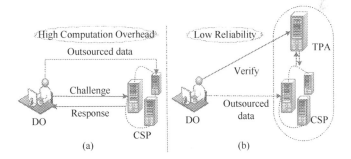

Fig. 1. Two examples of outsourced data environment. (a) Traditional outsourced data model; (b) outsourced data model by utilizing TPA

In order to solve above problem, we need a decentralized solution for data integrity verification in untrusted environment. Whereas proposing such a solution is not straightforward and will require addressing the following critical challenges. Firstly, since there is no central node, we need an effective storage strategy to store the verification records for data integrity in decentralized scenario. Moreover, in order to ensure the verification records are not maliciously tampered with, the storage strategy needs an effective mechanism to improve the security of records. Finally, in order to avoid overly relying on a central auditor, we should utilize a distributed verification approach for data integrity.

The blockchain is first introduced by the bitcoin [24], which is considered to be the new innovation of computer technology. In addition, it is viewed as the basis of the next generation of cloud computing. Blockchain is expected to transform the traditional Internet from information-based model to trust-based model [32]. The core idea of the blockchain is the storage and verification of data in an untrusted environment. More significantly, the features of blockchain include decentralization, redundancy storage, collective maintenance and tamper resistant. These features motivate us to bring the blockchain into data integrity verifications in untrusted environment.

In this paper, we design a novel decentralized model in untrusted environment based on blockchain, namely DCOM (Decentralized COllaborative verification Model). Based on our proposed model, we present DIV (Decentralized Integrity Verification) algorithm which including two phases, i.e., WriteBlock and Check-Block. To the best of our knowledge, DCOM is the first formally to build a decentralized model based on blockchain technology for verifying data integrity in untrusted environment. The contributions of this paper outline as follows.

- We present a decentralized model based on blockchain. The model includes a collaborative network consisting of some verification peers, and each peer maintains verification records by a blockchain form.
- We propose a verification algorithm for data integrity which divides into two phases. In WriteBlock phase, DO can task verification peers to store entired verification records formed by a blockchain. Additionally, in Check-Block phase, DO can request collaborative network to obtain the verification results by retrieving the local blockchain.
- We evaluate our proposed approach on a real outsourced data service scenario. Experimental results demonstrate that our proposed approach is efficient and effective.

The rest of this paper is organized as follows. In Sect. 2, we give preliminaries and define the problem. Section 3 describes decentralized collaborative model based on blockchain. In Sect. 4, we discuss the decentralized data integrity verification algorithm. We show the experimental evaluations in Sect. 5. We overview related works in Sect. 6, and conclude the paper in Sect. 7.

2   Preliminaries and Problem Statement

2.1   Preliminaries

Proofs of Retrievability. POR is a *Challenge/Response* protocol which enables the CSP proves that the raw file can be retrieved without downloading the entire data. Note that, POR only guarantees that a fraction p of the file can be retrieved. For that reason, POR is usually performed on a file which has been *erasure-coded*. In such a way, the file can be recovered by any fraction p of stored data. More precisely, POR assumes a model consisting of a user, and a service provider that stores a file uploaded by the users. A POR protocol includes four key steps [29] which are *Setup*, *Store*, *Challenge*, and *Response*.

- *Setup.* This algorithm generates public keys pub_k and private keys pri_k. Each user distributes pub_k to all parties, and keeps pri_k secretively.
- *Store.* This algorithm takes pri_k as input keys of the DO and a raw file. The file F gets processed and outputs the produced $F*$ which will be stored on the CSP. Additionally, a file tag τ generated which contains additional information about F, such as metadata and secret information.
- *Challenge, Response.* The challenge and response algorithms define a procedure for proving the retrievability of outsourced data. At any point of time, DO can *Challenge* service providers for proving the retrievability of F. The *Response* algorithm executed by service providers returns TRUE if the verification succeeds, meaning that the file can be retrieved, and FALSE otherwise.

Blockchain. Blockchain, also named distributed ledger, is an *append-only* data model maintained by a set of peers who do not fully trust each other. The features of blockchain include decentralization, redundancy storage, collective maintenance and tamper resistant. Firstly, blockchain is a decentralized model. The procedures of storing, transmitting and verifying data in blockchain are entirely based on a decentralized model. In such a model, there is no central node, which means that all nodes in the blockchain have identical obligation. Secondly, blockchain can be viewed as a reliable database. Blockchain stores data redundantly, i.e., each node has a complete replication of blockchain. Hence, in order to modify the data, malicious nodes must control more than half of computation power of the entire blockchain network, which is great difficult to achieve. Finally, data stored by blockchain are safe and trustworthy. Blockchain use asymmetric cryptography methods to preserve privacy of data. Additionally, blockchain adopts hash algorithms to ensure data cannot be tampered with, and deploys consensus algorithms to guarantee consistency of all blockchain replications. Figure 2 shows an example of blockchain.

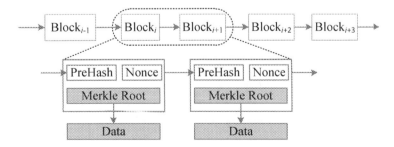

Fig. 2. The data structure of blockchain. Data are packed into blocks which are linked to previous blocks.

2.2 Problem Statement

System Model. We build on the definition from [8]. More precisely, we assume that, in order to reduce the computational workload, the DO outsources verification work of data integrity to the TPA. However, the DO usually can not determine the confidence of TPA before outsourcing the verification work, that is, DO outsources data integrity verification to untrusted TPA. It is worth noting that we assume the CSP safely stores DO's data. Unless otherwise specified, in the following of this paper, we suppose the CSP is trustworthy and the TPA is untrusted. Our model includes four components including data owner, cloud service provider, collaborative network and blockchain. In the following section, we give details of each component of proposed model.

Threat Model. To analyze data integrity of outsourced in untrusted environment, we need to define the threat model. More specifically, in this paper, we assume the TPA is untrusted. The risks of using untrusted TPA to verify the integrity of outsourced data is due to the following reasons. First, since DO loses the supervision of verifying process and record, the untrusted TPA maybe arbitrarily modify the verification records and return the incorrect results to DO. In this case, in order to verify the data integrity, DO can only download the entire data. Second, in the public auditing of outsourced data, the DO fully relies on a centralized TPA for verification. However it is possible that the TPA is inquisitive about the cloud data. The existing technology usually outsources the verification records of outsourced data and the signature of DO to the TPA simultaneously. This case makes the untrusted TPA can use DO's signature to query the data stored in the CSP, resulting in reduced data security. More seriously, the untrusted TPA can delete the data and forge the verification records.

Design Goals. The reason for the above threat model is that DO, in order to reduce the computational workload, hands over the integrity verification to an untrusted centralized TPA. The main motivation of this work is to develop a decentralized model for verifying data integrity in untrusted environment. Unlike traditional approaches, this work proposes to achieve the following goals. Firstly, to avoid above security issues, our verifying model is decentralized, that is, there is no centralized party for controlling the whole processing for data integrity verification. And then, in our context, the security of data should be guarantee by decentralized ways. Moreover, the efficiency of verification process should efficient as traditional approaches, such as POR and OPOR.

3   Decentralized Collaborative Model

3.1   Architecture Overview

As shown in Fig. 3, DCOMB consists of four key components which including data owner, cloud service provider, collaborative network, and blockchain. In the following, we introduce the details of each component. **Data Owner (DO).** DO outsources data in order to reduce local storage overhead for massive data by utilizing "infinite" storage capacity, and adopt computation resources provided by the service providers. **Cloud Service Providers (CSP).** CSP controls a cluster consisted of numerous quantity hardware and software resources to provide outsourced data services. We argue that CSP has its own distributed storage system, such as HDFS [2] and Ceph [30], to manage massive data and metadata for replications. **Collaborative Network (CN).** CN consists of a number of verification peers (VPs) which keep communications with each other to form a P2P network. Each VP can join or leave the network at any time. When DO outsources data to CSP, all VP challenge the CSP to obtain POR records for each file, which reducing computation workload of POR. **Blockchain (BC).** BC stores the POR records of each outsourced file in a blockchain. In order to avoid reliable issues by utilizing TPA, each VP maintains a entire replication of blockchain locally.

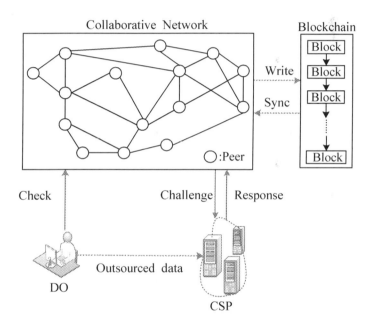

Fig. 3. Architecture of DCOMB model.

3.2 Data Model

As already elaborated, in traditional blockchain, each block includes two key parts including block header and block body. The block header maintains the digest calculated by the previous block, and the block body stores data such as financial transactions [24] and smart contracts [13]. This combination manner makes that each block in the blockchain only depends on its previous block. Specifically, in DCOMB model, each block body stores POR records formed by a Merkle Tree [21].

As shown in Fig. 4, the leaf node of the POR Merkle Tree consists of a two-tuple, where ρ_i represents the POR records of user i, and $Sig_S()$ is a signature function. According to the definition of Merkle Tree, the POR records can be verified by given intermediate hash values, i.e., verification path [11]. For example, the gray entities of Fig. 4 consist a verification path for verifying $(\rho_2, \text{Sig}_S(\rho_2))$

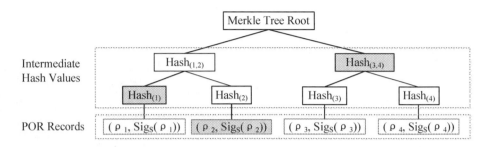

Fig. 4. Sketch of POR Merkle Tree.

4   Decentralized Integrity Verification Algorithm

In this section, we propose DIV algorithm based on DCOMB model. The algorithm includes three key steps, that is **Setup**, **WriteBlock**, and **CheckBlock**. In the following, we introduce the specification of each step.

4.1   Setup

We first define some functions for setting up the DIV algorithm. These functions are the prerequisites for the DIV algorithm, including $KeyGen()$, $Split()$, and $PORGen()$.

- $(pub_k, pri_k) \leftarrow KeyGen()$. Similar to traditional POR protocols, each DO needs to execute the key generation function for receiving public key pub_k and private key pri_k. All DO distribute their pub_k to the collaborative network and keep the pri_k secretively.
- $F^* \leftarrow Split(F)$. Given the file F, the function split the F by coding method, such as *erasure code*, which contains n blocks, each s sectors long (i.e. n and s are reconstruction thresholds).
- $\rho_i \leftarrow PORGen(pri_k, F^*)$. The function gets POR records ρ_i for F^* by using the method in [29].

After generating the required parameters, DO signs ρ_i using his pri_k which denoted by ρ_i^*, and transmits the ρ_i^* to the collaboration network CN. For the sake of simplicity, in the following of this paper, we use ρ_i to denote ρ_i^*.

Algorithm 1. WriteBlock Algorithm

Input: DO set $S_{DO} = \{DO_1, DO_2, \cdots, DO_m\}$, outsourced data set $S_F = \{F_1, F_2, \cdots, F_m\}$, verification peers set $S_{VP} = \{VP_1, VP_2, \cdots, VP_n\}$

1 Outsourced all data in S_F to CSP;
2 **for** *each DO in S_{DO}* **do**
3      $(pub_k, pri_k) \leftarrow KeyGen()$;
4      $F^* = Split(F)$;
5   gets and signs POR records ρ computed by CSP using $PORGen(pri_k, F^*)$;
6 $S_P = \{(\rho_1, sig_1), (\rho_2, sig_2), \cdots, (\rho_m, sig_m)\}$;
7 Transmit S_P to all VPs S_{VP};
8 **for** *each VP in S_{VP}* **do**
9   **if** *not receives* message$_{succ}$ **then**
10        $result = $ VerifySig(S_P);
11   **if** result **then**
12   Generates a POR Merkle Tree for S_P and builds a *block*;
13   Broadcasts *message$_{succ}$* to other VPs and gets write permission to blockchain;
14   Writes block into local blockchain;
15   Broadcasts *message$_{sync}$* to other *vps*;
16   **else**
17   **if** *receives* message$_{sync}$ **then**
18   Synchronizes the state of local blockchain;

4.2 WriteBlock Algorithm

In this subsection, we give the details of WriteBlock algorithm for storing the POR records based on DCOM model. When receiving the POR records, all VPs validate whether the signatures are valid by utilizing the corresponding pub_k. If the process passed, all VPs cache the POR records in their memory, otherwise discard it. Then, the VPs construct a POR Merkle Tree and adopt the POW [24] consensus mechanism to contest for writing the blockchain BC. Finally, synchronizing the local blockchain state of all VP. Consider a DO set $S_{DO} = \{DO_1, DO_2, \cdots, DO_m\}$, $S_F = \{F_1, F_2, \cdots, F_m\}$ represents the data of each DO respectively, and $S_{VP} = \{VP_1, VP_2, \cdots, VP_n\}$ are verification peers consisting a collaborative network. The procedure of *WriteBlock* is shown as follow.

- Each DO in S_{DO} gets the cryptographic keys and encode the file according to Sect. 4.1. Then, sending the processed file to the CSP;
- The CSP compute the POR records set $S_\rho = \{\rho_1, \rho_2, \cdots, \rho_m\}$ for each F_i, where $0 < i \leq m$. All VPs in S_{VP} validate each ρ in S_ρ, and construct the POR Merkle Tree using the valid POR records. Moreover, each VP calculates a random value for a given difficulty simultaneously. VP* denotes the peers who first finishing all above calculations, and send *message$_{succ}$* to other peers.

– VP* gets the *write permission* for linking the constructed block to the global blockchain, and broadcasts message $message_{sync}$ to other VPs for synchronizing their local blockchain.
– When received $message_{sync}$, VPs synchronize the local state of blockchain to match the global blockchain state.

4.3   CheckBlock Algorithm

In this subsection, we discuss CheckBlock algorithm for verifying the POR records stored in blockchain. Based on the tamper-resistant feature of blockchain, the data stored in the blockchain cannot be modified maliciously. Firstly, DO sends request to the collaborative network. Then, each VP checks whether the state of local blockchain is consistent with global state. The VPs who have inconsistent state need to synchronize the local block by reading global blockchain. Finally, returning the corresponding POR records by retrieving local blockchain.

Consider a DO sends request r to CN to check the POR records. $S_{VP} = \{VP_1, VP_2, \cdots, VP_n\}$ denotes a verification peer set, and sig represents corresponding signature for DO. The procedure of *CheckBlock* algorithm is shown as follow.

– DO Sends $\{r, sig\}$ to VP_i in S_{VP}, where $0 < i \leq n$.
– Each VP checks whether the state of local metadata blockchain, which represents by VP_i. BC_L, matches the global blockchain BC_G. If inconsistent, the VP_i must synchronize the local state first.
– Moreover, all VP with consistent state retrieve the local blockchain to obtain the corresponding POR records ρ by utilizing sig. The verification peer who finishing the retrieve processing, denoted by VP*, broadcast the $message_{succ}$ to other VPs.
– VPs stop current retrieving process, only if receiving quorum amount (more than half of S_{VP}, for simplicity) of $messages_{succ}$, and VP* returns the *result*.

4.4   Security Analysis

In this subsection, we analyze how our purposed algorithm avoids the threats proposed in Sect. 2.2.

In writeBlock phase, all VP validate the signatures for each DO. The invalid POR records cannot be stored in the blockchain. Moreover, if the signature of DO changed such as replacing the signature algorithm or cryptographic device, the new POR records must be re-signed by the new signature. In addition, only the newer timestamps block can be retrieved. According to the Sect. 3.2, the data in DCOM model constructed a Merkle Tree which consisting of $\{\rho_i, sig_i\}$, where ρ_i and sig_i represent the POR records and signature of DO respectively. Moreover, according to the definition of Merkle Tree, the Root changes along with ρ_i or sig_i which ensures the integrity of the POR records. Additionally, according to Algorithm 1, only one VP in the collaborative network can write new block

---

Algorithm 2. CheckBlock Algorithm

    Input: DO, corresponding signature sig, and verification peers set $S_{VP} = \{VP_1,$
         $VP_2, \cdots, VP_n\}$
    Output: verification result set $result$

1 DO sends verification request and signature to all VP in S_{VP};
2 **for** each VP in S_{VP} **do**
3   **while** not receives exceed quorum of $message_{succ}$ **do**
4   checks the state of local blockchain MBC_L;
5   **if** the state of local MBC_L is inconsistent **then**
6   synchronizes the state of local blockchain;
7   **else**
8            $result$ = retrievePORrecords(sig);
9   **if** $result$ **then**
10   Broadcasts $message_{succ}$ to other VPs;

11 Return $result$;

---

to the blockchain. Furthermore, all VP periodically synchronize the local state in order to maintain consistency of blockchain. The blockchain synchronization ensures that the data stored in each VP are consistent. Therefore, in any point of time, the POR records returned by VP in collaborative network are accurate for any given signature of DO. In the DIV algorithm, DO outsources the data to the CSP who responsible for computing the POR records. The POR records are then sent to the collaboration network where all verification peers store the POR records for all DOs in a form of blockchain. Further, we can get POR records as long as a certain number of verification peers.

5 Experiments

5.1 Experimental Setting

We performed our experiments on a machine with Intel (R) Xeon E5-2620 CPU @ 2.0 GHz, 32 GB RAM, and Ubuntu Linux OS. All proposed model and algorithm were implemented in Golang. In addition, we simulate some VP to constitute a collaborative network. The datasets are raw data and the experimental parameters are shown as Table 1.

Table 1. Experimental parameters

Parameters	Default value	Variation range
Data size (MB)	100	50 (D_1), 100 (D_2), 168 (D_3)
Quantity of VP	10	5, 10, 20
Quantity of DO	20	10, 20, 30, 50

5.2   Performance Evaluations

In this subsection, we show experimental results of our proposed approach. Firstly, we evaluate the concurrent performance of DIV algorithm. Then, we compare DIV algorithm to state-of-art algorithms including POR [29] and OPOR [8], which are widely used in outsourced data services.

In Fig. 5 (where DIV-X, X represents the quantity of VP), we evaluate the time required by the DO to perform DIV algorithm, when compared to POR and OPOR schemes. For this purpose, we vary in the x-axis the fraction of challenged blocks of the total number of blocks of the file. Figure 5(a), (b) and (c) show the comparison results of the algorithm when using the three datasets. It can be seen that the DIV algorithm needs more time to store the POR records than the two algorithms. This is because that the DIV algorithm needs to synchronize and write valid POR records into the blockchain, resulting in a waste of time. Furthermore, the storage efficiency of the DIV algorithm is proportional to the quantity of VP. The reason for this is that the more VP we use, the more messages need to be broadcasted. We can argue that the storage efficiency of the DIV algorithm is similar to the traditional algorithms, and the additional overheads are acceptable.

In Fig. 6, we evaluate the verification by varying in the x-axis the fraction of verification blocks of the total number of blocks of entire file. From the results we can argue that, as the amount of data increases, the time for verifying the entire data grows. The verification efficiency of DIV algorithm is similar to two traditional algorithms. At the same time, it can be seen that the verification time increases along with the quantity of VPs. The reason for this is that VP first check state of local blockchain and synchronize the inconsistent blocks. This makes the verification time is proportional to the quantity of VPs, but the growth trend is slow. We compared the concurrent performance to the two traditional algorithms. First, we simulate some DOs to send storage requests simultaneously. More precisely, we assume that these DOs store the same quantity of data. Figure 7(a), (b) and (c) show the store latency for each algorithm at three datasets. From the experimental results we can see that the store latency of DIV algorithm is proportional to the quantity of data, and the efficiency of our proposed algorithm is similar to the traditional ones.

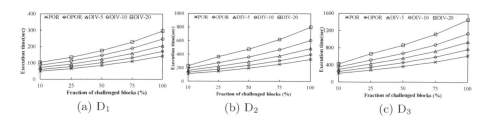

Fig. 5. Store performance with three datasets.

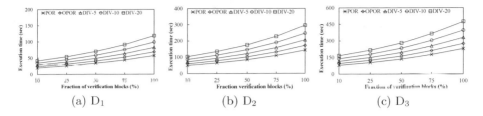

Fig. 6. Verification performance with three datasets.

In addition, we conduct experiments for evaluating the concurrent verification performance. We also simulate several DOs to send verification requests simultaneously. Figure 8(a), (b) and (c) show that the traditional algorithms have low concurrent capability for verifying three datasets, and the POR algorithm is slightly worst than the OPOR algorithm which introduces TPA to improve the verification efficiency. Additionally, the concurrent verification capability of DIV algorithm is inversely proportional to the quantity of VPs. The reason for the results is that all VPs retrieve the local block at the same time, meaning that more VPs can reduce the response delay.

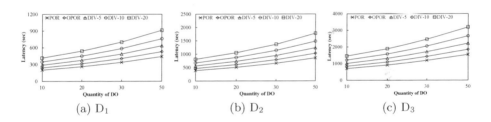

Fig. 7. Concurrent store performance.

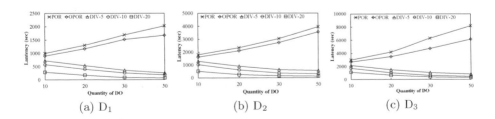

Fig. 8. Concurrent verification performance.

6 Related Work

Integrity Verification Scheme. Juels [18] introduce a POR scheme, and present a sentinels-based algorithm for user to challenge the server to guarantee that the file blocks can be retrieved. Additionally, Shacham and Waters [26] purpose specific POR algorithms including MAC-based and RSA-based signatures, and give detailed security analysis for the algorithms. Ateniese et al. [9] propose a variant of POR scheme called proofs of data possession (PDP), which supports an unbounded number of challenges and enables public verifiability of the scheme. All above schemes required the users to challenge the server, which increasing the computational overload. To remedy that, Armknecht et al. [8] purpose a outsourced POR scheme, which enables an external party to execute a POR protocol with the server.

Untrusted Data Management. Feldman et al. [16] introduce a generic framework for building a wide variety of collaborative applications with untrusted servers. Shraer et al. [27] present a service for securing user interaction with untrusted cloud storage, which guaranteeing integrity and consistency of outsourced data. Brenner et al. [12] add a transparent encryption layer to ZooKeeper by means of a privacy proxy, which establishing confidential coordination for distributed applications. To process queries on the untrusted environment, Chen et al. [14] present novel schemes for verifiable skyline queries via untrusted CSPs. TrustedDB [10] is an outsourced database builded and run on actual hardware, which allows users to execute SQL queries with privacy without having to trust the CSP.

Blockchain. The blockchain technology recently gains increasingly consideration due to its success for trusted and secure mechanism. Blockstack [7] adds features of blockchain into traditional DNS service, and proposes a secure global naming and storage system to query and modify the data through Virtual Chain [7,17,23]. In order to solve the data integrity problem, [6] proposes a decentralized data security sharing network system, which effectively improves the efficiency and security of data. In [25], the authors propose a decentralized data storage model based on POR, which improving the data reliability. In [20], the authors propose a decentralized metadata storage model, which improves the security of cloud storage services. To ensure the data integrity, the authors in [31] provide a block-based storage network by encrypting and signing the raw data, and save blocks into P2P file system. The authors in [33] present a decentralized preserve privacy system by using blockchain. BigchainDB [22] combines traditional distributed database with blockchain, and it improves the security of data and solve the capacity of blockchains simultaneously. In [15], the authors present a framework, namely BLOCKBENCH, to analysis the private blockchain which includes consistency algorithms, data models, execution engines, and applications on the chain. EtherQL [19] proposed an flexible efficient approach for retrieving blockchain, which including range queries and top-k queries. In Beihang chain [28], the authors summarize the application development method based on blockchain, and give the key problems to resolve for developing blockchain applications.

7   Conclusion

In this paper, we introduced a decentralized model based on blockchain, namely DCOM, to avoid the security issues causing by utilizing TPA to run POR protocols. Moreover, we proposed an efficient algorithm named DIV for data integrity verification based on DCOM model. Our experimental results show that our proposal is efficient and effective. We argue that the proposed solution brings the features of blockchain into outsourced data management in untrusted environment. In terms of future work, we plan to explore more efficient mechanisms to optimize the verification procedures, to design an extended scheme which supports dynamic verification, and to improve existing blockchain systems based on the proposed models and algorithms to support data verification in untrusted environment.

Acknowledgement. This research was partially supported by the National Natural Science Foundation of China (Nos. 61472069, 61402089, and U1401256), the Fundamental Research Funds for the Central Universities (Nos. N161602003, N171607010, N161904001, and N160601001), the Natural Science Foundation of Liaoning Province (No. 2015020553).

References

1. Amazon. http://aws.amazon.com
2. Apache hadoop. http://hadoop.apache.org/
3. Baidu. http://cloud.baidu.com
4. Google. http://www.google.com
5. Microsoft azure. http://www.microsoft.com/windowsazure
6. 1e96a1b27a6cb85df68d728cf3695b0c46dbd44d: Filecoin: a cryptocurrency operated file storage network (2014)
7. Ali, M., Nelson, J., Shea, R., Freedman, M.J.: Blockstack: a global naming and storage system secured by blockchains. https://www.zurich.ibm.com/dccl/papers/nelson_dccl_slides.pdf. Accessed 13 Dec 2016
8. Armknecht, F., Bohli, J.M., Karame, G.O., Liu, Z., Reuter, C.A.: Outsourced proofs of retrievability. In: ACM SIGSAC Conference on Computer and Communications Security, pp. 831–843 (2014)
9. Ateniese, G., Burns, R., Curtmola, R., Herring, J., Kissner, L., Peterson, Z., Song, D.: Provable data possession at untrusted stores. In: ACM Conference on Computer and Communications Security, pp. 598–609 (2007)
10. Bajaj, S., Sion, R.: TrustedDB: a trusted hardware based database with privacy and data confidentiality. In: ACM SIGMOD International Conference on Management of Data, SIGMOD 2011, Athens, Greece, June 2011, pp. 205–216 (2011)
11. Becker, G.: Merkle signature schemes, merkle trees and their cryptanalysis. Technical report. Ruhr-University Bochum (2008)
12. Brenner, S., Wulf, C., Kapitza, R.: Running zookeeper coordination services in untrusted clouds (2014)
13. Buterin, V.: A next-generation smart contract and decentralized application platform (2014)

14. Chen, W., Liu, M., Zhang, R., Zhang, Y., Liu, S.: Secure outsourced skyline query processing via untrusted cloud service providers. In: IEEE International Conference on Computer Communications, INFOCOM 2016, pp. 1–9. IEEE (2016)
15. Dinh, T.T.A., Wang, J., Chen, G., Liu, R., Ooi, B.C., Tan, K.L.: Blockbench: a framework for analyzing private blockchains (2017)
16. Feldman, A.J., Zeller, W.P., Freedman, M.J., Felten, E.W.: SPORC: group collaboration using untrusted cloud resources. In: Usenix Conference on Operating Systems Design and Implementation, pp. 337–350 (2010)
17. Jiye, W., Lingchao, G., Aiqiang, D.: Block chain based data security sharing network architecture research. J. Comput. Res. Dev. **54**(4), 742–749 (2017)
18. Juels, A.: PORs: proofs of retrievability for large files. In: ACM Conference on Computer and Communications Security, pp. 584–597 (2007)
19. Li, Y., Zheng, K., Yan, Y., Liu, Q., Zhou, X.: EtherQL: a query layer for blockchain system. In: Candan, S., Chen, L., Pedersen, T.B., Chang, L., Hua, W. (eds.) DASFAA 2017. LNCS, vol. 10178, pp. 556–567. Springer, Cham (2017). https://doi.org/10.1007/978-3-319-55699-4_34
20. Lowry, S., Wilkinson, J.: Metadisk: blockchain-based decentralized file storage application. https://storj.io/metadisk.pdf. Accessed 2 March 2017
21. Mao, J., Zhang, Y., Li, P., Li, T., Wu, Q., Liu, J.: A position-aware merkle tree for dynamic cloud data integrity verification. Soft. Comput. **21**(8), 2151–2164 (2017)
22. McConaghy, T., Marques, R.: BigchainDB: a scalable blockchain database. https://www.bigchaindb.com/whitepaper/bigchaindb-whitepaper.pdf. Accessed 11 Jan 2017
23. Miller, A., Juels, A., Shi, E., Parno, B., Katz, J.: Permacoin: repurposing bitcoin work for data preservation. In: IEEE Symposium on Security and Privacy, pp. 475–490 (2014)
24. Nakamoto, S.: Bitcoin: a peer-to-peer electronic cash system (2008)
25. Sengupta, B., Bag, S., Ruj, S., Sakurai, K.: Retricoin: bitcoin based on compact proofs of retrievability. In: International Conference on Distributed Computing and Networking, p. 14 (2016)
26. Shacham, H., Waters, B.: Compact proofs of retrievability. J. Cryptol. **26**(3), 442–483 (2013)
27. Shraer, A., Cachin, C., Cidon, A., Keidar, I., Yan, M., Shaket, D.: Venus: verification for untrusted cloud storage. In: ACM Workshop on Cloud Computing Security Workshop, pp. 19–30 (2010)
28. Tsai, W.T., Yu, L., Wang, R., Liu, N., Deng, E.Y.: Blockchain application development techniques (2017)
29. Waters, B.: Compact proofs of retrievability. J. Cryptol. **26**(3), 442–483 (2008)
30. Weil, S.A., Brandt, S.A., Miller, E.L., Long, D.D., Maltzahn, C.: Ceph: a scalable, high-performance distributed file system. In: Proceedings of the 7th Symposium on Operating Systems Design and Implementation, pp. 307–320. USENIX Association (2006)
31. Wilkinson, S., Boshexski, T.: Metadisk: blockchain-based decentralized file storage application. https://storj.io/storj.pdf. Accessed 11 January 2017
32. Yong, Y., Feiyue, W.: The development status and prospects of blockchain technology. Acta Autom. Sin. **42**(4), 481–494 (2016)
33. Zyskind, G., Nathan, O., Pentland, A.S.: Decentralizing privacy: using blockchain to protect personal data. In: IEEE Security and Privacy Workshops, pp. 180–184 (2015)

Enabling Concurrency on Smart Contracts Using Multiversion Ordering

An Zhang$^{(\boxtimes)}$ and Kunlong Zhang

School of Computer Science and Technology, Tianjin University, Tianjin, China
{zhangan,zhangkl}@tju.edu.cn

Abstract. Blockchain-based platforms, such as Ethereum, allow transactions in blocks to call user-defined scripts named *smart contracts*. In the blockchain network, after being generated by a miner, a block will be validated many times by the peers who accept it. Hence by enabling concurrency on smart contracts, especially validation, we can improve the efficiency and the throughput of those platforms.

By introducing multiversion transaction ordering, this paper presents a concurrent scheme called *MVTO* to run smart contracts concurrently. First, the miners are able to use any concurrency control technique to discover a conflict-serializable schedule. Then, validators use MVTO to verify the block by replaying this schedule concurrently and deterministically. The evaluation shows that this mechanism achieves approximately 2.5x speedup in the block validation using a thread pool with 3 threads.

Keywords: Blockchain · Smart contract · Concurrency
Multiversion transaction ordering

1 Introduction

Platforms like Ethereum are essentially instances of distributed Byzantine-fault-tolerant database. They are built on a decentralized, peer-to-peer network where peers do not fully trust each other. Generally, there are two kinds of peer nodes in the network: miner and validator. We briefly describe their work as follows: Miners repeatedly collect transactions in the network and package them into new *blocks*. When creating a new block, the miner incorporates the cryptographic hash of the preceding block of the new block, *i.e.*, the most recent block in its local storage, into the header of the newly generated block. The cryptographic hash in each block's header acts as the pointer to its preceding block and thus form a chain of blocks, called *blockchain*. The newly generated block is then published to other validators. Validators follow a consensus protocol to decide whether to accept a newly received block or not and how to synchronize with other peers to reach consensus of blockchain states across the network. In a word, the blockchain is a shared immutable database for recording the history of transactions.

© Springer International Publishing AG, part of Springer Nature 2018
Y. Cai et al. (Eds.): APWeb-WAIM 2018, LNCS 10988, pp. 425–439, 2018.
https://doi.org/10.1007/978-3-319-96893-3_32

Ethereum introduces *smart contract* into blockchain. A smart contract is a collection of code (its functions) and data (its states) which are stored on the blockchain with a unique account and address [1]. A contract will execute when it is triggered by a transaction sent to its address. The functions called can be written in several Turing-complete languages such as Solidity [2]. Thus miner will charge fees from target contract's account for every computational step to ensure that the execution will finish. This fee refers to the *gas* in Ethereum.

Smart Contracts execute on blockchain in two scenarios.

1. Mining: When a miner proposes a new block, it starts to execute contracts according to the order of the transactions in the block. The merkle root of final states is then stored in the block.
2. Validating: When a validator receives a new block, it re-executes smart contract with the exact same order adopted by the miner when this block was generated. Then the validator checks the consistency of the resulting states by using merkling techniques.

Despite having the advantages such as tamper-proof and Byzantine-fault-tolerant provided by blockchain, smart contract platforms suffer from the limitation of throughput. This limitation is partly prompted by the lack of concurrent mechanisms in existing smart contracts designs. When miners and validators deal with a block, they execute smart contracts serially to produce a deterministic result. Furthermore, Ethereum is planning to change its consensus protocol from *proof of work* (POW) to an energy saving protocol called *proof of stake* (POS). After switching to POS, Ethereum can significantly save time originally needed in the POW phase. In this trend, Ethereum can execute smart contracts that are more complicated and time-consuming. However, one needs to apply concurrency to fully exploit those saved computational powers.

There are three reasons why smart contracts can not employ naive concurrent solution. First, a smart contract can be called several times by different transactions during a block's execution and therefore race conditions may occur when those calls of the same contract execute in parallel. Second, smart contracts need to execute transactionally. In other words, if Ethereum executes multiple transactions concurrently, it must produce a *conflict-serializable* schedule where the final states can be produced by a serial schedule[1]. Third, validating a block requires deterministic execution which can not be provided by naive concurrency approach.

In the mentioned two scenarios where smart contracts are executed, the computational power spent on these scenarios is unbalanced. A block only gets executed one time when created by miners. If this block is accepted, every node in the network will validate this block. Thus, improving the efficiency of block validation is more important than block generation in mining phase.

We propose a concurrent scheme for smart contracts. In this scheme, when a miner proposes a new block, it can employ any concurrent control technique, such

---

[1] A serial schedule is a schedule where transaction are executed serially and do not interleave each other.

as 2PL or timestamp, as long as it can produce a *conflict-serializable* schedule. During the execution, the miner needs to record the *write set* of every transaction in the block, *i.e.*, the set which contains the data items that the transaction tries to write. After the miner finishes the block's execution, it stores the write sets into the block. Then, the miner adjusts the transaction order of the block to match the serial order of the resulting schedule and publish the new block. Before the validator executes the newly received block, it constructs a "write chain" on the conflicting data items using the write sets and the transaction order in the block. The "write chain" pre-determines the contention relationships among block's transactions and the priority of these transactions. The proposed mechanism, called *multiversion transaction ordering* (MVTO), uses "write chain" to resolve conflicts at runtime and then produces deterministic results. Meanwhile, by using the multiversion technique in the "write chain", MVTO further reduces the conflict at runtime.

This paper makes following contributions:

1. a scheme to run smart contracts called by transactions in a block concurrently where miner can employ any concurrent control technique as long as it produces conflict-serializable schedule.
2. a multiversion concurrency control mechanism to validate a block concurrently and deterministically. The evaluation shows that this mechanism achieves approximately 2.5x speedup in the block validation using a thread pool with 3 threads.

The rest of the paper is organized as follows. Section 2 introduces a simplified smart contract model and the notions used in this paper. Section 3 presents the details of the proposed mechanism called MVTO. Section 4 proves the correctness of MVTO. Section 5 illustrates the experiments and summarizes the results. Section 6 reviews the related work. Section 7 concludes the paper.

2   Background

This paper uses a simplified smart contract model to illustrate the proposed mechanism. This section will introduce the notions that related to this paper.

2.1   Smart Contract

Smart contracts can be seen as a collection of self-defined states maintained in blockchain and functions that manipulate these states.

A simplified proxy ballot contract is shown in Fig. 1. The contract defines two persistent states: voters (line 6) and proposals (line 7). "Proposals" is an array of "Proposal". Each "proposal" contains the number of votes it owns (line 3). The "voters" maps a unique memberID to the data structure "Voter" (line 2). Voters can vote to a specific proposal and they can only vote once.

Client firstly collects votes to the same proposal. Then it calls the function "proxyVote" (line 10–16) to cast these votes by sending a transaction to this

```
1 ▾ contract ProxyBallot {
2       //define Voter<int voteTo, int weight, int voted>
3       //define Proposal<int voteCount>
4
5       //self-define state
6       mapping(memberID -> Voter) voters;
7       Proposal[] proposals;
8
9       /// Cast votes to proposal ${toProposal} by Proxy.
10      function proxyVote(uint[] voters, uint8 toProposal) {
11          for(int senderID in voters){
12              Voter sender = voters[msg.sender];
13              if (sender.voted || toProposal >= proposals.length) throw;
14              sender.voted = true;
15              sender.voteTo = toProposal;
16              proposals[toProposal].voteCount += sender.weight;
17          }
18      }
19      //Other functions ...
20  }
```

Fig. 1. Simplified ballot smart contract

ballot contract's address. Voters cast their votes by adding their weight to target candidate's "voteCount". Most of those state changes are vulnerable to race conditions and may result in inconsistent states. Furthermore, functions can use **throw** statement to handle exceptions such as double voting (line 13). The **throw** statement can abort the contract, discarding the transient variables and undoing any state changes.

2.2   Data Action

A data action is a primitive operation (read or write) on a state's data item [3]. For example, in Fig. 1, voters cast their votes by performing an update action[2] on the target proposal in the state variable "proposals" (line 15). We refer to t_i as the i-th transaction in block. Notation $r_i(x)$ and $w_i(x)$ are the read and write action on data item x executed by t_i respectively.

2.3   Conflict

Two data actions *conflict* with each other if executing them in either order yields different results. For example, there are two situations where two actions conflict in the single-version concurrency control scenario:

1. Two actions of the same transaction, *e.g.*, $r_i(x)$ conflict with $r_i(y)$.
2. Two actions of different transactions and one of them is a write action, *e.g.*, $w_i(x)$ conflict with $r_j(x)(i \neq j)$ and $w_i(x)$ conflict with $w_j(x)(i \neq j)$.

2.4   Schedule

A *schedule* is the sequence of data actions that transactions actually performed. A *serial schedule* is a schedule that meets the following conditions: (i) the actions that belong to the same transaction preserve the order in transaction; (ii) the

---

[2] The update action can be divided into a read and a write action.

actions that belong to different transactions don't interleave with each other. For example, a serial schedule is shown in Example 1.

Example 1 (serial schedule).

$$r_1(x); w_1(x); r_1(y); w_2(x); w_2(y); r_3(x).$$

Conflict-Serializable Schedule. Two schedules are *conflict-equivalent* if they can turn into each other by swapping adjacent non-conflicting data actions. A schedule is *conflict-serializable* if it is conflict-equivalent to a serial schedule. The *serial order* of a conflict-serializable schedule is the transaction order of the serial schedule which it conflict-equivalent with.

Non-conflicting actions can be fully parallelized because the results won't change. However, swapping the execution order of conflict actions might change results and violate the correctness of transactional execution. In Example 2, we can observe that conflict-serializable schedules with the same serial order will produce the same results for each action that belongs to them.

Example 2 (conflict-equivalent). Following schedules are conflict-equivalent with the serial schedule in Example 1 and therefore have the same serial order $t_1 < t_2 < t_3$.

- $r_1(x); w_1(x); w_2(x); r_1(y); w_2(y); r_3(x);$
- $r_1(x); w_1(x); w_2(x); r_1(y); r_3(x); w_2(y);$

Hence, in order to ensure the correctness, smart contracts need concurrency control mechanism to produce conflict-serializable schedule.

Recoverable Schedule. When a transaction writes a data item, the write action is tentative and may be reverted if the transaction aborts. For the schedule "$w_2(x); r_3(x)$", $r_3(x)$ reads the value of x which is previously written by $w_2(x)$. The effect of $r_3(x)$ will not be recoverable and become a dirty data if t_2 aborts after t_3 commits. We refer to t_2 as a *read-from transaction* of t_3.

A schedule needs to be recoverable to prevent this dirty data issue. A *recoverable schedule* is a schedule where each transaction commits only after all its read-from transactions have committed. In this way, the changes made by the transaction that read dirty data are still revertable because the transaction must wait for its read-from transactions to commit.

2.5   Multiversion Concurrency Control

When using single-version concurrency control methods, transactions may have to abort because of non-serializable action. This paper employs multiversion concurrency control (MVCC) to reduce conflict. Instead of overwriting the data item when performing write action, MVCC preserves each version of write actions to avoid aborting transactions due to reasons like *read too late*. Since result of each write action is recorded along with its timestamp, a read action can know which version it should read. We refer to the version of the data item x written by transaction t_i as notation x_i.

In MVTO, multiversion reduces the conflict among the data actions. Write actions on the same data item don't conflict because they write to different versions. The conflict happens between the read action and write action on the same version of the data item where the read action must wait for the write action finishes.

3   Proposed Mechanism

3.1   Basic Idea

Mining Phase. In mining phase, since correctness of smart contracts' execution requires conflict-serializability, any concurrency control mechanism used in DBMS can be applied to produce a conflict-serializable schedule. The block generating process can be divided into two steps: (i) contracts' execution and (ii) consensus protocol. After block's execution (step 1), the miner can acquire the serial order of transactions according to the produced schedule. Then, the miner adjusts the transaction order of the block to match the serial order. At this point, the miner can compute the hash of the reorganized block and step into consensus protocol (step 2). Figure 2 illustrates this block's reorganization process. In this example, the original order, which is decided by the miner, is $t_1 < t_2 < t_3$ (as shown in the left-hand figure). After the execution, the resulting schedule shows that the serial order is $t_1 < t_3 < t_2$ and therefore the miner reorganizes the block according to the serial order (as shown in the right-hand figure).

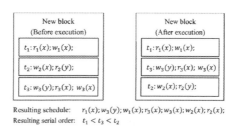

Fig. 2. When mining: reorganize transactions according to schedule

Validation Phase. In the concurrent scenario, verifying the transactions that call smart contracts need three conditions to ensure the correctness:

1. Conflict-serializability: the resulting schedule of validation must be *conflict-serializable*.
2. Schedule replayability: the resulting schedule must be *conflict-equivalent* with the schedule produced by the miner of this block.

3. Execution recoverability: the resulting schedule must be *recoverable* because the transactions can abort at any time.

We refer to the proposed mechanism as *multiversion transaction ordering* (MVTO). MVTO is similar to timestamp concurrency control because they both decide the presumptive transactions' serial order. Timestamp concurrency control presume the serial order using timestamp and abort those transactions which violate the presumptive order. Timestamps are assigned at runtime, which means the presumptive serial order is determined at runtime and the final order may change due to the abortion of transactions. On the contrary, MVTO presumes the serial order to be the transaction order[3] of the block before the execution so that it can validate deterministically. Therefore the validator can know the priority of transactions before the smart contracts' execution. Hence, MVTO can make sure that the data actions never violate the presumptive order.

In order to achieve such functionalities, MVTO needs transactions to provide their *write set* to the validation. Transaction's write set is the set which contains the data items that this transaction will write. Since smart contracts are written in Turing-compete languages, it is impossible in general to statically determine which data actions a smart contract will perform. However, every smart contract called in the block has already been executed when the block is created by the miner. Since mining phase and validation phase share the same initial states from the common blockchain history, these two phases also share the same write set of each transaction as long as their final serial order are identical. Therefore, mining phase can provide the write set of transactions to validation phase by recording all the write actions it has performed. By using these write sets, the validator can pre-determine all the versions that a data item will have and resolve the conflicts at runtime by letting the transaction with smaller index in the presumptive serial order execute first. Hence, the conflict-serializability and replayability of schedule can hold.

When the transaction t_i reads the data item's value which is previously written by another transaction t_j in this block, t_i will record the handler of t_j as its *read-from transaction*. In order to produce recoverable schedules, t_i cannot commit until all its *read-from transactions* has committed.

3.2   Data Structure

Transaction. The transaction in the block has four possible status: "Init", "Active", "Aborted" and "Committed". "Init" is the initial status of transaction. "Active" means there is a thread executing the transaction. A transaction are "Committed" when it completes all its tasks. A transaction will be "Aborted" when it uses up the *gas* of the contract or **throw** other exceptions. We refer to $status(t_i)$ as the status of t_i.

---

[3] This transaction order is also the serial order in the mining phase.

Read Set and Write Set. Transactions use *read set, i.e.,* the set of their read-from transactions, to ensure their recoverability. Read set is constructed at runtime and will be initialized into an empty set if the transaction restarts. After read action $r_i(x)$ reads the value written by another transaction t_j in the block, a tuple like $\langle t_j, x_j^i \rangle$ is inserted into read set of t_i. In this tuple, t_j is the read-from transaction of t_i and x_j^i is the value which t_i read from t_j. Note that we use x_j^i to indicate x_j^i may be inconsistent with x_j at the time t_i tries to commit, this will be discussed later. We refer to $rs(t_i)$ as the read set of t_i.

Every transaction has a *write set* that contains all the write actions it will perform. The write set of a transaction is generated during the first execution when the block is created by the miner. We refer to $ws(t_i)$ as the write set of transaction t_i. The elements in write set are data locators for the data items. With the write set, Validator can determine the conflict relation among transactions on the specific data items and construct the "write chain" for these data items accordingly.

Write Chain on Data Item. By using the write set, validator can construct *write chain* on every data item that will be accessed by the transactions in the block.

Figure 3 shows this write chain under an example. There are 5 data items (a, b, c, d, e). A validator receives a new block containing 4 transactions (Fig. 3a). According to the data actions of these 4 transactions, their write sets is shown in Fig. 3b.

$$t_1 : r_1(a); w_1(a); r_1(b); w_1(b); r_1(e); w_1(e);$$
$$t_2 : r_2(b); w_2(b); r_2(d); w_2(d); r_2(d);$$
$$t_3 : r_3(e); w_3(e); r_3(b); w_3(b);$$
$$t_4 : r_4(a); r_4(b); r_4(c); r_4(d); r_4(e);$$

$$ws(t_1) : \{a, b, e\}$$
$$ws(t_2) : \{b, d\}$$
$$ws(t_3) : \{b, e\}$$

(a) Transactions   (b) Write set

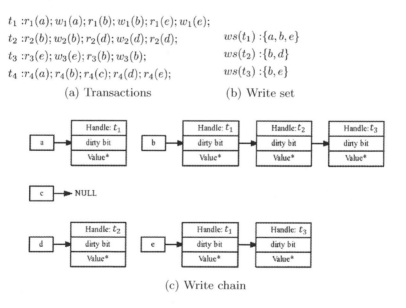

(c) Write chain

Fig. 3. Extended data structure on data item

Before the validator replays those transactions, it uses the write sets to construct the write chain. As shown in Fig. 3c, the write chain contains all the versions the data item will have during the block's execution. Elements in the write chain of a data item are 3-tuples: $\langle handle, value, write\ bit \rangle$, each of them represents a version of this data item. These elements are ordered by the index of their writer transaction, $i.e.$, the presumptive serial order. The "handle" is the reference to the transaction which is going to write the value into this element. Read actions will use this handle to access the index and the status of the writer transaction. When the transaction writes its version, it writes into the "value" of the element and set the "write bit" to **true**. For the read actions that want to read this version, the "value" is readable only when "write bit" is **true**. Otherwise, they have to wait for the writer transaction to write this version. We refer to $\langle t_i, x_i, wb(x_i) \rangle$ as the $target\ element$ of action $w_i(x)$.

When a read action $r_i(x)$ tries to read x, it must decide which version of x to read. Starting from x_0 which is the original value of x, the $r_i(x)$ will find the version as follows:

Step 1. Find x_j that $j < i$, and there is no other version x_k where $j < k < i$.
Step 2. Determine whether $status(t_j) = aborted$.
      (a) If $status(t_j) \neq aborted$, then x_j is the proper version to read.
      (b) If $status(t_j) = aborted$, then trace back through the chain to find the nearest preceding version x_m where $status(t_m) \neq aborted$.

3.3 Concurrency Control Mechanism for Validation

Rules for Scheduling. The $scheduler$ deals with the requests of data actions in MVTO. After the validator constructs all the write chains for data items, scheduler must follow certain rules to make sure the resulting schedule is conflict-serializable and the serial order is identical to the presumptive serial order. The rules are described as follows:

Rule 1. Suppose scheduler receives a request of $w_i(x)$.
      (a) If $w_i(x)$ is legal, $i.e.$, scheduler can find the element of t_i on write chain of x, then scheduler grants the request. Let t_i write x_i into the target element and then set $wb(x_i) = true$.
      (b) If $w_i(x)$ is illegal, then abort t_i, the newly received block fails the validation.
Rule 2. Suppose scheduler receives a request of $r_i(x)$. The scheduler will find out the right version x_j for $r_i(x)$.
      (a) If $j = 0$, grant the request and return the original value of x.
      (b) If $j \neq 0$, then check the status of the read-from transaction t_j.
          i. If $status(t_j) = Aborted$, then find the proper version again and restart from rule 2.
          ii. If $status(t_j) = Committed$ then grant the request, return the value of x_j and add $\langle t_j, x_j^i \rangle$ to $rs(t_i)$.
          iii. If $status(t_j) = Active$
             A. If $wb(x_j) = true$, grant $r_i(x)$ as in 2(b)ii.
             B. If $wb(x_j) = false$, delay t_i until $wb(x_j)$ is set to $ture$, then grant $r_i(x)$ as in 2(b)ii.

Rules of Commit. After transaction t_i finishes all its tasks, t_i will try to commit. The rules for committing transaction, which is shown in Algorithm 1, will maintain the recoverability of schedule.

Algorithm 1. Commit t_i

---

1 **foreach** $\langle t_j, val \rangle \in rs(t_i)$ **do**
2   **while** $status(t_j) \neq Committed$ **do**
3   **if** $status(t_j) = Aborted$ **then**
4   **go to:** RestartPoint
5   **end**
6   sleep()
7   **end**
8   **if** $checkConsistency(\langle t_j, val \rangle) = \textbf{\textit{false}}$ **then**
9   **go to:** RestartPoint
10   **end**
11 **end**
12 $status(t_i) = Committed$

As discussed earlier, t_i can't commit until all its read-from transactions (line 1) have committed (line 2) and some of its read-from transactions that are active may abort at any time. Aborting a transaction will discard all its changes to the states and therefore causing the transactions which have read from this aborted transaction have to restart (line 3). When a transaction t_j restarts, $rs(t_j)$ will be cleared and for $\forall x \in ws(t_j)$, $wb(x_j)$ in x's write chain must be set to **false**. The results of the restarted transaction's data actions might change because the read set of transaction will change when its read-from transaction aborts. Thus, when a transaction t_i tries to commit, the original value x_j^i in $rs(t_i)$ may be inconsistent with the newest version x_j because t_j may have restarted and overwritten the original value with new value x_j^*. So t_i must check the consistency between $\forall x_j^i \in rs(t_i)$ and the newest x_j, and t_i must restart when finding the inconsistency (line 8–10). The effect of restart can spread through the write chain after the element of the aborted transaction, making more active transactions restart due to the inconsistency.

Note that the consistency between the committing transaction and its committed read-from transactions will hold because the committed transactions won't restart and change the versions they have written.

4   Correctness

Observation 1. *For conflicting actions $r_i(x)$ and $w_j(x)(i \neq j)$, if $r_i(x)$ read the version x_j which is written by $w_j(x)$, MVTO will ensure $j <= i$ and scheduling rules grant $r_i(x)$ only after $w_j(x)$ finishes.*

Lemma 1. *MVTO will produce conflict-serializable schedule.*

Proof. According to the conflict relation among data actions in final schedule, we can build a precedence-graph [3] which shows the dependency among the transactions in the block. The conflict notion in MVTO which is presented in Sect. 2.5 shows that the conflict happens between read actions and write action on the same version of the data item. With Observation 1, we can conclude that the precedence-graph is acyclic. As proved elsewhere [3], the schedule with an acyclic precedence graph is conflict-serializable.

Lemma 2. *Final schedule produced by MVTO is conflict-equivalent to serial schedule that matches the transaction order in the block.*

Proof. The precedence-graph of the final schedule produced by MVTO is a directed acyclic graph (Lemma 1). We can observe that the result of the topological sort of this precedence-graph is the serial order of the final schedule. In this graph, for all edges that are pointing from t_i to t_j, MVTO makes sure that $i < j$ (Observation 1), where i and j are also the index of transactions in the block. Therefore, the serial order of schedule produced by MVTO is equivalent to transaction order in the block. Thus we can conclude that the final schedule is conflict-equivalent to serial schedule that matches the transaction order in the block.

Lemma 3. *Validation will succeed if the block is legal.*

Proof. Concurrent schedules of a block's validation in MVTO will produce the same results for each data action as long as the (i) initial states are identical and (ii) these schedules are *conflict-equivalent.*

Blockchain makes sure the honest miners and validators share the same history of blocks when creating or validating the new block. Therefore, the initial states of the new block in validation phase are identical to those initial states in mining phase.

The final schedule of the validation is *conflict-equivalent* to serial schedule that matches the transaction order in the block (Lemma 2). Meanwhile, since the miner adjusts the transaction order in the block to match the serial order of the produced concurrent schedule, the final schedule of the validation and the concurrent schedule produced by miner are both *conflict-equivalent* to the same serial schedule. Therefore these two schedules are *conflict-equivalent* to each other.

Hence, we can conclude that mining and validation can produce the same result as long as the block is legal.

5   Evaluation

MVTO aims to improve the throughput for blocks validation by executing smart contracts in parallel. We use a benchmark for MVTO that vary the workload of a transaction, the percentage of abortion, the percentage of conflict and the size of the thread pool to evaluate this approach.

5.1   Benchmark

The benchmark is the block that only contains the "ProxyBallot" contract which is shown in Fig. 1. Each block contains 200 transactions. In a transaction that calls "ProxyBallot" contract, the workload is the number of the voters which are collected by this transaction. The conflict percentage is defined to be the percentage of transactions that vote to "proposals[0]". Other non-conflicting transactions will vote to different proposals. To control the abort rate of a block, some transactions are set to abort after they finish.

We set up four experiments where we vary the workload, percentage of abortion, percentage of conflict and size of thread pool respectively. (1) Experiment 1 varies the workload for each transaction from 2000 to 20000 voters with 15% data conflict and 10% abort. (2) Experiment 2 varies the percentage of data conflict in the block from 0% to 100% with 20000 workload and 10% abort. (3) Experiment 3 varies the percentage of abortion in the block from 0% to 100% with 20000 workload and 15% data conflict. All the above three experiments use a thread pool with 3 worker threads. (4) Experiment 4 varies the size of the thread pool from 3 to 15 and the conflict percentage from 5% to 15% with 20000 workload and 10% abort.

5.2   Results

We use C++ implementation to run the evaluation on a machine with 4-core 4.00 GHz CPU. All the results are the mean of 100 times executions. Results are shown in Fig. 4, each contains the results of serial and concurrent validation on the same block. Results of serial validation serve as the baseline when showing MVTO's speedup.

Figure 4a shows the speedup of MVTO over serial validation when varying the workload per transaction with 15% conflict and 10% abort. Because of the overhead of multithreading, MVTO is slower than serial validation when workload is lower than 4000. MVTO achieves speedup when workload is higher than 4000 and achieves 2.5x speedup when workload is 20000.

Figure 4b shows the result when varying the conflict percentage of the block with 20000 workload per transaction and 10% abort. As the conflict percentage raise, the speedup of MVTO keeps dropping from 2.5x to nearly 0.5x. When conflict percentage reaches 60%, MVTO becomes slower than serial validation.

Figure 4c shows the speedup of MVTO when varying the abort percentage of the block with 20000 workload per transaction and 15% conflict. We can observe that serial execution is significantly slower when there are more transactions abort during validation. The reason is the difference between single-version and multiversion when dealing with transaction's abortion. To abort transactions in single-version, the undo logs are commonly needed to roll back every tentative change that made on state variables. While MVTO can just discard the version which is written by the aborted transaction. Hence MVTO shows lower cost of restarting transactions compared to single-version technique.

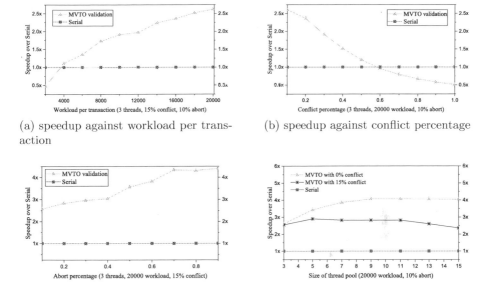

(a) speedup against workload per transaction

(b) speedup against conflict percentage

(c) speedup against abort percentage

(d) speedup against size of thread pool

Fig. 4. Evaluation results

Figure 4d shows that how the size of thread pool affect the speedup of MVTO under blocks with or without conflict. When the block admit no conflict, the speedup is generally higher with lager thread pool and MVTO cannot achieve more than 4x speedup due to the limitation of 4-core CPU. However, when the block contains 15% conflict, the effect of speedup begins to drop when using thread pool larger than 5 threads. This is because more active threads can lead to more possibility to have synchronization on the conflicting data item between active transactions at runtime.

5.3 Discussion

As we mentioned in Sect. 1, Ethereum can deal with more complicated and time-consuming contracts when it switch to POS. Hence, despite MVTO can only achieve desirable speedup when executing smart contracts with enough workload due to the overhead of multithreading (Fig. 4a), it can bring speedup to Ethereum and any other platforms which will deal with complicated smart contracts.

The results in Fig. 4c shows the advantage of multiversion technique used in MVTO compared to single-version implementation when dealing with data roll back which caused by aborting and restarting transactions. Results in Fig. 4d shows that MVTO may slow down when using a thread pool with too much worker threads because of the overhead of the synchronization between conflicting data action. In this evaluation, the transactions in benchmark only calls the same smart contract so that the conflict percentage can be very high (as shown in Fig. 4b). In practice, miner will receive many transactions that trigger unrelated smart contracts, thus the blocks in reality might face less data conflict than the blocks used in this evaluation.

6   Related Work

Dickerson et al. [4] treat every smart contract invocation as a speculative atomic action. Therefore the miner can discover a serializable schedule when creating the new block by using locks in a 2PL manner. When executing, the miner records the schedule by storing the trace of the lock on every data item into the block, so that validator can retrieve and replay that same schedule deterministically. Serge and Hobor [5] present the similarities between multi-transactional behaviors of smart contract and problems of shared-memory concurrency. Bocchino et al. [6] survey many techniques for replaying a concurrent schedule deterministically.

Ethereum [1] may be the most popular platform among all the smart contract platforms on public blockchain. Its recent project Plasma [7] tries to remission the low throughput by employing the sharding techniques. Plasma split the state space into multiple partitions where each runs on a different child blockchain, forming a multiple blockchains ecosystem with tree hierarchy. Other ongoing platforms such as Polkadot [8], EOS [9] and Aelf [10] also adopt similar techniques to parallelize contracts that working on different state space.

Garcia-Molina et al. [3] introduces concurrency control mechanism that employ multiversion techniques. Many software transactional memory (STM) techniques [11,12] fit well with smart contract because smart contracts must be executed transactionally. Ghosh et al. [13] uses "update chain" that similar to the write chain in MVTO to implement a concurrency control mechanism using multiversion timestamp under STM. This mechanism aims to reduce aborts when facing conflicting update transactions. The "update chain" is constructed at runtime while MVTO constructs write chain before the execution.

7   Conclusion

We have proposed a concurrent scheme called MVTO to increase the throughput of blockchain-based smart contract platform. In this scheme, the miner can use any concurrency control technique to discover a conflict-serializable schedule. The write set of each transaction is recorded into the newly generated block along with the final states. The order of transactions in block is adjusted by the miner to match the serial order of the resulting schedule thereby validators can know which serial order they should replay. Before the validation, validators use those write sets and the transaction order to construct "write chain" on conflicting data items, therefore they can pre-determine the dependency among transactions in the block. In summary, validators can validate a block in a concurrent and deterministic manner. The Evaluation shows that MVTO can achieve approximately 2.5x speedup when validating a block with conflicting input data using a thread pool with 3 worker threads.

Furthermore, MVTO can be integrated into existing systems without compromising their original architecture. Developers just need to implement the concurrency control and multiversion on the underlying basic data items. Existing platforms mostly plan to increase throughput by employing sharding. MVTO is compatible with the sharding techniques where sharding scale through the multi-chains architecture and MVTO enable concurrency within each child chain. Hence, MVTO can achieve significant scalability by using sharding techniques to divide the states into different partitions.

In conclusion, MVTO can speed up the block validation and increase the throughput of smart contract platforms.

References

1. Wood, G.: Ethereum: a secure decentralised generalised transaction ledger. Ethereum Project Yellow Paper, vol. 151 (2014)
2. Ethereum: Solidity documentation. http://solidity.readthedocs.io/en/develop
3. Garcia-Molina, H., Ullman, J.D., Widom, J.: Database System Implementation, pp. 883–940. Prentice-Hall, Upper Saddle River (2000)
4. Dickerson, T.D., Gazzillo, P., Herlihy, M., Koskinen, E.: Adding concurrency to smart contracts. In: Proceedings of the ACM Symposium on Principles of Distributed Computing, PODC 2017, Washington, DC, USA, 25–27 July 2017, pp. 303–312 (2017)
5. Sergey, I., Hobor, A.: A concurrent perspective on smart contracts. In: FC 2017 International Workshops on Financial Cryptography and Data Security, WAHC, BITCOIN, VOTING, WTSC, and TA, Sliema, Malta, 7 April 2017, pp. 478–493 (2017). Revised Selected Papers
6. Bocchino, R., Adve, V., Adve, S., Snir, M.: Parallel programming must be deterministic by default. In: Proceedings of the First USENIX Conference on Hot Topics in Parallelism, p. 4 (2009)
7. Poon, J., Buterin, V.: Plasma: scalable autonomous smart contracts. White Paper (2017)
8. Wood, G.: Polkadot: vision for a heterogeneous multi-chain framework. White Paper (2016)
9. EOS: EOS.IO technical white paper. https://eos.io/
10. AELF: a multi-chain parallel computing blockchain framework. https://aelf.io/
11. Herlihy, M., Luchangco, V., Moir, M., Scherer III., W.N.: Software transactional memory for dynamic-sized data structures. In: Proceedings of the Twenty-Second ACM Symposium on Principles of Distributed Computing, PODC 2003, Boston, Massachusetts, USA, 13–16 July 2003, pp. 92–101 (2003)
12. Zhang, D., Dechev, D.: Lock-free transactions without rollbacks for linked data structures. In: Proceedings of the 28th ACM Symposium on Parallelism in Algorithms and Architectures, SPAA 2016, Asilomar State Beach/Pacific Grove, CA, USA, 11–13 July 2016, pp. 325–336 (2016)
13. Ghosh, A., Chaki, R., Chaki, N.: A new concurrency control mechanism for multi-threaded environment using transactional memory. J. Supercomput. **71**(11), 4095–4115 (2015)

ElasticChain: Support Very Large Blockchain by Reducing Data Redundancy

Dayu Jia[1], Junchang Xin[1,2(✉)], Zhiqiong Wang[3],
Wei Guo[4], and Guoren Wang[5]

[1] School of Computer Science and Engineering,
Northeastern University, Shenyang, China
xinjunchang@mail.neu.edu.cn
[2] Key Laboratory of Big Data Management and Analytics,
Shenyang, Liaoning Province, China
[3] School of Sino-Dutch Biomedical and Information Engineering,
Northeastern University, Shenyang, China
[4] School of Computer, Shenyang Aerospace University, Shenyang, China
[5] School of Computer Science and Technology,
Beijing Institute of Technology, Beijing, China

Abstract. Blockchains are secure by design and they have been widely used in digital asses, trade finance, information security and many other fields. However, the current blockchain protocol requires that each full node must contain the complete chain. When the storage capacity of a full node is less than that of the complete chain, this node cannot be a member of blockchain system. With the input data increasing, the number of full nodes in blockchains would decrease. The security of blockchains would significantly reduce. Therefore, we provide the ElasticChain, which can improve storage scalability under the premise of ensuring blockchain data safety. The full nodes in ElasticChain store the part of the complete chain based on the duplicate ratio regulation algorithm. Meanwhile, the node reliability verification method was used for increasing the stability of full nodes and reducing the risk of data imperfect recovering caused by the reduction of duplicate number. The experimental results on real datasets show that ElasticChain has the same stability, fault tolerance and security with the current blockchain system and it improves the storage scalability extremely.

1 Introduction

With the increasing popularity of digital encryption currency such as Bitcoin, blockchain technology is gaining more and more attention. The blockchain is a kind of new decentralized protocol that can safely store digital currency and digital assets. Blockchains can effectively solve the consensus problem in Byzantine agreement [1] by using digital encryption [2], timestamp, distributed consensus [3] and economic incentive. It realizes decentralized point to point transaction

© Springer International Publishing AG, part of Springer Nature 2018
Y. Cai et al. (Eds.): APWeb-WAIM 2018, LNCS 10988, pp. 440–454, 2018.
https://doi.org/10.1007/978-3-319-96893-3_33

when nodes do not need to believe each other in a distributed system. Hence, blockchains can effectively reduce the trust cost in the real economy and redefine the property rights in the internet age.

Although blockchain technology can improve the data security and reliability significantly, a lot of bottlenecks still exist. It is worthwhile to research on improving throughput and reducing mining cost, while the shortage of storage scalability is one of the most serious problems as well. Taking Bitcoin as an example, there are two kinds of nodes in the network, the full nodes and the lightweight nodes. Running a full node is the only way you can use Bitcoin in a trustless way. They do not suffer from many attacks that affect lightweight wallets. By September 28, 2017, the number of certified addresses in Bitcoin system is 9892723 [4] and the Bitcoin system contains 484,490 blocks with the storage capacity of 124.47 GB. The blockchain protocol requirements that each full node retains the complete blockchain, and each node is in the same import. If a node wants to ensure its bitcoins safety in utmost, this node should be a full node. In that way, there will be nearly 10 million nodes contributing more than 100 GB of disk space to store the blockchain data. In other words, the Bitcoin system only saves about 100 GB data with nearly 1000 PB storage space. This leads to a great waste of storage space.

Moreover, the capacity of the Bitcoin and the number of participating nodes will increase rapidly with the time continuing. Bitcoin system will consume larger storage space. When the storage capacity of a full node in system is less than the capacity of the blockchain, it cannot continue being a member of this blockchain system. It greatly limits the joining of nodes with small storage capacity. It also enables the nodes to exit their own blockchain system. When the full nodes in the blockchain system become fewer, the total computing power of the blockchain system will be reduced accordingly. And it will indirectly reduce the security of blockchain that based on the POW (proof-of-work) [5]. If it is not universal for the full nodes to join in blockchain system, the development and application of database system based on blockchain technologies will be greatly constrained.

In order to solve the above problems, ElasticChain is presented in this paper. The major contributions of the paper are the followings:

(1) We present a duplicate ratio regulation algorithm. Nodes store parts of the complete chain under the premise of ensuring blockchain data safety by algorithms. It decreases the memory burden for the full nodes in blockchain.

(2) We present a node reliability verification method. We design three roles for nodes: user role, storage role and verification role. The verification nodes record and update the stability values of storage nodes in real time. Then, the high stability nodes are chosen to store the duplicates of each block. It improves the data stability for blockchain.

(3) The benefits of the ElasticChain are evaluated, which shows that it has the same stability, fault tolerance and security as the current blockchain system and greatly improves the storage capacity of blockchain nodes.

2   Background

A novel method [6] for decentralised peer-to-peer software license validation using cryptocurrency blockchain technology was proposed to ameliorate software piracy, and to provide a mechanism for all software developers to protect their copyrighted works. Zyskind [7] et al. described a decentralized personal data management system that ensures users own and control their data. They implemented a protocol that turns a blockchain into an automated access-control manager without a third party. Some security models have also been proposed, such as Blockstack [8] and HotNets [9]. Blockstack a global naming and storage system secured by blockchains, enables the introduction of new functionality without modifying the underlying blockchain. HotNets is a novel quantitative framework to analyse the security and performance implications of various consensus and network parameters of POW blockchains. Some of the methods [10,11] were employed to establish the quantization standard of security and scalability for the systems based on blockchain. Some research propose several consensuses, not only POW, but also POS (Proof of Stake) [12] and PBFT (Practical Byzantine Fault Tolerance) [13]. In POS-based cryptocurrencies, the creator of the next block is chosen via various combinations of random selection and wealth or age. So there is little cost in working on several chains (unlike in proof-of-work systems). It saves lots of resources, but anyone can abuse this rule to attempt to double-spend. Byzantine consensus is a fundamental and well-studied problem in the area of distributed system, and PBFT efficiently solves this problem.

When someone attempt to attack the honest blockchain, they must meet the requirements of consensus. Take the POW as an example, POW's security relies on the principle [11] that no entity should gather more than 50% of the processing power because such an entity can effectively control the system by sustaining the longest chain. The attacks on POW include double-spending attacks and selfish mining. Double-spending is the same single coin spent more than once. Recent studies [14] have shown that double-spending attacks on fast payments succeed with overwhelming probability and can be mounted at low cost. Selfish mining is considered that some miners want to increase their relative mining share in the blockchain. Recent studies [15] show that, as a result of these attacks, a selfish miner equipped with originally 33% mining power can effectively earn 50% of the mining power.

3   Duplicate Ratio Regulation Algorithm

In the POW-based blockchain system, if a attacker wants to modify the data in a block, he needs to compute the hash value of all blocks behind this block quickly. The attack is successful only if the attacker calculates faster than the actual blockchain. Therefore, modifying a new block costs less computational pride than modifying a former block. According to the security of each block in blockchain, ElasticChain stores each block using the duplicate ratio regulation algorithm. In

ElasticChain, the new blocks store a large number of duplicates because of their weak safety, while the former blocks are stored a small number of duplicates due to their strong security. The duplicate ratio regulation algorithm ensures the POW's security and reduces the storage capacity for nodes in blockchain system.

In the duplicate ratio regulation algorithm, firstly, it calculates the minimum number of duplicates for each block according to the security analysis. Then, the algorithm calculates the number of contiguous blocks that need to be grouped together. The security of these contiguous blocks is similar and they are stored in the same number of duplicates. Finally, we calculate the number of duplicates that need to be saved.

3.1   Security Analysis

Nakamoto [16], the founder of blockchain, supposed some attackers produce a new parallel chain to replace the honest chain, and only successful if the production of the parallel chain is faster than the honest one. The probability of an attacker catching up from a given deficit is analogous to a Gambler's Ruin problem. We supposed that p is the probability an honest node finds the next block and q is probability the attacker finds the next block. When z blocks have been linked after the attack begins, the attacker's potential progress will be a Poisson distribution and the expected value is calculated by Formula 1.

$$\lambda = z(q/p) \tag{1}$$

The probability of catching up the honest chain is P_z:

$$P_z = \sum_{k=0}^{\infty} \frac{\lambda^k e^{-\lambda}}{k!} \times \begin{cases} (\frac{q}{p})^{(z-k)} & if \quad k \le z \\ 1 & if \quad k > z \end{cases} \tag{2}$$

To avoid summation of infinite sequences, we convert the Formula (2) to:

$$P_z = 1 - \sum_{k=0}^{z} \frac{\lambda^k e^{-\lambda}}{k!} \times (1 - (\frac{q}{p})^{(z-k)}) \tag{3}$$

We wrote code in the Java language to calculate P_z when $q = 0.1$, $p = 0.9$ and z is from 0 to 30. Then, we plotted the cruve of P_z using matlab, as shown in Fig. 1. The probability of catching an honest chain declines rapidly with the number of blocks increasing. Moreover, the probability is a big number at the beginning of attack. It means attackers are likely to attack successfully. However, it is impossible to complete a successful attack as the number of blocks increased.

Because the decrease rate of P_z is very fast with the increase of z, the value of P_z cannot be obviously expressed in Fig. 1 when z is larger than 5. Therefore, when z takes 10 to 15, the cruve of P_z is drawn in Fig. 2.

It can be seen from Figs. 1 and 2 that attackers are becoming more and more vulnerable to catch up with the honest chain as the number of blocks growing.

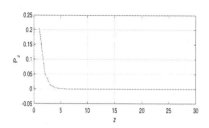

Fig. 1. Cruve of P_z

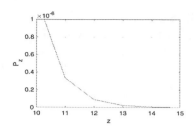

Fig. 2. Cruve of P_z when z takes 10 to 15

The more primitive blocks are, the less likely blockchain can be tampered and the higher security it is. Therefore, the number of duplicates of each block is determined by their location. We store a small number of duplicates of the original blocks and store enough number of duplicates of the new blocks in the blockchain system. The function relations are shown in Formula (4). M is the total number of nodes in the blockchain. i is the sequence number of each block. n is the currently total number of the block and m_i is the number of duplicates to store. P_{n-i} is the probability that the block i is caught by an attacker. It also can be considered as a security factor for block i.

$$m_i = \lceil P_{n-i} \times M \rceil \tag{4}$$

However, the blockchain consensus insists that if more than 50% of the nodes store the same data, the data is treated as the real one. In other word, if more than half of the nodes in the network are controlled, the data in the entire network will be controlled. Therefore, we cannot set the number of duplicates for each block very small. According to the different security of blockchain system, we set k is the minimum number of duplicates for each block.

3.2   The Number of Duplicates

Borel's Law [17] defines that any probability below 1 in 10^{50} is automatically zero. According to the Formula (3), we calculated the probability of P_z until it reduces to 10^{-50} as z increasing by integer value. At this point, it is impossible to catch up with the honest node for a attacker. Therefore, each z blocks are considered as a set of data fragment to store the same number of duplicates.

Finally, the number of duplicates of per block is determinated. The number of duplicates for block i is named m_i, and the minimum number of duplicates is k. The duplicate ratio regulation algorithm is shown in Algorithm 1.

3.3   Example and Optimization

Here, we take an example, when $q = 0.1$, we calculate P_z according to Formula (3) at first. In order to simplify the grouping process, the value of z is an integer multiple of ten. When $z \geq 100$, P_z is smaller than 10^{-50}. Each 100 blocks are

---

Algorithm 1. Duplicate ratio regulation algorithm

Input: $P_z = 1$, $z = 1$, the total number of nodes in blockchain M, the number of blocks in the current blockchain n.

Output: duplicate allocation method

1 Estimating q

2 **for** each block P_z

3   **if** $(P_z > 10^{-50})$

4          $P_z = 1 - \sum_{k=0}^{z} \frac{\lambda^k e^{-\lambda}}{k!} \times (1 - (\frac{q}{p})^{(z-k)})$

5          $z = z + 1$

6   **end if**

7 **end for**

8 $z_{min} = z$

9 Estimating k (according to M and n)

10 **for** each block i

11      $m_i = \lceil P_{n-i} \times M \rceil$

12   **if** $(m_i < k)$

13   Splitting the blockchain. Each z_{min} blocks are split into a fragment, and each block i in a same fragment is saved as the same duplicates m_i.

14   **else**

15   Splitting the blockchain. Each z_{min} blocks are split into a fragment, and the block i is saved as k duplicates.

16   **end if**

17 **end for**

---

saved in the same number of duplicates as a set of data fragment. Then the Formula (4) is used to calculate the number of duplicates in each fragment. The P_z in Formula (4) is calculated by Formula (3), but the Formula (3) is complex. Therefore, the Weibull function was adopted to fit the cruve of P_z using MATLAB. We choose Weibull function to fit Formula (3), because its fitting result is the closest to the cruve of P_z comparing with other functions. The fitting result is shown as the Formula (5).

$$f(x) = a \times b \times x^{(b-1)} \times exp(-a \times x^b) \qquad (5)$$

a $= 1.905$ (1.886, 1.924), b $= 0.723$ (0.7154, 0.7307). The fitting variance (SSE) is $1.215e^{-5}$, and the R-square is 0.9997.

It can be seen from the fitting result that P_z has negative exponential relation with z. Therefore, in order to simplify the calculation in segmentation process, we modify the Formula (3) to the Formula (6) to calculate the number of duplicates. The allocation scheme of duplicates is shown in Fig. 3.

$$m = 2^{-\lceil \frac{(n-i)}{100} \rceil} \times M \qquad (6)$$

4   Node Reliability Verification Method

The nodes in blockchain can be arbitrarily added, and some nodes may always fail and produce DATM (data missed). However, ElasticChain proposes the

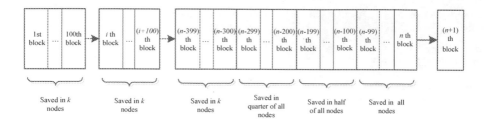

Fig. 3. The allocation scheme of duplicates

duplicate ratio regulation algorithm in which a relatively small number of duplicates of the former blocks are stored in the network because of their strong security. When most of the nodes with the former blocks fail, it will have a great impact on the recovery of former blocks. Therefore, ElasticChain uses a verification method of node reliability to improve the stability of nodes and reduce the risk of data imperfect recovering.

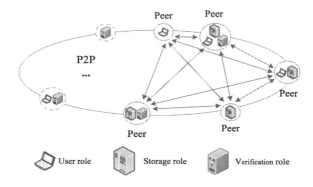

Fig. 4. The ElasticChain architecture

The framework for node reliability verification method is shown in Fig. 4. The nodes in network include three roles: the user node, the storage node and the verification node. A node in network would have one, two or three roles at the same time. The user node is the owner of the original data, and it can upload and query blockchain data. The storage node is the holder of the duplicates and the verification node is the verifier of the stability for storage nodes. And we establish two new blockchains: the P (Position) chain and the POR (Proofs of Reliability) chain, as shown in Fig. 5. The P chain is stored in the user nodes to record the location of the data duplicates. The POR chain is stored in the verification nodes to record the reliability of each storage node.

The implementation of P chains and POR chains are all based on blockchain technology. It guarantees the security of location information of duplicates and the reliability evaluation of storage nodes.

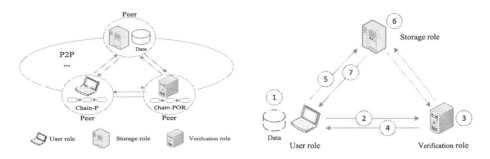

Fig. 5. The nodes in ElasticChain   **Fig. 6.** ElasticChain stored

4.1   Store

When the node reliability verification method is used for data storage, Elastic-Chain uses the POR (Proofs of Retrievability) method [18,19] to encrypt the blockchain data of the user nodes, and obtains the corresponding ciphertext and key. POR are cryptographic proofs that prove the retrievability of non-local data. More precisely, POR assume a model comprising of a user, and a service provider that stores a file pertaining to the user. POR consist basically of a challenge-response protocol in which the service provider proves to the user that its file is still intact and retrievable. In ElasticChain, a user node stores the ciphertext in storage nodes, the verification nodes can check the integrity of the data at any time. While checking, the storage node will be randomly selected a portion of the ciphertext data and return it to the verification node. The verification node calculates the received ciphertext with the key generated by POR. Then, we can find out whether the data in the storage node is complete. Thus, the POR method can be used to verify the data integrity in real-time with a little communication cost.

In the process of data storage, firstly, ElasticChain uses the POR method to encrypt each block which belongs to the user nodes, and obtains the corresponding ciphertext and key. Secondly, the user nodes calculate the number of duplicates for each block based on the duplicate ratio regulation algorithm. Thirdly, the user nodes store the key generated by POR method into the local memory, and send one copy of the key to the verification nodes. Finally, the storage nodes store the ciphertext.

At this step, the node reliability verification method will access the reliability information of storage nodes which is stored in the verification nodes, and find out a few storage nodes with higher reliable values to store the data of each block. In order to ensure the reliability information of storage nodes avoiding being tampered with maliciously, the verification nodes store it into POR chain.

Meanwhile, in order to insure the read speed for user nodes, the storage nodes' addresses are returned to the user nodes and saved in the P chain. The P chain ensures the security of these addresses. However, a P chain from a user node only store the addresses which keep the ciphertext produced by this user node. The other addresses of storage nodes are not stored in this user node. So, the user node can read its own data quickly. The process of ElasticChain storing data is shown in Fig. 6, and the details of the process are shown in Algorithm 2.

Algorithm 2. ElasticChain store
<hr/>
1 Use POR method to encrypt each block.
2 The user node calculates the number of duplicates for each block.
3 The user node stores the key generated by the POR method into the local memory.
4 The user node sends one copy of the key to the verification nodes.
5 The verification node accesses the reliability of each storage node in the POR chain.
6 Return the storage nodes with the highest reliability to the user node.
7 Store each block in these storage nodes.
8 Return the addresses of storage nodes to the user node and store it in the P chain.

4.2   Retrieve

When a user node reads the data, the user node accesses the P chain in the local disk to find out the storage location of the data. Then, ElasticChain system finds the corresponding storage nodes according to the location information, and asks them to return the ciphertext data to the user node. Finally, The user node recoveries the ciphertext according to the key which is saved locally and generated by POR method, and then obtains the initial data. The process of ElasticChain retrieving is shown in Fig. 7, and the details of the process are shown in Algorithm 3.

Algorithm 3. ElasticChain Retrieve
<hr/>
1 The user node accesses the P chain to find out the addresses of storage nodes.
2 Storage nodes return the ciphertext data to user node.
3 The user node retrieves the ciphertext according to the key generated by POR, and obtains the initial data.

4.3   Storage Node Reliability Verification

In ElasticChain, the blocks are saved in storage nodes. However, storage nodes may fail and produce DATM in some conditions. In order to reduce the instability of storage nodes, the verification nodes verify the partial ciphertext data in real time. The validation method requires storage nodes to send the randomly partial ciphertext back at any time. After that, the verification nodes detect the storage status of the storage nodes and write the real-time status into the POR chain. When the user nodes apply for storing data, the verification nodes provide the latest reliability value of storage node for the user nodes. Then, user nodes can select the most stable storage nodes to store the block data.

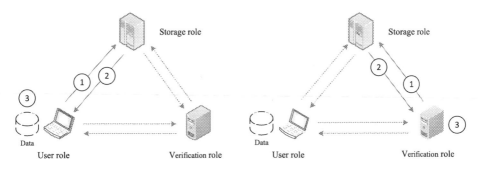

Fig. 7. ElasticChain retrievable

Fig. 8. Storage node reliability verification

The process of storage node reliability verification is shown in Fig. 8. Firstly, ElasticChain sets the same reliability values to each storage node. Then, the verification nodes check the reliability of data in storage nodes at every same period of time. If the data in the storage nodes is complete, the reliability value remains unchanged. If the storage node data is modified or lost, the verification nodes will reduce its reliability value and store it in the POR chain. The ElasticChain uses the reliability values of each storage node in the POR chain as a standard to select the highly reliable storage nodes.

4.4 Incentive Mechanism

In bitcoin system, the miners calculate the hash value of the next block, and the large numbers of calculations ensure the security of bitcoin. Thus, the bitcoin system will award each successful miner a number of bitcoins. This has inspired hundreds of miners to mine new bitcoins by consuming their calculation ability of CPU and large amount of power. In ElasticChain, storage nodes and verification nodes provide their own large disk space, which guarantees the data security of the user nodes. For stimulating storage nodes and verification nodes, they can be user nodes to store data safely or be paid by user nodes in ElasticChain. The more storage space they provide, the more data they can store in ElasticChain or the more payments they can get.

5   Evaluation

The experimental environment is a computer with IntelCore i5-6500, 3.20 GHz of CPU and 16 GB of memory. Experimental nodes are created using VMware Workstation 12.5.2. Each node has an ubuntu16.04 system with 1 GB of memory and 60 GB of hard disk space. We built ElasticChain, P chain and POR chain blockchain projects by use of the open source Hyperledge fabric v0.6.

The experiment established four, eight, twelve and sixteen nodes, respectively. All nodes are storage nodes, user nodes and verification nodes. The exper-

iments run a transaction code named chaincode_example02.go. When each transaction is completed, a 5.39 KB broadcast message is generated.

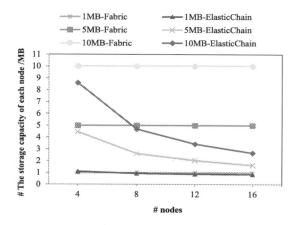

Fig. 9. The average storage space occupied by per node

5.1   Storage Space

Firstly, we experimented on the storage space occupied by ElasticChain. In this
section, we designed 6 experiments, which compare with the ElasticChain and
Hyperledge fabric with the different number of nodes and processing different
amount of data.

When all nodes are running normally and are not attacked, each 500 KB
data is fragmented into a group of slices. The minimum number of duplicates for
each slice is 2, and the number of duplicates is calculated by Formula (6). When
the transaction completes 186 times, 930 times and 1860 times, the broadcast
data 1.00 MB, 5.00 MB, and 10.00 MB are generated, respectively. Figure 9 shows
the average storage space occupied by per node of ElasticChain and blockchain
system based on Hyperledge fabric. We can get the following conclusions.

(1) When few nodes join the network, the average storage space occupied by
    each node in the ElasticChain is similar to that of the fabric blockchain.
    However, when the number of nodes increases, the average storage space
    occupied by ElasticChain nodes is reduced significantly.
(2) When the amount of data stored is small, the average storage space occupied
    by the ElasticChain nodes is similar to that of the fabric blockchain. This
    is because the location information of the storage nodes is saved in the P
    chain, and the reliability evaluation information of storage nodes is saved in
    the POR chain. The size of each data in the P chain and POR chain are
    both fixed values. Therefore, when the amount of stored data is increasing
    continuously, the average storage space occupied by the ElasticChain nodes
    is reduced significantly compared with the fabric blockchain system.

(3) As the stored data increasing, the increment of average storage space of ElasticChain nodes tends to be flat.

Therefore, ElasticChain has good storage scalability in the multi-node and large data applications.

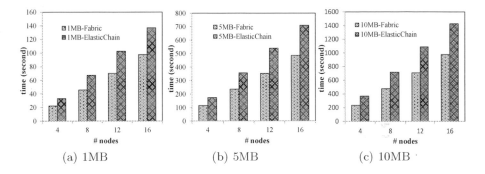

(a) 1MB   (b) 5MB   (c) 10MB

Fig. 10. The processing time of ElasticChain and Hyperledge fabric

Then, the processing time of ElasticChain and fabric are shown in Fig. 10. The processing time refers to the time from when a transaction was started to when it finished confirmation and write operation. We can get the following conclusions.

(1) The processing time of ElasticChain is slightly longer than the time of Hyperledge fabric. It is because that ElasticChain divides the blockchain into slices, it will take some time to process. And in ElasticChain, the operations on P chain and POR chain also take a period of time.
(2) With the number of nodes and storage data increasing, the processing time of ElasticChain increases basically linearly. It is because when ElasticChain stores each transaction, it will do the same work. ElasticChain will increase the same length of time when it deal with the new transaction.

5.2 Fault Tolerance

In the practical applications, it is very common that some peers in blockchain system go down, and the data in these peers cannot be recovered. The integrity of the data would be affected. In Hyperledge fabric, the data is stored in each node. When some peers go down, the user can download data from other nodes. However, the duplicates of data in ElasticChain are less than that in Hyperledge fabric, and ElasticChain will be more affected than Hyperledge fabric on the integrity of data.

Our experiment set up 8, 12 and 16 storage nodes, and there were four nodes of them are unstable nodes. These four unstable nodes were not verification nodes, and the failure probability of them were 0.8, 0.6, 0.4 and 0.2, respectively.

When the experiment had completed the transaction 930 times and 1860 times, we got 5.00 MB and 10.00 MB of data, and the duplicates allocation strategy was as same as the above experiment. Figure 11 shows the recovery of Elastic-Chain, the blockchain system which only based on the duplicate ratio regulation algorithm and Hyperledge fabric.

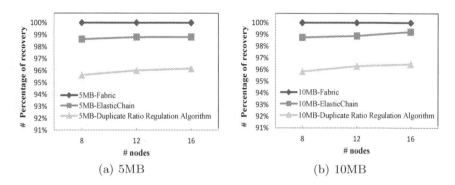

(a) 5MB   (b) 10MB

Fig. 11. The fault tolerance of ElasticChain, the blockchain system only based on the duplicate ratio regulation algorithm and Hyperledge fabric

It can be seen from Fig. 11 that the unstable nodes had a negligible effect on Hyperledge fabric. The blockchain system which only based on the duplicate ratio regulation algorithm was more affected, and the ElasticChain was less affected. It is because that ElasticChain chose the better stability of the nodes to store data through the reliability verification method. It can be seen from the experiment that as the number of nodes are increased, the data recovery ratio of ElasticChain increases, and the fault tolerance of the system is enhanced.

5.3 Security

We tested the security of ElasticChain refering to the Blockbench method [20]. When an attacker intentionally modifies the data in storage nodes, the blockchain will produce a bifurcation. The security of the system can be judged by the number of blocks generated by the bifurcation blocks. The smaller number of bifurcation blocks are generated, the safer this system is. In practice, there are many nodes to join the blockchain and we want to design the simulation in a pragmatic way. In our experiments, we just did the experiment with 16 nodes, and did not establish 4 nodes, 8 nodes and 12 nodes. When running Hyperledge fabric v0.6 and ElasticChain, the attack appeared at 100 s after the system beginning and ended at 250 s. The running results of the two systems are shown in Fig. 12.

The experiment shows that when Hyperledge fabric and ElasticChain are attacked, no bifurcation chains are created. It is because ElasticChain is also based on the Hyperledger fabric system. The consensus of Hyperledger guarantees the security of the blocks when the chains are attacked. However, when the

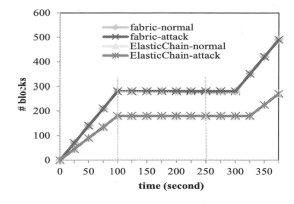

Fig. 12. The security of ElasticChain

attack stopped, Hyperledge fabric and ElasticChain needed a period of time to recover from the attack. As we can see from Fig. 12, ElasticChain has a longer recovery time than the Hyperledger fabric.

The experiments above show that when the fabric-based ElasticChain is attacked, the system is of high security, though it needs more processing time.

6   Conclusion

In our study, we present ElasticChain, which can improve storage scalability under the premise of ensuring blockchain data safety. In ElasticChain, the duplicate ratio regulation algorithm implements that the full nodes with small storage capacity only store parts of the blockchain instead of the complete chain. The reliability verification method was used for increasing the stability of storage nodes and reducing the risk of data imperfect recovering caused by the reduction of duplicate number. In the future, we can improve the duplicate ratio regulation algorithm to compute the number of duplicates more accurate and reduce more storage space under the premise of data security. Moreover, ElasticChain can be applied to other blockchain systems, such as Ethereum and Parity.

Acknowledgement. This research was partially supported by the National Natural Science Foundation of China (Nos. 61472069, 61402089, 61402298, and U1401256), the Fundamental Research Funds for the Central Universities (Nos. N161602003, N171607010, N161904001, and N160601001), the Natural Science Foundation of Liaoning Province (No. 2015020553, and 20170540702). Junchang Xin is the corresponding author.

References

1. Eyal, I., Gencer, A.E., Renesse, R.V.: Bitcoin-NG: a scalable blockchain protocol, pp. 45–59 (2015)

2. Bonneau, J., Miller, A., Clark, J., Narayanan, A., Kroll, J.A., Felten, E.W.: Research perspectives and challenges for bitcoin and cryptocurrencies, pp. 104–121 (2015)
3. Yuan, Y., Wang, F.Y.: Blockchain: the state of the art and future trends. Acta Automatica Sinica (2016)
4. Blockmeta: The Blockchain Data of Bitcoin. https://blockmeta.com/. Accessed 28 Sept 2017
5. Li, J., Wolf, T.: A one-way proof-of-work protocol to protect controllers in software-defined networks. In: Symposium on Architectures for NETWORKING and Communications Systems, pp. 123–124 (2016)
6. Herbert, J., Litchfield, A.: A novel method for decentralised peer-to-peer software license validation using cryptocurrency blockchain technology. In: Australasian Computer Science Conference, pp. 27–35 (2015)
7. Zyskind, G., Nathan, O., Pentland, A.S.: Decentralizing privacy: using blockchain to protect personal data. In: IEEE Security and Privacy Workshops, pp. 180–184 (2015)
8. Ali, M., Nelson, J., Shea, R., Freedman, M.J.: Blockstack: a global naming and storage system secured by blockchains, pp. 181–194 (2016)
9. Hari, A., Lakshman, T.V.: The internet blockchain: a distributed, tamper-resistant transaction framework for the internet. In: ACM Workshop on Hot Topics in Networks, pp. 204–210 (2016)
10. Gervais, A., Karame, G.O., Glykantzis, V., Ritzdorf, H., Capkun, S.: On the security and performance of proof of work blockchains. In: ACM Sigsac Conference on Computer and Communications Security, pp. 3–16 (2016)
11. Karame, G.: On the security and scalability of bitcoin's blockchain. In: ACM SIGSAC Conference on Computer and Communications Security, pp. 1861–1862 (2016)
12. Bentov, I., Lee, C., Mizrahi, A., Rosenfeld, M.: Proof of activity: extending bitcoins proof of work via proof of stake. ACM SIGMETRICS Perform. Eval. Rev. **42**(3), 34–37
13. Distler, T., Cachin, C., Kapitza, R.: Resource-efficient byzantine fault tolerance. IEEE Trans. Comput. **65**(9), 2807–2819 (2016)
14. Karame, G.O., Androulaki, E., Capkun, S.: Double-spending fast payments in bitcoin. In: ACM Conference on Computer and Communications Security, pp. 906–917 (2012)
15. Eyal, I., Sirer, E.G.: Majority is not enough: bitcoin mining is vulnerable. In: International Conference on Financial Cryptography and Data Security, pp. 436–454 (2014)
16. Nakamoto, S.: Bitcoin: a peer-to-peer electronic cash system. Consulted (2008)
17. Borel, E.: Probabilities and Life. Dover Publications Inc., New York (1962)
18. Juels, A.: PORs: proofs of retrievability for large files. In: ACM Conference on Computer and Communications Security, pp. 584–597 (2007)
19. Armknecht, F., Bohli, J.M., Karame, G.O., Liu, Z., Reuter, C.A.: Outsourced proofs of retrievability. In: ACM SIGSAC Conference on Computer and Communications Security, pp. 831–843 (2014)
20. Dinh, A., Wang, J., Chen, G., Ooi, B.C., Tan, K.-L.: Blockbench: a framework for analyzing private blockchains (2017)

A MapReduce-Based Approach
for Mining Embedded Patterns from
Large Tree Data

Wen Zhao and Xiaoying Wu[✉]

Computer School, Wuhan University, Wuhan, China
{wenzhao,xiaoying.wu}@whu.edu.cn

Abstract. Finding tree patterns hidden in large datasets is an important research area that has many practical applications. Unfortunately, previous contributions have focused almost exclusively on extracting patterns from a set of small trees on a centralized machine. The problem of mining embedded patterns from large data trees has been neglected. However, this pattern mining problem is also important for many modern applications that arise naturally and in particular with the explosion of big data. In this paper, we propose a novel MapReduce approach to mine embedded patterns from a single large tree which can handle situations when either the tree itself or intermediate mining results at low frequency thresholds cannot fit in the memory of any individual computer node. Furthermore, we come up with a set of optimizations to minimize internode communication. Experimental evaluation shows that our algorithm can scale well to trees with over ten million vertices.

Keywords: Tree pattern · MapReduce · Holistic twig-join algorithm

1 Introduction

Nowadays, huge amounts of data are represented, exported and exchanged between and within organizations in tree-structure form, e.g., XML and JSON files, RNA sequences, and software traces. Finding interesting tree patterns that are hidden in tree datasets has many practical applications. The goal is to capture the complex relations that exist among the data entries. Because of its importance, tree mining has been the subject of extensive research.

The Problem. Previous contributions have focused almost exclusively on mining patterns from a set of small trees. The problem of mining embedded patterns from large data trees has been neglected.

This can be explained by the increased complexity of this task due mainly to three reasons: (a) embeddings generate a larger set of candidate patterns and this substantially increases their computation time; (b) the problem of finding an unordered embedding of a tree pattern to a data tree is NP-Complete [3].

© Springer International Publishing AG, part of Springer Nature 2018
Y. Cai et al. (Eds.): APWeb-WAIM 2018, LNCS 10988, pp. 455–462, 2018.
https://doi.org/10.1007/978-3-319-96893-3_34

This renders the computation of the frequency of a candidate embedded pattern difficult; and (c) mining a large data tree is more complex than mining a set of small data trees. Indeed, the single large tree setting is more general than the set of small trees, since the latter can be modelled as a single large tree rooted at a virtual unlabeled node.

2   Proposed Approach

As a common manner of many existing distributed pattern mining approaches, our approach: EtpmLtd (Embedded Tree Pattern Miner on Large Tree Data) iterates between the local mining phase and the global summary phase (Fig. 1. shows the main framework of our approach).

We present the pesudo code of EtpmLtd with its local mining phase and global summary phase in Algorithm 1. We assume the input data are preprocessed to a list of occurrences of nodes in order of their depth-first position in the tree. And inverted lists of each label are also extracted during the preprocessing procedure. We use the list of occurrences of nodes as the input of our algorithm.

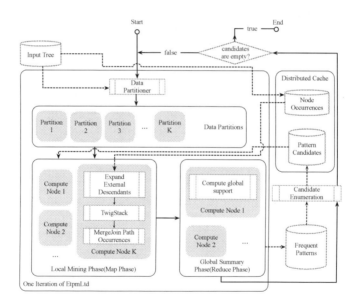

Fig. 1. Framework of our proposal.

2.1   Candidate Generation

In order to systematically generate candidate patterns, we adopt the equivalence class based pattern generation method introduced in [9] outlined next. To minimizing the redundant generation of the isomorphic representation of the same pattern, we use a canonical form for tree patterns.

Algorithm 1. EtpmLtd

---

1 **Input**   : data partitions: $T = \{T_i, \cdots\}$, minimum support threshold: $minsup$
2 **Output**   : frequent pattern $\mathcal{P}$
3 add $T \rightarrow C$, $size \leftarrow 1, candidates \leftarrow \emptyset, \mathcal{P} \leftarrow \emptyset$;
4 **repeat**
5 $\quad$ **if** $size \neq 1$ **then**
6 $\quad\quad$ add Enumerate$(patterns) \rightarrow C$,
7 $\quad$ $\Omega^P \leftarrow$ LocalMiningPhase$(candidates, T)$;
8 $\quad$ $patterns \leftarrow$ GlobalSummaryPhase(Ω^P);
9 $\quad$ $\mathcal{P} \leftarrow \mathcal{P} \cup patterns$;
10 **until** $patterns$ is $empty$;
11 **return** $\mathcal{P}$;
12 **LocalMiningPhase** $T_i, candidates$:
13 $\quad$ **if** $size = 1$ **then**
14 $\quad\quad$ Report$(label\ l,\ number\ of\ nodes\ with\ label\ l)$;
15 $\quad$ **else**
16 $\quad\quad$ **foreach** $P \in candidates$ **do**
17 $\quad\quad\quad$ $\Omega^P_{c_i} \leftarrow$ MergeJoin$(\ T_i \cup$ Expand$(T_i, P, C),\ P)$;
18 $\quad\quad\quad$ Report$(P, \Omega^P_{c_i})$;

19 **GlobalSummaryPhase** $\Omega^P, minsup$:
20 $\quad$ **if** $[size = 1$ & CountSupport$(\Omega^P) \geq minsup]$ or IsFrequent$(\Omega^P, minsup)$ **then**
21 $\quad\quad$ Report(P);

Equivalence Class Expansion Based Candidate Generation. Let P be a pattern of size $k - 1$. Each node of P is identified by its depth-first position in the tree. The rightmost leaf of P, denoted rml, is the node with the highest depth-first position. The immediate prefix of P is the sub-pattern of P obtained by deleting the rml from P. The equivalence class of P is the set of all the patterns of size k that have P as the immediate prefix. We denote the equivalence class of P as $[P]$.

Let $[P]$ be a prefix equivalence class of size k patterns, and let pair (x, i) denote the pattern in the class where x represents the label of the rml, and i represents the depth-first position of its father node in the pattern. We also use P^i_x represent the pattern. We can join any two patterns P^i_x and any other pattern P^j_y (including itself) in $[P]$ by adding the rml of P^j_y to the right most path of P^i_x to produce new patterns. The join operation $\otimes$ is defined as follows:

(1) If $i = j$, then $p^i_x \otimes p^j_y = (p^i_x)^j_y$, only if P is not an empty immediate prefix;
(2) If $i \geq j$, then $p^i_x \otimes p^j_y = (p^i_x)^k_y$.

At the beginning of local mining phase of iteration $k + 1$, we obtain all possible frequent pattern candidates of size $k + 1$ (denoted as C_{k+1}) by performing equivalence class expansion of the size k frequent patterns (denoted as F_k) mined from iteration k (function $Enumerate$).

2.2 Local Mining Phase

Local mining phase extracts embeddings rooted at each partition for each pattern. It corresponds to the map phase of MapReduce framework. Given a pattern

P, we use $P.al$, $P.rl$ and $P.ol$ to represent the label(s) of all nodes, root node and other nodes (exclude root) of the pattern respectively. We denote the occurrences list of label l from data partition distributed to compute node c_i as $O_{c_i}^l$, and the occurrences list of label l from data partition distributed to any other compute nodes except c_i as $O_{c-c_i}^l$ (c is the set of all compute nodes). One challenge of mining a partitioned tree is that a globally frequent pattern P can be missed due to the fact that certain edges involved in the tree isomorphisms span different partitions, which will results in false negatives.

Eliminating False Negatives via External Descendants. To prevent false negatives, we propose a technique called external descendant expansion (function *Expand*). The main idea is that before computing support for pattern P of size k, we expand the partition T_i (the tree partition on compute node c_i) by requesting from other partitions the descendant nodes of root nodes of P. That is, compute node c_i has to obtain the occurrences list of $P.ol$ on all other compute nodes except c_i, namely $O_{c-c_i}^{P.ol}$. So that compute node c_i will be able to extract all embeddings that the root of the pattern occurs on this data partition. To minimize the occurrences list that compute node c_i read from distributed cache, for each pattern P and any occurrence $o_{c-c_i}^{P.ol}$ from $O_{c-c_i}^{P.ol}$, we read it into memory iff $o_{c-c_i}^{P.ol}$ is a descendant of any occurrence $o_{c_i}^{P.rl}$ from $O_{c_i}^{P.rl}$.

A Holistic Twig-Join Approach for Computing Path Occurrences. We use a holistic twig-join algorithm TwigStack [2], the state-of-art algorithm for computing all the occurrences of tree-pattern queries on tree data. Algorithm TwigStack joins multiple inverted lists at a time to avoid generating intermediate join results. And finally for any two path occurrences o_{P_i} and $o_{P_{l_j}}$ of path P_{l_i} and P_{l_j}, they can be merge joined iff the data nodes do not obey the relations of their corresponding pattern nodes.

2.3   An Improvement of EtpmLtd: Algorithm EtpmLtd+

Pattern Pruning via Local Support. For a globally frequent pattern P, let $v^* \in P.V$ denote the node with the minimum number of mappings. That is $\sigma(P) = |\Phi(v^*)|$. Now, for each partition T_i, let $O_{c_i}^P$ be the set of occurrences, and let $\Phi_i(v)$ be the corresponding set of mappings for any $v \in P.V$. We could define the local support of P in partition T_i to be $\sigma_i(P) = |\Phi_i(v^*)| = \min_{v \in P.V}\{|\Phi_i(v)|\}$. And further let v_i^* denote the node with the minimum number of mappings in partition T_i. We define the maximum local frequency of P as $\theta(P) = \max_{v \in P.V}\{|\Phi_i(v)|\}$. And a pattern is locally frequent iff its maximum local frequency satisfies the condition that $\theta(P) \geq minsup/K$. Note that suppose P is not locally frequent, which is $\theta(P) < minsup/K$, thus: $minsup = \sum_{i=1}^{K} minsup/K > \sum_{i=1}^{K} \theta(P) = \sum_{i=1}^{K} \max_{v \in V}\{|\Phi_i(v)|\} \geq \max_{v \in V}\{\sum_{i=1}^{K} |\Phi_i(v)|\} \geq \min_{v \in V}\{\sum_{i=1}^{K} |\Phi_i(v)|\} = \min_{v \in V}\{|\Phi(v)|\} = \sigma(P)$. Which is $\sigma(P) < minsup$. So that a pattern P could be globally frequent only it's locally frequent as the primary condition.

Minimizing Communication via Support Bounding. To eliminate false negatives, we came up with an external descendant expansion procedure. Now we divide the occurrence lists of pattern P on partition T_i as two parts: $\bar{O}_{c_i}^P$ and $\widetilde{O}_{c_i}^P$, denoting the occurrence lists that can be found before the expansion and the occurrences lists that only can be found after the expansion procedure. It is obvious that $O_{c_i}^P = \bar{O}_{c_i}^P \cup \widetilde{O}_{c_i}^P$, and $O_{c_i}^P \cap \widetilde{O}_{c_i}^P = \emptyset$. And on the basis of $\bar{O}_{c_i}^P$ and $\widetilde{O}_{c_i}^P$, we define the before-expansion local support and after-expansion local support as $\bar{\sigma}_i(P)$ and $\widetilde{\sigma}_i(P)$. By the relations of $O_{c_i}^P$, $\bar{O}_{c_i}^P$ and $\widetilde{O}_{c_i}^P$, we can get $\sigma_i(P) = \bar{\sigma}_i(P) + \widetilde{\sigma}_i(P)$. Then global frequency of pattern P satisfies the following condition: $\sigma_i(P) = \sum_{i=1}^{K} (\bar{\sigma}_i(P) + \widetilde{\sigma}_i(P)) \geq \sum_{i=1}^{K} \bar{\sigma}_i(P)$. Hence we can derive a lower bound estimation for $\sigma(P)$ as: $\bar{\sigma}(P) = \sum_{i=1}^{K} \min_{v \in P} \{|\bar{\Phi}_i(v)|\} = \sum_{i=1}^{K} \bigcup_{|O_{c_i}^P|} \bar{\phi}(v)$. In which $\bar{\Phi}_i(v)$ and $\bar{\phi}(v)$ represent the set or an instance of unique mappings for a node in P respectively. For a pattern P, if $\bar{\sigma}(P) \geq minsup$, we can say that the pattern P is globally frequent.

3 Experimental Evaluation

Experimental Settings. All our distributed experiments are performed on a Hadoop cluster with up to 20 nodes. Each node consists of a 12-core 2.6 GHz processor, with 256 GB memory for namenode and 128 GB memory for datanodes. We conduct experiments on the following three algorithms and compare their performance in terms of running time and communication cost: (1) EtpmLtd; (2) EtpmLtd+; (3) A baseline approach that partitions pattern candidates among compute nodes and each compute node reads a complete copy of tree data, which we call BaselineLtd in short. Baseline approach also works in an iterative way. It uses the same algorithm to enumerate pattern candidates and count support with EtpmLtd and EtpmLtd+.

Datasets. We used two main datasets in our experiments, namely, Treebank[1], and xml. The properties of the datasets are shown in Table 1. The xml datasets (xmlf1, xmlf5, xml10 and xmlf20) are generated by XMark [6], which is a XML document generator that produces scaled documents according to the DTD specified in The XML Benchmark Project. The factor f decides the scaling of the document. We set f to 1, 5, 10 and 20 to obtain XML documents and preprocessed them to single large tree datasets by their xml elements.

Performance Results. We study the performance of the distributed algorithms by varying one of the two different parameters, namely user defined parameter minimum support ($minsup$) and the number of partitions (K), while keeping the other fixed. In the plots which X axis represents the parameter $minsup$, the value s on the X axis indicates $minsup$ of $s\%$ with respect to number of nodes $|V|$, i.e., we are using relative support for each pattern.

---

[1] http://www.cis.upenn.edu/~treebank.

Table 1. Datasets properties.

Dataset	# V	# Labels	# Paths	Size (KB)
Treebank	2,437,666	250	1,392,231	122,949
xmlf1	1,666,315	74	1,211,774	82,351
xmlf5	8,353,141	74	6,073,932	427,654
xmlf10	16,703,050	74	12,147,169	874,926
xmlf20	33,423,024	74	24,305,687	1,777,761

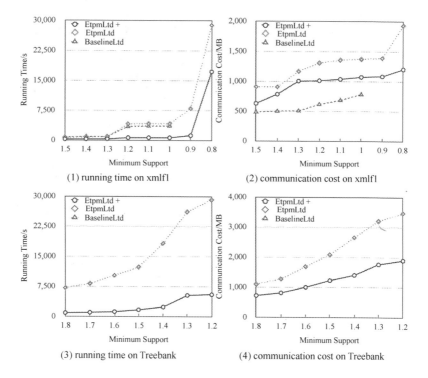

Fig. 2. Running time and communication cost on 2 datasets under different thresholds.

Varying Minsup. In this section, the number of partitions K is set as 130. Figure 2 shows the running time and communication cost for xmlf1 and Treebank. We calculate the communication cost by the HDFS communication by all compute nodes. We observe that our algorithms EtpmLtd and EtpmLtd+ can run even at a very low *minsup*. On xmlf1, BaselineLtd would crash down for threshold lower than 1%. And on both datasets, we can conclude that EtpmLtd+ can run faster than EtpmLtd, and with far more less communication cost. For the reason that xmlf1 are relatively small tree data, it can be fit into an individual node memory, so that BaselineLtd, which need no more information once the data partitions are assigned, can run faster than our algorithms. But it would turn

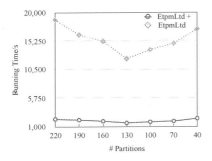

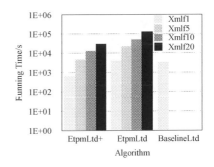

Fig. 3. Running time on Treebank under different number of partitions.

Fig. 4. Running time of xmlf* datasets on three algorithms.

out to be infeasible for lower threshold setting or larger scale data. And we can see that for Treebank, BaselineLtd crashes down even at the highest support level in the experimental settings.

Varying Number of Partitions. In this section, we will compare the influence on EtpmLtd and EtpmLtd+ by setting different number of data partitions on Treebank. We set the number of partitions from 220 to 40, with 30 step size. Figure 3 shows the running time. Results reveal that too many or too little data partitions are both not good for the performance of our MapReduce approaches.

Scalability. We use xml1, xmlf5, xmlf10 and xmlf20 to evaluate the scalability of EtpmLtd, EtpmLtd+ and BaselineLtd. The numbers in their names are almost equal to the ratio of their dataset scale. The minimum support threshold setting is 1.0%. We conduct an experiment on a centralized machine with our sequential version of EtpmLtd, and results show that running on xmlf5, xmlf10 and xmlf20 would crash down with an OutOfMemory Exception. Figure 4 shows the running time of EtpmLtd, EtpmLtd+ and BaselineLtd on the four datasets. Even for a dataset with scale of more than 10 million vertices, EtpmLtd and EtpmLtd+ can work well. But BaselineLtd would fail for the lack of memory.

4 Related Work

The problem of mining tree patterns has been studied since the last decade. Among many such algorithms, sleuth [9] is the representative one for mining unordered embedded patterns. And EmbTPMBit [8] is the first sequential algorithm for mining unordered embedded patterns from a large single tree. It compactly encodes embedded occurrences of the patterns into lists of occurrences for the nodes of the patterns. The experimental results of [8] show that EmbTPM-Bit greatly outperforms sleuth. We employee the same idea of EmbTPMBit for computing pattern support.

DistGraph [7] is a distributed approach for mining induced patterns from large single graph data. The distributed system it uses can directly do communication between compute nodes. The optimizations designed in this paper are partially inspired by the work of DistGraph. Several other algorithms are mining patterns from a set of small graphs. For instance, FSM-H [1] and MRFSE [5] are two iterative approach based on MapReduce to mine unordered patterns. MRFSM [4] is a two-step filter-and-refinement MapReduce framework for frequent subgraph mining.

5 Conclusion

In this paper, we proposed and studied a MapReduce-based approach for frequent embedded pattern mining: EtpmLtd and its optimized version EtpmLtd+. They ensure no false negatives by expand local data to obtain external descendants. EtpmLtd+ uses local support bounding to minimize the communication cost, which can result in an evident speedup.

We conduct experiments to compare the properties of these two algorithms and evaluate the performance against the BaselineLtd approach that uses the pattern candidates partition scheme. The result shows that our two algorithms both can work well on large datasets with *minsup* set extremely low while BaselineLtd crashes.

References

1. Bhuiyan, M.A., Hasan, M.A.: An iterative mapreduce based frequent subgraph mining algorithm. IEEE Trans. Knowl. Data Eng. **27**(3), 608–620 (2015). https://doi.org/10.1109/TKDE.2014.2345408
2. Bruno, N., Koudas, N., Srivastava, D.: Holistic twig joins: optimal XML pattern matching. In: Proceedings of ACM SIGMOD International Conference on Management of Data, pp. 310–321 (2002)
3. Kilpeläinen, P., Mannila, H.: Ordered and unordered tree inclusion. SIAM J. Comput. **24**(2), 340–356 (1995). https://doi.org/10.1137/S0097539791218202
4. Lin, W., Xiao, X., Ghinita, G.: Large-scale frequent subgraph mining in mapreduce. In: 2014 IEEE 30th International Conference on Data Engineering, pp. 844–855, March 2014. https://doi.org/10.1109/ICDE.2014.6816705
5. Lu, W., Chen, G., Tung, A.K.H., Zhao, F.: Efficiently extracting frequent subgraphs using mapreduce. In: 2013 IEEE International Conference on Big Data, pp. 639–647, October 2013. https://doi.org/10.1109/BigData.2013.6691633
6. Schmidt, A.: Xmark—an XML benchmark project. https://projects.cwi.nl/xmark/. Accessed 28 June 2003
7. Talukder, N., Zaki, M.J.: A distributed approach for graph mining in massive networks. Data Min. Knowl. Disc. **30**(5), 1024–1052 (2016). https://doi.org/10.1007/s10618-016-0466-x
8. Wu, X., Theodoratos, D.: Leveraging homomorphisms and bitmaps to enable the mining of embedded patterns from large data trees. In: Renz, M., Shahabi, C., Zhou, X., Cheema, M.A. (eds.) DASFAA 2015. LNCS, vol. 9049, pp. 3–20. Springer, Cham (2015). https://doi.org/10.1007/978-3-319-18120-2_1
9. Zaki, M.J.: Efficiently mining frequent embedded unordered trees. Fundam. Inf. **66**, 33–52 (2004)

Author Index